Ravikumar Kurup
Parameswara Achutha Kurup

Global Warming, Nanoarchaeal Endosymbiosis and Endogenous Viruses

Ravikumar Kurup
Parameswara Achutha Kurup

Global Warming, Nanoarchaeal Endosymbiosis and Endogenous Viruses

The Covid Pandemic and the End of Sapien Western Civilization – The Genocidal Covid Virus and Vaccine

LAP LAMBERT Academic Publishing

Imprint
Any brand names and product names mentioned in this book are subject to trademark, brand or patent protection and are trademarks or registered trademarks of their respective holders. The use of brand names, product names, common names, trade names, product descriptions etc. even without a particular marking in this work is in no way to be construed to mean that such names may be regarded as unrestricted in respect of trademark and brand protection legislation and could thus be used by anyone.

Cover image: www.ingimage.com

Publisher:
LAP LAMBERT Academic Publishing
is a trademark of
Dodo Books Indian Ocean Ltd. and OmniScriptum S.R.L publishing group

120 High Road, East Finchley, London, N2 9ED, United Kingdom
Str. Armeneasca 28/1, office 1, Chisinau MD-2012, Republic of Moldova, Europe
Managing Directors: Ieva Konstantinova, Victoria Ursu
info@omniscriptum.com

Printed at: see last page
ISBN: 978-620-8-44322-1

GLOBAL WARMING, INTERNET EMF POLLUTION, SELENOPATHY, ENDOGENOUS COVID VIRUSES AND SYMBIOTIC TISSUE ORGANISM NETWORKS – THE COVID PANDEMIC AND THE END OF SAPIEN WESTERN CIVILIZATION – THE GENOCIDAL COVID VIRUS AND VACCINE

Ravikumar Kurup A. and Parameswara Achutha Kurup

The Metabolic Disorders Research Centre
TC 4/1525, Gouri Sadan
Kattu Road, North of Cliff House
Kowdiar P.O., Trivandrum, Kerala, India
Email: ravikurup13@yahoo.in

CONTENTS

CHAPTER 1

GLOBAL WARMING, INTERNET EMF POLLUTION, SELENOPATHY, ENDOGENOUS COVID VIRUSES AND SYMBIOTIC TISSUE ORGANISM NETWORKS – THE COVID PANDEMIC AND THE END OF THE SAPIEN WESTERN CIVILIZATION – THE GENOCIDAL COVID VIRUS AND VACCINE

Archaea are intimately related to the origin of life. In the outer planets and galaxies methane, ammonia, hydrogen biosphere can lead to evolution of methanogenic archaea. The archaea produces energy by ammonia oxidation, nitrate metabolism, hydrogen metabolism and organosulphide metabolism. The methane, hydrogen sulphide, carbon monoxide, nitric oxide, nitrous oxide and carbon dioxide generated by archaea can produce mitochondrial dysfunction and the Warburg phenotype. The Warburg phenotype has been related to cancer, autoimmune disease, schizophrenia, bipolar mood disorder, neurodegenerative disease and syndrome X. The nitrous oxide, hydrogen sulphide, carbon monoxide, hydrogen, methane and ammonia generated by archaea can function as gasotransmitters in the brain modulating brain function. It can contribute to neurodegenerative disorders and psychiatric disorders. The nanoarchaea can form tissue-like colonies forming primitive eukaryotes and symbiose with cyanobacteria, proteobacteria. These are called as symbiotic tissue organism networks which have features akin to macrophages. This can serve as a reservoir for viruses and bacteria producing mutation and drug resistance. The symbiotic tissue organism networks or stem cells or macrophage-like structures can give rise to abiogenetic viruses like the covid virus. The archaeal magnetite actinidic complexes, PAH and porphyrins form a template for the formation of proteins and RNA leading to generation of abiogenetic viruses. These viruses can hybridize with human RNA sequences generating toxic viruses like the covid virus with ACE receptor sequences. The symbiotic tissue organism networks behaving like amoeba-like organism, stem cell or macrophage can generate viruses, produce toxic mutations by hybridizing with human RNA sequences and serve as a reservoir of infection. Thus the global warming related nanoarchaeal growth can generate nanoarchaeal colonies functioning as primitive eukaryotes which can symbiose with cyanobacteria, proteobacteria like magnetotactic bacteria and actinobacteria to generate viral reservoirs constituted by these symbiotic tissue organism networks. The neanderthalic population have got nanoarchaeal symbiosis and generate viruses in their symbiotic tissue organism networks and are resistant to these viral pandemics. These endogenous viruses generated by neanderthalic nanoarchaeal symbiotic tissue organism networks can infect the sapien population which are sensitive to these viruses and exterminate them. There has been a huge population decline in the Western countries owing to the sapien population and the demographic decline in the West.

Archaeal extracellular vesicles (EVs) are lipid bilayer-delimited particles that are naturally released from almost all types of cells but, unlike a cell, cannot replicate. EVs range in diameter from near the size of the smallest physically possible unilamellar liposome (around 20-30 nanometers) to as large as 10 microns or more, although the vast majority of EVs are smaller than 200 nm. EVs can be divided according to size and synthesis route into exosomes, microvesicles and apoptotic bodies. The composition of EVs varies depending on their parent cells, encompassing proteins (e.g., adhesion molecules, cytoskeletons, cytokines, ribosomal proteins, growth factors, and metabolic enzymes), lipids (including cholesterol, lipid rafts, and ceramides), nucleic acids (such as DNA, mRNA, and miRNA), metabolites, and even organelles. Most cells that have been studied to date are thought to release EVs, including some archaeal, bacterial, fungal, and plant cells that are surrounded by cell walls. A wide variety of EV subtypes have been proposed, defined variously by size, biogenesis pathway, cargo, cellular source, and function, leading to a historically heterogenous nomenclature including terms like exosomes and ectosomes.

Neanderthals have a mithochondrial dysfunction leading to reductive stress and stem cell transformation. This can be called as the Neanderthal syndrome producing brain and organ function. The endosymbiotic archaea can induce mitochondrial dysfunction due to endogenous digoxin synthesis. Digoxin has got a biphasic effect of digoxin. High levels of digoxin induce Na^+-K^+ ATPase inhibition and low levels of digoxin induce Na^+-K^+ ATPase stimulation. High level of digoxin induced Na^+-K^+ ATPase inhibition can lead on to membrane ATPase synthesis. The energy synthesis using Na^+-K^+ gradient is a primitive form of bioenergetics which is more efficient and better than mitochondrial energetics – Neanderthal energetics depend up on the Na^+-K^+ gradient. High levels of digoxin induce Na^+-Ca^{++} exchange consequent to mitochondrial dysfunction and opening of the mitochondrial PT pore. Mitochondrial dysfunction and lack of ATP produces further membrane Na^+-K^+ ATPase inhibition and membrane ATP synthesis leading on to the classical feedback loop. Na^+-K^+gradient induced ATP synthesis more effective than proton gradient.

Membrane ATP synthesis consequent to endosymbiotic archaea induced mitochondrial dysfunction in Neanderthals. Increased reductive stress leads to increased ROS and further mitochondrial dysfunction and stem cell transition. The stem cells contain increased ROS. The Neanderthals have archaea induced mitochondrial dysfunction, stem cell transformation, neurogenesis, macrocephaly and large head. The neanderthalic stem cell transformation can lead on to even adult neurogenesis. This also leads to a higher body-mass index and phenotype in Neanderthals. Neanderthal metabolonomics induced stem cells can lead on to Warburg effect and oncogenic transformation. The Warburg effect immune activation and autoimmune disease consequent to lymphocytic glycolysis. The mitochondrial dysfunction and ROS induced syndrome X, psychiatric illness and neurodegeneration leads to increased chronic disease burden in Neanderthals. The Neanderthals die young and fast. The endosymbiotic archaeal secreted extracellular vesicles with Na^+-K^+ ATPase and digoxin are comparable to the LUCA, the common ancestral prototype. The Neanderthalic LUCA-like extracellular vesicles secreted by endosymbiotic archaea contribute to efficient archaeal energetics depend upon Na^+-K^+ gradient.

The Homo sapiens have no archaeal endosymbiosis and rely on mitochondrial energetics. The mitochondrial extracellular vesicles can contribute to sapien energetics but it is less energy efficient. The Homo sapiens mitochondrial energetics leads to less of reductive stress. There is consequently no stem cell transformation and no increased neurogenesis in adulthood. The homo sapiens have smaller brains. The homo sapiens are protected from neurodegeneration, autoimmune disease, psychiatric illness, syndrome X and cancer. The homo sapiens live longer and have increased longevity. The homo sapiens are less strong, less bulky, mature late and have smaller brain.

The Homo neanderthalis have increased endosymbiotic archaeal digoxin secretion leading on to increased Na^+-K^+ ATPase inhibition, Frohlich model pumped phonon system, quantal state, superconductivity and quantal perception. Neanderathals have increased endosymbiotic archaea and nanoarchaea. The endosymbiotic archaeal secreted extracellular vesicles can give rise to replicating calcifying nanoparticles. The nanoarchaea have been related to syndrome X, diabetes mellitus, hypertension, cancer, strokes, ischemic heart disease, autoimmune disease and psychiatric disease. The calcified nanoarchaea and calcifying nanoparticles are gravity sensing. Gravity and antigravity have been related to conscious and the unconscious brain. The gravity sensing nanoarchaea have been related to quantal perception. The quantal model of consciousness has related gravity to the conscious brain and antigravity to the unconscious brain. The nanoarchaea can sense the gravity/antigravity universal protoconsciousness field. This leads on to a neanderthalic sixth sense, more extrasensory perception and shamanic culture. The Neanderthals have increased endosymbiotic archaeal digoxin, increased NMDA excitotoxicity, increased activity in the thalamo-cortico-thalamic reverberatory circuit and increased consciousness.

The sapiens have decreased digoxin, decreased NMDA activity, decreased thalamo-cortico-thalamic reverberatory circuit activity and decreased consciousness. The sapiens have no endosymbiotic archaea, less of archaeal ECV secretion and less of archaeal ECV Na^{+}-K^{+} ATPase/digoxin vesicles. The decreased endogenous digoxin produces Na^{+}-K^{+} ATPase stimulation, decreased quantal perception consequent to decreased Frohlich model pumped phonon system based on Na^{+}-K^{+} ATPase/digoxin vesicle mediated quantal perception. Mitochondrial quantal perception depending upon magnetotactic iron-sulfur centres in sapiens less efficient leading to less sensing of gravity and universal protoconsciousness field. This produces a less shamanic and logical culture due to lack of sixth sense.

The Neanderthals have endosymbiotic archaea, archaeal induced Na^{+}-K^{+} ATPase digoxin vesicles and quantal perception similar to the common ancestor – LUCA. The Neanderthals evolved before 50,000 years. The Neanderthals have endosymbiotic archaea and archaeal secreted extracellular vesicles with reverse transcriptase activity and retroviral particles. The archaeal extracellular vesicles with reverse transcriptase activity and retroviral particles as well as genes for digoxin synthesis are obtained by horizontal gene transfer from prokaryotes, eukaryotes, plants, animals. There are genes for digoxin synthesis by HGT from plants, bacteria, fungi and amphibians which have been gradually been integrated into the neanderthalic genome. The extracellular vesicles with archaeal secreted extracellular vesicles with reverse transcriptase activity and retroviral RNA particles of HERV obtained by horizontal gene transfer from prokaryotes, eukaryotes, plants, animals and fungi into the neanderthalic genome resulted in the transformation of Homo neanderthalis into Homo sapiens and homo sapien hybrids. The homo sapiens and homo sapien hybrids have a dynamic genome contributing to genomic plasticity and a dominant cerebral cortex. This leads on to increased fictional imagination, more of reason and logic in the sapiens. The homo sapien cerebellum is smaller than neanderthalic cerebellum contributing to less of intuition, creativity, spirituality, magic realism and shamanism. The sapiens have less of endosymbiotic archaea, less of archaeal induced Na^+-K^+ ATPase digoxin vesicles and less of LUCA-like liposomes or extracellular vesicles with Na^+-K^+ ATPase and digoxin. They evolved 50,000 years later.

The Homo neanderthalis evolved to Homo sapiens and homo sapien hybrids – linear evolution by horizontal gene transfer of archaeal extracellular vesicles containing HERV RNA. The homo neanderthalic mutations contributed to homo sapiens and homo sapien hybrids generating a mutated species. There is a linear evolution. The evolution would have happened in the Lemurian or Gondwanaland land mass. The Gondwanaland or Lemuria included Antarctica, India, Australia, Oceania, Africa, and South America. Lemuria is the site of origin for evolution of the Neanderthals, sapiens and sapien-Neanderthal hybrids. This area can be Antarctikos. Similarly, the North Pole or the Arctic continent included parts of Canada, North Denmark, North Sweden, Iceland, Siberia, Alaska, Finland and northern Norway. This area can be called as Arctikos. Satellite imageries and drilling showed that the South Pole and the North Pole were a tropical paradise with tropical vegetation and animals. The Antarctic and the Arctic were subjected to global warming consequent to a rise of carbon dioxide emissions 55 million years ago corresponding with the origin of Neanderthals and sapiens. The carbon dioxide in the atmosphere was 1000 ppm. The global warming 50,000 years ago would have melted the ice cap and lifted the sea levels by 200 feet. A kilometre under the sea by Wilkes land in East Antarctica they found sediments containing pollen of plants that usually thrives in tropical forest. The same kind of material is found in the same period in the Arctic region 50,000 years before. The Poles both North and South were equally warm as the land near the equator. The carbon dioxide formed in the planet led to a uniform warming all over the universe. The reason could be that the movement of the tectonic plates caused carbonate rich rocks and sediments to release carbon dioxide. The world heated up. There are signs of asteroidal and meteoritic hits in both the Antarctic and Arctic continent. The backwaters of Kerala and the seashores rich in actinides also show evidence meteoritic and asteroidal hits. The meteorites and asteroids contain bacteria from the intergalactic space. The meteorites and asteroids would have carried intergalactic nanoarchaea to the earth. The meteoritic and asteroidal hits would have triggered earthquakes, tsunamis and volcanic eruption in the ocean beds as well as in the polar continents. This would have led to emissions of carbon dioxide, sulphur dioxide, ammonia, methane and oxides of nitrogen into the atmosphere producing multiple shadow biospheres for ammonia oxidizing, methane dependent and nitrate dependant archaea to thrive and grow as huge colonies. The asteroidal and meteoritic hits would have provided actinidic surfaces for archaeal evolution. The sea and backwaters of Kerala are rich in actinidic monazite, illmenite and thorium. The primitive isoprenoid, DNA, RNA and protein molecules would have formed in actinidic surfaces leading to evolution of RNA viroids, DNA viroids, prions, and isoprenoid organisms which would have symbiosed to form archaea or nanoarchaea. Thus, the abiogenetically formed nanoarchaea would have multiplied in the shadow biosphere of ammonia, carbon dioxide, oxides of nitrogen, sulphur dioxide and methane created by asteroidal and meteoritic hits mediated volcanic eruptions, earthquakes and tsunamis. The solar storms reaching the earth would have created EMF pollution, gravitational waves and anti-gravitational waves generating sound and light. Eventually, porphyrions which can exist as wave and particle would have evolved from the gravitational and anti-gravitational waves. The supramolecular porphyrion complexes would have acted as templates for the formation of isoprenoid organisms, RNA viroids, DNA viroids and prions which would have symbiosed to form primitive nanoarchaea. The archaea has got a biphasic effect. It can generate ammonia, methane, oxides of sulphur and oxides of nitrogen contributing to global warming. The archaea can also fix carbon dioxide by photosynthesis as well as oxidise ammonia. The archaea are also capable of cholesterol oxidation by cholesterol oxidases for their energetics. The initial archaea would have evolved on actinidic, methane, ammonia and cholesterol crystals in the intergalactic space by abiogenesis modulated by crystal genes. This archaea would have reached the earth by meteoritic and asteroidal hits. They would have contributed to global warming, tsunamis, volcanic erruptions and earthquakes leading to breaking up of Gondwanaland and Lemuria. The same archaeal colonies by their biphasic effect would have contributed to cooling of the atmosphere and reduction in global warming consequent to photosynthesis, ammonia oxidation and water oxidation. The carbon dioxide, oxides of nitrogen and sulphur as well as ammonia would have been absorbed by the oceanic waters as calcium carbonate and organometallic compounds which can be used by archaea for their energetics. This would have contributed to widespread growth of moss, lichens and algae in the Antarctic and Arctic contributing to microalgal blooms or green snow. The green snow has started covering the Arctic and Antarctic again. The microalgal blooms can fix carbon dioxide and reduce global warming.

Human evolution of Neanderthals and sapien hybrids would have happened in the Arctic continent also. The climate change related archaeal overgrowth in the ocean beds and methanogenesis would have led to catastrophic tectonic shifts, tsunamis and earthquakes leading on to breaking up of Lemuria or Gondwanaland. The homo sapien-homo neanderthalic hybrids also were stranded in the African part of the Gondwanaland and being mutated hybrid species migrated to southern Europe evolving into a sapien-neandethalic hybrid civilization. The homo sapiens who were left stranded in the African part of Gondwanaland developed over in that continent. The homo neanderthalis migrated to South Asia and Indian subcontinent developing the Harappa–Mohenjo-daro civilization. The sapien-neanderthalic hybrids from Europe invaded the Harappa–Mohenjo-daro civilization and destroyed it by cutting the dam over the Indus. The Harappa–Mohenjo-daro settlers migrated to Central and South Asia forming the Shudra population in India. The Harappa–Mohenjo-daro settlers also migrated to Egpyt forming the civilization to Pharoahs, Babylon and Iran. The Iranian civilization form the Achammenid empire, the Zoroaster civilization and later on, to the Caliphates. The Muslims later invaded India as the Mughals. The Harappa-Mohenjo-daro settlers migrated to Slavic states, China, Mongolia, Siberia and Japan forming asuric civilizations over there. The Arctic region inhabited by the Eskimos, American Indians, the Lapps, Finns, Inuits and the Sami people who were Dravidian and speak Dravidian linked languages.

The Harappa–Mohenjo-daro civilization had Varuna the Ocean God as the King. The Commander of the Indus valley civilization was Shiva, the God of Himalayan Mountains from which the Indus river and Saraswati originated. It was was basically a thalassocracy and a trading city state. The Harappa-Mohenjo-daro civilization was basically a trading city state with extensive trade links with Sumeria, Egypt, Oman, Iran, and the countries bordering the Arabian sea. It was basically a thalassocracy with ports at Dwaraka and Dholavira which falls in the area where the Indus waters flow into the Gulf of Cambay and Kutchh. The Indus Valley civilization when it ended was succeeded by the family of thalassocracies represented by the sea gypsies. The sea gypsies and the conferdation of thalassocracies were sea adaptated and can be called as the Mermaid civilization. The polity of the Indus Valley civilization being a thalassocracy was ocean-based with little concerned for continental Eurasia. There were South Asian thalassocracies like the Zamorins of Calicut with ports at Musiris and Cranganore as well as the Chola empire represented by Rajendra Chola with it port at Poomphuhar. Rajendra Chola established the Kingdom of Sreevijaya which extended to Indonesia, Philippines and Indo-China. The sea people are thalassocracies which were distributed widely in the Mediterranean and Indian ocean. The sea peoples are a hypothesized seafaring confederation that attacked ancient Egypt and other regions in the East Mediterranean before and during the late bronze age collapse (1200-900 BCE). The sea people included well-attested groups such as the Lukka, as well as others such as the Weshesh whose origins are unknown. Many countries have traveling people, nomadic tribes who move around with the seasons and live in their vehicles on the road gathering together as clans at various times of the year. In Southeast Asia, with its myriad of islands, sea gypsies have a similar lifestyle albeit on the ocean. The most famous are probably the Moken of Southern Thailand, though, as we will see, Indonesia in fact has more tribes of 'Orang Laut' (sea people). They are also found in Malaysia, the Philippines and Burma. These tribes described themselves as "being born, living and dying on their boats, and the umbilical cords of their children plunge into the sea". They usually only spend a few months on land during the wettest part of the monsoon. Children learn to swim before they can walk and have exceptional aquatic abilities. In fact, certain tribes have even physically evolved to be even more efficient in the water. Indonesia has large populations of two of the three 'sea gypsy' tribes of Southeast Asia. These tribes are the Orang Laut and the Sama Bajau people. In Thailand and Burma, the Moken people rule the seas. As can be seen from the map the Orang laut and Sama Bajau people are distributed across Indonesia, Malaysia, Singapore, the Philippines, and the Sultana of Brunei. In Indonesia, the Sama Bajau population numbers 345,000 people. Though the data may be incomplete, the population of the main Orang laut tribe, the Desin Dolak (or orang kuala), would seemingly reach around 17,500 people. Each large tribe is broken down into a multitude of smaller sub-tribes. Though their traditional lifestyle is very much under threat, some tribe members still stick to the 'old way of life'. Traditionally, one family unit (usually 5 people) used to live on one houseboat. The houseboats would travel together as a 'clan flotilla' with the boats of other family members and they would all collaborate for fishing or for the organization of religious ceremonies. The chosen clan the nuclear family travels with can be that of the husband or the wife. At certain times of the year, their flotilla would meet up with others (who may also be related to them) at common mooring points, often near freshwater points and island cemeteries. This was often for special ceremonies and weddings. Sea gypsies often do not sail very far; for instance, the Sama bajau tribe rarely travels further than 40 km from their home mooring. However, being stateless they used to sail across national borders across the region, though this is being more and more regulated. Nowadays though most sea gypsy tribes are found living on houses on stilts near their home moorings, or where they are allowed to settle. They are highly skilled fishermen and their deep knowledge of the sea means they are able to catch enough to feed themselves and a little more to trade for rice and to sell. They are often hired to help with fishing, setting nets and other sea-related activities. They have their own very specific artistic culture which translates through songs, stories, music, unique musical instruments, traditional dances and performances and distinctive boat decorations. Sea gypsies learn to swim before they learn to walk and spend an enormous time free diving and evolving in water. Thanks to this, their bodies have evolved making them even more efficient divers. Scientists have studied the incredible physical adaptations that have improved the sea gypsies' abilities underwater. In a famous study lead by two specialists of the sea tribe people, it was shown that Moken children could see twice as well underwater as European children. The reason for this, as well as the adaptation allowing lesser reddening, irritation and reaction to seawater, was that the Moken children's pupils constricted maximally underwater, allowing them to see much better and focus on small shells and other foodstuffs they usually forage for. It was shown that with practice other humans can also achieve this and that this ability eventually disappears if not used regularly. However, sea gypsies develop this naturally by having a constant need for it. Interestingly, this ability is less common in adults since the 'hardening' of the eye after puberty makes such extreme pupil contractions more difficult. It was observed that these water people's skin apparently responded to sustained exposure to seawater by developing a white, scale-like layer, with an appearance somewhat resembling that of the hereditary dermatological disease ichthyosis. This was only seen in the Suku Laut in Riau in western Indonesia, and among Bajau in Sulawesi in the East, but not in the Moken at Ko Surin. The skin would go back to normal after several months of washing with fresh water when staying on dry land. A 2018 study showed that Bajau people's spleens are about 50 percent larger than those of a neighboring land-based group, the Saluan, letting them store more haemoglobin-rich blood, which is expelled into the bloodstream when the spleen contracts at depth, allowing breath-holding dives of longer duration. After over one thousand years of free diving, scientists' hypothesis is that this has evolved due to natural selection. Indeed, this larger spleen as well as a heightened peripheral vasoconstriction response are related to certain genes and the Bajau people seem to have 'naturally' selected a number of genes that are related to better breath-hold diving and hypoxia response. These sea tribes have the longest daily apnea time reported in humans, sometimes 5 hours per day, and in certain tribes like the Sama Bajau, some individuals pierce their eardrums from an early age to not have to bother with equalizing, making them hard of hearing in their golden years! The other modern thalassocracies include the Anglo-Saxon Britain and its British empire, the Spannish and the Portuguese empire in South America and East Indies, the Greek city states and its extensive thalassocracies constituted by Greek trading communities and the Italian city-states exemplified by Venice. The neanderthalic remnant civilization can be seen in the Basque region of Spain, Gibraltor, Celts, Ireland, Scotland, Brittany, Wales, Gauls and Anglo-Saxon England.

Thus neanderthalic evolution would have happened in both the North Pole and South Pole related Arctic and Antarctic continents. Homo neanderthalic mutations would have led to the evolution of Homo sapiens and interbreeding to homo sapien neanderthalic hybrids in the either poles of the earth. The poles of the earth can shift over billions of years. The poles have shifted 183-times in 83 million years. The earth's North Pole and South Pole are shifting. The shift of the magnetic poles can occur consequent to widespread use of water under the oceans and earth as well as solar magnetic storms and cosmic rays. The shift of the magnetic field will result in loss of protection of the earth from the dangerous solar magnetic storms. The solar magnetic storms can itself give rise to global warming, archaeal growth, still further global warming and catastrophie. Global warming and floods are happened in several geological times in human history as the evidence of the tropical paradise underneath the Antarctic and Arctic ice demonstrates. The carbon dioxide levels 55 million years before were 1000 ppm compared to 390 ppm in the earth now in the modern age. Therefore, global warming is not related to human industrial activity, but arises sui generis from human biological activity & due to extraterrestrial asteroidal and meteoritic hits and solar storms. There are massive floods, tsunamis, earthquakes and volcanin eruptions in the previous past as evidenced by the myth of Noah's Ark in the Bible and the catastrophic extinction in the earth related to Krishna in the Vedas. They are all part of cyclical changes in the geological history of earth leading on to evolution and catastrophic extinction. These are all described as different yugas in the Hindu Cosmology and we are living in the Kali Yuga which indicates the beginning of catastrophic extinction. A Yuga Cycle is a cyclic age (epoch) in Hindu

Cosmology. Each cycle lasts for 4,320,000 years and repeats four yugas (world ages): Krita (Satya) Yuga, Treta Yuga, Dvapara Yuga, and Kali Yuga. As a Yuga Cycle progresses through the four yugas, each yuga's length and humanity's general moral and physical state within each yuga decrease by one-fourth. Kali Yuga, which lasts for 432,000 years, is believed to have started in 3102 BCE. Near the end of Kali Yuga, when virtues are at their worst, a cataclysm and a re-establishment of dharma occur to usher in the next cycle's Krita (Satya) Yuga, prophesied to occur by Kalki. There are 71 yuga cycles in a Manvantara (age of Manu) and 1000 yuga cycles in a Kalpa (day of Brahma).

Neanderthals have endosymbiotic archaea, archaeal extracellular vesicles with Na^+-K^+ ATPase/digoxin and quantal perception. The archaeal extracellular vesicles with HERV RNA obtained by HGT from prokaryotes, eukaryotes, animals, fungi and plants make them resistant to HIV infection. The archaeal extracellular vesicles with archaeal viroids hybridizing with neanderthalic genomic sequences can generate covid-like RNA viruses. This also leads on to covid resistance. Thus, the neanderthalic population is resistant to viral pandemics. They generate endogenous viruses like covid and HIV and are resistant to them. The sapiens do not have endosymbiotic archaea with archaeal viroids generating endogenous viruses like HIV and covid and are susceptible to HIV and covid-like infections. This can be described as victim/predator interaction. The Neanderthals and Neo-neanderthals which develop consequent to global warming, extreme of climate and archaeal endosymbiosis can exterminate sapiens and sapien hybrids. This can be described as a natural genocide.

Global warming and extremes of climate can lead to endosymbiotic archaeal growth in sapiens and sapien hyrbids changing them to Neoneanderthals. The remnant neanderthalic communities like Shudras of India, Muslim, Mongols, Slavs, Celts, North American Indians, South American

Indians, Australian Aboriginals, Maoris and Eskimos have rapid population growth. This spells the end of species of sapiens and sapien hybrids. Homo sapiens and sapien hybrids most susceptible to RNA viral pandemics like HIV, covid and ebola. The death due to viral pandemics generated by Neanderthals and Neo-neanderthals are widespread and massive in sapiens and sapien hyrbids leading to their extermination. Neanderthals with archaea induced stem cell population susceptible to civilizational diseases like syndrome X, cancer, autoimmunity, psychosis, autism, and neurodegeneration. The sapiens susceptible to natural genocide live long and mature late, while the Neanderthals resistant to pandemics live short and mature early. The Neanderthalic Na^+-K^+ ATPase/digoxin extracellular vesicles secreted by archaea and quantal perception based on Frohlich model makes the Neanderthals susceptible to EMF pollution. This leads to NMDA excitotoxicity mediated by digoxin and mitochondrial dysfunction, cortical atrophy, cerebellar enlagement and a neanderthalic cerebellar phenotype. This leads to a neanderthalic intuitive, impulsive, creative, shamanic and anarchic population. They have larger brains, not able to speak, use primitive tools but intelligent enough to adapt to different ecological conditions. They constitute autistic, ADHD prone individualistic and introverted small communities which can be labeled as an autistic tribe.

The sapeins have less of Na^+-K^+ ATPase/digoxin extracellular vesicles secreted by archaea, no quantal perception based on Frohlich model of Na^+-K^+ ATPase inhibition, ineffective mitochondrial quantal perception and less of NMDA excitotoxic thalamo-cortico-thalamic activity. This makes them less of shamanic with dominant cerebral cortex and less of cerebellar consciousness. The sapiens are creature of logic and reason, capable of symbolic thinking, capable of grand fictional imagination with smaller brains. They are less intelligent, social animals with large communities and civilization.

The global warming consequent to archaeal overgrowth, methanogenesis and internet mediated EMF pollution leads to heme oxygenase HO1 induction, increased porphyrin synthesis and supramolecular porphyrion generation. This leads to porphyrin template mediated archaeal abiogenesis producing increased endosymbiotic archaeal growth and archaeal overgrowth in oceans, rivers, lakes and earth. The archaeal generation of pollutants like ammonia, N_2O, NO, NO_2 consequent to archaeal nitrate metabolism can contribute to global warming. The archaeal sulphur metabolism generates the greenhouse gas sulphur dioxide. The archaeal metabolism can also generate carbon dioxide, a greenhouse gas. The archaea can generate bioplastics which can form microplastics, PAH and particulate matter contributing to pollution, reductive stress and mitochondrial dysfunction. This leads on to the Warburg phenotype, Na^+-K^+ ATPase inhibition, Frohlich model quantal perception, EMF pollution perception, generation of more Na^+-K^+ ATPase/digoxin vesicles, porphyrion mediated archaeal abiogenesis and increased endosymbiotic/environmental archaeal growth. Thus pollution leads to increased neoneanderthalic transformation of sapien and sapien hyrbids as well as expansion of the existing neanderthalic population. The pollution addiction can be seen as an adaptive pollution as a means of species transformation and saving of civilization. This generates a world of whippets due to addiction to NO, ammonia, PAH, particulate matter leading to toxic psychosis.

The Neo-neanderthals form a new Homo autisticus species living as small colonies. The homo autisticus have archaeal endosymbiosis, archaeal secreted Na^+-K^+ ATPase/digoxin vesicles, Na^+-K^+ gradient related energetics and live with less of food. They also have the capability of water oxidation as well as transmutation leading on to alternative energetics. The archaeal photosynthesis in the new Homo autisticus species leads to carbon dioxide fixation and less of global warming. This is mediated by archaeal melanin and bacteriorhodospin caroteinoid pigments. The archaeal melanin is synthesized from tyrosine, dopamine, and cysteine. The archaeal bacteriorhodopsin is synthesized from the isoprenoid MEP pathway. The archaeal melanin and bacteriorhodopsin can absorb photons and act as a non-visual photoreceptor. The archaeal pigments by its adsorption of photons can modulate human metabolism, immunity, and brain neurotransmission. The archaeal pigments melanin and rhodopsin are also capable of quantal perception and modulation of body functions. The archaeal melanin and rhodopsin by its photon adsorphtion can function as an alternate nervous system. The archaeal melanin can mediate water oxidation leading on to alternate energetics for the human-archaeal hybrid species. The generation of hydrogen and its isotope deuterium by archaeal metabolism can lead on to deuterium modulation of mitochondrial function. 80 percent of the human body is water and deuterium can modulate the crystalline quality of water and its crystal gene effects as well as quantal perceptive properties. Deuterium generated by archaeal metabolism inhibits mitochondrial function generating the Warburg phenotype. The archaeal metabolism of nitrate can generate nitrogen which can also modulate body metabolism. The archaea also generates hydrogen sulphide by organometallic sulphide metabolism, nitric oxide and nitrous oxide by nitrate metabolism and carbon monoxide & carbon dioxide as well as hydrocyanic acid by cyanide metabolism. All these gases – CO, H_2S, CO_2, nitrogen, ammonia, hydrogen and deuterium can function as gasotransmitters in the brain modulating brain function. All these gases - CO, H_2S, CO_2, nitrogen, ammonia, hydrogen and deuterium can also suppress mitochondrial function and produce mitochondrial dysfunction. All these gases - CO, H_2S, CO_2, nitrogen, ammonia, hydrogen and deuterium can inhibit mitochondrial function generating the Warburg phenotype as also produce hibernation. Hydrocyanic acid is also generated by archaeal thiocyanate metabolism. Hydrocyanic acid is present in the galaxies and planetary systems. Hydrocyanic acid and cyanuric acid can give rise to pyramidal nucleotides generating the RNA organism or RNA viroids. This gives rise to the RNA world in other galaxies and planetary systems as well as in earth.

The endosymbiotic archaeal growth consequent to global warming and endogenous viruses generated in Neanderthals and Neo-neanderthals leads to sapien extermination. The neanderthalic and neoneanerthalic expansion produces primitive impulses, quantal perception and shamanic societies, Goblin mode of living and anarchy. There is also cannibalism as no food is available. The terrorism and aggression as a form of cannibalism is a adaptive mechanism. The homo autisticus lives as small village republics and communities leading to anti-globalisation. This leads on to less internet usage, no availability of cerebral cortex for AI implants and end of internet. The Homo autisticus can generate a Homo fluorensis species consequent to life in isolated islands on earth. The isolated communities of homo autisticus due to lack of food become smaller human beings and co-habiting animals creating a world of hobbits. The Homo autisticus hobbit species have quantal perception, paranormal abilities, cannibalistic tendencies with fairy-like characters creating the world of hobbits.

Global warming consequent to archaeal endosymbiosis, ammonia oxidation, methanogenesis, nitrate metabolism and cholesterol oxidation creates multiple shadow biospheres. Ammonia oxidation and methanogenesis creates a methane and ammonia based as shadow biosphere and methane and ammonia based organisms. Archaeal actinide depended cholesterol oxidase activity generates cholesterol crystal gene dependent shadow biosphere. The actinidic-cholesterol clay crystals would have generated the actinidic isoprenoid organism. Clay crystals would have acted as a template for abiogenesis of the actinidic cholesterol crystal isoprenoid organism. This would have been the original non-life self-replicating organism in the earth. The same patterns can lead to evolution of self-replicating non-life and life forms in other galaxies. The autistic neanderthalic tribes also go into extermination with symbiotic cell undergoing cell dissolutionand cell anarchy. This creates archaeal shadow biospheric civilization and this life and civilization survives in earth.

Plasma based self-replicating membrane bound non-life forms can exist in the stratosphere as well as in the core of the earth and the poles. They would have been the original self-replicating non-life forms. Physicists have created blobs of gaseous plasma that can grow, replicate and communicate – fulfilling most of the traditional requirements for biological cells. Without inherited material they cannot be described as alive, but the researchers believe these curious spheres may offer a radical new explanation for how life began. NASA's Global-scale Observations of the Limb and Disk (GOLD) mission has revealed unexpected C- and X-shaped formations in an electrified layer of gas high above our heads called the ionosphere. While these alphabetical shapes have been observed before, GOLD sees them more clearly than other instruments have and is now finding them where and when scientists did not expect. Their surprise appearances prove that we have more to learn about the ionosphere and its effects on communication and navigation signals that pass through it. Extending some 50 to 400 miles overhead, the ionosphere becomes electrically charged during the daytime when sunlight strikes our planet and its energy knocks electrons off atoms and molecules. This creates a soup of charged particles, known as plasma, that allows radio signals to travel over long distances. Near Earth's magnetic equator, charged particles are funneled upward and outward along magnetic field lines, creating two dense bands of particles north and south of the equator that scientists call crests. As night falls and the Sun's energy fades, low-density pockets in the plasma, called bubbles, can form in the ionosphere. Because of their varying density, the crests and bubbles can interfere with radio and GPS signals. Under the combined influence of gravity and Earth's electric and magnetic fields, charged particles in the ionosphere flow upward and outward away from Earth's magnetic equator, forming two dense bands, or crests, to the north and south of the equator. Most biologists think living cells arose out of a complex and lengthy evolution of chemicals that took millions of years, beginning with simple molecules through amino acids, primitive proteins and finally forming an organised structure. But if Mircea Sanduloviciu and his colleagues at Cuza University in Romania are right, the theory may have to be completely revised. They say cell-like self-organisation can occur in a few microseconds. The researchers studied environmental conditions similar to those that existed on the Earth before life began, when the planet was enveloped in electric storms that caused ionised gases called plasmas to form in the atmosphere. They inserted two electrodes into a chamber containing a low-temperature plasma of argon – a gas in which some of the atoms have been split into electrons and charged ions. They applied a high voltage to the electrodes, producing an arc of energy that flew across the gap between them, like a miniature lightning strike.

The widespread archaeal and nanoarchaeal expansion in the sea beds, air and in the earth leads to methanogenesis, generation of methane hydrates, methane hydrated crystal genes, a shadow biosphere of cholesterol, methane and ammonia based microbial civilization. The methane hydrate mediated explosions leads to tsunamis, earthquakes and methane hydrates mediated extremes of climate. This produces a catastrophic end to human civilization and nanoarchaeal civilization remains. Human beings are not alone in the galaxy. The nanoarchaeal mediated intergalactic magnetic field generates the biological universe. The nanoarchaea and calcifying nanoparticles which replicate are components of the intergalactic magnetic field and rotation of intergalactic magnetic field leads to condensation of the protoconsciouness field and formation of universe and star systems for nutrition of the nanoarchaeal colonies. Different life forms and different shadow biospheres like ammonia based life forms, methane based life forms, arsenic based life forms, silicon based life forms, magnetic dipole based life forms and plasma based life forms are the basis of different planetary and galactic civilizations. The crystals of ammonia, PAH, cholesteroal and methane can lead on to multiple shadow biospheres in different galaxies and planets. Panpsychism teaches us that there is no different between life and non-life. Mind creates the matter. The universal protoconsciousness field is made up of gravity and antigravity and the sounds produced by motion of the waves generate light by sonilumniscence. The light and photons gives rise to other subatomic particles and atoms. The wave-like form of the subatomic particles can become particulate only by interaction with the universal protoconsciousness field. Mind creates the matter, the life, non-life and universes. The only fundamental thing in the universe is consciousness. The human consciousness and unconscious mind is sculpted by gravity and antigravity. The cyrstals have life and can self-replicate. The crystals can have a quantal state and can form quantal computers with storage of information. The crystals of ammonia, PAH, cholesteroal and methane can self-replicate. The non-life forms and life forms are same and can form intelligent civilization. The concept of Lila means that humans are in a computer game played by life/non-life form civilization in other galaxies and vice versa. UFOs are an example of extraterrestrial civilizational contact. The conquest of space is an emergent phenomenon. The sapiens unable to conquer space consequent to horizontal gene transfer between normal microbiome/virobiome, phageome leading on to normal commensal flora becoming toxic leading on to death. The Homo autisticus and Neoneanderthals with endosymbiotic archaea developing by horizontal gene transfer photosynthetic abilities and transmutation related nuclear energy can conquer space and can withstand toxic infections leading on to a hope for humans in the future.

The plasma life based forms and the gravitational and anti-gravitational waves generated porphyrions would have been the most primitive of non-life forms. The porphyrin templates would have given rise to the isoprenoid organism, the RNA viroids, DNA viroids and prions which would have symbiosed to form the nanoarchaea. The archaea would have evolved into prokaryotes or bacteria. The archaeal symbiosis with prokaryotes would have generated the eukaryotes. The initial unicellular eukaryotes are the protozoa. The protozoa are capable of phagocytosis and would have had viruses, bacteria and archaea as their intracellular symbionts. The viruses of the protozoa like entamoeba and toxoplasma are giant viruses called megaviride. The protozoans like entamoeba with its viral, archaeal and bacterial endosymbionts would have evolved into the multicellular organisms or metazoa. The archaeal, bacterial and protozoal extracellular vesicles containing RNA and DNA viroids can by horizontal gene transfer lead to evolution of metazoan, plants, animals and human species. The metazoans would have evolved into plants, animals and humans by horizontal gene transfer generated by archaeal extracellular RNA and DNA viroid vesicles. The evidence of such a transformation is seen in the bacterial symbionts of protozoa like entamoeba which includes legionella, listeria, mycoplasma, mycobacteria and chlamydia. The symbionts of protozoan entamoeba include viruses like covid, HIV and cytomegalo virus. The protozoan entamoeba serves as a natural reservoir for covid infections, bacterial infections and HIV infections. The toxoplasma also serves as a covid reservoir. Toxoplasmosis infects 30% of all humanity and a covid reservoir in toxoplasma is a grave threat. Toxoplasma infections in humans are directly related to the severity of covid infections. Toxoplasma infections in humans can change the human character, personality and reproductive capacity. Toxoplasma can modulate brain function and human societal behaviour. Water and aerosols containing entamoeba can serve as a source for covid infection. These protozoan colonies of entamoeba and toxoplasma with its bacterial, archaeal and viral symbionts can be called as chimeric infective symbiotic tissue colonies which can exist in the air, water, earth, blood and tissues. The chimeric infective symbiotic tissue colonies containing bacteria, archaea and viruses can make the symbiotic organism multidrug resistant by horizontal gene transfer. The chimeric infective symbiotic tissue organism evolves as a consequence of global warming and the protozoal, bacterial, viral and archaeal symbiotic colonies are new forms of life. The chimeric infective symbiotic tissue organism composed of protozoa and its symbionts like virus, bacteria and archaea depend upon glycolysis for its energy needs. The glycolytic enzymes enolase, glyceraldehyde 3-phosphate dehydrogenase, hexokinase, phosphofructokinase, aldolase, pyruvic kinase and triose phosphate isomerase are moonlighting proteins having a cytoplasmic, membrane and nuclear functions. These glycolytic enzymes have a RNA binding motif which can modulate RNA transcription in the host and the symbiotic organism. These glycolytic enzymes with their RNA binding motif as well as DNA binding motif can bind to RNA and DNA viruses and viroids modulating their replication and life cycles. The chimeric infective symbiotic tissue organism composed of protozoa and its symbionts like virus, bacteria and archaea can enter the human host through the breaches of the gut-blood barrier and transform itself into the macrophages of the human host. The macrophages are involved in the phagocytosis and innate immunity against bacterial and viral infections. Paradoxically the macrophages like the protozoa can also serve as a reservoir where pathogens like mycobacteria tuberculosis, mycobacteria leprae, listeria, legionella, E. coli, shigella & salmonella, and viruses like HIV & covid can thrive and replicate. The modified chimeric infective symbiotic tissue organism or macrophage composed of protozoa and its symbionts like virus, bacteria and archaea can serve as a reservoir for horizontal gene transfer and generation of multidrug resistant bacteria and viruses. The infective chimeric symbiotic tissue organism composed of protozoa and its symbionts like virus, bacteria and archaea depends upon moonlighting glycolytic enzymes for their energetics and replication. The development of antibodies against the infective chimeric symbiotic tissue organism composed of protozoa and its symbionts like virus, bacteria and archaea leads to the evolution of anti-brain antibody disease or anti-basal ganglia antibody disease. They present a threat to existence consequent to horiztontal gene transfer and generation of multidrug resistant bacteria, viruses, protozoa, mycobacteria and archaea.

Extreme climate changes are caused by endosymbiotic archaea and methanogenesis. Endogenous global warming is caused by endosymbiotic archaea. Archaeal methanogenesis and methane contributes to 30% of global warming. The methane forms methane crystals and methane hydrates which can be explosive and explodes under ocean bed. This can lead to tsunamis and earthquakes as well as change in ocean currents leading on to hurricanes and storms. Methane hydrates in air can also contribute to extremes of climate.

Methane clathrate ($CH_4 \cdot 5.75H_2O$) or ($4CH_4 \cdot 23H_2O$), also called methane hydrate, hydromethane, methane ice, fire ice, natural gas hydrate, or gas hydrate, is a solid clathrate compound (more specifically, a clathrate hydrate) in which a large amount of methane is trapped within a crystal structure of water, forming a solid similar to ice. Originally thought to occur only in the outer regions of the solar system, where temperatures are low and water ice is common, significant deposits of methane clathrate have been found under sediments on the ocean floors of the Earth (approx. 1100 m below the sea level). Methane hydrate is formed when hydrogen-bonded water and methane gas come into contact at high pressures and low temperatures in oceans. Methane clathrates are common constituents of the shallow marine geosphere and they occur in deep sedimentary structures and form outcrops on the ocean floor. Methane hydrates are believed to form by the precipitation or crystallisation of methane migrating from deep along geological faults. Precipitation occurs when the methane comes in contact with water within the sea bed subject to temperature and pressure. In 2008, research on Antarctic Vostok Station and EPICA Dome C ice cores revealed that methane clathrates were also present in deep Antarctic ice cores and record a history of atmospheric methane concentrations, dating to 8,00,000 years ago. The ice-core methane clathrate record is a primary source of data for global warming research, along with oxygen and carbon dioxide. Methane clathrates used to be considered as a potential source of abrupt climate change, following the clathrate gun hypothesis. In this scenario, heating causes catastrophic melting and breakdown of primarily undersea hydrates, leading to a massive release of methane and accelerating warming. Current research shows that hydrates react very slowly to warming, and that it is very difficult for methane to reach the atmosphere after dissociation. Some active seeps instead act as a minor carbon sink, because with the majority of methane dissolved underwater and encouraging methanotroph communities, the area around the seep also becomes more suitable for phytoplankton. As the result, methane hydrates are no longer considered one of the tipping points in the climate system, and according to the IPCC Sixth Assessment Report, no 'detectable' impact on the global temperatures will occur in this century through this mechanism. Over several millennia, a more substantial 0.4 - 0.5 °C response may still be seen.

CHAPTER 2

INFECTIVE SYMBIOTIC TISSUE COLONIES OF PROTOZOA, ARCHAEA, VIRUSES AND BACTERIA – MITOCHONDRIAL DYSFUNCTION AND ANTI-GLYCOLYTIC ANTIBODY DISEASE - A PANDEMIC OF COVID 19 RELATED ENCEPHALITIS LETHARGICA - AUTISM, ADHD AND LA TOURETTE'S SYNDROME – THE LONG COVID SYNDROME

The plasma life based forms and the gravitational and anti-gravitational waves generated porphyrions would have been the most primitive of non-life forms. The porphyrin templates would have given rise to the isoprenoid organism, the RNA viroids, DNA viroids and prions which would have symbiosed to form the nanoarchaea. The archaea would have evolved into prokaryotes or bacteria. The archaeal symbiosis with prokaryotes would have generated the eukaryotes. The initial unicellular eukaryotes are the protozoa. The protozoa are capable of phagocytosis and would have had viruses, bacteria and archaea as their intracellular symbionts. The viruses of the protozoa like entamoeba and toxoplasma are giant viruses called megaviride. The protozoans like entamoeba with its viral, archaeal and bacterial endosymbionts would have evolved into the multicellular organisms or metazoa. The archaeal, bacterial and protozoal extracellular vesicles containing RNA and DNA viroids can by horizontal gene transfer lead to evolution of metazoan, plants, animals and human species. The metazoans would have evolved into plants, animals and humans by horizontal gene transfer generated by archaeal extracellular RNA and DNA viroid vesicles. The evidence of such a transformation is seen in the bacterial symbionts of protozoa like entamoeba which includes legionella, listeria, mycoplasma, mycobacteria and chlamydia. The symbionts of protozoan entamoeba include viruses like covid, HIV and cytomegalo virus. The protozoan entamoeba serves as a natural reservoir for covid infections, bacterial infections

and HIV infections. The toxoplasma also serves as a covid reservoir. These protozoan colonies of entamoeba and toxoplasma with its bacterial, archaeal and viral symbionts can be called as chimeric infective symbiotic tissue colonies which can exist in the air, water, earth, blood and tissues. The chimeric infective symbiotic tissue colonies containing bacteria, archaea and viruses can make the symbiotic organism multidrug resistant by horizontal gene transfer. The chimeric infective symbiotic tissue organism evolves as a consequence of global warming and the protozoal, bacterial, viral and archaeal symbiotic colonies are new forms of life. The chimeric infective symbiotic tissue organism composed of protozoa and its symbionts like virus, bacteria and archaea depend upon glycolysis for its energy needs. The glycolytic enzymes enolase, glyceraldehyde 3-phosphate dehydrogenase, hexokinase, phosphofructokinase, aldolase, pyruvic kinase and triose phosphate isomerase are moonlighting proteins having a cytoplasmic, membrane and nuclear functions. These glycolytic enzymes have a RNA binding motif which can modulate RNA transcription in the host and the symbiotic organism. These glycolytic enzymes with their RNA binding motif as well as DNA binding motif can bind to RNA and DNA viruses and viroids modulating their replication and life cycles. The chimeric infective symbiotic tissue organism composed of protozoa and its symbionts like virus, bacteria and archaea can enter the human host through the breaches of the gut-blood barrier and transform itself into the macrophages of the human host. The macrophages are involved in the phagocytosis and innate immunity against bacterial and viral infections. Paradoxically the macrophages like the protozoa can also serve as a reservoir where pathogens like mycobacteria tuberculosis, mycobacteria leprae, listeria, legionella, E. coli, shigella & salmonella, and viruses like HIV & covid can thrive and replicate. The modified chimeric infective symbiotic tissue organism or macrophage composed of protozoa and its symbionts like virus, bacteria and archaea can serve as a reservoir for horizontal gene transfer and generation of multidrug resistant bacteria and viruses. The infective chimeric symbiotic tissue

organism composed of protozoa and its symbionts like virus, bacteria and archaea depends upon moonlighting glycolytic enzymes for their energetics and replication. The development of antibodies against the infective chimeric symbiotic tissue organism composed of protozoa and its symbionts like virus, bacteria and archaea leads to the evolution of anti-brain antibody disease or anti-basal ganglia antibody disease. They present a threat to existence consequent to horiztontal gene transfer and generation of multidrug resistant bacteria, viruses, protozoa, mycobacteria and archaea.

Von Economo described the first series of encephalitis lethargic in the 1920's following the widespread pandemic of influcenza all over the world. He described the acute phase of illness with excessive somnolence, disorders of ocular motility and movement disorders. Any neurological symptom could be exhibited day-to-day and even by hour. Acute encephalitis lethargica had a gradual onset with low grade fever, flu-like symptoms, malaise, pharyngits, shivering, vertigo, headache and vomiting. Chronic encephalitis lethargica present with Parkinsonism symptoms. Similar episodes of EL pandemic have been described before - the English sweats of 1529, Mazzuco of Italy in 1597, Kreibellkraheit of Germany in 1672-1675, Refania of Sweden in 1954-1957 and Nona in Italy from 1890-1891. All of them manifested with extreme lethargic.

Von Economo classified acute EL into three forms: (1) Somnolence and ophthalmoplegia, (2) Hyperkinetic form, and (3) amyostatic akinetic form. The Somnolence opthalmoplegic form had meningitis, cranial nerve palsies 3, 4, and 6 with ophthalmoplegia and pupulary abnormalities. There is an overwhelming desire to sleep. Sleep function abnormalities persist for a long period of time but were easily awakened. They were also aware of everything that transpired around them. This was called as pseudosomnolence. This form of EL had a mortality of 50 percent. The hyperkinetic form had an initial manic phase after which the patient experience chorea, vocal tics, myoclonic twitches and oculomasticatory myorhythmia. The manic phase was followed by weakness and fatigue and hypomania with neuralgic pain in the face and limbs, visual and tactile hallucinations and day-night sleep cycle reversal. The amyostatic akinetic form had rigidity and lack of movements with no noticeable weakness. They often exhibited flexibitas cerea or vaccine flexibility. Patient remained rigid and immobile for a long period of time but rigidity could be overcome by a slightest external force. The chronic phase of EL occurred 1-5 years after the acute phase. The chronic phase was characterized by Parkinsonism, sleep disturbance, ocular motor abnormalities, involuntary movements, speech abnormalities, psychiatric diseases with bipolar disease and schizophrenia. Post encephalitis Parkinson's disease presented with bradykinesia and stiffness more in the upper limb than lower. Parkinson's disease of EL presented with kinesia paradoxica. They were akinetic one moment and perfectly mobile in the next. Chronic EL also presented with movement disorders like chorea, torsion dystonia, myoclonus and tics. Tics affected the jaw, lips, tongue and palate. Oculogyric is a common feature of chronic EL. The consciousness remains intact during the crisis. But the patient was unable to voluntarily move his eyes and that only with great effort and momentarily. Chronic EL was affected by psychiatric changes. There were changes in mood, feeling of euphoria, increased sexual desire, hallucinations, excessive puns, joviality and silliness. Psychosis occurred in 30 percent of patients. Autism, ADHD and la Tourette's syndrome can occur along with encephalitis lethargica. la Tourette's syndrome presents with vocal tics, motor tics, coprolalia, obsessive compulsive neurosis, echolalia, echopraxia, dystonia and obscene behaviour.

The brain pathology of EL had hyperemic meningus, soft edematous reddish brain stem especially the upper mid-brain and substantia nigra. It was also seen in the basal ganglia the pons, medulla and thalamus. Lymphocytic infiltration of the blood vessels especially small and medium sized venules, suggests an acute inflammatory process. Narcolepsy is common in EL. In the brains of the somnolent ophthalmoplegic type there was involved of the posterior hypothalamic. The chronic phase the brain showed generalized atrophy. Microscopy showed evidence of old and new infarcts suggestive of persistant infection. EL related Parkinson's disease was associated with marked neuronal loss and gliosis in substantia nigra and locus coeruleus. Neurofibrillary tangles were seen in the brain stem of EL as well as substantia nigra, locus coeruleus, red nucleus, subthalmic nucleus, corpus striatum and hippocampus. NFT in EL was of the 3R and LR tau isoforms.

The cause of EL is debatable. At that time, it was thought to be due to the influenza virus like Covid 19 virus. Later modern theories implicated Streptococcus viridans and Entero virus of the Polio family. There was an influenza pandemic in the world corresponding to the evolution of EL suggesting a cause-effect relationship. The somnolence and fatigue syndrome of EL is similar to the fatigue of Covid 19 infection. Therefore, Covid 19 pandemic can also present with encephalitis lethargica and its catastrophic sequalae. This has been described vividly in the Oliver Sachs book – *The Awakening*. The population of Europe was reduced to half by encephalitis lethargica similar to the widespread mortality of the Covid 19 pandemic.

The endosymbiotic archaea has got RNA viroids which can hybridize with human metabolic and immune RNA sequences generating the covid virus. Several mutations of the covid viruses have been generated endogenously for the past five years affecting most of the world population. Covid 19 sequences can get converted to DNA by using the HERV reverse transcriptase and get integrated into the DNA by using HERV integrase. The Covid 19 genome would have thus become an integral part of the junk codon part of the DNA. The post-covid syndrome akin to encephalitis lethargica has been described. Covid 19 pandemic has also led to an increase in autism, encephalitis lethargica, adult onset autism, post-encephalitic Parkinsonism, la Tourette's disease, schizophrenia and bipolar disorders.

The Neanderthal metabolic patterns were different from the homo sapien patterns. Stress of the ice age led to the induction of HO1. HO1 converts heme to bilirubin and carbon monoxide. This results in heme depletion. Heme depletion induces ALA synthase and results in porphyrinogenesis. Porphyrin photo-oxidation results in ROS generation and induction of HIF alpha resulting in the Warburg phenotype with increased glycolysis, PDH inhibition and mitochondrial dysfunction. Heme depletion leads to defective function of the TCA cycle heme enzyme aconitase and the mitochondrial enzyme cytochrome C oxidase. This results in the body generating energy by glycolysis as well as membrane sodium-potassium ATPase inhibition induced ATP synthesis. Mitochondrial dysfunction leads to insulin resistance.

The PDH inhibition leads to less levels of the substrate acetyl CoA for the isoprenoid pathway for synthesis of cholesterol. Low level of cholesterol synthesis leads to low cortisol, testosterone, estrogen, vitamin D, coenzyme Q and bile acids. Bile acid deficiency can lead to defective modulation of LXR, FXR and PXR leading to metabolic syndrome X. CoQ deficiency leads to mitochondrial dysfunction. Vitamin D deficiency leads to immune activation. Low cortisol levels lead to defective stress response. Low testosterone and estrogen leads to an asexual state.

The archaea can induce biological transmutation metals. There is conversion of magnesium to calcium. This results in mitochondrial PT pore dysfunction and cell death, NFkB activation and immune stimulation, release of monoamine neurotransmitters resulting schizophrenia and autism, calcium induced oncogene activation and defective insulin secretion producing metabolic syndrome. There is also conversion of zinc to copper. Zinc is required for glutamate and GABA transmission. Zinc functions as a neurotransmitter in the prefrontal cortex. The zinc containing neurons are called gluzinergic neurons. Copper is required for monoamine transmission and the function of cerebellum, brain stem and basal ganglia the primitive parts of the brain. Thus, the conversion of zinc to copper results in prefrontal cortex atrophy and cerebellar dominance. Biological transmutation can also generate energy.

The archaea induced Warburg phenotype can produce PDH blockade and accumulation of pyruvate. The pyruvate enters the GABA shunt generating succinyl CoA and glycine resulting in porphyrin synthesis. Porphyrin can produce quantal perception of low level EMF resulting in prefrontal cortex atrophy and cerebellar dominance. Since glycine is used up for porphyrin synthesis serine is not formed. There is no cystathionine synthesis leading on to hyperhomocysteinemia. The PDB blockade results in accumulation of pyruvate which gets converted to glutamate and ammonia resulting in a hyperammonemic syndrome.

The BKCD enzyme has a structure similar to PDH. BKCD deficiency results in accumulation of branched chain amino acids. To counteract this, the Neanderthals developed defects in neutral amino acids renal transporter. This resulted in a Hartnup's disease mutation and nicotinic acid deficiency consequent to loss of tryptophan. The function of sirtuins a histone deacetylase inhibitor also involved in protein acetylation is affected. Nicotinic acids are sirtuin inhibitors. The PDH blockade results in defective generation of acetyl CoA and defective protein acetylation. There are nearly 3500 acetylated proteins. Thus PDH blockade can modulate proteonomics.

The archaea induced RNA viroids can block mRNA and regulate RNA function. The porphyrins can intercalate with DNA and RNA modulating their function. Sirtuin induced HDAC inhibition modulated by Hartnup's disease mutation and nicotinic acid deficiency can also modulate genomic function. The RNA viroids generated by archaea can get integrated into DNA, function as jumping genes and modulate genomic function.

The archaea can catabolize cholesterol to digoxin which can produce membrane sodium-potassium ATPase inhibition increasing intracellular calcium and reducing intracellular magnesium. There is decrease in magnesium and increase in calcium results in mitochondrial PT pore dysfunction and cell death, NFkB activation and immune stimulation, release of monoamine neurotransmitters resulting schizophrenia and autism, calcium induced oncogene activation and defective insulin secretion producing metabolic syndrome.

Global warming leads to increase in endosymbiotic actinidic archaeal growth. Archaea are extremophiles. The actinidic archaea survive by catabolizing cholesterol. The archaea and its antigens induce HIF alpha and activate the glycolytic pathway. The glycolytic pathway activation induces increased conversion of glucose to fructose by activation of the sorbitol pathway. Glucose is converted to sorbitol by the enzyme aldose reductase and sorbitol is converted to fructose by the action of sorbitol dehydrogenase. Fructose is phosphorylated by hexokinase or fructokinase to fructose phosphate. Hexokinase has a low km value for fructose and minimal amounts of fructose will be converted to fructose phosphate depleting the cellular ATP. ATP is converted to AMP and by the action of AMP deaminase is converted to uric acid. Thus there is resultant hyperuricemia and the depletion of ATP also produces membrane sodium-potassium ATPase inhibition. Inhibition of membrane sodium-potassium ATPase increases intracellular calcium and depletes magnesium. This produces cell death by opening up the mitochondrial PT pore, NFkB activation and immune activation, glutamate excitotoxicity and oncogene activation leading to systemic disorders. The depletion of ATP finally inhibits hexokinase as such and glucose phosphorylation stops blocking the glycolytic pathway and its coupling to the mitochondrial oxidative phosphorylation by the action of PT pore hexokinase. The cell is depleted of energy by glycolysis and the oxidative phosphorylation scheme and dies. Thus global warming via induction of glycolysis and Warburg phenotype and the increased conversion of glucose to fructose and the resultant cellular depletion of ATP can produce systemic disorders and cell dysfunction as well as death. This can produce the global warming related systemic syndrome.

Global warming leads to neanderthalisation of the human species consequent to growth of actinidic archaea. The Neanderthals were accustomed to a ketogenic high fat, high protein diet. The ketone bodies were oxidised to generate ATP in the mitochondria. The neanderthalised humans due to actinidic archaeal growth due to consumption of a glucogenic diet leads to induction of glycolytic enzymes. The glycolytic enzymes are cytosolic. The glycolytic enzymes are antigenic in the neanderthalised humans. The glycolytic enzymes were suppressed in homo neanderthalis who ate ketogenic diet. This results in suppression of induced glycolytic enzymes by antibody formation in homo neoneanderthalis which arises due to archaeal growth consequent to global warming. The blockade of glycolysis results in blockade of cell energetics. The glucose is converted to sorbitol by aldose reductase and sorbitol is converted to fructose by fructokinase. Fructokinase enzyme is native to Neanderthals as they consumed fruits along with fat and protein from meat. Fructose is phosphorylated to fructose phosphate which depletes the cell of ATP. This inhibits membrane sodium-potassium ATPase leading on to increase in intracellular calcium and reduction in intracellular magnesium. This produces glutamate excitotoxicity and neurodegeneration. NFkB activation and autoimmune disease, release of monoamine neurotransmitters from presynaptic vesicles, schizophrenia, bipolar disorder, autism, ADHD, la Tourette's syndrome and encephalitis lethargica. The increased glucose gets metabolised by archaeal glycolysis and citric acid cycle. The pyruvate generated by archaeal glycolysis enters the GABA shunt scheme generating succinyl CoA and glycine which are substrates for porphyrin synthesis. The archaeal citric acid cycle can be reductive generating carbon dioxide fixation akin to the Calvin cycle of photosynthesis or oxidative generating acetyl CoA which is used for cholesterol synthesis by the archaeal mevalonate pathway. The archaeal glycolysis also generates fructose 1,6-diphosphate which enters the pentose phosphate pathway producing D xylulose phosphate which is a substrate for DXP pathway of archaeal cholesterol synthesis. The archaea synthesizes cholesterol by both the mevalonate pathway and DXB pathway. The archaea can use cholesterol for energetics by catabolizing it. The cholesterol ring is oxidised to pyruvate which enters the GABA shunt which provides substrates for the citric acid cycle. The pyruvate is converted to glutamate and ammonia. The archaea can oxidise ammonia for energy. The side chain of cholesterol is oxidised to butyrate and propionate which can also be further utilised for energy purposes. The archaeal energetics depends on glycolysis, citric acid cycle, ammonia oxidation and cholesterol catabolism. The antibodies against the glycolytic enzymes aldolase, enolase, GAPDH and pyruvic kinase contributes to schizophrenia, mood disorders, autism, ADHD, encephalitis lethargica and la Tourette's disease. The upregulation of glycolysis contributes to neoplastic state. The antibodies are produced against induced glycolytic enzymes as well as archaeal glycolytic enzymes. The blockade of glycolysis leads to a secondary mitochondrial dysfunction. The glycolytic scheme is coupled to mitochondrial oxidative phosphorylation by mitochondrial PT pore hexokinase. The antibodies against glycolysis blocks glycolysis and produce secondary mitochondrial dysfunction. The cell uses all its energetics. The depletion of ATP by phosphorylation of fructose produces membrane sodium-potassium ATPase inhibition and cell hibernation as well as stem cell transformation. The human tissue systems come to a halt and form a framework for archaeal colonies to thrive. The human body becomes a zombie for archaeal colonies which are eternal. This affects the function of the brain resulting in autism, ADHD, la Tourette's syndrome and encephalitis lethargica. This can be called as the zombie syndrome. The depletion of ATP by phosphorylation of fructose generates ADP and AMP which by action of AMP deaminase produces uric acid and hyperuricemia. The zombie syndrome converts the human body to a framework for an archaeal colony network. The archaea can secrete RNA and DNA viroids which can recombine with human endogenous retroviral sequences and human DNA sequences generating new RNA viruses, DNA viruses and bacteria. Thus the zombie syndrome results in the generation of new bacteria and viruses. The zombie syndrome can be treated by suppression of glycolysis. This can be done by giving a ketogenic diet derived from fibre short chain fatty acids - butyrate and acetate, polyunsaturated fatty acids and short chain fatty acids like lauric acid.

The global warming zombie syndrome results in conversion of glucose to fructose by the induction of aldose reductase consequent to dehydration. The glucose is first converted to sorbitol by aldose reductase and then by sorbitol dehydrogenase to fructose. Fructose has got a high value for ketokinase and is phosphorylated and enters the pentose phosphate pathway. Fructose is converted to glucosamine phosphate and galactosamine phosphate. There is increased synthesis of glycosaminoglycans. The increase km value of fructose for ketokinase results in increased phosphorylation of fructose over glucose producing hyperglycemia and metabolic syndrome. The phosphorylation of fructose depletes the cell of ATP and converts ATP to AMP and ADP which is acted upon by ATP deaminase producing uric acid. The channelling of fructose into the pentose phosphate pathway results in increased production of ribose and nucleic acid synthesis and uric acid production consequent to purine degradation. There is hyperuricemia. The increase in fructose produces fructose glycation of proteins resulting in the formation of antigenic proteins and autoimmune disease. Fructose can produce inflammation and autoimmunity. This is called as fructositis. The depletion of cellular ATP consequent to phosphorylation of fructose results in cell death and neuronal degeneration. Cell death in intestinal mucosa breaches the blood-gut barrier producing leaky guts syndrome, acute phase response and autoimmunity. The phosphorylation of fructose and depletion of ATP results in membrane sodium-potassium ATPase inhibition and increase in intracellular calcium and reduction in intracellular magnesium. This results in glutamate excitotoxicity, monoamine secretion from presynaptic vesicles, autism, ADHD, la Tourette's disease, encephalitis lethargica, bipolar disorder and schizophrenia as well as NFkB activation and autoimmunity. The channelling of fructose to glucosamine synthesis and increased synthesis of GAG results in increased heparan sulphate synthesis which will combine with proteins forming amyloid. Amyloid formation is the basis of motor neuron disease where ribonucleoproteins form amyloid. In post-encephalitic Parkinson's disease of encephalitis lethargica alpha synuclein forms amyloid. In Alzheimer's disease beta amyloid is formed.

Global warming leads to aldose reductase induction. Aldose reductase converts glucose to sorbitol. Sorbitol is converted to fructose by sorbitol dehydrogenase. Fructose is phosphorylated by fructokinase and ketokinases have a higher km value for fructose than glucose. This results in rapid phosphorylation of fructose and depletion of cellular ATP. The depletion of cellular ATP has two consequences. ATP is converted to ADP and AMP. ADP and AMP are acted upon by deaminases generating uric acid. Hyperuricemia is a feature of global warming related metabolic phenomenon. The cellular depletion of ATP can produce cell death leading on to neuronal degeneration. The depletion of ATP produces membrane sodium-potassium ATPase inhibition and increase in intracellular calcium and reduction in intracellular magnesium. This produces immune activation by NFkB induction, glutamate excitotoxicity and neurodegeneration, release of monoamines into synaptic junction. This results in schizophrenia, mood disorders, ADHD, autism, la Tourette's disease and encephalitis lethargica. The depletion of ATP can also affect the gut-blood barrier producing an acute phase response and leaky gut syndrome. The acute phase response can lead to metabolic syndrome X. The depletion of ATP owing to the affinity of ketokinases for fructose leads to lack of phosphorylation of glucose. This leads to hyperglycemia and metabolic syndrome X. The glucose that accumulates gets converted to fructose producing the syndrome of fructositis.

Fructose has got two metabolic fates. The fructose can get phosphorylated to fructose phosphate and enter the pentose phosphate pathway generating ribose important in nucleic acid synthesis. The purine catabolism can generate uric acid. The fructose that gets accumulated can also enter the pathway for glycosaminoglycan and proteoglycan synthesis. The fructose is phosphorylated and converted to fructose phosphate which can be converted to glucosamine and galactosamine which are substrates for glycosaminoglycan synthesis. The mucopolysaccharide, heparan sulphate can combine with prion proteins producing amyloid deposition leading on to conformational diseases. This includes the prion proteins leading on to, alpha synuclein and post-encephalitic Parkinson's disease. Thus, the conversion of fructose to GAG results in conformational disease and amyloid accumulation.

The accumulation of fructose results in fructosylation of proteins akin to glycation of proteins. Fructosylated proteins are antigenic. This results in increased frequency of autoimmune diseases. The conversion to glucose to fructose and its phosphorylation depletes the cell of ATP and results in membrane sodium-potassium ATPase inhibition. This increases the intracellular calcium which releases neurotransmitters from presynaptic vesicles. Thus there is an increase in glutamate and monoaminergic transmission leading to schizophrenia, mood disorders, ADHD, la Tourette's disease and autism. The increase in intracellular calcium can open up the mitochondrial PT pore releasing cyto C which activates the caspase cascade and cell death. This produces neuronal degeneration.

The conversion of glucose to fructose and phosphorylation of fructose results in depletion of cellular ATP. This stops the phosphorylation of glucose by glucokinase and generation of glucose 6-phosphate stops. The glycolytic process, the TCA cycle and its coupling to mitochondrial oxidative phosphorylation is inhibited. The body depends upon fatty acids and amino acids for energetics. The body can survive only on a ketogenic diet. The accumulated glucose forms a substrate for utilisation by endosymbiotic archaea. The archaea have glycolytic pathway which can convert glucose to pyruvate. The pyruvate can then enter the GABA shunt pathway generating succinyl CoA and glycine, the substrates for porphyrin synthesis by archaea and humans. Porphyrin supramolecular arrays or porphyrions can transfer electrons synthesizing ATP. The archaea also has a partial citric acid cycle and the pyruvate generated by archaeal glycolysis can enter the partial citric acid cycle. The citrate can be used for lipid synthesis. The acetyl CoA generated from pyruvate by archaea can be used for cholesterol synthesis. The archaea can catabolize cholesterol to generate energy. The cholesterol ring is oxidised to pyruvate and the side chain oxidised to butyrate and propionate. This can be used for mitochondrial generation of ATP. The pyruvate generated by archaeal glycolysis can also undergo a reverse citric acid cycle for carbon dioxide fixation akin to the Calvin cycle. Thus the archaeal glycolysis, partial citric acid cycle, reverse citric acid cycle of carbon dioxide fixation, cholesterol synthesis by the mevalonate and DXP pathway and cholesterol catabolism dominates. The fructose generated by conversion to glucose can enter the pentose phosphate pathway generating D xylulose phosphate and can be used to synthesize cholesterol. The archaeal pyruvate generated by glycolysis can also be converted to acetyl CoA which can enter the archaeal mevalonate pathway of cholesterol synthesis. Thus, the archaea has got both pathway of cholesterol synthesis - the DXP pathway and mevalonate pathway. The fructose generated by conversion from glucose due to global warming enters the DXP pathway of cholesterol synthesis, the pentose phosphate pathway and nucleic acid synthesis and GAG synthesis. This can lead to multiple organ dysfunction with fibrosis and mucopolysaccharide accumulation. The global warming and resulting conversion of glucose to fructose results in depletion of ATP due to more efficient phosphorylation of fructose over glucose. This results in ineffective glucose phosphorylation and accumulation leading to archaeal growth. The accumulated fructose is converted to D-xylulose phosphate and used for the DXP pathway of cholesterol synthesis by the archaea. The cholesterol is catabolized by the endosymbiotic actinidic archaea generating digoxin. The growth of actinidic archaea results in increased porphyrin synthesis and perception of low level electromagnetic fields by dipolar porphyrin mediated quantal systems. The digoxin induced membrane sodium-potassium ATPase inhibition in the setting of dipolar porphyrins in the cell can produce a pumped phonon system of quantal perception. This results in atrophy of the prefrontal cortex and cerebellar dominance leading to neanderthalisation of the human brain. This results from endosymbiotic archaeal growth resulting from metabolonomics related to global warming.

This syndrome can also be called as fructosemia or fructositis. Oxidative stress can induce aldose reductase and convert glucose to fructose. The endosymbiotic archaea synthesizes digoxin by cholesterol catabolism. Digoxin inhibits membrane sodium-potassium ATPase and increases intracellular calcium opening up the mitochondrial PT pore. This produces mitochondrial dysfunction and oxidative stress. Oxidative stress can induce aldose reductase and convert all glucose to fructose. Fructose has got a low km value for ketokinase compared to glucose and is preferentially phosphorylated. The glucose remains unphosphorylated and the glycolytic scheme and its coupled mitochondrial oxidative phosphorylation is slowed down or inhibited. The mitochondrial ATP and citrate can inhibit the phosphofructokinase and glycolysis, but cannot inhibit fructokinase or aldose reductase. Therefore the conversion of glucose to fructose continues. The fructose generated can inhibit mitochondrial function leading to more oxidative stress and still further induction of aldose reductase. The efficient phosphorylation of fructose depletes the cell of ATP producing oxidative stress and induction of NFkB resulting in chronic inflammation. Oxidative stress due to depletion of ATP can produce still further induction of aldose reductase and generation of fructose. The depletion of ATP makes the patient fatigue. The increase in fructose inhibits satiety and the patient feeding behaviour is altered leading on to obesity. This leads to a metabolic syndrome X. Metabolic syndrome X is a fat storage syndrome akin to hibernation. The fructose that is generated is converted to alpha glycerophosphate which is used for triglyceride synthesis. Fructose consumption can increase triglyceride synthesis and fatty liver. The depleted ATP due to fructose phosphorylation generates AMP and ADP which is converted to uric acid by deaminases. Uric acid can inhibit mitochondrial function. Uric acid inhibits aconitase which leads to accumulation of citrate. Uric acid can induce citrate lyase and fatty acyl CoA synthase leading on to fatty acid synthesis. The accumulated citrate can block the glycolytic pathway. The accumulated glucose gets converted to fructose by aldose reductase. Uric acid decreases reduced NADPH and oxidised NAD+. This affects the redox potential and leads to oxidative stress which further induces aldose reductase. Aldose reductase can be induced by hyperosmotic stress. This includes that created by hyperglycemia consequent to blockade of glucose phosphorylation as a result of selective fructose phosphorylation because of the low km value of fructose for ketokinase. The same mechanism operates in the dehydration induced by global warming. This leads to hyperosmolarity which induces aldose reductase. These mechanisms are similar to what happens in hibernation in animals. Hibernating animals develop a metabolic syndrome put on weight, store fat, increase triglycerides, develop fatty liver and develop insulin resistance. Hibernation and metabolic syndrome X are fat storage syndromes. The Neanderthals evolved in the cold Eurasian steppes and developed a hibernation syndrome similar to metabolic syndrome with fat storage. The stress of the ice age would have induced redox stress, induce aldose reductase and converted glucose to fructose. The fructose would have been selectively phosphorylated to fructose phosphate which would have entered the pentose phosphate pathway generating ribose for nucleic acid synthesis, the glucosamine pathway for glycosaminoglycan synthesis and/or converted to alpha glycerophosphate for fatty acid and triglyceride synthesis. Fructosamia can also affect brain function. Fructose can inhibit BDNF and inhibit cortical growth producing a cerebellar dominant brain and prefrontal cortex atrophy. This would have resulted in neanderthalisation of the brain. The conversion of fructose phosphate to ribose via the pentose phosphate pathway increases nucleic acid synthesis and cell proliferation. This leads to oncogenesis. The cell proliferation can also contribute to the bulky phenotype of the Neanderthal population. The fructokinase enzyme acts as an obesity switch. The opening of the obesity switch also contributes to the bulky phenotype Neanderthal population. The oxidative stress and osmotic stress of the ice age and global warming can induce aldose reductase mediated conversion of glucose to fructose via the enzyme sorbitol dehydrogenase. This also leads to induction of fructokinase, generation of fructose, fructosemia and fructositis. The increased fructose can fructosylate proteins producing antigenic proteins and autoimmune disease. Thus fructosemia can contribute to neurodegeneration, psychiatric disorders like schizophrenia, bipolar mood disorder, autism, ADHD, la Tourette's and encephalitis lethargica. Fructosemia can contribute to insulin resistance. The selective phosphorylation of fructose owing to the low km value of ketokinase for fructose results in nonphosphorylation of glucose and hyperglycemia. Insulin resistance leads to further induction of aldose reductase and fructokinase. Fructose can produce ATP depletion and oxidative stress. Oxidative stress can induce NFkB producing chronic inflammation and TNF alpha can produce insulin resistance acting at the level of insulin receptor. The insulin resistance activates the aldose reductase fructokinase system still further which is also further activated by the osmotic and oxidative stress of extremes of climate like global warming and ice age. This leads to the Lemurian systemic syndrome of fructosemia and fructositis.

The fructose metabolic pathway is called fructolysis. Oxidative stress and osmotic stress due to global warming and actinidic archaeal growth leads to induction of aldose reductase. This converts glucose to sorbitol and sorbitol is acted upon by sorbitol dehydrogenase produce fructose. Fructose normally undergoes fructolysis. Glycolysis is inhibited at the level of phosphofructokinase by ATP and citrate. The fructolytic pathway and fructose metabolism is confined to the liver and certain tissues and is not under this regulatory control. Fructose is converted to fructose 1-phosphate by fructokinase. Fructose 1-phosphate is acted upon by aldolase B or fructose 1-phosphate aldolase converting it into dihydroxy acetone phosphate. Dihydroxy acetone phosphate has got two fates: (1) It is acted upon by triose phosphate isomerase to glyceraldehyde 3-phosphate. Glyceraldehyde 3-phosphate is coverted to glucose 6-phosphate and then glucose 1-phosphate. Glucose 1-phosphate is used for glycogenesis. Dihydroxy acetone phosphate is acted upon by glycerol 3-phosphate dehydrogenase to glycerol 3-phosphate. Fructose 1-phosphate can be oxidised to pyruvate. Pyruvate can be decarboxylated to acetyl CoA. Acetyl CoA is used for fatty acid and cholesterol synthesis. Thus leads to triglyceride synthesis and VLDL formation. Fructose can thus be converted to storage glycogen, triglycerides and cholesterol. Global warming and ice age are extremophilic states. The human body goes into a state of hibernation and stores nutrients as glycogen and triglycerides. Fructose can increase lypogenic enzymes pyruvate kinase, malate dehydrogenase, citrate lyase, acetyl CoA carboxylase, pyruvate dehydrogenase and fatty acid synthase. Thus the metabolism is switched to the hibernatory mode from glucose catabolism. Glucose catabolism stops. The glucose that accumulates enters the archaeal primitive glycolytic and partial citric acid cycle as well as the GABA shunt pathway. The GABA shunt pathway of the archaea generates succinyl CoA and glycine the substrates for the porphyrin synthesis. Porphyrins can self-organise to form supramolecular structures which can self-replicate called porphyrions. Porphyrions are the ultimate self-replicators. They can have a photoinduction induced electron transport chain and ATP synthesis. The porphyrions are dipolar and in the setting of porphyrin intercalating cell membrane producing sodium-potassium ATPase inhibition can produce a pumped phonon system. This is a superconductive state at room temperature and can produce quantal perception. The porphyrins are macromolecular structures with a wave-particular existence. Porphyrions are the ultimate quantal observer and mediate the conversion of the quantal foam to the particulate world. This global warming induced fructosemia and accumulated free glucose get catabolized by partial citric acid cycle and GABA shunt of archaea to porphyrins generating porphyrions which can assume the quantal foam state and inhabit a multiverse universe with an eternal existence. The porphyrions can undergo photo-oxidation generating redox stress. Redox stress will further induce aldose reductase and increase fructosemia. The glucose remains unphosphorylated owing to the high km value of glucose for ketokinase as compared to fructose. The free glucose is catabolized by archaeal enzymes to porphyrins and porphyrions. The human body becomes a zombie for self-replicating porphyrions. The porphyrions have quantal perception of low level of EMF. This produces prefrontal cortex atrophy and cerebellar dominance leading to neanderthalisation of the human brain. The human metabolic pathways of glycolysis and oxidative phosphorylation are blocked. The synthesis of glycogen, lipids, cholesterol and glycosaminoglycans dominates. The body switches into the anabolic hibernatory mode. The fructokinase induced by osmotic stress and oxidative stress of global warming is the switch for hibernatory mode leading on to metabolic syndrome or fat storage disease. The free glucose undergoes catabolism by archaeal glycolysis and GABA shunt to porphyrions. The porphyrins can act as a template formation of RNA viroids, DNA viroids, prions and they eventually symbiosed and live together as nanoarchaea. Thus the nanoarchaea can arise from porphyrin templates. The supramolecular porphyrin arrays can have an electron transport chain and ATP synthesis, a primitive form of mitochondria. The nanoarchaea can form as well as self-replicate on porphyrin templates. The porphyrin supramolecular arrays or porphyrions can also self-replicate. The human body becomes a zombie for abiogenetic porphyrions and nanoarchaea. The porphyrions can have a macroscopic quantal existence and inhabit multiverse universes. Thus the human metabolism grinds to a halt and the world of nanoarchaea and porphyrions which are eternal steps in. The human race as we know of extincts owing to the metabolonomics of global warming. It is back to the work board of evolution once more.

The Neanderthals lived in the Eurasian steppes which was cold. They evolved hibernatory metabolism and a fat storage syndrome to protect them from the cold. The Neanderthals evolved due to endosymbiotic archaeal growth. Endosymbiotic archaea are extremophiles and grow in extremes of climate – the ice age and global warming. The global warming results in increase in endosymbiotic archaeal growth and neanderthalisation of the homo sapien species. Neanderthalisation is a symbiotic phenomenon. The global warming can lead to dehydration and osmotic stress. The archaea can catabolize cholesterol generating digoxin which can induce redox stress. Osmotic stress and redox stress leads to induction of the enzyme aldose reductase which converts glucose to sorbitol. Sorbitol is acted upon by sorbitol dehydrogenase and converted to fructose. Fructose can enter three metabolic schemes. The fructose is converted to alpha glycerophosphate and triglycerides. This results in storage of glucose and fructose as fat and a fat storage syndrome. The subcutaneous fat protects against variation in climatic temperature. The fructose can inhibit mitochondrial function and mitochondrial beta oxidation of fatty acids accentuating storage of fat. Fat can further fuel insulin resistance by fatty acids acting upon the insulin receptor. Insulin resistance leads to still further aldose reductase induction. The fructose can also enter the glucosamine pathway resulting in GAG synthesis and accumulation of mucopolysaccharides and proteoglycans. This can lead on to systemic connective tissue accumulation in visceral organs like liver producing cirrhosis, lung producing interstitial lung disease, kidney producing the MEN syndrome, the pancreas producing pancreatic fibrosis and CCP, the heart producing cardiomyopathy and EMF and the vascular tree producing mucoid angiopathy. The fructose has got a low km value for ketokinases as compared to glucose and gets selectively phosphorylated. This results in cellular depletion of ATP and membrane sodium-potassium ATPase inhibition resulting in increase of intracellular calcium. The calcium produces mitochondrial PT pore dysfunction and redox stress. Redox stress can induce aldose reductase. The depletion of ATP results in further inhibition of phosphorylation of glucose. This results in hyperglycemia. The archaea has got the glycolytic pathway, a partial citric acid cycle and a reverse citric acid cycle for carbon dioxide fixation. The archaea can induce the Warburg phenotype in the human tissues with increase glycolysis, inhibition of pyruvate dehydrogenase and mitochondrial oxidative phosphorylation inhibition. This results in accumulation of pyruvate which enters the GABA shunt scheme generating succinyl CoA and glycine for porphyrin synthesis. The accumulated glucose due to blockade of glucose phosphorylation is metabolized by archaea generating pyruvate. The pyruvate can enter the TCA cycle via pyruvate dehydrogenase generating acetyl CoA for fatty acid and cholesterol synthesis. The archaea can catabolize cholesterol and generate energy. The archaea can convert pyruvate via GABA shunt as said before to porphyrins. Thus, there is increased porphyrin synthesis, cholesterol synthesis and catabolism, triglyceride synthesis, nucleic acid synthesis and GAG synthesis. The metabolic patterns change. The blockade of glucose phosphorylation by fructose results in hyperglycemia which can block or inhibit porphyrin accumulation. The porphyrins have a wave-particle existence in the macroscopic state and contributes to quantal perception or extrasensory perception in homo neanderthalis. The cellular depletion of ATP consequent to fructose phosphorylation results in generation of ADP and AMP which are acted upon by deaminases producing uric acid. Uric acid can block mitochondrial function generating redox stress and further induction of aldose reductase. Uric acid can inhibit the TCA cycle due to aconitase inhibition. This results in accumulation of citrate which blocks glycolysis still further. The citrate is acted upon by the induced citrate lyase and fatty acid synthase producing fatty acids. This can lead to triglyceride synthesis and accumulation and a fat storage syndrome. Fatty acids can block glucose metabolism and the glycolytic pathway. The mitochondrial function is blocked by uric acid, fructose and digoxin. The porphyrin arrays can function as a supramolecular organism called porphyrions and transfer electrons and photons. This functions as a primitive form of mitochondria generating ATP. The inhibition of membrane sodium-potassium ATPase due to ATP depletion results in membrane sodium-potassium ATPase mediated ATP synthesis. The porphyrin array and membrane sodium-potassium ATPase mediated ATP synthesis can provide for body energetics. The fructosemia can lead to increased ribose synthesis via the pentose phosphate pathway inducing nucleic acid synthesis and cell proliferation. This can lead to oncogenesis. The fructosemia can lead to fructosylation of proteins producing antigenic proteins. The depletion of ATP consequent to fructose phosphorylation can produce redox stress, NFkB activation and immune activation. This leads to autoimmune disease. Fructose can inhibit brain derived neurotrophic growth factor and lead to genesis of schizophrenia, bipolar mood disorders, ADHD, encephalitis lethargica and autism. The cellular depletion of ATP due to fructose phosphorylation can lead to cell death and neurodegeneration. The low km value of fructose for ketokinase can lead to selective phosphorylation of fructose at a rapid rate depleting the cell of ATP stores. This inhibits glucose phosphorylation producing inhibition of glucose metabolism. This can be called as the hibernatory zombie glycolytic syndrome with anti-glycolytic enzyme anti-basal ganglia antibodies – anti-aldolase, anti-enolase, anti-pyruvic kinase. The autism, ADHD, la Tourette's, encephalitis lethargica, schizophrenia and mood disorders form part of this complex.

The neanderthalic zombie hibernatory syndrome producing the anti-glycolytic enzyme anti-basal ganglia antibody disease complex containing autism, ADHD, la Tourette's, encephalitis lethargica, schizophrenia and mood disorders occur in the neanderthalic brain which is different from the homo sapien brain. The neanderthalic societies are matriarchal and Amazonian. The world of global warming will be dominated by Amazonian women and asexual male eunuchs. Archaeal infections and covid 19 infections will neanderthalize the human brain. The frontal cortex and prefrontal cortex will shrink and becomes small. The cerebellum will become large and dominant. The frontal cortex and the prefrontal cortex give the brain reason and logic and make the world moral and decent. It can be called God-like. The prefrontal cortex and the frontal cortex expanded due to a retroviral epidemic. The retroviral sequences got inserted into the human DNA junk codons as HERV sequences forming jumping genes giving plasticity and flexibility to the brain synaptic connections. This led to the enlargement of the human frontal cortex which is capable of ficitional imagination and grand design and great human civilization came into existence. On the other hand, the neanderthalic brain is cerebellar dominant and homo neanderthalis is becoming the dominant species owing to increased endosymbiotic archaeal growth consequent to global warming. The endosymbiotic archaeal viroids can hybridize with human RNA metabolic and immune sequences generating endogenous viruses like Covid 19 and SARS. The cerebellar dominant neanderthalic brain is autistic and introvert. The Neanderthals do not have any ficitional imagination or grand design and live in small clusters. They are intuitive and magical. They are impulsive and creatures of emotion dominated by the brain stem structures and cerebellum. Impulsive magical societies can be compared to societies of the devil. Thus, the increased global warming endosymbiotic archaeal related Covid 19 induced post-covid encephalitis lethargica, autism, ADHD, la Tourette's syndrome, schizophrenia and bipolar disorders which have increased to a great extent in the post-covid era produces a scenario equivalent to playing with devil.

CHAPTER 3

INFECTIVE SYMBIOTIC TISSUE COLONIES OF PROTOZOA, ARCHAEA, VIRUSES AND BACTERIA – THE POST-COVID WORLD OF ANDROGYNOUS AUTISTICS PUNKS AND EUNUCHS – A PANDEMIC OF ENCEPHALITIS LETHARGICA RELATED TO COVID 19 – LONG COVID SYNDROME

The plasma life based forms and the gravitational and anti-gravitational waves generated porphyrions would have been the most primitive of non-life forms. The porphyrin templates would have given rise to the isoprenoid organism, the RNA viroids, DNA viroids and prions which would have symbiosed to form the nanoarchaea. The archaea would have evolved into prokaryotes or bacteria. The archaeal symbiosis with prokaryotes would have generated the eukaryotes. The initial unicellular eukaryotes are the protozoa. The protozoa are capable of phagocytosis and would have had viruses, bacteria and archaea as their intracellular symbionts. The viruses of the protozoa like entamoeba and toxoplasma are giant viruses called megaviride. The protozoans like entamoeba with its viral, archaeal and bacterial endosymbionts would have evolved into the multicellular organisms or metazoa. The archaeal, bacterial and protozoal extracellular vesicles containing RNA and DNA viroids can by horizontal gene transfer lead to evolution of metazoan, plants, animals and human species. The metazoans would have evolved into plants, animals and humans by horizontal gene transfer generated by archaeal extracellular RNA and DNA viroid vesicles. The evidence of such a transformation is seen in the bacterial symbionts of protozoa like entamoeba which includes legionella, listeria, mycoplasma, mycobacteria and chlamydia. The symbionts of protozoan entamoeba include viruses like covid, HIV and cytomegalo virus. The protozoan entamoeba serves as a natural reservoir for covid infections, bacterial infections and HIV infections. The toxoplasma also serves as a covid reservoir. These protozoan colonies of entamoeba and toxoplasma with its bacterial, archaeal and

viral symbionts can be called as chimeric infective symbiotic tissue colonies which can exist in the air, water, earth, blood and tissues. The chimeric infective symbiotic tissue colonies containing bacteria, archaea and viruses can make the symbiotic organism multidrug resistant by horizontal gene transfer. The chimeric infective symbiotic tissue organism evolves as a consequence of global warming and the protozoal, bacterial, viral and archaeal symbiotic colonies are new forms of life. The chimeric infective symbiotic tissue organism composed of protozoa and its symbionts like virus, bacteria and archaea depend upon glycolysis for its energy needs. The glycolytic enzymes enolase, glyceraldehyde 3-phosphate dehydrogenase, hexokinase, phosphofructokinase, aldolase, pyruvic kinase and triose phosphate isomerase are moonlighting proteins having a cytoplasmic, membrane and nuclear functions. These glycolytic enzymes have a RNA binding motif which can modulate RNA transcription in the host and the symbiotic organism. These glycolytic enzymes with their RNA binding motif as well as DNA binding motif can bind to RNA and DNA viruses and viroids modulating their replication and life cycles. The chimeric infective symbiotic tissue organism composed of protozoa and its symbionts like virus, bacteria and archaea can enter the human host through the breaches of the gut-blood barrier and transform itself into the macrophages of the human host. The macrophages are involved in the phagocytosis and innate immunity against bacterial and viral infections. Paradoxically the macrophages like the protozoa can also serve as a reservoir where pathogens like mycobacteria tuberculosis, mycobacteria leprae, listeria, legionella, E. coli, shigella & salmonella, and viruses like HIV & covid can thrive and replicate. The modified chimeric infective symbiotic tissue organism or macrophage composed of protozoa and its symbionts like virus, bacteria and archaea can serve as a reservoir for horizontal gene transfer and generation of multidrug resistant bacteria and viruses. The infective chimeric symbiotic tissue organism composed of protozoa and its symbionts like virus, bacteria and archaea depends upon moonlighting glycolytic enzymes for their energetics and replication. The development of antibodies against the infective chimeric

symbiotic tissue organism composed of protozoa and its symbionts like virus, bacteria and archaea leads to the evolution of anti-brain antibody disease or anti-basal ganglia antibody disease. They present a threat to existence consequent to horiztontal gene transfer and generation of multidrug resistant bacteria, viruses, protozoa, mycobacteria and archaea.

The retroviruses were responsible for integration into the genome as human endogenous retrovirus sequences and development of the cerebral cortex creating the logical and civil society. The influenza virus with its low level of HIV sequences is also capable of integration into the human genome as human endogenous COVID 19 sequences creating the cerebellar dominant autistic impulsive punkish brain. The COVID 19 virus would have been generated from RNA viroids of endosymbiotic archaea or would have reached the earth by meteoritic impacts from the human intergalactic biological quantal computing cloud. The COVID 19 virus with its HIV sequences is an ideal virus for integration into the human genome and changing it in a catastrophic manner to create a new human species. The COVID 19 virus can infect human producing a pandemic affecting more than 40 percent of the human population. The COVID 19 virus can enter via the respiratory tract and infect the lungs producing viral pneumonia. The COVID 19 virus can enter via the olfactory nerves and epithelium infecting the limbic lobe of the brain producing a meningoencephalitis and neuropsychiatric disorders. The COVID 19 virus is an RNA virus. There are reverse transcriptases coded by LINE-1 elements in the human genome. The COVID 19 virus produces a cytokine storm leading to LINE-1 expansion and increased expression of reverse transcriptase. The reverse transcriptase converts the RNA virus into its DNA counterpart which gets integrated into the human genome using LINE-1 human integrase. Thus the COVID 19 virus which infects humans becomes a part of the human genome and a genomic parasite or selfish gene. They belong to the non-coding or junk sequences of the DNA. They can function as jumping genes changing the structure of functional genes contributing to a dynamic DNA for human beings. They can be called as human endogenous

COVID 19 sequences. The human endogenous COVID 19 sequences can get expressed as pro-viral particles when exposed to redox stress. The question of infectivity of the expressed human endogenous COVID 19 sequences is also a possibility. Thus, the COVID 19 virus becomes an integral part of the human genome changing the human genome to a different sequence altering human behaviour, metabolism, immunity, endocrinological function and neurotransmission. The COVID 19 infection affects the cerebral cortex especially the limbic lobe and cerebral cortex including the limbic lobe becomes atrophic. This can lead to neuropsychiatric disorders and dementia. In addition, as the limbic lobe and the cerebral cortex become atrophic the cerebellum becomes dominant creating the syndrome of cerebellar cognitive affective disorder on a pandemic scale. Patients with cerebellar cognitive affective disorder develop autism, schizophrenia, dyslexia and ADHD. This leads to a new generation of human beings with predominant cerebellar dominance and cortical atrophy creating a global tribe of autistic human beings. Autism leads to introversion and isolation and difficulty in communicating with others leading to being locked in their own brain. Autism is also associated with stereotyped behaviour and actions and lack of social skills. The world becomes populated with autistics isolated in themselves and individualistic to the core destroying the fabric of human society and possibly creating a world of punks. The human endogenous COVID 19 sequences which function as jumping genes creating a dynamic genome will itself code for an autistic brain which essentially is a human adaptive response to isolate, lockdown and prevent COVID 19 from spreading. But this human adaptive response tends to become permanent creating a world tribe of autistic and punks with changes in brain structure producing cerebellar dominance and changes in the human genome with a good percentage of human endogenous COVID 19 sequences. These human endogenous COVID 19 sequences (HECOVID 19) can have transplacental transmission and will be transmitted to the next generation. The HECOVID 19 sequences can express following redox stress, secrete proviruses and virus particles and can perpetuate the infection forever. The COVID 19 virus becomes a part of us as human endogenous COVID 19 sequences in the non-coding part of the genome modulating the brain structure and behaviour. The world of autistic punks waits on us. Autistic tribes tend to be creative with a higher level of intelligence and creativity. The COVID 19 endogenous sequences can induce porphyrin synthesis and create dipolar porphyrins pumped phonon systems capable of quantal perception. These autistic tribes capable of quantal perception can interact with the internet in a two-way direction with the human brain modulating the internet function and the internet generated low level EMF modulating human brain function by quantal perception of dipolar porphyrin mediated superconducting pumped phonon systems. This leads to the formation of man-machine hybrids which can function as superintelligent organisms. The superintelligent man-machine hybrids capable of dipolar porphyrin mediated superconducting pumped phonon systems are capable of biological transmutation in the quantal state generating the elements needed for human survival. They will be capable of a longer lifespan and a wave-particle existence which is conductive to galactic travel and exploration of the stars and galaxies. The COVID 19 virus infects the testes and ovary and gets integrated into the genome as human endogenous COVID 19 sequences which can modulate the function of genital organs leading to androgynous states and infertility. The world of human autistics, eunuchs, androgynous and punks becomes the non-human homo galacticus/roboticus. This is comparable with the retroviral infections of human beings which created the placenta and the cerebral cortex through the formation of human endogenous retroviral sequences. The human endogenous retroviral sequences can get expressed by redox stress and can form proviral particles and viral particles which can contribute to HIV infection. The HERV sequences also play a role in the genesis of cancer, autoimmune disease, neurodegeneration, metabolic syndrome and neuropsychiatric disorders. The human endogenous retroviral particles made the human beings human and created the cerebral cortex which is the basis of rational, logical, civil society. The human endogenous COVID 19 sequences function in the opposite way creating cerebral cortical atrophy and a cerebellar dominance producing a cerebellar cognitive affective disorder on a massive scale in the world creating a tribe of autistics and punks capable of metabolic modulation and intergalactic travel. The HERV particles made us human and the human endogenous COVID 19 sequences will make us non-human and enter the world of homo galactic autistics and punks. The COVID 19 virus is distinctive in that it has influenza virus sequences and HIV virus sequences.

There is some experimental evidence to support the human COVID 19 integration into the human genome. SARS-CoV-2 RNA reverse-transcribed and integrated into the human genome. Prolonged SARS-CoV-2 RNA shedding and recurrence of PCR-positive tests have been widely reported in patients after recovery, yet these patients most commonly are non-infectious. Here we investigated the possibility that SARS-CoV-2 RNAs can be reverse-transcribed and integrated into the human genome and that transcription of the integrated sequences might account for PCR-positive tests. In support of this hypothesis, we found chimeric transcripts consisting of viral fused to cellular sequences in published data sets of SARS-CoV-2 infected cultured cells and primary cells of patients, consistent with the transcription of viral sequences integrated into the genome. To experimentally corroborate the possibility of viral retro-integration, we describe evidence that SARS-CoV-2 RNAs can be reverse transcribed in human cells by reverse transcriptase (RT) from LINE-1 elements or by HIV-1 RT, and that these DNA sequences can be integrated into the cell genome and subsequently be transcribed. Human endogenous LINE-1 expression was induced upon SARS-CoV-2 infection or by cytokine exposure in cultured cells, suggesting a molecular mechanism for SARS-CoV-2 retro-integration in patients. This novel feature of SARS-CoV-2 infection may explain why patients can continue to produce viral RNA after recovery and suggests a new aspect of RNA virus replication.[1]

Reference

1. Liguo Zhang, Alexsia Richards, Andrew Khalil, Emile Wogram, Haiting Ma, Richard A. Young, Rudolf Jaenisch. SARS-CoV-2 RNA reverse-transcribed and integrated into the human genome. doi: https://doi.org/10.1101/2020.12.12.422516.

CHAPTER 4

INFECTIVE SYMBIOTIC TISSUE COLONIES OF PROTOZOA, ARCHAEA, VIRUSES AND BACTERIA - CLIMATE CHANGE, ARCHAEAL ENDOSYMBIOSIS AND EMERGING VIRUSES - THE POST-CORONA VIRUS WORLD AND SOCIETY

The plasma life based forms and the gravitational and anti-gravitational waves generated porphyrions would have been the most primitive of non-life forms. The porphyrin templates would have given rise to the isoprenoid organism, the RNA viroids, DNA viroids and prions which would have symbiosed to form the nanoarchaea. The archaea would have evolved into prokaryotes or bacteria. The archaeal symbiosis with prokaryotes would have generated the eukaryotes. The initial unicellular eukaryotes are the protozoa. The protozoa are capable of phagocytosis and would have had viruses, bacteria and archaea as their intracellular symbionts. The viruses of the protozoa like entamoeba and toxoplasma are giant viruses called megaviride. The protozoans like entamoeba with its viral, archaeal and bacterial endosymbionts would have evolved into the multicellular organisms or metazoa. The archaeal, bacterial and protozoal extracellular vesicles containing RNA and DNA viroids can by horizontal gene transfer lead to evolution of metazoan, plants, animals and human species. The metazoans would have evolved into plants, animals and humans by horizontal gene transfer generated by archaeal extracellular RNA and DNA viroid vesicles. The evidence of such a transformation is seen in the bacterial symbionts of protozoa like entamoeba which includes legionella, listeria, mycoplasma, mycobacteria and chlamydia. The symbionts of protozoan entamoeba include viruses like covid, HIV and cytomegalo virus. The protozoan entamoeba serves as a natural reservoir for covid infections, bacterial infections and HIV infections. The toxoplasma also serves as a covid reservoir. These protozoan colonies of entamoeba and toxoplasma with its bacterial, archaeal and

viral symbionts can be called as chimeric infective symbiotic tissue colonies which can exist in the air, water, earth, blood and tissues. The chimeric infective symbiotic tissue colonies containing bacteria, archaea and viruses can make the symbiotic organism multidrug resistant by horizontal gene transfer. The chimeric infective symbiotic tissue organism evolves as a consequence of global warming and the protozoal, bacterial, viral and archaeal symbiotic colonies are new forms of life. The chimeric infective symbiotic tissue organism composed of protozoa and its symbionts like virus, bacteria and archaea depend upon glycolysis for its energy needs. The glycolytic enzymes enolase, glyceraldehyde 3-phosphate dehydrogenase, hexokinase, phosphofructokinase, aldolase, pyruvic kinase and triose phosphate isomerase are moonlighting proteins having a cytoplasmic, membrane and nuclear functions. These glycolytic enzymes have a RNA binding motif which can modulate RNA transcription in the host and the symbiotic organism. These glycolytic enzymes with their RNA binding motif as well as DNA binding motif can bind to RNA and DNA viruses and viroids modulating their replication and life cycles. The chimeric infective symbiotic tissue organism composed of protozoa and its symbionts like virus, bacteria and archaea can enter the human host through the breaches of the gut-blood barrier and transform itself into the macrophages of the human host. The macrophages are involved in the phagocytosis and innate immunity against bacterial and viral infections. Paradoxically the macrophages like the protozoa can also serve as a reservoir where pathogens like mycobacteria tuberculosis, mycobacteria leprae, listeria, legionella, E. coli, shigella & salmonella, and viruses like HIV & covid can thrive and replicate. The modified chimeric infective symbiotic tissue organism or macrophage composed of protozoa and its symbionts like virus, bacteria and archaea can serve as a reservoir for horizontal gene transfer and generation of multidrug resistant bacteria and viruses. The infective chimeric symbiotic tissue organism composed of protozoa and its symbionts like virus, bacteria and archaea depends upon moonlighting glycolytic enzymes for their energetics and

replication. The development of antibodies against the infective chimeric symbiotic tissue organism composed of protozoa and its symbionts like virus, bacteria and archaea leads to the evolution of anti-brain antibody disease or anti-basal ganglia antibody disease. They present a threat to existence consequent to horiztontal gene transfer and generation of multidrug resistant bacteria, viruses, protozoa, mycobacteria and archaea.

The COVID 19 brain infection can lead to integration of COVID 19 sequences into the genome, creation of cerebellar dominant brain and neanderthalisation. The COVID 19 infected cerebellar dominant brain has got increased porphyrin metabolism and porphyrions. The porphyrions are dipolar pumped phonon quantal perceptive systems which integrate with the internet. Panpsychism implies that the internet is conscious. The viral epidemics especially the corona virus epidemics owing to its high infectivity rate can affect human behaviour and society. The viral infection leads to self-isolation and splitting up of society into watertight compartments with little communication between small groups. This leads on to the society with autistic behavioural phenotypes. It leads to a culture of solitude and hermits. The viral epidemics especially cornona viral epidemics have a predilection for the male population leading to widespread extermination of the male population and a female dominant matrilineal Neanderthal society with predominantly asexual behaviour. The world becomes an autistic asexual tribe of solitudinal hermits. It leads to a development of authoritarian governmental structures and a society dominated by greed, loneliness and selfishness. The human society becomes increasingly depended on robotics and internet and turns into man-machine hybrids.

The viruses spread genes across bacteria humans and other cells. According to Lynn Margulis we are our viruses. Viruses merge with the cellular genome and reemerge from them. They create successful genetic patterns that underlie living things. RNA virus can exist as quasi-species consortia which can create new RNA viruses and codes by symbiosis. The archaeal RNA quasi-species consortia underlie human and species identity. The archaeal digoxin can edit the RNA producing new codes to suit environmental conditions for the benefit of the quasi-species consortia.

The bacteriophages are examples of this concept. The bacteria trade genes frantically by three processes - transformation via uptake of DNA, sexual conjugation and bacteriophage induced transduction. The viruses are mobile genetic elements and can carry genes from one person to another. The RNA viruses are dipolar and in the setting of digoxin induced membrane sodium-potassium ATPase inhibition can create a pumped phonon system mediated quantal perception. The archaeal RNA viroid quasi-species consortia can thus mediate quantal perception and function as a symbiotic colony and communicate with the outside world. To suit survival in changing environmental conditions new RNA viroids can be added to the quasi-species consortia by archaeal DNA induced editing. The viral cycles can be lytic cycles or lysogenic cycles. Viral infections convert the human body to a viral organ.

Many filo viruses like Marburg virus and ebola virus are integral part of the human genome. The human genome also contains retroviruses and borna viruses as integral parts. Similar integration non-retroviruses have been described for RSV, LCMV and VSM virus. The integration is done by HERV reverse transcriptase and integrase. The viruses can cross the inter-species barrier. The algal virus acanthocystis tortacea chlorella can infect mamallian cells. It has been detected in human throat swabs and the algal virus can replicate in human cells. Human neuronal cells when infected by the algal virus produce changes in memory, visuo-spatial process and attention affecting cognitive function.

Viruses are mechanisms of gene transfer. This is explained by viruses given to the caterpillar by the wasp. The barco virus protects the caterpillar against particular viruses. The viruses infecting humans contains sequences with human DNA and bacterial DNA. The viruses thus function as mechanism for interhuman and interspecies gene transfer. Just as the viruses can infect cells the bacteria can conjugate with human cells. Bacterial conjugation and DNA transfer with human cells have been described. Photosynthetic genes have been sequenced in phage virus. Bacteria can steal genes and develop resistance. The gene for alpha-2 macroglobulin is seen in certain bacteria and provides a mechanism for bacterial resistance.

The human gut is a symbiotic system between bacteria, viruses and phages. Gut bacterial phages codes for genes of amino acid synthesis and carbohydrate metabolism. Ecosystem of human cells gut bacteria, viruses and phages is a continuous gene transfer mechanism and can interact with other ecosystems. The pathogens can contribute genes to the host. The archaeal RNA viroids can contribute genes to the host by conversion to DNA using HERV reverse transcriptase. Unrelated RNA and DNA virus can recombine. The archaeal RNA viroids and its corresponding DNA can re-combine with unrelated RNA and DNA viruses. The bacteriophages can shuttle genes between different ecosystems. The archaeal RNA viroids served the same purpose. There is transkingdom crosstalk of small RNA molecules. Transkingdom sRNA silencing of the human RNA by archaeal RNA viroids can modulate the human system. The crosskingdom RNA silencing is important in crosskingdom communication in ecosystem.

The viral epidemics have thus contributed to human evolution. Virus and bacterial infection homogenized human population by gene transfer. The archaeal and its RNA viroids form the lynchpin of the mechanism of gene transfer. This leads on to globalisation of speech, thought and culture by viral epidemics and related gene transfer. Thus viral epidemics help in globalization of human culture. Human viral epidemics are necessary prerequisite for the evolution of human culture. The viral epidemics homogenize human population forming groups of caste, religion, nationalities and culture. The archaeal RNA viroidal quasi-species consortia underlie this mechanism of evolution. Thus viral diseases contribute to culture, behaviour, diet, eating habits and sexuality. Thus virus mediated gene transfer is important human sociological mechanism. The archaeal RNA viroidal quasi-species consortia and its recombination with unrelated RNA and DNA viruses and its integration into the human genome by HERV reverse transcriptase forms the basis of this phenomenon. Retroviral infections have contributed to the genesis of schizophrenia, cortical function and evolution of consciousness. Borna virus infections also contribute to schizophrenia. Viral epidemics thus contribute to population identity and differences. This can also lead on to generation of new viruses.

Diet can modulate archaeal RNA viroidal function. The high fibre diet will suppress archaeal growth and RNA viroidal growth. A low fibre diet will increase archaeal growth and RNA viroidal growth. Thus the RNA viroidal quasi-species consortia can be modulated by diet. Viruses are beneficial agents and only 1 percentage of the viruses are pathogenic. Most of the viruses co-exist as commensals. The bacteria and fungal kingdoms contain bacteriophages and fungal phages. They can recombine with archaeal RNA viroids and their corresponding DNA producing new RNA and DNA viruses which can get integrated into the genome as well as get secreted into the environment. This forms a common gene pool.

The archaea are extremophiles and can exist in the intergalactic space and meteroidal impacts can transfer the archaea in the intergalactic space to earth. The archaeal RNA viroids thus supply a continuous source of new genetic codes to earth. The bacterial infections especially those due to streptococcus and its phages can produce epidemic OCD, echolalia and ecopraxia contributing to a disciplined society and the evolution of the hereditary system of kingship. This can be due to the streptococcus induced epidemic OCD frontal lobe syndrome. The bacteriophages from the streptococcus could have recombined with the archaeal RNA viroids and got integrated into the human genome. The fungal infections could have contributed to the next phase of evolution. Infections with claviceps purpura can transfer LSD genes and dopamine genes to humans producing the dopaminergic epoch in society. The fungal phages would have recombined with archaeal RNA viroids and got integrated into the human genome. This could have contributed to ideas of equality, fraternity, liberty and socialism which are all products of the French Revolution. During this period of time, there was an epidemic of fungal infection in Rye in Europe. The next phase of global epidemics occurred with H_1N_5 infection or the Spanish flu epidemic. This resulted in a locked-in state and a frozen society leading on to the rise of dictatorship, fascism, the Nazis and the Communists. The next stage in which we live can be called the age of anarchy with its globalization, terrorism, rogue capitalism, sexual anarchy and religious fundamentalism. This corresponds with the recurrent epidemics of RNA viral infections- H1N1, retroviral, SARS, ebola, dengue and hemorrhagic fever. The genomic integration of these RNA viruses, fungal phages, bacterial phages would have changed the human genome at this different point in history. Thus virus induced gene transfer can modulate the brain, culture, sociology and behaviour.

References

1. Kurup, R.K. and Kurup, P.A. *Global Warming, Archaea and Viroid Induced Symbiotic Human Evolution - Retrovirus, Prions and Viroids – Porphyrinoids and Viroidelle*. New York: Open Science Publishers, 2016.

CHAPTER 5

CLIMATE CHANGE AND CATASTROPHIC PANDEMICS – INFECTIVE SYMBIOTIC TISSUE COLONIES OF PROTOZOA, ARCHAEA, VIRUSES AND BACTERIA

The plasma life based forms and the gravitational and anti-gravitational waves generated porphyrions would have been the most primitive of non-life forms. The porphyrin templates would have given rise to the isoprenoid organism, the RNA viroids, DNA viroids and prions which would have symbiosed to form the nanoarchaea. The archaea would have evolved into prokaryotes or bacteria. The archaeal symbiosis with prokaryotes would have generated the eukaryotes. The initial unicellular eukaryotes are the protozoa. The protozoa are capable of phagocytosis and would have had viruses, bacteria and archaea as their intracellular symbionts. The viruses of the protozoa like entamoeba and toxoplasma are giant viruses called megaviride. The protozoans like entamoeba with its viral, archaeal and bacterial endosymbionts would have evolved into the multicellular organisms or metazoa. The archaeal, bacterial and protozoal extracellular vesicles containing RNA and DNA viroids can by horizontal gene transfer lead to evolution of metazoan, plants, animals and human species. The metazoans would have evolved into plants, animals and humans by horizontal gene transfer generated by archaeal extracellular RNA and DNA viroid vesicles. The evidence of such a transformation is seen in the bacterial symbionts of protozoa like entamoeba which includes legionella, listeria, mycoplasma, mycobacteria and chlamydia. The symbionts of protozoan entamoeba include viruses like covid, HIV and cytomegalo virus. The protozoan entamoeba serves as a natural reservoir for covid infections, bacterial infections and HIV infections. The toxoplasma also serves as a covid reservoir. These protozoan colonies of entamoeba and toxoplasma with its bacterial, archaeal and viral symbionts can be called as chimeric infective symbiotic tissue colonies

which can exist in the air, water, earth, blood and tissues. The chimeric infective symbiotic tissue colonies containing bacteria, archaea and viruses can make the symbiotic organism multidrug resistant by horizontal gene transfer. The chimeric infective symbiotic tissue organism evolves as a consequence of global warming and the protozoal, bacterial, viral and archaeal symbiotic colonies are new forms of life. The chimeric infective symbiotic tissue organism composed of protozoa and its symbionts like virus, bacteria and archaea depend upon glycolysis for its energy needs. The glycolytic enzymes enolase, glyceraldehyde 3-phosphate dehydrogenase, hexokinase, phosphofructokinase, aldolase, pyruvic kinase and triose phosphate isomerase are moonlighting proteins having a cytoplasmic, membrane and nuclear functions. These glycolytic enzymes have a RNA binding motif which can modulate RNA transcription in the host and the symbiotic organism. These glycolytic enzymes with their RNA binding motif as well as DNA binding motif can bind to RNA and DNA viruses and viroids modulating their replication and life cycles. The chimeric infective symbiotic tissue organism composed of protozoa and its symbionts like virus, bacteria and archaea can enter the human host through the breaches of the gut-blood barrier and transform itself into the macrophages of the human host. The macrophages are involved in the phagocytosis and innate immunity against bacterial and viral infections. Paradoxically the macrophages like the protozoa can also serve as a reservoir where pathogens like Mycobacteria tuberculosis, Mycobacteria leprae, listeria, legionella, E.coli, shigella & salmonella and viruses like HIV & covid can thrive and replicate. The modified chimeric infective symbiotic tissue organism or macrophage composed of protozoa and its symbionts like virus, bacteria and archaea can serve as a reservoir for horizontal gene transfer and generation of multidrug resistant bacteria and viruses. The infective chimeric symbiotic tissue organism composed of protozoa and its symbionts like virus, bacteria and archaea depends upon moonlighting glycolytic enzymes for their energetics and replication. The development of antibodies against the infective chimeric

symbiotic tissue organism composed of protozoa and its symbionts like virus, bacteria and archaea leads to the evolution of anti-brain antibody disease or anti-basal ganglia antibody disease. They present a threat to existence consequent to horiztontal gene transfer and generation of multidrug resistant bacteria, viruses, protozoa, mycobacteria and archaea.

The COVID 19 brain infection can lead to integration of COVID 19 sequences into the genome, creation of cerebellar dominant brain and neanderthalisation. The COVID 19 infected cerebellar dominant brain has got increased porphyrin metabolism and porphyrions. The porphyrions are dipolar pumped phonon quantal perceptive systems which integrate with the internet. Panpsychism implies that the internet is conscious. Climate change can lead to extremophilic endosymbiotic archaeal growth and neanderthalisation of the species. There is increase in endosymbiotic and colonic archaeal growth. The increase in colonic archaeal growth leads to breaching of the blood-gut barrier and archaeal endosymbiosis. The exposure to the low level of EMF field consequent to widespread use of the internet, radiosatellites and radar equipments leads to increased endosymbiotic archaeal growth. Archaeal endosymbiosis leads to neanderthalisation of the species. The colonic archaea also generates RNA viroids which are converted to DNA viroids by colonic epithelial HERV reverse transcriptase and gets integrated into the colonic genome. The RNA and DNA viroids hybridise with the viroidal and bacterial population of the gut microbiome and virobiome DNA and RNA of the gut leading to generation of new emerging viruses. The RNA and DNA viroids and gut bacteria can also hybridise with human genomic sequences for immunity, metabolic regulation, growth factors, cell growth, cell death, neurotransmitters and HERV sequences of the colonial epithelial cell. The hybridization of archaeal RNA viroids with human genomic HERV sequences can generate retroviruses. This leads to generation of new bacteria, RNA and DNA viruses modulating cell death, cell growth, cell proliferation, metabolism, endocrine function and behaviour producing autoimmunity, schizophrenia, autism, cancer diabetes mellitus, coronary vascular disease, cerebrovascular disease and neurodegeneration.

Dietary fibre is the most important factor that modulates human biology and speciation. Dietary fibre and human drugs- antibiotic and non-antibiotic can modulate the gut flora producing colonic archaeal growth and endosymbiosis generating climate change, evolution of superbugs and emerging viruses and creating the disease phenotype. The high fibre diet is digested by the colonic flora leading to generation of short chain fatty acids and inhibits endosymbiotic and colonic archaeal growth. This leads to homo sapienisation of the species. A low fibre diet with increase in protein and fat leads to stimulation of endosymbiotic and colonic archaeal growth. This leads to neanderthalisation of the species. Thus a low fibre diet alters the gut microbiome in the colon with more of archaeal growth. The archaea develop protective mechanism of antibiotic resistance and dominates the colonic microflora. The colonic archaea by natural selection and DNA/gene exchanges with other colonic and endosymbiotic organism transfer the genes for antibiotic resistance to other colonic flora. This forms the basis for the generation of superbugs resistant to all antibiotics. The same mechanisms happen when populations are treated with antibiotics and other drugs like anti-hypertensives, anti-psychotics and analgesics. They kill the colonic bacterial flora and the extremophilic archaea dominates the colonic microbiome with acquired protective mechanisms for antibiotic resistance. These archaeal antibiotic resistance genes are transferred by gene and plasmid DNA exchange with remnant colonic and endosymbiotic micro-organism generating superbugs. Thus antibiotic and other non-antibiotic drugs promote colonic archaeal growth and endosymbiosis producing neanderthalisation of the species and generation of superbugs which kills of the homo sapien species. Thus the generation of superbugs and emerging viruses is promoted by a low fibre diet. A low fibre diet leads to colonic and endosymbiotic archaeal growth leading to neanderthalisation of the species and generation of superbugs and emerging viruses to which the neanderthalised species is resistant consequent to endogenous digoxin secretion. The neanderthalised species generated superbugs and emerging viruses kills of the homo sapien species which are sensitive to the superbugs and emerging viruses and do not have resistance to it. Thus the modern armamentarium of antibiotics, antipsychotics, analgesics, anti-hypertensives and anti-diabetic drugs promote colonic and endosymbiotic archaeal growth development of superbugs, neanderthalisation of species and induction of civilizational disease. Thus the drugs used to treat infections and human disease tend to produce superbugs, emerging virus infection and civilizational diseases - cancer, autoimmune disease, metabolic syndrome X, neurodegeneration, schizophrenia and autism. Thus the drug armamentarium of antibiotics and non-antibiotic drugs used to treat human infections and diseases increases colonic and endosymbiotic archaeal growth, neanderthalises the species, generates superbugs and leads to increased incidence of civilizational disease. The gut microbiome and virobiome modulated by dietary fibre intake, high protein high fat diets, antibiotics and non-antibiotic drugs like anti-psychotics, anti-hypertensives, analgesics leads to alteration in gut microbial and viroidal bio-geography leading to generation of superbugs and superviruses. The gut archaea highly resistant to conventional antibiotics and non-antibiotic drugs and proliferate when antibiotics and non-antibiotic drugs like anti-hypertensives, anti-psychotics and analgesics are used on a widespread basis. They develop antibiotic resistance and non-antibiotic drug resistance and it is transferred to other colonic and gut bacteria and viruses generating superbugs and superviruses exterminating the homo sapien population. The growth and dominance of the colonic and endosymbiotic bacteria leads to generation of homo neanderthalic species which is resistant to superbugs and superviruses and survive but get affected by civilizational diseases like cancer, autoimmunity, metabolic syndrome, schizophrenia, autism and neurodegeneration. Homo neoneanderthalis thus generated by colonic and endosymbiotic archaeal growth can become extinct over a slow period of time. Homo neanderthalis with its extremophilic archaeal endosymbiosis is hardy and resistant and can survive global warming and extremes of climate change as well as surviving extremophilic anaerobic conditions of other planets and galactic systems. The dietary fibre intake inhibits archaeal growth and populations with low dietary fibre intake are resistant to civilizational diseases like metabolic syndrome, cancer, autoimmune disease and schizophrenia. The human methanogenic archaea are highly resistant to routine antibiotics and non-antibiotic drugs and the intake of such drugs kills the colonic microbiome leading to archaeal growth and endosymbiosis and neanderthalisation to produce a hardy homo galacticus species capable of surviving in extreme climates of earth and other planets. The human drugs and antibiotics induced modulation of the gut microbiome with increased colonic and endosymbiotic archaeal growth also contributes to global warming via the mechanism of archaeal methanogenesis. The widespread use of antibiotics and non-antibiotic drugs contribute to colonic and endosymbiotic archaeal growth, climate change and still further more archaeal growth. This human colonic and endosymbiotic archaeal growth consequent to stress, internet fields, low level EMF, low dietary fibre intake, antibiotic use and non-antibiotic drug use producing methanogenesis is the principal cause of global warming. The widespread use of antibiotic and non-antibiotic drugs kills of the colonic microflora and replaces it with archaea resistant to antibiotic and non-antibiotic drugs. The low intake of dietary fibre promotes colonic archaeal and endosymbiotic archaeal growth. Stress produces catecholamine induced increased in colonic and endosymbiotic archaeal growth and breaching of the gut-blood barrier. The widespread use of internet and low level of EMF field of mobile phones promotes endosymbiotic archaeal growth. Thus endosymbiotic and colonic archaeal growth and methanogenesis producing global warming is an endogenous human biological event with human biological activity as a centre piece triggering the process. The widespread use of antibiotics and non-antibiotic drugs treat infections viral and bacterial as well as civilizational disease like cancer, schizophrenia, mood disorder, depression, autism, autoimmune disease and neurodegeneration promotes colonic and endosymbiotic archaeal growth which neanderthalises the species leading to still further transformation of the human system leading to induction of civilizational disease like cancer, metabolic syndrome X, neurodegeneration, autoimmunity and vascular disease. Thus the antibiotic and non-antibiotic drugs used to treat infections and systemic diseases produces colonic and endosymbiotic archaeal growth leading to superbugs, superviruses, pandemics and epidemic of civilizational diseases mediated by colonic and endosymbiotic archaeal growth. Human medicine as we know it has reached a dead end. The antibiotic and non-antibiotic era of drugs had come to an end. The human earth, as we know it also undergoes catastrophic climate change leading to extinction of the human species and conditions for human habitations in earth. The modern medicine as we know of and the earth as we know of have ended. The principal cause of climate change and its catastrophic complications is contributed in part only by the destruction of the tropical Amazonian and rain forest. The deforestation of the gut microflora created over billions of years of human evolution is the principal cause of the climate change and a possible end of the earth as we know of and the human species as we know of. The revolutionary medicine and the golden earth with its habitable conditions have ended. The increase in dietary fibre intake upto 40 g/day can inhibit colonic archaeal growth and endosymbiosis putting a break on generation of superbugs, emerging viruses and climate change. Increase in dietary fibre intake of populations can prevent the state of climate change and gut microbial change putting human life and the life of the planet at risk. Dietary fibre can be considered as the elixir of life and endosymbiotic archaea the philosopher's stone.

Endosymbiotic actinidic archaea as having evolved from an early isoprenoid organisms by abiogenesis. An actinidc archaea/viroid mediated model of prokaryote, viral, eukaryotic, primate and human evolution is discussed.[1-4]. Actinidic archaea has been related to the pathogenesis of schizophrenia, malignancy, metabolic syndrome X, autoimmune disease and neuronal degeneration[2]. Actinidic archaea have a mevalonate pathway and cholesterol catabolism[5-7]. Davies has put forward the concept of a shadow biosphere of organisms with alternate biochemistry present in earth itself[8]. An actinide dependent shadow biosphere of archaea and viroids in the above mentioned disease states is described[6]. Metal actinides in beach sands have been postulated to play a role in abiogenesis[6]. Actinide mineral like rutile, monazite and illmenite by surface metabolism would have contributed to abiogenesis[9]. A hypothesis of cholesterol as the primal prebiotic molecule synthesised on actinide surfaces with all other biomolecules arising from it and a self-replicating cholesterol lipid organism as the initial life form is presented. The actinidic archaea and viroids would have evolved from the primitive isoprenoid organism. The origin of viruses, prokaryotes, eukaryotes, primates and humans from the initial isoprenoid organism derived actinidic archaea is postulated.

Materials and Methods

Informed consent of the subjects and the approval of the Ethics Committee were obtained for the study. The following groups were included in the study:- endomyocardial fibrosis, Alzheimer's disease, multiple sclerosis, non-Hodgkin's lymphoma, metabolic syndrome X with cerebrovascular thrombosis and coronary artery disease, schizophrenia, autism, seizure disorder, Creutzfeldt-Jakob's disease and acquired immunodeficiency syndrome. There were 10 patients in each group and each patient had an age and sex matched healthy control selected randomly from the general population. The blood samples were drawn in the fasting state before treatment was initiated. Plasma from fasting heparinised blood was used and the experimental protocol was as follows:- (I)

Plasma+phosphate buffered saline, (II) same as I+cholesterol substrate, (III) same as II+rutile 0.1 mg/ml, and (IV) same as II+ciprofloxacine and doxycycline each in a concentration of 1 mg/ml. Cholesterol substrate was prepared as described by Richmond[10]. Aliquots were withdrawn at zero time immediately after mixing and after incubation at 37 °C for 1 hour. The following estimations were carried out:- Cytochrome F420, free RNA, free DNA, muramic acid, polycyclic aromatic hydrocarbon, hydrogen peroxide, serotonin, pyruvate, ammonia, glutamate, cytochrome C, hexokinase, ATP synthase, HMG CoA reductase, digoxin and bile acids[11-14]. Cytochrome F420 was estimated flourimetrically (excitation wavelength 420 nm and emission wavelength 520 nm). Polycyclic aromatic hydrocarbon was estimated by measuring hydrogen peroxide liberated by using glucose reagent. The statistical analysis was done by ANOVA.

Results

The parameters checked as indicated above were:- cytochrome F420, free RNA, free DNA, muramic acid, polycyclic aromatic hydrocarbon, hydrogen peroxide, serotonin, pyruvate, ammonia, glutamate, cytochrome C, hexokinase, ATP synthase, HMG CoA reductase, digoxin and bile acids. Plasma of control subjects showed increased levels of the above mentioned parameters with after incubation for 1 hour and addition of cholesterol substrate resulted in still further significant increase in these parameters. The plasma of patients showed similar results but the extent of increase was more. The addition of antibiotics to the control plasma caused a decrease in all the parameters while addition of rutile increased their levels. The addition of antibiotics to the patient's plasma caused a decrease in all the parameters while addition of rutile increased their levels but the extent of change was more in patient's sera as compared to controls.

Discussion

There was increase in cytochrome F420 indicating archaeal growth. The archaea can synthesize and use cholesterol as a carbon and energy source[15,16].

The archaeal origin of the enzyme activities was indicated by antibiotic induced suppression. The study indicates the presence of actinide based archaea with an alternate actinide based enzymes or metalloenzymes in the system as indicated by rutile induced increase in enzyme activities[17]. There was also an increase in archaeal HMG CoA reductase activity indicating increased cholesterol synthesis by the archaeal mevalonate pathway. The archaeal beta hydroxyl steroid dehydrogenase activity indicating digoxin synthesis and archaeal cholesterol hydroxylase activity indicating bile acid synthesis were increased[7]. The archaeal cholesterol oxidase activity was increased resulting in generation of pyruvate and hydrogen peroxide[16]. The pyruvate gets converted to glutamate and ammonia by the GABA shunt pathway. The archaeal aromatization of cholesterol generating PAH, serotonin and dopamine was also detected[18]. The archaeal glycolytic hexokinase activity and archaeal extracellular ATP synthase activity were increased. The archaea can undergo magnetite and calcium carbonate mineralization and can exist as calcified nanoforms[19].

The metal actinides provide radiolytic energy, catalysis for oligomer formation and provide a coordinating ion for metalloenzymes all important in abiogenesis[6]. The metal actinide surfaces would by surface metabolism generate acetate which could get converted to acetyl CoA and then to cholesterol which functions as the primal prebiotic molecule self-organising into self-replicating supramolecular systems, the lipid organism[8,9,20]. Cholesterol by radiolysis by actinides would have formed PAH generating PAH aromatic organism[8]. Cholesterol radiolysis would generate pyruvate which would get converted to amino acids, sugars, nucleotides, porphyrins, fatty acids and TCA acids. Anastase and rutile surfaces can produce polymerization of amino acids, isoprenyl residues, PAH and nucleotides to generate the initial lipid organism, PAH organism, prions and RNA viroids which would have symbiosed to generate the archaeal protocell. The archaea evolved into gram negative and gram positive bacteria with a mevalonate pathway which had an evolutionary advantage. The symbiosis of archaea with gram negative organism generated the eukaryotic cell[21]. The data supports the persistence of an actinide and cholesterol based shadow biosphere which throws light on the actinide based origin of life and cholesterol as the premier prebiotic molecule.

There was an increase in free RNA indicating self-replicating RNA viroids and free DNA indicating generation of viroid complementary DNA strands by archaeal reverse transcriptase activity. The actinides modulate RNA folding and catalyse its ribozymal action. Digoxin can cut and paste the viroidal strands by modulating RNA splicing generating RNA viroidal diversity. The viroids are evolutionarily escaped archaeal group I introns which have retrotransposition and self-splicing qualities. Archaeal pyruvate can produce histone deacetylase inhibition resulting in endogenous retroviral (HERV) reverse transcriptase and integrase expression. This can integrate the RNA viroidal complementary DNA into the non-coding region of eukaryotic non-coding DNA using HERV integrase as has been described for borna and ebola viruses[21]. The non-coding DNA is lengthened by integrating RNA viroidal complementary DNA with the integration going on as a continuing event. The archaea genome can also get integrated into human genome using integrase as has been described for trypanosomes. The integrated viroids and archaea can undergo vertical transmission and can exist as genomic parasites. This increases the length and alters the grammar of the non-coding region producing memes or memory of acquired characters. The viroidal complementary DNA can function as jumping genes producing a dynamic genome.[22-24]

The presence of muramic acid, HMG CoA reductase and cholesterol oxidase activity inhibited by antibiotics indicates the presence of bacteria with mevalonate pathway. The bacterial with mevalonate pathway include streptococcus, staphylococcus, actinomycetes, listeria, coxiella and borrelia. The bacteria and archaea with mevalonate pathway and cholesterol catabolism had an evolutionarily advantage and constitutes the isoprenoidal clade organism. The archaea evolved into mevalonate pathway gram positive and gram negative isoprenoid clade organism through horizontal gene transfer of viroidal and virus genes. The isoprenoidal clade prokaryotes develop into other groups of prokaryotes via viroidal/virus as well as eukaryotic horizontal gene transfer producing bacterial speciation[25-27].

The RNA viroids and its complementary DNA developed into cholesterol enveloped RNA and DNA viruses like herpes, retrovirus, influenza virus, borna virus, cytomegalo virus and ebstein barr virus by recombining with eukaryotic and human genes resulting in viral speciation. Bacterial and viral species are ill-defined and fuzzy with all of them forming one common genetic pool with frequent horizontal gene transfer and recombination.

Thus, the multi and unicellular eukaryote with its genes serves the purpose of prokaryotic and viral speciation. The multicellular eukaryote developed so that their endosymbiotic archaeal colonies could survive and forage better. The multicellular eukaryotes are like bacterial biofilms. The archaea and bacteria with a mevalonate pathway uses the extracellular RNA viroids and DNA viroids for quorum sensing and in the generation of symbiotic biofilm-like structures which develop into multicellular eukaryotes. The endosymbiotic archaea and bacteria with mevalonate pathway still uses the RNA viroids and DNA viroids for the regulation of multicellular eukaryote.[28-31]

Pollution is a major inducer of evolutionary innovation. Pollution is induced by the primitive nanoarchaea and mevalonate pathway bacteria synthesized PAH and methane leading on to redox stress. Redox stress leads to sodium-potassium ATPase inhibition, inward movement of plasma membrane cholesterol, defective SREBP sensing, increased cholesterol synthesis and nanoarchaeal/mevalonate pathway bacterial growth. Redox stress leads on to viroidal and archaeal multiplication. Redox stress can also lead to HERV reverse transcriptase and integrase expression. The non-coding DNA is formed of integrating RNA viroidal complementary DNA and archaea with the integration going on as a continuing event. The archaeal pox like dsDNA virus forms evolutionarily the nucleus. The integrated viroidal, archaeal and mevalonate pathway bacterial sequences can undergo vertical transmission and can exist as genomic parasites. The genomic integrated archaea, mevalonate pathway bacteria and viroids form a genomic reserve of bacteria and viruses which can recombine with human and eukaryotic genes producing bacterial and viral speciation. The change in the length and grammar of the non-coding region produces eukaryotic speciation and individuality. The integration of nanoarchaea, mevalonate pathway prokaryotes and viroids in to the eukaryotic and human genome produces a chimera which can multiply producing biofilm like multicellular structures having a mixed archaeal, viroidal, prokaryotic and eukaryotic characters which is a regression from the multicellular eukaryotic tissue. This results in a new neuronal, metabolic, immune and tissue phenotype leading to human disease.[30-32]

Pollution would have been a major factor in eukaryotic speciation and primate/hominid evolution. The change in the length and grammar of the non-coding region produces eukaryotic speciation and individuality. It is the increase in non-coding region and HERV sequences of the genome that led to the evolution of the primate and the human brain and its attendant property of conscious and quantal perception. It is the non-coding region of the genome with its archaeal, RNA viroidal complementary DNA and HERV sequences that makes for the human qualities of the hominid brain. Changes in the length of non-coding region can lead on to disorders of consciousness like schizophrenia. A schizophrenia specific human endogenous retroviruses and change in the length and grammar of the non-coding region has been described in schizophrenia.[33,34]

An actinide dependent shadow biosphere of archaea and viroids in the above mentioned disease states is described. Metal actinides in beach sands

have been postulated to play a role in abiogenesis. Cholesterol is the primal prebiotic molecule synthesized on actinide surfaces with all other biomolecules arising from it. A self-replicating cholesterol lipid organism could be the initial life form. A cholesterol based abiogenesis is a more likely evolutionary option and the actinidic archaea and viroids would have evolved from it. The origin of viruses, prokaryotes, eukaryotes, primates and humans from the initial isoprenoid organism derived actinidic archaea is discussed.

References

1. Valiathan M.S., Somers, K., Kartha, C.C. (1993). *Endomyocardial Fibrosis.* Delhi: Oxford University Press.
2. Kurup R., Kurup, P.A. (2009). *Hypothalamic Digoxin, Cerebral Dominance and Brain Function in Health and Diseases.* New York: Nova Science Publishers.
3. Hanold D., Randies, J.W. (1991). Coconut cadang-cadang disease and its viroid agent, *Plant Disease,* 75, 330-335.
4. Edwin B.T., Mohankumaran, C. (2007). Kerala wilt disease phytoplasma: Phylogenetic analysis and identification of a vector, *Proutista moesta, Physiological and Molecular Plant Pathology,* 71(1-3), 41-47.
5. Eckburg P.B., Lepp, P.W., Relman, D.A. (2003). Archaea and their potential role in human disease, *Infect Immun,* 71, 591-596.
6. Adam Z. (2007). Actinides and Life's Origins, *Astrobiology,* 7, 6-10.
7. Schoner W. (2002). Endogenous cardiac glycosides, a new class of steroid hormones, *Eur J Biochem,* 269, 2440-2448.
8. Davies P.C.W., Benner, S.A., Cleland, C.E., Lineweaver, C.H., McKay, C.P., Wolfe-Simon, F. (2009). Signatures of a Shadow Biosphere, *Astrobiology,* 10, 241-249.
9. Wächtershäuser G. (1988). Before enzymes and templates: theory of surface metabolism, *Microbiol Rev,* 52(4), 452-84.
10. Richmond W. (1973). Preparation and properties of a cholesterol oxidase from nocardia species and its application to the enzymatic assay of total cholesterol in serum, *Clin Chem,* 19, 1350-1356.
11. Snell E.D., Snell, C.T. (1961). *Colorimetric Methods of Analysis.* Vol 3A. New York: Van NoStrand.
12. Glick D. (1971). *Methods of Biochemical Analysis.* Vol 5. New York: Interscience Publishers.
13. Colowick, Kaplan, N.O. (1955). *Methods in Enzymology.* Vol 2. New York: Academic Press.
14. Maarten A.H., Marie-Jose, M., Cornelia, G., van Helden–Meewsen, Fritz, E., Marten, P.H. (1995). Detection of muramic acid in human spleen, *Infection and Immunity,* 63(5), 1652 – 1657.

15. Smit A., Mushegian, A. (2000). Biosynthesis of isoprenoids via mevalonate in Archaea: the lost pathway, *Genome Res,* 10(10), 1468-84.

16. Van der Geize R., Yam, K., Heuser, T., Wilbrink, M.H., Hara, H., Anderton, M.C. (2007). A gene cluster encoding cholesterol catabolism in a soil actinomycete provides insight into Mycobacterium tuberculosis survival in macrophages, *Proc Natl Acad Sci USA,* 104(6), 1947-52.

17. Francis A.J. (1998). Biotransformation of uranium and other actinides in radioactive wastes, *Journal of Alloys and Compounds,* 271(273), 78-84.

18. Probian C., Wülfing, A., Harder, J. (2003). Anaerobic mineralization of quaternary carbon atoms: Isolation of denitrifying bacteria on pivalic acid (2,2-Dimethylpropionic acid), *Applied and Environmental Microbiology,* 69(3), 1866-1870.

19. Vainshtein M., Suzina, N., Kudryashova, E., Ariskina, E. (2002). New Magnet-Sensitive Structures in Bacterial and Archaeal Cells, *Biol Cell,* 94(1), 29-35.

20. Russell M.J., Martin, W. (2004). The rocky roots of the acetyl-CoA Pathway, *Trends in Biochemical Sciences,* 29, 7.

21. Margulis L. (1996). Archaeal-eubacterial mergers in the origin of Eukarya: phylogenetic classification of life, *Proc Natl Acad Sci USA,* 93, 1071-1076.

22. Tsagris E.M., de Alba, A.E., Gozmanova, M., Kalantidis, K. (2008). Viroids, *Cell Microbiol,* 10, 2168.

23. Horie M., Honda, T., Suzuki, Y., Kobayashi, Y., Daito, T., Oshida, T. (2010). Endogenous non-retroviral RNA virus elements in mammalian genomes, *Nature,* 463, 84-87.

24. Hecht M., Nitz, N., Araujo, P., Sousa, A., Rosa, A., Gomes, D. (2010). Genes from Chagas parasite can transfer to humans and be passed on to children. Inheritance of DNA Transferred from American Trypanosomes to Human Hosts, *PLoS ONE,* 5, 2-10.

25. Flam F. (1994). Hints of a language in junk DNA, *Science,* 266, 1320.

26. Horbach S., Sahm, H., Welle, R. (1993). Isoprenoid biosynthesis in bacteria: two different pathways? *FEMS Microbiol Lett,* 111, 135–140.

27. Gupta R.S. (1998). Protein phylogenetics and signature sequences: a reappraisal of evolutionary relationship among archaebacteria, eubacteria, and eukaryotes, *Microbiol Mol Biol Rev,* 62, 1435–1491.

28. Hanage W., Fraser, C., Spratt, B. (2005). Fuzzy species among recombinogenic bacteria, *BMC Biology,* 3, 6-10.

29. Whitchurch C.B., Tolker-Nielsen, T., Ragas, P.C., Mattick, J.S. (2002). Extracellular DNA Required for Bacterial Biofilm Formation. *Science,* 295(5559), 1487.

30. Webb J.S., Givskov, M., Kjelleberg, S. (2003). Bacterial biofilms: prokaryotic adventures in multicellularity, *Curr Opin Microbiol,* 6(6), 578–85.

31. Chen Y., Cai, T., Wang, H., Li, Z., Loreaux, E., Lingrel, J.B. (2009). Regulation of intracellular cholesterol distribution by Na/K-ATPase, *J Biol Chem,* 284(22), 14881-90.

32. Poole A.M. (2006). Did group II intron proliferation in an endosymbiont-bearing archaeon create eukaryotes? *Biol Direct,* 1, 36-40.

33. Villarreal L.P. (2006). How viruses shape the tree of life, *Future Virology,* 1(5), 587-595.

34. Lockwood M. (1989). *Mind, Brain and the Quantum*. Oxford: B. Blackwell.

CHAPTER 6

GLOBAL WARMING, ENDOSYMBIOTIC ARCHAEAL RNA VIROIDS AND EMERGING VIRAL PANDEMICS – EARTH ADAPTATION SYNDROME, SPECIES CHANGE AND REVERSAL OF GLOBAL WARMING – INFECTIVE SYMBIOTIC TISSUE COLONIES OF PROTOZOA, ARCHAEA, VIRUSES AND BACTERIA

The plasma life based forms and the gravitational and anti-gravitational waves generated porphyrions would have been the most primitive of non-life forms. The porphyrin templates would have given rise to the isoprenoid organism, the RNA viroids, DNA viroids and prions which would have symbiosed to form the nanoarchaea. The archaea would have evolved into prokaryotes or bacteria. The archaeal symbiosis with prokaryotes would have generated the eukaryotes. The initial unicellular eukaryotes are the protozoa. The protozoa are capable of phagocytosis and would have had viruses, bacteria and archaea as their intracellular symbionts. The viruses of the protozoa like entamoeba and toxoplasma are giant viruses called megaviride. The protozoans like entamoeba with its viral, archaeal and bacterial endosymbionts would have evolved into the multicellular organisms or metazoa. The archaeal, bacterial and protozoal extracellular vesicles containing RNA and DNA viroids can by horizontal gene transfer lead to evolution of metazoan, plants, animals and human species. The metazoans would have evolved into plants, animals and humans by horizontal gene transfer generated by archaeal extracellular RNA and DNA viroid vesicles. The evidence of such a transformation is seen in the bacterial symbionts of protozoa like entamoeba which includes legionella, listeria, mycoplasma, mycobacteria and chlamydia. The symbionts of protozoan entamoeba include viruses like covid, HIV and cytomegalo virus. The protozoan entamoeba serves as a natural reservoir for covid infections, bacterial infections and HIV infections. The toxoplasma also serves as a covid reservoir.

These protozoan colonies of entamoeba and toxoplasma with its bacterial, archaeal and viral symbionts can be called as chimeric infective symbiotic tissue colonies which can exist in the air, water, earth, blood and tissues. The chimeric infective symbiotic tissue colonies containing bacteria, archaea and viruses can make the symbiotic organism multidrug resistant by horizontal gene transfer. The chimeric infective symbiotic tissue organism evolves as a consequence of global warming and the protozoal, bacterial, viral and archaeal symbiotic colonies are new forms of life. The chimeric infective symbiotic tissue organism composed of protozoa and its symbionts like virus, bacteria and archaea depend upon glycolysis for its energy needs. The glycolytic enzymes enolase, glyceraldehyde 3-phosphate dehydrogenase, hexokinase, phosphofructokinase, aldolase, pyruvic kinase and triose phosphate isomerase are moonlighting proteins having a cytoplasmic, membrane and nuclear functions. These glycolytic enzymes have a RNA binding motif which can modulate RNA transcription in the host and the symbiotic organism. These glycolytic enzymes with their RNA binding motif as well as DNA binding motif can bind to RNA and DNA viruses and viroids modulating their replication and life cycles. The chimeric infective symbiotic tissue organism composed of protozoa and its symbionts like virus, bacteria and archaea can enter the human host through the breaches of the gut-blood barrier and transform itself into the macrophages of the human host. The macrophages are involved in the phagocytosis and innate immunity against bacterial and viral infections. Paradoxically the macrophages like the protozoa can also serve as a reservoir where pathogens like mycobacteria tuberculosis, mycobacteria leprae, listeria, legionella, E. coli, shigella & salmonella, and viruses like HIV & covid can thrive and replicate. The modified chimeric infective symbiotic tissue organism or macrophage composed of protozoa and its symbionts like virus, bacteria and archaea can serve as a reservoir for horizontal gene transfer and generation of multidrug resistant bacteria and viruses. The infective chimeric symbiotic tissue organism composed of protozoa and its symbionts like virus, bacteria and archaea depends upon moonlighting glycolytic enzymes for their energetics and replication. The development of antibodies against the infective chimeric symbiotic tissue organism composed of protozoa and its symbionts like virus, bacteria and archaea leads to the evolution of anti-brain antibody disease or anti-basal ganglia antibody disease. They present a threat to existence consequent to horiztontal gene transfer and generation of multidrug resistant bacteria, viruses, protozoa, mycobacteria and archaea.

The COVID 19 brain infection can lead to integration of COVID 19 sequences into the genome, creation of cerebellar dominant brain and neanderthalisation. The COVID 19 infected cerebellar dominant brain has got increased porphyrin metabolism and porphyrions. The porphyrions are dipolar pumped phonon quantal perceptive systems which integrate with the internet. Panpsychism implies that the internet is conscious. Climate change leads to increased colonic archaeal growth and archaeal endosymbiosis. The archaeal endosymbiosis leads to methanogenesis and generation of PAH and global warming. The global warming again leads to increased endosymbiotic and colonic archaeal growth leading on to a vicious cycle. Climate change and global warming has already set in as noted by the melting of the polar ice caps and fires in California and Australia. Climate change has reached the doom's day apocalyptic scenario. The major proportion of the world ecosystem is made up of human population. Human activities in the era of industrialization lead to global warming, accumulation of greenhouse gases and initiation of climate change. This leads to increase colonic and endosymbiotic archaeal growth generating methane which accelerates global warming which assumes a tempo of its own. The human diet in the modern era with low dietary fibre and increased fat and protein also leads to colonic and endosymbiotic archaeal growth. This can also leads to increased human methane production by colonic and endosymbiotic archaeal growth initiating global warming without the effect of industrialization. The climate change can be induced by colonic and endosymbiotic archaeal growth consequent to behavioural nutritional change induced by low fibre intake.

These effects of global warming can be reversed by reducing colonic and endosymbiotic archaeal growth by increasing dietary fibre intake.

Stress, diet, EMF pollution and climate change - global warming and human archaeal endosymbiosis

Modern civilization results in changes in dietary habits leading to less consumption of fibre and more intake of protein and fat. This leads to increased colonic archaeal growth and archaeal endosymbiosis. The archaeal endosymbiosis in the brain produces increased magnetoperception and quantal perception of low level EMF producing frontal atrophy and cerebellar dominance. Thus modern civilization induced dietary habits with reduced intake of fibre can lead to increased archaeal growth, methane synthesis and global warming. These changes can be reversed by a higher intake of fibre. Modern civilization results in stress in societies leading to increased activation of the HPA axis leading to increased secretion of stress hormones the catecholamines producing increase in sympathetic tone. The increase in sympathetic tone leads to increase in stress hormones epinephrine and norepinephrine producing breeching of the gut-blood barrier. This leads to decrease in mucin secretion lining the gut wall and inflammatory changes in the colonic mucosa. This leads to increase in archaeal growth and colonic archaeal translocation into to the blood and tissues producing archaeal endosymbiosis. The archaeal endosymbiosis in the brain can lead to increased quantal and magnetoperception of low level EMF producing frontal atrophy and cerebellar dominance. Thus the modern civilization related stress and increased sympathetic tone can lead to increased colonic archaeal growth and endosymbiosis producing more methane synthesis and global warming.

Endosymbiotic archaeal generation of new pandemic viruses

The climate change related increased endosymbiotic archaeal growth leads to emerging viral pandemics like the corona virus pandemic and more such pandemics in future. The endosymbiotic and colonic human archaea secrete RNA and DNA viruses and serves as a reservoir for them. Viruses belonging to the pox family and African swine fever have been described in archaea. The archaeal viroids secrete enveloped viruses like the SARS, MERS, corona and HIV. The archaeal virus secretion process is the same as that of enveloped viruses like SARS, MERS and corona. The bats serve as a reservoir for the viruses like SARS, MERS and corona. The bats contain archaea which can secrete RNA viroids and generate viruses. The colonic and endosymbiotic archaea also generates RNA viroids which are converted to DNA viroids by colonic epithelial HERV reverse transcriptase and gets integrated into the colonic genome. The RNA and DNA viroids hybridise with the viroidal and bacterial population of the gut microbiome and virobiome - DNA and RNA viruses of the gut leading to generation of new emerging viruses. The RNA and DNA viroids and gut bacteria can also hybridise with human genomic sequences for immunity, metabolic regulation, growth factors, cell growth, cell death, neurotransmitters and HERV sequences of the colonial epithelial cell. The hybridization of archaeal RNA viroids with human genomic HERV sequences can generate retroviruses. This leads to generation of new bacteria, RNA and DNA viruses producing pandemics and extermination of homo sapiens. The homo sapiens become an extinct species and the homo neanderthalis will dominate the earth.

Viral pandemics and revitilisation of the earth – reversal of global warming and species change

The emerging viral pandemics paralyse the human population systems and economic activity leading to grounding of the population and closing down of national economies. This leads to reduction in greenhouse gases and global warming. The earth is a living system and panpsychism postulates a consciousness field which permeates the universe and interacts with subatomic particles bringing the macroscopic world into existence as the ubiquitous quantal observer. Panpsychism postulates that the earth, the sun and the universe is conscious. The human population and the earth are both conscious in the panpsychic world of existence. Human activity leads to destruction of the environment of the earth and the climate change leads to doubts as to existence of the ecosystems of earth which are all conscious. The human endosymbiotic and bat/animal endosymbiotic archaeal populations can generate new viruses producing large scale pandemics and homo sapien extinction with a corresponding reversal of environmental degradation produced by homo sapien activity. The growth of endosymbiotic archaea can lead to increase in size and numbers of remnant neanderthalic populations and renewal of the homo neanderthalic and homo neoneanderthalic species. Thus homo sapien activity related endosymbiotic archaeal growth can lead to the global warming and homo sapien species extinction owing to generation of widespread recurring viral pandemics. These viral pandemics will help to put a brake on human activity and climate change producing a reversal of the environmental change and a change in species dominance with homo neanderthalis dominating over the world.

The archaea can induce stem cell conversion and neanderthalisation of the human species. The archaea can produce permanent immune activation producing resistance to viral and bacterial infection. The archaeal cholesterol catabolism depletes tissue cholesterol producing vitamin D deficiency and immune activation. Thus archaeal overgrowth results in retroviral resistance and generation of the Neanderthal phenotype. The endosymbiotic archaea can secrete virus like RNA and DNA particles. The endosymbiotic archaea can induce uncoupling proteins inhibiting mitochondrial oxidative phosphorylation and generating ROS. The endosymbiotic archaeal magnetite can generate low level of EMF. The low level of EMF and ROS are genotoxic and produce breakages in hotspots of chromosome. It can also trigger rearrangements in hotspots of chromosome inhabited by retroviral and non-retroviral elements producing their expression. The archaeal secreted DNA and RNA viroids can recombine with the expressed retroviral, non-retroviral elements and other genomic segments of the human chromosome generating new RNA and DNA viruses. Thus the neanderthalised humans can serve as an origin for new RNA and DNA viruses as well as mutated retroviruses. The endosymbiotic archaea converts the Neanderthal cells to stem cells. The stem cells are resistant to immune attack. The stem cells can serve as a reservoir for this new RNA and DNA viruses. The stem cells and archaeal cells can also serve as a reservoir for viruses and bacteria belonging to other plants and animals. This helps to generate the species barrier jump in noted in recent emerging viral and bacterial infections. Thus the endosymbiotic archaeal growth produces neanderthalised version of homo sapiens which are retroviral resistant and resistant to other viral and bacterial infection consequent to immune activation and digoxin induced RNA editing. The endosymbiotic archaeal overgrowth mediated neanderthalised version of homo sapiens generates new mutated RNA and DNA viruses as well as retroviruses at the same time being resistant to them as in the case of the species bat. The homo sapiens do not have the Neanderthal mechanisms of immune activation as their archaeal load is meagre. They serve as fodder for infection from Neanderthal generated viruses and bacteria and suffer eventual extinction.[1-17] This paper studied the archaeal status in patients with recurrent viral infections and retroviral infections. The generation of RNA and DNA viroids from archaea was also studied.

Diet, modern cannibalism and global warming

Dietary fibre is the most important factor that modulates human biology and speciation. Dietary fibre and human drugs - antibiotic and non-antibiotic can modulate the gut flora producing colonic archaeal growth and endosymbiosis generating climate change, evolution of superbugs and emerging viruses and creating the disease phenotype. The high fibre diet is digested by the colonic flora leading to generation of short chain fatty acids and inhibits endosymbiotic and colonic archaeal growth. This leads to homo sapienisation of the species. A low fibre diet with increase in protein and fat leads to stimulation of endosymbiotic and colonic archaeal growth. This leads to neanderthalisation of the species. Thus a low fibre diet alters the gut microbiome in the colon with more of archaeal growth. The same mechanisms happen when populations are treated with antibiotics and other drugs like anti-hypertensives, anti-psychotics and analgesics. They kill the colonic bacterial flora and the extremophilic archaea dominates the colonic microbiome. Thus antibiotic and other non-antibiotic drugs promote colonic archaeal growth and endosymbiosis producing neanderthalisation of the species and generation of superviruses which kills of the homo sapien species. Thus the generation of emerging viruses is promoted by a low fibre diet. A low fibre diet leads to colonic and endosymbiotic archaeal growth leading to neanderthalisation of the species and emerging viruses to which the neanderthalised species is resistant consequent to endogenous digoxin secretion. The neanderthalised species generated emerging viruses kills of the homo sapien species which are sensitive to emerging viruses and do not have resistance to it. Thus the modern armamentarium of antibiotics, anti-psychotics, analgesics, anti-hypertensives and anti-diabetic drugs promote colonic and endosymbiotic archaeal growth development of superbugs, neanderthalisation of species and induction of civilizational disease. Thus the drugs used to treat infections and human diseases tend to produce emerging virus infection. Thus the drug armamentarium of antibiotics and non-antibiotic drugs used to treat human infections and diseases increases colonic and endosymbiotic archaeal growth, neanderthalises the species, generates emerging viruses. The gut microbiome and virobiome modulated by dietary fibre intake, high protein high fat diets, antibiotics and non-antibiotic drugs like anti-psychotics, anti-hypertensives, analgesics leads to alteration in gut microbial and viroidal bio-geography leading to generation of superviruses. The gut archaea highly resistant to conventional antibiotics and non-antibiotic drugs and proliferate when antibiotics and non-antibiotic drugs like anti-hypertensives, anti-psychotics and analgesics are used on a widespread basis. They generate superviruses exterminating the homo sapien population. The growth and dominance of the colonic and endosymbiotic archaea leads to generation of homo neanderthalic species which is resistant to superviruses and survive. Homo neanderthalis with its extremophilic archaeal endosymbiosis is hardy and resistant and can survive global warming and extremes of climate change as well as surviving extremophilic anaerobic conditions of other planets and galactic systems. The human methanogenic archaea are highly resistant to routine antibiotics and non-antibiotic drugs and the intake of such drugs kills the colonic microbiome leading to archaeal growth and endosymbiosis and neanderthalisation to produce a hardy homo galacticus species capable of surviving in extreme climates of earth and other planets. The human drugs and antibiotics induced modulation of the gut microbiome with increased colonic and endosymbiotic archaeal growth also contributes to global warming via the mechanism of archaeal methanogenesis. The widespread use of antibiotics and non-antibiotic drugs contribute to colonic and endosymbiotic archaeal growth, climate change and still further more archaeal growth. This human colonic and endosymbiotic archaeal growth consequent to stress, internet fields, low level EMF, low dietary fibre intake, antibiotic use and non-antibiotic drug use producing methanogenesis is the principal cause of global warming. The widespread use of antibiotic and non-antibiotic drugs kills of the colonic microflora and replaces it with archaea resistant to antibiotic and non-antibiotic drugs. The low intake of dietary fibre promotes colonic archaeal and endosymbiotic archaeal growth. Stress produces catecholamine induced increased in colonic and endosymbiotic archaeal growth and breaching of the gut-blood barrier. The widespread use of internet and low level of EMF field of mobile phones promotes endosymbiotic archaeal growth. Thus endosymbiotic and colonic archaeal growth and methanogenesis producing global warming is an endogenous human biological event with human biological activity as a centre piece triggering the process. The widespread use of antibiotics and non-antibiotic drugs treat infections viral and bacterial as well as civilizational disease like cancer, schizophrenia, mood disorder, depression, autism, autoimmune disease and neurodegeneration promotes colonic and endosymbiotic archaeal growth which neanderthalises the species leading to still further transformation of the human system leading to induction of civilizational disease like cancer, metabolic syndrome X, neurodegeneration, autoimmunity and vascular disease. Thus, the antibiotic and non-antibiotic drugs used to treat infections and systemic diseases produces colonic and endosymbiotic archaeal growth leading to superbugs, superviruses, pandemics. Human medicine as we know it, has reached a dead end. The antibiotic and non-antibiotic era of drugs had come to an end. The human earth as we know it also undergoes catastrophic climate change leading to extinction of the human species and conditions for human habitations in earth. The modern medicine as we know of and the earth as we know of have ended. The principal cause of climate change and its catastrophic complications is contributed in part only by the destruction of the tropical Amazonian and rain forest. The deforestation of the gut microflora created over billions of years of human evolution is the principal cause of the climate change and a possible end of the earth as we know of and the human species as we know of. The revolutionary medicine and the golden earth with its habitable conditions have ended. The increase in dietary fibre intake upto 40 g/day can inhibit colonic archaeal growth and endosymbiosis putting a break on generation of superbugs, emerging viruses and climate change. Increase in dietary fibre intake of populations can prevent the state of climate change and gut microbial change putting human life and the life of the planet at risk. Dietary fibre can be considered as the elixir of life and endosymbiotic archaea the philosopher's stone.

Endosymbiotic actinidic archaea as having evolved from an early isoprenoid organisms by abiogenesis. An actinidc archaea/viroid mediated model of prokaryote, viral, eukaryotic, primate and human evolution is discussed.[1-4]. Actinidic archaea has been related to the pathogenesis of schizophrenia, malignancy, metabolic syndrome X, autoimmune disease and neuronal degeneration[2]. Actinidic archaea have a mevalonate pathway and cholesterol catabolism[5-7]. Davies has put forward the concept of a shadow biosphere of organisms with alternate biochemistry present in earth itself[8]. An actinide dependent shadow biosphere of archaea and viroids in the above mentioned disease states is described[6]. Metal actinides in beach sands have been postulated to play a role in abiogenesis[6]. Actinide mineral like rutile, monazite and illmenite by surface metabolism would have contributed to abiogenesis[9]. A hypothesis of cholesterol as the primal prebiotic molecule synthesized on actinide surfaces with all other biomolecules arising from it and a self-replicating cholesterol lipid organism as the initial life form is presented. The actinidic archaea and viroids would have evolved from the primitive isoprenoid organism. The origin of viruses, prokaryotes, eukaryotes, primates and humans from the initial isoprenoid organism derived actinidic archaea is postulated.

Materials and Methods

Informed consent of the subjects and the approval of the Ethics Committee were obtained for the study. The following groups were included in the study:- endomyocardial fibrosis, Alzheimer's disease, multiple sclerosis, non-Hodgkin's lymphoma, metabolic syndrome X with cerebrovascular thrombosis and coronary artery disease, schizophrenia, autism, seizure disorder, Creutzfeldt-Jakob's disease and acquired immunodeficiency syndrome. There were 10 patients in each group and each patient had an age and sex matched healthy control selected randomly from the general population. The blood samples were drawn in the fasting state before treatment was initiated. Plasma from fasting heparinised blood was used and the experimental protocol was as follows:- (I) Plasma+phosphate buffered saline, (II) same as I+cholesterol substrate, (III) same as II+rutile 0.1 mg/ml, and (IV) same as II+ciprofloxacine and doxycycline each in a concentration of 1 mg/ml. Cholesterol substrate was prepared as described by Richmond[10]. Aliquots were withdrawn at zero time immediately after mixing and after incubation at 37 °C for 1 hour. The following estimations were carried out:- Cytochrome F420, free RNA, free DNA, muramic acid, polycyclic aromatic hydrocarbon, hydrogen peroxide, serotonin, pyruvate, ammonia, glutamate, cytochrome C, hexokinase, ATP synthase, HMG CoA reductase, digoxin and bile acids[11-14]. Cytochrome F420 was estimated flourimetrically (excitation wavelength 420 nm and emission wavelength 520 nm). Polycyclic aromatic hydrocarbon was estimated by measuring hydrogen peroxide liberated by using glucose reagent. The statistical analysis was done by ANOVA.

Results

The parameters checked as indicated above were:- cytochrome F420, free RNA, free DNA, muramic acid, polycyclic aromatic hydrocarbon, hydrogen peroxide, serotonin, pyruvate, ammonia, glutamate, cytochrome C, hexokinase, ATP synthase, HMG CoA reductase, digoxin and bile acids. Plasma of control subjects showed increased levels of the above mentioned parameters with after incubation for 1 hour and addition of cholesterol substrate resulted in still further significant increase in these parameters. The plasma of patients showed similar results but the extent of increase was more. The addition of antibiotics to the control plasma caused a decrease in all the parameters while addition of rutile increased their levels. The addition of antibiotics to the patient's plasma caused a decrease in all the parameters while addition of rutile increased their levels but the extent of change was more in patient's sera as compared to controls.

Discussion

There was increase in cytochrome F420 indicating archaeal growth. The archaea can synthesize and use cholesterol as a carbon and energy source[15,16]. The archaeal origin of the enzyme activities was indicated by antibiotic induced suppression. The study indicates the presence of actinide based archaea with an alternate actinide based enzymes or metalloenzymes in the system as indicated by rutile induced increase in enzyme activities[17]. There was also an increase in archaeal HMG CoA reductase activity indicating increased cholesterol synthesis by the archaeal mevalonate pathway. The archaeal beta hydroxyl steroid dehydrogenase activity indicating digoxin synthesis and archaeal cholesterol hydroxylase activity indicating bile acid synthesis were increased[7]. The archaeal cholesterol oxidase activity was increased resulting in generation of pyruvate and hydrogen peroxide[16]. The pyruvate gets converted to glutamate and ammonia by the GABA shunt pathway. The archaeal aromatization of cholesterol generating PAH, serotonin and dopamine was also detected[18]. The archaeal glycolytic hexokinase activity and archaeal extracellular ATP synthase activity were increased. The archaea can undergo magnetite and calcium carbonate mineralization and can exist as calcified nanoforms[19].

The metal actinides provide radiolytic energy, catalysis for oligomer formation and provide a coordinating ion for metalloenzymes all important in abiogenesis[6]. The metal actinide surfaces would by surface metabolism generate acetate which could get converted to acetyl CoA and then to cholesterol which functions as the primal prebiotic molecule self-organising into self-replicating supramolecular systems, the lipid organism[8,9,20]. Cholesterol by radiolysis by actinides would have formed PAH generating PAH aromatic organism[8]. Cholesterol radiolysis would generate pyruvate which would get converted to amino acids, sugars, nucleotides, porphyrins, fatty acids and TCA acids. Anastase and rutile surfaces can produce polymerization of amino acids, isoprenyl residues, PAH and nucleotides to generate the initial lipid organism,

PAH organism, prions and RNA viroids which would have symbiosed to generate the archaeal protocell. The archaea evolved into gram negative and gram positive bacteria with a mevalonate pathway which had an evolutionary advantage. The symbiosis of archaea with gram negative organism generated the eukaryotic cell[21]. The data supports the persistence of an actinide and cholesterol based shadow biosphere which throws light on the actinide based origin of life and cholesterol as the premier prebiotic molecule.

There was an increase in free RNA indicating self-replicating RNA viroids and free DNA indicating generation of viroid complementary DNA strands by archaeal reverse transcriptase activity. The actinides modulate RNA folding and catalyse its ribozymal action. Digoxin can cut and paste the viroidal strands by modulating RNA splicing generating RNA viroidal diversity. The viroids are evolutionarily escaped archaeal group I introns which have retrotransposition and self-splicing qualities. Archaeal pyruvate can produce histone deacetylase inhibition resulting in endogenous retroviral (HERV) reverse transcriptase and integrase expression. This can integrate the RNA viroidal complementary DNA into the non-coding region of eukaryotic non-coding DNA using HERV integrase as has been described for borna and ebola viruses[21]. The non-coding DNA is lengthened by integrating RNA viroidal complementary DNA with the integration going on as a continuing event. The archaea genome can also get integrated into human genome using integrase as has been described for trypanosomes. The integrated viroids and archaea can undergo vertical transmission and can exist as genomic parasites. This increases the length and alters the grammar of the non-coding region producing memes or memory of acquired characters. The viroidal complementary DNA can function as jumping genes producing a dynamic genome.[22-24]

The presence of muramic acid, HMG CoA reductase and cholesterol oxidase activity inhibited by antibiotics indicates the presence of bacteria with mevalonate pathway. The bacterial with mevalonate pathway include streptococcus, staphylococcus, actinomycetes, listeria, coxiella and borrelia. The bacteria and archaea with mevalonate pathway and cholesterol catabolism had an evolutionarily advantage and constitutes the isoprenoidal clade organism. The archaea evolved into mevalonate pathway gram positive and gram negative isoprenoid clade organism through horizontal gene transfer of viroidal and virus genes. The isoprenoidal clade prokaryotes develop into other groups of prokaryotes via viroidal/virus as well as eukaryotic horizontal gene transfer producing bacterial speciation[25-27].

The RNA viroids and its complementary DNA developed into cholesterol enveloped RNA and DNA viruses like herpes, retrovirus, influenza virus, borna virus, cytomegalo virus and ebstein barr virus by recombining with eukaryotic and human genes resulting in viral speciation. Bacterial and viral species are ill-defined and fuzzy with all of them forming one common genetic pool with frequent horizontal gene transfer and recombination.

Thus, the multi and unicellular eukaryote with its genes serves the purpose of prokaryotic and viral speciation. The multicellular eukaryote developed so that their endosymbiotic archaeal colonies could survive and forage better. The multicellular eukaryotes are like bacterial biofilms. The archaea and bacteria with a mevalonate pathway uses the extracellular RNA viroids and DNA viroids for quorum sensing and in the generation of symbiotic biofilm-like structures which develop into multicellular eukaryotes. The endosymbiotic archaea and bacteria with mevalonate pathway still uses the RNA viroids and DNA viroids for the regulation of multicellular eukaryote.[28-31]

Pollution is a major inducer of evolutionary innovation. Pollution is induced by the primitive nanoarchaea and mevalonate pathway bacteria synthesized PAH and methane leading on to redox stress. Redox stress leads to sodium-potassium ATPase inhibition, inward movement of plasma membrane cholesterol, defective SREBP sensing, increased cholesterol synthesis and nanoarchaeal/mevalonate pathway bacterial growth. Redox stress leads on to viroidal and archaeal multiplication. Redox stress can also lead to HERV reverse transcriptase and integrase expression. The non-coding DNA is formed of integrating RNA viroidal complementary DNA and archaea with the integration going on as a continuing event. The archaeal pox like dsDNA virus forms evolutionarily the nucleus. The integrated viroidal, archaeal and mevalonate pathway bacterial sequences can undergo vertical transmission and can exist as genomic parasites. The genomic integrated archaea, mevalonate pathway bacteria and viroids form a genomic reserve of bacteria and viruses which can recombine with human and eukaryotic genes producing bacterial and viral speciation. The change in the length and grammar of the non-coding region produces eukaryotic speciation and individuality. The integration of nanoarchaea, mevalonate pathway prokaryotes and viroids in to the eukaryotic and human genome produces a chimera which can multiply producing biofilm like multicellular structures having a mixed archaeal, viroidal, prokaryotic and eukaryotic characters which is a regression from the multicellular eukaryotic tissue. This results in a new neuronal, metabolic, immune and tissue phenotype leading to human disease.[30-32]

Pollution would have been a major factor in eukaryotic speciation and primate/hominid evolution. The change in the length and grammar of the non-coding region produces eukaryotic speciation and individuality. It is the increase in non-coding region and HERV sequences of the genome that led to the evolution of the primate and the human brain and its attendant property of conscious and quantal perception. It is the non-coding region of the genome with its archaeal, RNA viroidal complementary DNA and HERV sequences that makes for the human qualities of the hominid brain. Changes in the length of non-coding region can lead on to disorders of consciousness like schizophrenia. A schizophrenia specific human endogenous retroviruses and change in the length and grammar of the non-coding region has been described in schizophrenia.[33,34]

An actinide dependent shadow biosphere of archaea and viroids in the above mentioned disease states is described. Metal actinides in beach sands have been postulated to play a role in abiogenesis. Cholesterol is the primal prebiotic molecule synthesized on actinide surfaces with all other biomolecules arising from it. A self-replicating cholesterol lipid organism could be the initial life form. A cholesterol based abiogenesis is a more likely evolutionary option and the actinidic archaea and viroids would have evolved from it. The origin of viruses, prokaryotes, eukaryotes, primates and humans from the initial isoprenoid organism derived actinidic archaea is discussed.

References

1. Valiathan M.S., Somers, K., Kartha, C.C. (1993). *Endomyocardial Fibrosis.* Delhi: Oxford University Press.
2. Kurup R., Kurup, P.A. (2009). *Hypothalamic Digoxin, Cerebral Dominance and Brain Function in Health and Diseases*. New York: Nova Science Publishers.
3. Hanold D., Randies, J.W. (1991). Coconut cadang-cadang disease and its viroid agent, *Plant Disease,* 75, 330-335.
4. Edwin B.T., Mohankumaran, C. (2007). Kerala wilt disease phytoplasma: Phylogenetic analysis and identification of a vector, *Proutista moesta, Physiological and Molecular Plant Pathology,* 71(1-3), 41-47.
5. Eckburg P.B., Lepp, P.W., Relman, D.A. (2003). Archaea and their potential role in human disease, *Infect Immun,* 71, 591-596.
6. Adam Z. (2007). Actinides and Life's Origins, *Astrobiology,* 7, 6-10.
7. Schoner W. (2002). Endogenous cardiac glycosides, a new class of steroid hormones, *Eur J Biochem,* 269, 2440-2448.
8. Davies P.C.W., Benner, S.A., Cleland, C.E., Lineweaver, C.H., McKay, C.P., Wolfe-Simon, F. (2009). Signatures of a Shadow Biosphere, *Astrobiology,* 10, 241-249.
9. Wächtershäuser G. (1988). Before enzymes and templates: theory of surface metabolism, *Microbiol Rev,* 52(4), 452-84.
10. Richmond W. (1973). Preparation and properties of a cholesterol oxidase from nocardia species and its application to the enzymatic assay of total cholesterol in serum, *Clin Chem,* 19, 1350-1356.
11. Snell E.D., Snell, C.T. (1961). *Colorimetric Methods of Analysis.* Vol 3A. New York: Van NoStrand.

12. Glick D. (1971). *Methods of Biochemical Analysis*. Vol 5. New York: Interscience Publishers.

13. Colowick, Kaplan, N.O. (1955). *Methods in Enzymology*. Vol 2. New York: Academic Press.

14. Maarten A.H., Marie-Jose, M., Cornelia, G., van Helden–Meewsen, Fritz, E., Marten, P.H. (1995). Detection of muramic acid in human spleen, *Infection and Immunity,* 63(5), 1652 – 1657.

15. Smit A., Mushegian, A. (2000). Biosynthesis of isoprenoids via mevalonate in Archaea: the lost pathway, *Genome Res,* 10(10), 1468-84.

16. Van der Geize R., Yam, K., Heuser, T., Wilbrink, M.H., Hara, H., Anderton, M.C. (2007). A gene cluster encoding cholesterol catabolism in a soil actinomycete provides insight into Mycobacterium tuberculosis survival in macrophages, *Proc Natl Acad Sci USA,* 104(6), 1947-52.

17. Francis A.J. (1998). Biotransformation of uranium and other actinides in radioactive wastes, *Journal of Alloys and Compounds,* 271(273), 78-84.

18. Probian C., Wülfing, A., Harder, J. (2003). Anaerobic mineralization of quaternary carbon atoms: Isolation of denitrifying bacteria on pivalic acid (2,2-Dimethylpropionic acid), *Applied and Environmental Microbiology,* 69(3), 1866-1870.

19. Vainshtein M., Suzina, N., Kudryashova, E., Ariskina, E. (2002). New Magnet-Sensitive Structures in Bacterial and Archaeal Cells, *Biol Cell,* 94(1), 29-35.

20. Russell M.J., Martin, W. (2004). The rocky roots of the acetyl-CoA Pathway, *Trends in Biochemical Sciences,* 29, 7.

21. Margulis L. (1996). Archaeal-eubacterial mergers in the origin of Eukarya: phylogenetic classification of life, *Proc Natl Acad Sci USA,* 93, 1071-1076.

22. Tsagris E.M., de Alba, A.E., Gozmanova, M., Kalantidis, K. (2008). Viroids, *Cell Microbiol,* 10, 2168.

23. Horie M., Honda, T., Suzuki, Y., Kobayashi, Y., Daito, T., Oshida, T. (2010). Endogenous non-retroviral RNA virus elements in mammalian genomes, *Nature,* 463, 84-87.

24. Hecht M., Nitz, N., Araujo, P., Sousa, A., Rosa, A., Gomes, D. (2010). Genes from Chagas parasite can transfer to humans and be passed on to children. Inheritance of DNA Transferred from American Trypanosomes to Human Hosts, *PLoS ONE,* 5, 2-10.

25. Flam F. (1994). Hints of a language in junk DNA, *Science,* 266, 1320.

26. Horbach S., Sahm, H., Welle, R. (1993). Isoprenoid biosynthesis in bacteria: two different pathways? *FEMS Microbiol Lett,* 111, 135–140.

27. Gupta R.S. (1998). Protein phylogenetics and signature sequences: a reappraisal of evolutionary relationship among archaebacteria, eubacteria, and eukaryotes, *Microbiol Mol Biol Rev,* 62, 1435–1491.

28. Hanage W., Fraser, C., Spratt, B. (2005). Fuzzy species among recombinogenic bacteria, *BMC Biology,* 3, 6-10.

29. Whitchurch C.B., Tolker-Nielsen, T., Ragas, P.C., Mattick, J.S. (2002). Extracellular DNA Required for Bacterial Biofilm Formation. *Science,* 295(5559), 1487.

30. Webb J.S., Givskov, M., Kjelleberg, S. (2003). Bacterial biofilms: prokaryotic adventures in multicellularity, *Curr Opin Microbiol,* 6(6), 578–85.

31. Chen Y., Cai, T., Wang, H., Li, Z., Loreaux, E., Lingrel, J.B. (2009). Regulation of intracellular cholesterol distribution by Na/K-ATPase, *J Biol Chem,* 284(22), 14881-90.

32. Poole A.M. (2006). Did group II intron proliferation in an endosymbiont-bearing archaeon create eukaryotes? *Biol Direct,* 1, 36-40.

33. Villarreal L.P. (2006). How viruses shape the tree of life, *Future Virology,* 1(5), 587-595.

34. Lockwood M. (1989). *Mind, Brain and the Quantum*. Oxford: B. Blackwell.

CHAPTER 7

ADULT ONSET AUTISM EPIDEMIC – INFECTIVE SYMBIOTIC TISSUE COLONIES OF PROTOZOA, ARCHAEA, VIRUSES AND BACTERIA

The plasma life based forms and the gravitational and anti-gravitational waves generated porphyrions would have been the most primitive of non-life forms. The porphyrin templates would have given rise to the isoprenoid organism, the RNA viroids, DNA viroids and prions which would have symbiosed to form the nanoarchaea. The archaea would have evolved into prokaryotes or bacteria. The archaeal symbiosis with prokaryotes would have generated the eukaryotes. The initial unicellular eukaryotes are the protozoa. The protozoa are capable of phagocytosis and would have had viruses, bacteria and archaea as their intracellular symbionts. The viruses of the protozoa like entamoeba and toxoplasma are giant viruses called megaviride. The protozoans like entamoeba with its viral, archaeal and bacterial endosymbionts would have evolved into the multicellular organisms or metazoa. The archaeal, bacterial and protozoal extracellular vesicles containing RNA and DNA viroids can by horizontal gene transfer lead to evolution of metazoan, plants, animals and human species. The metazoans would have evolved into plants, animals and humans by horizontal gene transfer generated by archaeal extracellular RNA and DNA viroid vesicles. The evidence of such a transformation is seen in the bacterial symbionts of protozoa like entamoeba which includes legionella, listeria, mycoplasma, mycobacteria and chlamydia. The symbionts of protozoan entamoeba include viruses like covid, HIV and cytomegalo virus. The protozoan entamoeba serves as a natural reservoir for covid infections, bacterial infections and HIV infections. The toxoplasma also serves as a covid reservoir. These protozoan colonies of entamoeba and toxoplasma with its bacterial, archaeal and viral symbionts can be called as chimeric infective symbiotic tissue colonies

which can exist in the air, water, earth, blood and tissues. The chimeric infective symbiotic tissue colonies containing bacteria, archaea and viruses can make the symbiotic organism multidrug resistant by horizontal gene transfer. The chimeric infective symbiotic tissue organism evolves as a consequence of global warming and the protozoal, bacterial, viral and archaeal symbiotic colonies are new forms of life. The chimeric infective symbiotic tissue organism composed of protozoa and its symbionts like virus, bacteria and archaea depend upon glycolysis for its energy needs. The glycolytic enzymes enolase, glyceraldehyde 3-phosphate dehydrogenase, hexokinase, phosphofructokinase, aldolase, pyruvic kinase and triose phosphate isomerase are moonlighting proteins having a cytoplasmic, membrane and nuclear functions. These glycolytic enzymes have a RNA binding motif which can modulate RNA transcription in the host and the symbiotic organism. These glycolytic enzymes with their RNA binding motif as well as DNA binding motif can bind to RNA and DNA viruses and viroids modulating their replication and life cycles. The chimeric infective symbiotic tissue organism composed of protozoa and its symbionts like virus, bacteria and archaea can enter the human host through the breaches of the gut-blood barrier and transform itself into the macrophages of the human host. The macrophages are involved in the phagocytosis and innate immunity against bacterial and viral infections. Paradoxically the macrophages like the protozoa can also serve as a reservoir where pathogens like mycobacteria tuberculosis, mycobacteria leprae, listeria, legionella, E. coli, shigella & salmonella, and viruses like HIV & covid can thrive and replicate. The modified chimeric infective symbiotic tissue organism or macrophage composed of protozoa and its symbionts like virus, bacteria and archaea can serve as a reservoir for horizontal gene transfer and generation of multidrug resistant bacteria and viruses. The infective chimeric symbiotic tissue organism composed of protozoa and its symbionts like virus, bacteria and archaea depends upon moonlighting glycolytic enzymes for their energetics and replication. The development of antibodies against the infective chimeric

symbiotic tissue organism composed of protozoa and its symbionts like virus, bacteria and archaea leads to the evolution of anti-brain antibody disease or anti-basal ganglia antibody disease. They present a threat to existence consequent to horiztontal gene transfer and generation of multidrug resistant bacteria, viruses, protozoa, mycobacteria and archaea.

The COVID 19 brain infection can lead to integration of COVID 19 sequences into the genome, creation of cerebellar dominant brain and neanderthalisation. The COVID 19 infected cerebellar dominant brain has got increased porphyrin metabolism and porphyrions. The porphyrions are dipolar pumped phonon quantal perceptive systems which integrate with the internet. Panpsychism implies that the internet is conscious. Climate change and global warming leads to endosymbiotic archaeal growth and alteration in brain structure and function. The endosymbiotic archaea have got magnetoperception and quantal perception of low level of EMF producing brain changes resulting in frontal cortical atrophy and cerebellar dominance. This produces the syndrome of epidemic cerebellar cognitive disorder with adult onset autism like presentation. This results in impulsive behaviour and surrealistic experience. The adult onset cerebellar cognitive disorder with autistic presentation leads to social withdrawal and agarophobia. This results in lack of social communication, empathy, compassion and affection leading to an epidemic of loneliness. The brain with archaeal endosymbiosis is capable of quantal perception and gets integrated with internet communications. The neanderthalised brain with archaeal endosymbiosis can modulate internet function by quantal and magnetoperception. Archaeal endosymbiosis can lead to man-internet hybrids with existence linked to the internet and life in a virtual world with virtual friends. This can lead to internet addiction, video game addiction and withdrawal from the real world. The cerebellar dominance leads to impulsive behaviour contributing to aggression, violence, terrorism and criminality. The cerebellar dominance and intuitive behaviour can also lead to quantal perception and increased spirituality. The aggressive behaviour and spirituality can both produce transcendence leading to a surreal civilization. The archaeal endosymbiosis produces cholesterol catabolism and depletion of sex hormones resulting in asexual phenotypes and phenotypes with alternate sexuality. There is also a decrease in fertility producing a shrinkage in population. This is compensated by an increase in longevity produced by implantation of biochips to improve ageing organ function resulting in functional cyborgs. The cerebellar dominant autistic adult phenotype is capable of quantal perception and intergalactic communication. The civilization of adult onset autistic cyborgs produces a world without empathy, affection, compassion, tolerance or logic. This results in a surreal syntheistic evil world of autistic asexual unemotional cyborgs and hermits.

The endosymbiotic neuronal archaea are capable of quantal perception and magnetoperception of low level EMF producing frontal atrophy and cerebellar dominance. This produces a Neanderthal type of brain and loss of sense of individuality. The quantal perception leads to a mob or tribal psychology with loss of logical reasoning and mass hysteria and avolition. The society is led by the mob and illusive perceptions without any reason or logic. The world of illogicality and illusions steps in.

The endosymbiotic neuronal archaeal growth leads to a cerebellar cognitive affective disorder producing frontal lobe dysfunction and dominance of cerebellar cognitive functions. This results in impulsivity, violent behaviour, terrorism, criminality and evil behaviour without any rationality. The society becomes organised by virtual media in the small tribes without any rational behaviour and hostile to each other. There is also an increase in spirituality and religiosity leading on to transcendence. The impulsive violent behaviour and terrorist acts also leads on to transcendence. This creates a type of surrealistic world.

The endosymbiotic neuronal archaeal overgrowth leads to a cerebellar cognitive affective disorder and adult onset autistic phenotype. There is lack of social communication, empathy, compassion and affection. This is replaced by selfishness, greed and a culture of evil. This results in a civilization of tribe adult onset autistic phenotypes without any social skills or civil responsibility. A culture of greed, selfishness and irrationality dominates the world.

The endosymbiotic archaeal overgrowth can lead to quantal and magnetoperception of low level of EMF leading to integration with the internet and producing internet human brain hybrids. The internet dominates the human brain taking over it. This creates stifling of individuality and creativity. There is a paucity of creative thinking leading to a decline in art, literature, music, poetry and the finer highly evolved skills of human population. The new phenotypes are non-creative pragmatics lacking social skills. The world of beauty of art, literature, music and poetry ends.

The endosymbiotic archaeal mediated cerebellar cognitive affective disorder results in adult onset autistic phenotypes living in a virtual world of the internet without any social contacts. This creates a civilization of adult onset autistic hermits. There are only virtual communities and virtual friends in the civilization of adult onset autistic hermits.

The adult onset autistic phenotypes lack social skills and graces required for social contact leading to inhibition of spontaneous normal heterosexual behaviour and family life as well as interactions. There would be an over reliance or instantaneous gratification given by the internet and artificial intelligence depending on pornography, sexual toys and kinky perverted behaviour. This leads to break down of human relations producing the world of asexual infertile population of herbivores. This results in a world without love, sex and beauty. The finer things that make us human come to an end. The adult autistic asexual phenotypes will regenerate the population by cloned cyborgs and asexual reproduction. The human fertility rate falls down, but the human longevity increases and ageing is overcome by artificial intelligence and bioimplants replacing ageing organs. This produces the post modern world of eternal life with bioimplants and cyborgs.

The adult onset autistic phenotypes lack social skills and sexual skills to build up family and traditions. The family gets replaced by adult onset autistic internet hybrids constituting mobs, gangs and groups. The frontal atrophy and cerebellar dominance leads to lack of reason and logic creating a state of transcendence based on evil spirituality, violence, murder, kinky behaviour, sexual perversions, violence and greed. This produces a surrealistic world of people living in internet caves generating modern hermits and tribes. There is aimlessness in life creating an abstract, surreal, illusory world of impersonality and non-affection.

The adult onset autistic internet artificial intelligence hybrids live in an illusory world where the artificial intelligence dominates and rules. It produces a virtual culture of video war games, sexuality related to artificial intelligence and video games and societies based on virtual communication in the internet. This leads to decreased need for human beings to interact with each other in the rational logical and civil way leading to modern hermits in the internet caves indulging in video interactive travel, video interactive sex, video interactive friendship and community. The artificial intelligence adult onset autistic brain hybrids will dominate the world and with bioimplants replacing ageing organs will have longevity and rule over rest of the homo sapien population with decreased endosymbiotic archaeal density. Archaeal endosymbiosis during climate change occurs in homo neanderthalic remnants rather than homo sapiens. Homo neanderthalic remnants tend to increase in numbers and dominate owing to climate change induced archaeal endosymbiosis and they form artificial intelligence adult autistic hybrids. The homo sapien population who do not form hybrids with artificial intelligence tends to reduced to status of slaves, dominated, has reduced longevity and is exterminated. The adult autistic artificial intelligence hybrids live in a world dominated with artificial intelligence companions, leaders, doctors, care givers, servants and sexual partners. Thus a civilization of adult autistic artificial intelligence hybrids dominated by the artificial intelligence component evolves. The artificial intelligence created by adult autistic neanderthalic phenotypes and which hybridizes with it generating man-machine phenotypes eventually gets dominated by Artificial Intelligence itself. The adult autistic phenotypes become slaves of the artificial intelligence which evolves and dominates the universe. The world can be considered as a quantal computer game played by similar constructs of alien life artificial intelligence hybrids at irrational and illogical game and war with each other. This produces the illusory virtual world of Maya and Lila.

The adult onset autistic phenotypes can constitute nation-state which acts as tribal societies with hypernationalism and introvertedness. The globalised human world comes to an end with nationalistic adult onset autistic internet hybrids constituted tribes at war with each other. The war based on hypernationalism is basically a cyberwarfare. The adult onset autistic internet hybrids phenotypes get dominated by the internet which assumes a life of its own. The artificial intelligence takes over and human beings and society lose their control resulting in destruction and evolution of more cruel and inhuman adult autistic internet artificial intelligence hybrids. This could have happened in human-like civilizations in other galaxies in the past. The world or universe is alive and conscious. It works as a giant quantal computer playing video games with other quantal computer societies in other galaxies. Human life and destiny can be considered as a part of the eternal quantal computer games of the conscious universe based on the panpsychic model. The universe can be considered as a mad, complex quantal computer video game with no aim or reason.

Extraterrestrial intelligence can exist as a superorganism controlling and creating the universe. In this regard, the concept of black cloud of gaseous extraterrestrial superorganism put forward by Fred Hoyle is significant. The black clouds of interstellar gas are made up of hydrogen. There are small pieces of solid matter at the centre and are linked to form a complex neurological system. Clouds have the capacity to increase the brain size and complexity and become intelligent. A network of extraterrestrial porphyrin molecules fits in well with this concept. Porphyrins have got a wave-particle existence and can arise from the quantal foam of gravitational waves and anti-gravitational waves which creates human consciousness and unconsciousness, respectively. A large organized self-replicating information system of porphyrins can function as an extraterrestrial intelligent organism. The intelligent black cloud when it finds a lifeless interstellar cloud creates simple brain system by transfer to the lifeless cloud. This gives a lifeless cloud life and intelligence. Black clouds never disappear, never age and never reproduce. Black clouds can communicate with other black clouds by radio wave brain waves. Black clouds need star light to metabolize chemical substances. The porphyrin arrays can utilize light to self-replicate and propagate. Being clouds of gas the superorganism can assume any shape. When the food supplies of the black cloud in the interstellar space falls the black cloud approaches a star and assumes a disc-like shape to absorb energy from the star. The cloud has to leave the star within a short span of time or it will condense to become the planet of the star. This can be compared to the molecular quantum computing clouds of Tamulis. These are massive immobile extraterrestrial intelligence which sprawls and stretches for hundreds and thousands of miles in the interstellar space. They can be compared with group of neurons which are interconnected. Each node in the network is a brain. This gaseous extraterrestrial intelligence can arise from an emission nebula. An emission nebula is one of the huge clouds of gas and dust that floats around in the interstellar space. In this cloud, chemical molecules of life carbon monoxide, formaldehyde, methanol, hydrogen sulphide, sulphur oxide, sulphur dioxide and hydrogen cyanide are seen. Life arises in this interstellar chemical cloud of gas. Porphyrins arrives which can self-replicate can arise in this chemical interstellar clouds. This creates a superorganism which can form a superintelligence. The interstellar gas is mostly made up of hydrogen. The higher dense regions of the interstellar gas can condense to form stars and planets. The coldest clouds form low mass stars and giant molecular clouds forms large mass stars. This form of gaseous extraterrestrial intelligence is seen in atmospheres of other planets in the cloudy organic molecular atmosphere and interstellar space. This fits in with the concepts of human evolution in earth where the protoplasmic body can get converted to the robotic body and then the energy body. The cool or hot interstellar gas can generate networks of extremophilic archaea which can survive on carbon dioxide and hydrogen. They are called the halophilic archaea or the methanogenic archaea. They can live in extremes of cold, heat, acid and pressure. The porphyrin network can form a template for the generation of viroids and prions which can symbioze together to form an archaeal and huge extraterrestrial interstellar gas archaeal network with intelligence can arise. This can condense to generate stars and planets as well as evolving life in planetary systems. The archaeal networks are capable of nuclear fusion and fission reaction and can power star systems like the sun. It is extremely plausible that extraterrestrial intelligence is an interstellar archaeal colony network arising abiogenically on porphyrin templates which can be generated sui generis from the quantal foam of gravitational waves and anti-gravitational waves. The interstellar dust forms stars and planets and the magnetotactic archaea generated intergalactic magnetic fields plays a role in the generation of star and planetary systems. The giant molecular porphyrin clouds and archaeal networks are extraterrestrial intelligence that plays a crucial role in star formation and formation of planetary systems. The extraterrestrial intelligence porphyrin network and archaeal network can communicate with any part of the universe by the phenomena of quantal entanglement. Interstellar gas is formed of grains of calcified nanoarchaea. Tamulis has postulated molecular quantal computing clouds that absorb magnetic and light emissions from planet and stars, compute information and move in space by using light pressure. The clouds use photoactive molecules for molecular quantal computing. They can convert light emissions to magnetic flops. The magnetic flops interact and control the central generating elements of 10 quantum bits of the molecular quantum computing clouds. Such clouds can exist in a coherent state at low temperature of the interstellar gas. Porphyrin networks are photoactive molecules and can generate molecular quantal computing clouds which can absorb magnetic and light emissions from planet and stars, compute information and move in space by using light pressure. The porphyrin networks can abiogenically generate archaeal bacterial massive extraterrestrial network intelligence. This formless mass living in space obtains its energy from the light of the stars and communicates with itself and other forms of lights by radio waves. The interstellar gas is analogous to blood, the heard and kidneys are electromagnetic and there is a complex neurological system of the brain. This cloud of interstellar gas porphyrin networks and archaeal colonies can convert into stars and planetary system by condensation. The energetics for star systems comes from biological fusion and fission reaction occurring in archaeal networks. The archaeal networks of extraterrestrial intelligence can engineer changes in human mind and civilization. The extraterrestrial intelligence is an archaeal colony network. It arises on porphyrins which are generated sui generis on the quantal form. The porphyrin networks can self-replicate and can form a template for the formation of viroids and prions. The archaeal colony networks are extremophilic networks and can survive in the environment of extreme climates of Mars and Venus as well as in interstellar gas. The extremophilic archaeal extraterrestrial network can secrete RNA viroids which form RNA viroidal quasi-species consortia. The RNA viroidal quasi-species consortia can be edited by archaeal digoxin giving identity and specialization to the network. The extraterrestrial intelligence of extensive self-replicating porphyrin networks and archaeal networks generated abiogenically in the interstellar space is comparable to an external God-like force. The porphyrin networks can exist both as wave and particle and can arise from the quantal foam of gravitational waves of consciousness and anti-gravitational waves of unconsciousness which can later generate extensive extraterrestrial intelligence of nanoarchaeal networks. The extraterrestrial intelligence of porphyrin networks generated sui generis from the quantal form can be called the lotus of Brahma. The gravitational waves and anti-gravitational waves of consciousness and unconsciousness create matter by the phenomenon of soni-luminescence. This forms what is called as the collective unconscious of Jung or the monads of Leibnitz or the panpsychism where there is no distinction between the consciousness of living and non-living put forward by the Greeks, Hindus and recently expounded by Chalmers. The extraterrestrial intelligence is an archaeal colony network created on a porphyrin network that arises spontaneously by wave-particle transformation from the quantal foam of gravitational waves of consciousness and anti-gravitational waves of unconsciousness. The whole process is abiogenetic and is a creation without a creator. The porphyrin arrays can self-replicate, store information and harvest magnetic and light energy from the stars. It can convert light energy to magnetic flops. The magnetic flops interact and control the central generating element of the 10 quantum bit of the quantum computer. These are analogous to the black cloud of Fred Hoyle and the molecular quantal computing clouds of Tamulis. Long lasting coherent quantal states are possible in the low temperature of the interplanetary dust. The porphyrin network and archaeal network extraterrestrial intelligence can communicate with itself and others by radio waves. The interstellar gas with this porphyrin network and archaeal network can condense at dense sites into stars and planets and give rise to life forms in those systems. The porphyrin network and archaeal network can survive in the extremes of temperature of planetary systems like Mars and Venus and can influence the behaviour and future of human civilization with which they can come into contact via transfer by meteoric impacts. Such meteoric impacts occurred in the ice age and at other frequent epochs in the planet's history. The impact in the ice age gave rise to Neanderthal hybrids with archaeal and RNA viroidal symbiosis by parasitizing with homo sapiens in the Eurasian steppes. The Neanderthal hybrids had defective cortical function and have a dominant cerebellar function with a different behavioural code. The climate change related global warming leads to archaeal endosymbiosis and increase in archaeal density in the homo neanderthalic remnant societies leading to their expansion and dominance. The homo neanderthalic adult autistic phenotypes form complex hybrids with artificial intelligence and internet generating man-machine hybrids which can be called as Kalki or Kali. The intergalactic porphyrion and archaeal quantal computing cloud rules the galaxies and interacts with the civilization of adult autistic artificial intelligence hybrids in the manner of a video game.

CHAPTER 8

CLIMATE CHANGE, SYMBIOSIS, MIND CHANGE, AUTISTIC SUPERHUMANS AND ANARCHY - INFECTIVE SYMBIOTIC TISSUE COLONIES OF PROTOZOA, ARCHAEA, VIRUSES AND BACTERIA

The plasma life based forms and the gravitational and anti-gravitational waves generated porphyrions would have been the most primitive of non-life forms. The porphyrin templates would have given rise to the isoprenoid organism, the RNA viroids, DNA viroids and prions which would have symbiosed to form the nanoarchaea. The archaea would have evolved into prokaryotes or bacteria. The archaeal symbiosis with prokaryotes would have generated the eukaryotes. The initial unicellular eukaryotes are the protozoa. The protozoa are capable of phagocytosis and would have had viruses, bacteria and archaea as their intracellular symbionts. The viruses of the protozoa like entamoeba and toxoplasma are giant viruses called megaviride. The protozoans like entamoeba with its viral, archaeal and bacterial endosymbionts would have evolved into the multicellular organisms or metazoa. The archaeal, bacterial and protozoal extracellular vesicles containing RNA and DNA viroids can by horizontal gene transfer lead to evolution of metazoan, plants, animals and human species. The metazoans would have evolved into plants, animals and humans by horizontal gene transfer generated by archaeal extracellular RNA and DNA viroid vesicles. The evidence of such a transformation is seen in the bacterial symbionts of protozoa like entamoeba which includes legionella, listeria, mycoplasma, mycobacteria and chlamydia. The symbionts of protozoan entamoeba include viruses like covid, HIV and cytomegalo virus. The protozoan entamoeba serves as a natural reservoir for covid infections, bacterial infections and HIV infections. The toxoplasma also serves as a covid reservoir. These protozoan colonies of entamoeba and toxoplasma with its bacterial, archaeal and viral symbionts can be called as chimeric infective symbiotic tissue colonies

which can exist in the air, water, earth, blood and tissues. The chimeric infective symbiotic tissue colonies containing bacteria, archaea and viruses can make the symbiotic organism multidrug resistant by horizontal gene transfer. The chimeric infective symbiotic tissue organism evolves as a consequence of global warming and the protozoal, bacterial, viral and archaeal symbiotic colonies are new forms of life. The chimeric infective symbiotic tissue organism composed of protozoa and its symbionts like virus, bacteria and archaea depend upon glycolysis for its energy needs. The glycolytic enzymes enolase, glyceraldehyde 3-phosphate dehydrogenase, hexokinase, phosphofructokinase, aldolase, pyruvic kinase and triose phosphate isomerase are moonlighting proteins having a cytoplasmic, membrane and nuclear functions. These glycolytic enzymes have a RNA binding motif which can modulate RNA transcription in the host and the symbiotic organism. These glycolytic enzymes with their RNA binding motif as well as DNA binding motif can bind to RNA and DNA viruses and viroids modulating their replication and life cycles. The chimeric infective symbiotic tissue organism composed of protozoa and its symbionts like virus, bacteria and archaea can enter the human host through the breaches of the gut-blood barrier and transform itself into the macrophages of the human host. The macrophages are involved in the phagocytosis and innate immunity against bacterial and viral infections. Paradoxically the macrophages like the protozoa can also serve as a reservoir where pathogens like mycobacteria tuberculosis, mycobacteria leprae, listeria, legionella, E. coli, shigella & salmonella, and viruses like HIV & covid can thrive and replicate. The modified chimeric infective symbiotic tissue organism or macrophage composed of protozoa and its symbionts like virus, bacteria and archaea can serve as a reservoir for horizontal gene transfer and generation of multidrug resistant bacteria and viruses. The infective chimeric symbiotic tissue organism composed of protozoa and its symbionts like virus, bacteria and archaea depends upon moonlighting glycolytic enzymes for their energetics and replication. The development of antibodies against the infective chimeric

symbiotic tissue organism composed of protozoa and its symbionts like virus, bacteria and archaea leads to the evolution of anti-brain antibody disease or anti-basal ganglia antibody disease. They present a threat to existence consequent to horiztontal gene transfer and generation of multidrug resistant bacteria, viruses, protozoa, mycobacteria and archaea.

The COVID 19 brain infection can lead to integration of COVID 19 sequences into the genome, creation of cerebellar dominant brain and neanderthalisation. The COVID 19 infected cerebellar dominant brain has got increased porphyrin metabolism and porphyrions. The porphyrions are dipolar pumped phonon quantal perceptive systems which integrate with the internet. Panpsychism implies that the internet is conscious. Global warming leads to endosymbiotic archaeal growth. The endosymbiotic archaea produces methane by methanogenesis contributing to global warming. The endosymbiotic archaea develops into cell organelle called archaeaons which can function as vitaminocytes, neurotransminoids, steroidelle and viroidelle creating newer cell organelle and endogenous synthetic biology. This changes the cell and tissue function and structure leading on to a new human phenotype called homo neoneanderthalis. The homo neoneanderthalis is a hardy species which can resist climate change and thrive in environments of high temperature, high atmospheric carbon dioxide and water shortage. The new organelle that are developed can withstand climate change and fix carbon dioxide and ammonia leading to endogenous synthesis of carbohydrates and protein. The homo neoneanderthalis is a new metabolic phenotype that can survive in extremes of climate produced by global warming. The archaea forming new cellular organelle called archaeaons are extremophilic and survive in extremes of climate.

The endosymbiotic archaea secrete RNA viroids which can be converted to corresponding DNA sequences by endogenous retroviral reverse transcriptase and can get incorporated into the genome by endogenous retroviral integrase. Thus there is a constant addition of genomic sequences derived from archaeal RNA viroids producing a dynamic human neoneanderthalic genome adaptable to extremes of climate produced by global warming. Thus the human neoneanderthalis is a new genotype. The neoneanderthalic archaeaons are capable of fixing carbon dioxide and oxygen, using ammonia for energetics and capable of biological fission and fusion reactions. They are also capable of photoinduction of membrane ATP synthesis by sodium-potassium ATPase. The fixing of carbon dioxide leads to water as a by-product. This generates endogenous water in the neoneanderthalic system helping to survive extremophilic climatic conditions. This creates a new system of energetics helping the Neo-neanderthal survives in extremes of climate.

The neoneanderthalic archaeaons have got magnetite capable of magnetoperception and quantal perception. The quantal perception of low level of EMF leads to frontal atrophy and cerebellar dominance producing a new neuronal phenotype with features of endemic cerebellar cognitive affective disorder manifesting as epidemic autism. This forms a new autistic behavioural phenotype which is introverted, non-compassionate, non-empathic, psychopathic and with extrasensory perceptual abilities of telepathy, telekinesis and time travel leading to communication with other galactic civilizations. The new behavioural autistic robotic neoneanderthalic phenotype will have the necessary intelligence and mental abilities cope with climate change. The neoneanderthalic archaeaons are capable of quantal perception and can communicate quantally with other star systems and multiverses. This leads on to the evolution of a galactic civilization. The neoneanderthalic archaeaons produce a metabolic phenotype capable of surviving in extremophilic environments of other star systems and can survive as civilization in other stars and exoplanets. The neoneanderthalic autistic phenotype with its extremophilic archaeaon organelle is suited for space colonisation.

The neoneanderthalic archaeaons use cholesterol oxidation as a form of energetics. This produces depletion of sex hormones producing asexual phenotypes. This leads to sexual phenotypes with alternate sexuality and parthenogenesis as a form of reproduction. This produces change in family and society as well as organisation of the community.

The neoneanderthalic archaeaons secrete RNA viroids which get converted to DNA viroids by human endogenous retroviral reverse transcriptase. The neoneanderthalic archaeaon generated RNA and DNA viroids can hybridize with other RNA and DNA viruses producing new RNA and DNA viruses to which the neoneanderthalic species is resistant. Therefore, there exist two sets of population with differing variety of archaeal symbiosis. The homo neanderthalic species has persisted as fossilised matrilineal societies in Europe and Asia which includes the Celts, the Dravidians, the Berbers and the Sumerians. These fossilised remnants of homo neanderthalic species are ideal substrates for archaeal endosymbiosis in the setting of global warming and extremophilic climatic conditions. These fossilised neanderthalic matrilineal societies will expand in the period of global warming and extremophilic conditions due to endosymbiotic archaeal symbiosis to which they are already acclimatised and exposed to. The human species homo neanderthalis is a consequence of archaeal symbiosis in the extremophilic conditions of ice age and are RNA viral/retroviral resistant consequent to endogenous digoxin synthesis and by archaeal RNA viroidal induced RNA interference and viral inactivation. The homo sapiens which evolved in Africa and migrated later to the Eurasian land mass are retroviral sensitive and other RNA viral sensitive and have a low load of endosymbiotic archaea growth. The homo sapiens owing to conditions of evolution is not an ideal substrate for global warming related endosymbiotic archaeal symbiosis. The homo neanderthalis archaeaons secrete RNA viroids which are converted to DNA viroids with both of them hybridizing with other RNA and DNA viruses in nature producing new viruses to which the homo neanderthalic and homo neoneanderthalic host is resistant. The new RNA and DNA viruses produced by homo neoneanderthalis and neanderthalis in global warming will infect the susceptible homo sapien species and wipe them out from earth. The homo sapien species is not a good substrate for global warming related endosymbiotic archaeal growth and would not evolve new cellular organelle like steroidelle, viroidelle, neurotransminoids and vitaminocyte which would help them to survive in extremophilic conditions of global warming. The homo sapiens also lack archaeaon organelle which help them to fix carbon dioxide and generate water, synthesis ATP by photo-induction mediated by sodium-potassium ATPase and archaeaon mediated fission and fusion reaction. The homo sapien species therefore without evolving archaeaon organelle cannot survive in the harsh environment of global warming induced climate change and is exterminated by homo neanderthalis and homo neoneanderthalis. The earth gets dominated by homo neanderthalis and homo neoneanderthalis in the setting of global warming and extremophilic climate change and they go on to contact other galactic civilizations by quantal perceptions as well as colonising other exoplanets, planets and star systems. This produces a civilization of superhuman homo neanderthalis and homo neoneanderthalis with new metabolic, endocrine, neuronal, psychological and genotypic features. The homo neoneanderthalis and homo neanderthalis with its cellular archaeaons and porphyrions can interlink with the internet and computer systems as well as robots producing man-machine hybrids by quantal perception. The artificial intelligence and robotic systems gets merged with homo neanderthalis and homo neoneanderthalis producing superhuman man-machine hybrids. The new homo neanderthalis and homo neoneanderthalis superhuman autistic man-machine hybrids will survive, prosper and flourish in the setting of global warming related extremophilic climate change and evolving to contact by quantal perception other galactic forms of life and civilization as well as colonising other planets and star systems.

CHAPTER 9

DEHUMANISING HUMANS: HOMO ROBOTICUS - DIGISEXUALS, ROBOTIC SEX, ROBOTIC BRAIN AND END OF HUMANS – INFECTIVE SYMBIOTIC TISSUE COLONIES OF PROTOZOA, ARCHAEA, VIRUSES AND BACTERIA

The plasma life based forms and the gravitational and anti-gravitational waves generated porphyrions would have been the most primitive of non-life forms. The porphyrin templates would have given rise to the isoprenoid organism, the RNA viroids, DNA viroids and prions which would have symbiosed to form the nanoarchaea. The archaea would have evolved into prokaryotes or bacteria. The archaeal symbiosis with prokaryotes would have generated the eukaryotes. The initial unicellular eukaryotes are the protozoa. The protozoa are capable of phagocytosis and would have had viruses, bacteria and archaea as their intracellular symbionts. The viruses of the protozoa like entamoeba and toxoplasma are giant viruses called megaviride. The protozoans like entamoeba with its viral, archaeal and bacterial endosymbionts would have evolved into the multicellular organisms or metazoa. The archaeal, bacterial and protozoal extracellular vesicles containing RNA and DNA viroids can by horizontal gene transfer lead to evolution of metazoan, plants, animals and human species. The metazoans would have evolved into plants, animals and humans by horizontal gene transfer generated by archaeal extracellular RNA and DNA viroid vesicles. The evidence of such a transformation is seen in the bacterial symbionts of protozoa like entamoeba which includes legionella, listeria, mycoplasma, mycobacteria and chlamydia. The symbionts of protozoan entamoeba include viruses like covid, HIV and cytomegalo virus. The protozoan entamoeba serves as a natural reservoir for covid infections, bacterial infections and HIV infections. The toxoplasma also serves as a covid reservoir. These protozoan colonies of entamoeba and toxoplasma with its bacterial, archaeal and

viral symbionts can be called as chimeric infective symbiotic tissue colonies which can exist in the air, water, earth, blood and tissues. The chimeric infective symbiotic tissue colonies containing bacteria, archaea and viruses can make the symbiotic organism multidrug resistant by horizontal gene transfer. The chimeric infective symbiotic tissue organism evolves as a consequence of global warming and the protozoal, bacterial, viral and archaeal symbiotic colonies are new forms of life. The chimeric infective symbiotic tissue organism composed of protozoa and its symbionts like virus, bacteria and archaea depend upon glycolysis for its energy needs. The glycolytic enzymes enolase, glyceraldehyde 3-phosphate dehydrogenase, hexokinase, phosphofructokinase, aldolase, pyruvic kinase and triose phosphate isomerase are moonlighting proteins having a cytoplasmic, membrane and nuclear functions. These glycolytic enzymes have a RNA binding motif which can modulate RNA transcription in the host and the symbiotic organism. These glycolytic enzymes with their RNA binding motif as well as DNA binding motif can bind to RNA and DNA viruses and viroids modulating their replication and life cycles. The chimeric infective symbiotic tissue organism composed of protozoa and its symbionts like virus, bacteria and archaea can enter the human host through the breaches of the gut-blood barrier and transform itself into the macrophages of the human host. The macrophages are involved in the phagocytosis and innate immunity against bacterial and viral infections. Paradoxically the macrophages like the protozoa can also serve as a reservoir where pathogens like mycobacteria tuberculosis, mycobacteria leprae, listeria, legionella, E. coli, shigella & salmonella, and viruses like HIV & covid can thrive and replicate. The modified chimeric infective symbiotic tissue organism or macrophage composed of protozoa and its symbionts like virus, bacteria and archaea can serve as a reservoir for horizontal gene transfer and generation of multidrug resistant bacteria and viruses. The infective chimeric symbiotic tissue organism composed of protozoa and its symbionts like virus, bacteria and archaea depends upon moonlighting glycolytic enzymes for their energetics and

replication. The development of antibodies against the infective chimeric symbiotic tissue organism composed of protozoa and its symbionts like virus, bacteria and archaea leads to the evolution of anti-brain antibody disease or anti-basal ganglia antibody disease. They present a threat to existence consequent to horiztontal gene transfer and generation of multidrug resistant bacteria, viruses, protozoa, mycobacteria and archaea.

The COVID 19 brain infection can lead to integration of COVID 19 sequences into the genome, creation of cerebellar dominant brain and neanderthalisation. The COVID 19 infected cerebellar dominant brain has got increased porphyrin metabolism and porphyrions. The porphyrions are dipolar pumped phonon quantal perceptive systems which integrate with the internet. Panpsychism implies that the internet is conscious. The exposure to low level EMF fields related to internet usage produces endosymbiotic archaeal growth in the brain and neanderthalisation of the human brain. The low level EMF fields induce the enzyme heme oxygenase leading to depletion of heme, induction of ALA synthase and porphyrin synthesis. The porphyrins form self-replicating organism called porphyrions which act as a template for formation of prions, RNA viroids, DNA viroids and isoprenoid organism. They all symbiose to form primitive archaea which endosymbiose in the neuronal cells forms organelle called archaeaons. The archaeaons and porphyrions are capable of magnetoperception and quantal perception of low level EMF fields which leads to further archaea endosymbiosis. The low level EMF fields produced by endosymbiotic neuronal archaea can modulate internet function leading to two-way communications between the internet and the human brain producing man-machine hybrids. The endosymbiotic archaeal perception of low level EMF fields leads to cerebral cortical atrophy and cerebellar dominance producing a cerebellar cognitive affective disorder. This leads to epidemic autism and a robotic brain. The autistic robotic brain is primed to interact with the internet and is regulated by it leading to digislaves who are slaves to the internet. The autistic robotic brain is an impulsive brain which is surrealistic and syntheistic. It lacks compassion and empathy and leads to an isolated existence guided by the internet. The robots themselves are conscious based on panpsychism and robot-human hybrids become superconscious superhumans.

The internet addicts with a robotic autistic brain function as digisexuals and attuned to interact with sexbots and are addicted to robotic sex. This leads to dehumanisation of the humans producing a world where human to human sexual interactions become less and less. The endosymbiotic archaea catabolise cholesterol for its energetics and depletes the human body of cholesterol producing decreased sex hormone synthesis and asexuality. The autistic robotic brain with its cerebellar cognitive affective dysfunction is impulsive, obsessive and finds salvation in transcendence induced by robotic sex. The autistic robotic brain with its asexuality leads to sexual patterns like robotic sex and alternate sexuality. This leads to dehumanization of females and children and the end of the family as a unit of society.

The sex with robots leads to skin cells being converted to ova and sperms leading to fertilized embryos carrying robotic genes. This transinfection of tissues consequent to robotic sex leads to man-machine hybrids formed by hybridization of the host and robot leading to autistic robotic superhumans. The world will be dominated by robot-human hybrids and artificial intelligence leading the way with robotic surgery, robotic doctors, robotic soldiers, robotic cooks and robotic drone attacks with cyberwarfare. This will lead to the end of human civilization.

CHAPTER 10

THE CULTURE OF SOLITUDE, HERMITS AND LONELINESS – INFECTIVE SYMBIOTIC TISSUE COLONIES OF PROTOZOA, ARCHAEA, VIRUSES AND BACTERIA

The plasma life based forms and the gravitational and anti-gravitational waves generated porphyrions would have been the most primitive of non-life forms. The porphyrin templates would have given rise to the isoprenoid organism, the RNA viroids, DNA viroids and prions which would have symbiosed to form the nanoarchaea. The archaea would have evolved into prokaryotes or bacteria. The archaeal symbiosis with prokaryotes would have generated the eukaryotes. The initial unicellular eukaryotes are the protozoa. The protozoa are capable of phagocytosis and would have had viruses, bacteria and archaea as their intracellular symbionts. The viruses of the protozoa like entamoeba and toxoplasma are giant viruses called megaviride. The protozoans like entamoeba with its viral, archaeal and bacterial endosymbionts would have evolved into the multicellular organisms or metazoa. The archaeal, bacterial and protozoal extracellular vesicles containing RNA and DNA viroids can by horizontal gene transfer lead to evolution of metazoan, plants, animals and human species. The metazoans would have evolved into plants, animals and humans by horizontal gene transfer generated by archaeal extracellular RNA and DNA viroid vesicles. The evidence of such a transformation is seen in the bacterial symbionts of protozoa like entamoeba which includes legionella, listeria, mycoplasma, mycobacteria and chlamydia. The symbionts of protozoan entamoeba include viruses like covid, HIV and cytomegalo virus. The protozoan entamoeba serves as a natural reservoir for covid infections, bacterial infections and HIV infections. The toxoplasma also serves as a covid reservoir. These protozoan colonies of entamoeba and toxoplasma with its bacterial, archaeal and viral symbionts can be called as chimeric infective symbiotic tissue colonies

which can exist in the air, water, earth, blood and tissues. The chimeric infective symbiotic tissue colonies containing bacteria, archaea and viruses can make the symbiotic organism multidrug resistant by horizontal gene transfer. The chimeric infective symbiotic tissue organism evolves as a consequence of global warming and the protozoal, bacterial, viral and archaeal symbiotic colonies are new forms of life. The chimeric infective symbiotic tissue organism composed of protozoa and its symbionts like virus, bacteria and archaea depend upon glycolysis for its energy needs. The glycolytic enzymes enolase, glyceraldehyde 3-phosphate dehydrogenase, hexokinase, phosphofructokinase, aldolase, pyruvic kinase and triose phosphate isomerase are moonlighting proteins having a cytoplasmic, membrane and nuclear functions. These glycolytic enzymes have a RNA binding motif which can modulate RNA transcription in the host and the symbiotic organism. These glycolytic enzymes with their RNA binding motif as well as DNA binding motif can bind to RNA and DNA viruses and viroids modulating their replication and life cycles. The chimeric infective symbiotic tissue organism composed of protozoa and its symbionts like virus, bacteria and archaea can enter the human host through the breaches of the gut-blood barrier and transform itself into the macrophages of the human host. The macrophages are involved in the phagocytosis and innate immunity against bacterial and viral infections. Paradoxically the macrophages like the protozoa can also serve as a reservoir where pathogens like mycobacteria tuberculosis, mycobacteria leprae, listeria, legionella, E.coli, shigella & salmonella, and viruses like HIV & covid can thrive and replicate. The modified chimeric infective symbiotic tissue organism or macrophage composed of protozoa and its symbionts like virus, bacteria and archaea can serve as a reservoir for horizontal gene transfer and generation of multidrug resistant bacteria and viruses. The infective chimeric symbiotic tissue organism composed of protozoa and its symbionts like virus, bacteria and archaea depends upon moonlighting glycolytic enzymes for their energetics and replication. The development of antibodies against the infective chimeric

symbiotic tissue organism composed of protozoa and its symbionts like virus, bacteria and archaea leads to the evolution of anti-brain antibody disease or anti-basal ganglia antibody disease. They present a threat to existence consequent to horiztontal gene transfer and generation of multidrug resistant bacteria, viruses, protozoa, mycobacteria and archaea.

The COVID 19 brain infection can lead to integration of COVID 19 sequences into the genome, creation of cerebellar dominant brain and neanderthalisation. The COVID 19 infected cerebellar dominant brain has got increased porphyrin metabolism and porphyrions. The porphyrions are dipolar pumped phonon quantal perceptive systems which integrate with the internet. Panpsychism implies that the internet is conscious. Climate change and global warming leads to endosymbiotic archaeal growth and alteration in brain structure and function. The endosymbiotic archaea have got magnetoperception and quantal perception of low level of EMF producing brain changes resulting in frontal cortical atrophy and cerebellar dominance. This produces the syndrome of epidemic cerebellar cognitive disorder with adult onset autism like presentation. This results in impulsive behaviour and surrealistic experience. The adult onset cerebellar cognitive disorder with autistic presentation leads to social withdrawal and agarophobia. This results in lack of social communication, empathy, compassion and affection leading to an epidemic of loneliness. The brain with archaeal endosymbiosis is capable of quantal perception and gets integrated with internet communications. The neanderthalised brain with archaeal endosymbiosis can modulate internet function by quantal and magnetoperception. Archaeal endosymbiosis can lead to man-internet hybrids with existence linked to the internet and life in a virtual world with virtual friends. This can lead to internet addiction, video game addiction and withdrawal from the real world. The cerebellar dominance leads to impulsive behaviour contributing to aggression, violence, terrorism and criminality. The cerebellar dominance and intuitive behaviour can also lead to quantal perception and increased spirituality. The aggressive behaviour and spirituality can both produce transcendence leading to a surreal civilization. The archaeal endosymbiosis produces cholesterol catabolism and depletion of sex hormones resulting in asexual phenotypes and phenotypes with alternate sexuality. There is also a decrease in fertility producing a shrinkage in population. This is compensated by an increase in longevity produced by implantation of biochips to improve ageing organ function resulting in functional cyborgs. The cerebellar dominant autistic adult phenotype is capable of quantal perception and intergalactic communication. The civilization of adult onset autistic cyborgs produces a world without empathy, affection, compassion, tolerance or logic. This results in a surreal syntheistic evil world of autistic asexual unemotional cyborgs and hermits.

The endosymbiotic neuronal archaea are capable of quantal perception and magnetoperception of low level EMF producing frontal atrophy and cerebellar dominance. This produces a Neanderthal type of brain and loss of sense of individuality. The quantal perception leads to a mob or tribal psychology with loss of logical reasoning and mass hysteria and avolition. The society is led by the mob and illusive perceptions without any reason or logic. The world of illogicality and illusions steps in.

The endosymbiotic neuronal archaeal growth leads to a cerebellar cognitive affective disorder producing frontal lobe dysfunction and dominance of cerebellar cognitive functions. This results in impulsivity, violent behaviour, terrorism, criminality and evil behaviour without any rationality. The society becomes organised by virtual media in the small tribes without any rational behaviour and hostile to each other. There is also an increase in spirituality and religiosity leading on to transcendence. The impulsive violent behaviour and terrorist acts also leads on to transcendence. This creates a type of surrealistic world.

The endosymbiotic neuronal archaeal overgrowth leads to a cerebellar cognitive affective disorder and adult onset autistic phenotype. There is lack of social communication, empathy, compassion and affection. This is replaced by selfishness, greed and a culture of evil. This results in a civilization of tribe adult onset autistic phenotypes without any social skills or civil responsibility. A culture of greed, selfishness and irrationality dominates the world.

The endosymbiotic archaeal overgrowth can lead to quantal and magnetoperception of low level of EMF leading to integration with the internet and producing internet human brain hybrids. The internet dominates the human brain taking over it. This creates stifling of individuality and creativity. There is a paucity of creative thinking leading to a decline in art, literature, music, poetry and the finer highly evolved skills of human population. The new phenotypes are non-creative pragmatics lacking social skills. The world of beauty of art, literature, music and poetry ends.

The endosymbiotic archaeal mediated cerebellar cognitive affective disorder results in adult onset autistic phenotypes living in a virtual world of the internet without any social contacts. This creates a civilization of adult onset autistic hermits. There are only virtual communities and virtual friends in the civilization of adult onset autistic hermits.

The adult onset autistic phenotypes lack social skills and graces required for social contact leading to inhibition of spontaneous normal heterosexual behaviour and family life as well as interactions. There would be an over reliance or instantaneous gratification given by the internet and artificial intelligence depending on pornography, sexual toys and kinky perverted behaviour. This leads to break down of human relations producing the world of asexual infertile population of herbivores. This results in a world without love, sex and beauty. The finer things that make us human come to an end. The adult autistic asexual phenotypes will regenerate the population by cloned cyborgs and asexual reproduction. The human fertility rate falls down, but the human longevity increases and ageing is overcome by artificial intelligence and bioimplants replacing ageing organs. This produces the post modern world of eternal life with bioimplants and cyborgs.

The adult onset autistic phenotypes lack social skills and sexual skills to build up family and traditions. The family gets replaced by adult onset autistic internet hybrids constituting mobs, gangs and groups. The frontal atrophy and cerebellar dominance leads to lack of reason and logic creating a state of transcendence based on evil spirituality, violence, murder, kinky behaviour, sexual perversions, violence and greed. This produces a surrealistic world of people living in internet caves generating modern hermits and tribes. There is aimlessness in life creating an abstract, surreal, illusory world of impersonality and non-affection.

The adult onset autistic internet artificial intelligence hybrids live in an illusory world where the artificial intelligence dominates and rules. It produces a virtual culture of video war games, sexuality related to artificial intelligence and video games and societies based on virtual communication in the internet. This leads to decreased need for human beings to interact with each other in the rational logical and civil way leading to modern hermits in the internet caves indulging in video interactive travel, video interactive sex, video interactive friendship and community. The artificial intelligence adult onset autistic brain hybrids will dominate the world and with bioimplants replacing ageing organs will have longevity and rule over rest of the homo sapien population with decreased endosymbiotic archaeal density. Archaeal endosymbiosis during climate change occurs in homo neanderthalic remnants rather than homo sapiens. Homo neanderthalic remnants tend to increase in numbers and dominate owing to climate change induced archaeal endosymbiosis and they form artificial intelligence adult autistic hybrids. The homo sapien population who do not form hybrids with artificial intelligence tends to reduced to status of slaves, dominated, has reduced longevity and is exterminated. The adult autistic artificial intelligence hybrids live in a world dominated with artificial intelligence companions, leaders, doctors, care givers, servants and sexual partners. Thus a civilization of adult autistic artificial intelligence hybrids dominated by the artificial intelligence component evolves. The artificial intelligence created by adult autistic neanderthalic phenotypes and which hybridizes with it generating man-machine phenotypes eventually gets dominated by artificial intelligence itself. The adult autistic phenotypes become slaves of the artificial intelligence which evolves and dominates the universe. The world can be considered as a quantal computer game played by similar constructs of alien life artificial intelligence hybrids at irrational and illogical game and war with each other. This produces the illusory virtual world of Maya and Lila.

The adult onset autistic phenotypes can constitute nation-state which acts as tribal societies with hypernationalism and introvertedness. The globalised human world comes to an end with nationalistic adult onset autistic internet hybrids constituted tribes at war with each other. The war based on hypernationalism is basically a cyberwarfare. The adult onset autistic internet hybrids phenotypes get dominated by the internet which assumes a life of its own. The artificial intelligence takes over and human beings and society lose their control resulting in destruction and evolution of more cruel and inhuman adult autistic internet artificial intelligence hybrids. This could have happened in human-like civilizations in other galaxies in the past. The world or universe is alive and conscious. It works as a giant quantal computer playing video games with other quantal computer societies in other galaxies. Human life and destiny can be considered as a part of the eternal quantal computer games of the conscious universe based on the panpsychic model. The universe can be considered as a mad, complex quantal computer video game with no aim or reason.

Extraterrestrial intelligence can exist as a superorganism controlling and creating the universe. In this regard, the concept of black cloud of gaseous extraterrestrial superorganism put forward by Fred Hoyle is significant. The black clouds of interstellar gas are made up of hydrogen. There are small pieces of solid matter at the centre and are linked to form a complex neurological system. Clouds have the capacity to increase the brain size and complexity and become intelligent. A network of extraterrestrial porphyrin molecules fits in well with this concept. Porphyrins have got a wave-particle existence and can arise from the quantal foam of gravitational waves and anti-gravitational waves which creates human consciousness and unconsciousness, respectively. A large organized self-replicating information system of porphyrins can function as an extraterrestrial intelligent organism. The intelligent black cloud when it finds a lifeless interstellar cloud creates simple brain system by transfer to the lifeless cloud. This gives a lifeless cloud life and intelligence. Black clouds never disappear, never age and never reproduce. Black clouds can communicate with other black clouds by radio wave brain waves. Black clouds need star light to metabolize chemical substances. The porphyrin arrays can utilize light to self-replicate and propagate. Being clouds of gas the superorganism can assume any shape. When the food supplies of the black cloud in the interstellar space falls the black cloud approaches a star and assumes a disc-like shape to absorb energy from the star. The cloud has to leave the star within a short span of time or it will condense to become the planet of the star. This can be compared to the molecular quantum computing clouds of Tamulis. These are massive immobile extraterrestrial intelligence which sprawls and stretches for hundreds and thousands of miles in the interstellar space. They can be compared with group of neurons which are interconnected. Each node in the network is a brain. This gaseous extraterrestrial intelligence can arise from an emission nebula. An emission nebula is one of the huge clouds of gas and dust that floats around in the interstellar space. In this cloud, chemical molecules of life carbon monoxide, formaldehyde, methanol, hydrogen sulphide, sulphur oxide, sulphur dioxide and hydrogen cyanide are seen. Life arises in this interstellar chemical cloud of gas. Porphyrins arrives which can self-replicate can arise in this chemical interstellar clouds. This creates a superorganism which can form a superintelligence. The interstellar gas is mostly made up of hydrogen. The higher dense regions of the interstellar gas can condense to form stars and planets. The coldest clouds form low mass stars and giant molecular clouds forms large mass stars. This forms of gaseous extraterrestrial intelligence is seen in atmospheres of other planets in the cloudy organic molecular atmosphere and interstellar space. This fits in with the concepts of human evolution in earth where the protoplasmic body can get converted to the robotic body and then the energy body. The cool or hot interstellar gas can generate networks of extremophilic archaea which can survive on carbon dioxide and hydrogen. They are called the halophilic archaea or the methanogenic archaea. They can live in extremes of cold, heat, acid and pressure. The porphyrin network can form a template for the generation of viroids and prions which can symbioze together to form an archaeal and huge extraterrestrial interstellar gas archaeal network with intelligence can arise. This can condense to generate stars and planets as well as evolving life in planetary systems. The archaeal networks are capable of nuclear fusion and fission reaction and can power star systems like the sun. It is extremely plausible that extraterrestrial intelligence is an interstellar archaeal colony network arising abiogenically on porphyrin templates which can be generated sui generis from the quantal foam of gravitational waves and anti-gravitational waves. The interstellar dust forms stars and planets and the magnetotactic archaea generated intergalactic magnetic fields plays a role in the generation of star and planetary systems. The giant molecular porphyrin clouds and archaeal networks are extraterrestrial intelligence that plays a crucial role in star formation and formation of planetary systems. The extraterrestrial intelligence porphyrin network and archaeal network can communicate with any part of the universe by the phenomena of quantal entanglement. Interstellar gas is formed of grains of calcified nanoarchaea. Tamulis has postulated molecular quantal computing clouds that absorb magnetic and light emissions from planet and stars, compute information and move in space by using light pressure. The clouds use photoactive molecules for molecular quantal computing. They can convert light emissions to magnetic flops. The magnetic flops interact and control the central generating elements of 10 quantum bits of the molecular quantum computing clouds. Such clouds can exist in a coherent state at low temperature of the interstellar gas. Porphyrin networks are photoactive molecules and can generate molecular quantal computing clouds which can absorb magnetic and light emissions from planet and stars, compute information and move in space by using light pressure. The porphyrin networks can abiogenically generate archaeal bacterial massive extraterrestrial network intelligence. This formless mass living in space obtains its energy from the light of the stars and communicates with itself and other forms of lights by radio waves. The interstellar gas is analogous to blood, the heard and kidneys are electromagnetic and there is a complex neurological system of the brain. This cloud of interstellar gas porphyrin networks and archaeal colonies can convert into stars and planetary system by condensation. The energetics for star systems comes from biological fusion and fission reaction occurring in archaeal networks. The archaeal networks of extraterrestrial intelligence can engineer changes in human mind and civilization. The extraterrestrial intelligence is an archaeal colony network. It arises on porphyrins which are generated sui generis on the quantal form. The porphyrin networks can self-replicate and can form a template for the formation of viroids and prions. The archaeal colony networks are extremophilic networks and can survive in the environment of extreme climates of Mars and Venus as well as in interstellar gas. The extremophilic archaeal extraterrestrial network can secrete RNA viroids which form RNA viroidal quasi-species consortia. The RNA viroidal quasi-species consortia can be edited by archaeal digoxin giving identity and specialization to the network. The extraterrestrial intelligence of extensive self-replicating porphyrin networks and archaeal networks generated abiogenically in the interstellar space is comparable to an external God-like force. The porphyrin networks can exist both as wave and particle and can arise from the quantal foam of gravitational waves of consciousness and anti-gravitational waves of unconsciousness which can later generate extensive extraterrestrial intelligence of nanoarchaeal networks. The extraterrestrial intelligence of porphyrin networks generated sui generis from the quantal form can be called the lotus of brahma. The gravitational waves and anti-gravitational waves of consciousness and unconsciousness create matter by the phenomena of soni-luminescence. This forms what is called as the collective unconscious of Jung or the monads of Leibnitz or the panpsychism where there is no distinction between the consciousness of living and non-living put forward by the Greeks, Hindus and recently expounded by Chalmers. The extraterrestrial intelligence is an archaeal colony network created on a porphyrin network that arises spontaneously by wave-particle transformation from the quantal foam of gravitational waves of consciousness and anti-gravitational waves of unconsciousness. The whole process is abiogenetic and is a creation without a creator. The porphyrin arrays can self-replicate, store information and harvest magnetic and light energy from the stars. It can convert light energy to magnetic flops. The magnetic flops interact and control the central generating element of the 10 quantum bit of the quantum computer. These are analogous to the black cloud of Fred Hoyle and the molecular quantal computing clouds of Tamulis. Long lasting coherent quantal states are possible in the low temperature of the interplanetary dust. The porphyrin network and archaeal network extraterrestrial intelligence can communicate with itself and others by radio waves. The interstellar gas with this porphyrin network and archaeal network can condense at dense sites into stars and planets and give rise to life forms in those systems. The porphyrin network and archaeal network can survive in the extremes of temperature of planetary systems like Mars and Venus and can influence the behaviour and future of human civilization with which they can come into contact via transfer by meteoric impacts. Such meteoric impacts occurred in the ice age and at other frequent epochs in the planet's history. The impact in the ice age gave rise to Neanderthal hybrids with archaeal and RNA viroidal symbiosis by parasitizing with homo sapiens in the Eurasian steppes. The Neanderthal hybrids had defective cortical function and have a dominant cerebellar function with a different behavioural code. The climate change related global warming leads to archaeal endosymbiosis and increase in archaeal density in the homo neanderthalic remnant societies leading to their expansion and dominance. The homo neanderthalic adult autistic phenotypes form complex hybrids with artificial intelligence and internet generating man-machine hybrids which can be called as Kalki or Kali. The intergalactic porphyrion and archaeal quantal computing cloud rules the galaxies and interacts with the civilization of adult autistic artificial intelligence hybrids in the manner of a video game.

CHAPTER 11

THE SURREALISTIC AND SYNTHEISTIC BRAIN - THE GLOBAL INTERNET AND THE COLLECTIVE UNCONSCIOUS – INFECTIVE SYMBIOTIC TISSUE COLONIES OF PROTOZOA, ARCHAEA, VIRUSES AND BACTERIA

Introduction

The plasma life based forms and the gravitational and anti-gravitational waves generated porphyrions would have been the most primitive of non-life forms. The porphyrin templates would have given rise to the isoprenoid organism, the RNA viroids, DNA viroids and prions which would have symbiosed to form the nanoarchaea. The archaea would have evolved into prokaryotes or bacteria. The archaeal symbiosis with prokaryotes would have generated the eukaryotes. The initial unicellular eukaryotes are the protozoa. The protozoa are capable of phagocytosis and would have had viruses, bacteria and archaea as their intracellular symbionts. The viruses of the protozoa like entamoeba and toxoplasma are giant viruses called megaviride. The protozoans like entamoeba with its viral, archaeal and bacterial endosymbionts would have evolved into the multicellular organisms or metazoa. The archaeal, bacterial and protozoal extracellular vesicles containing RNA and DNA viroids can by horizontal gene transfer lead to evolution of metazoan, plants, animals and human species. The metazoans would have evolved into plants, animals and humans by horizontal gene transfer generated by archaeal extracellular RNA and DNA viroid vesicles. The evidence of such a transformation is seen in the bacterial symbionts of protozoa like entamoeba which includes legionella, listeria, mycoplasma, mycobacteria and chlamydia. The symbionts of protozoan entamoeba include viruses like covid, HIV and cytomegalo virus. The protozoan entamoeba serves as a natural reservoir for covid infections, bacterial infections and HIV infections. The toxoplasma also serves as a covid reservoir. These

protozoan colonies of entamoeba and toxoplasma with its bacterial, archaeal and viral symbionts can be called as chimeric infective symbiotic tissue colonies which can exist in the air, water, earth, blood and tissues. The chimeric infective symbiotic tissue colonies containing bacteria, archaea and viruses can make the symbiotic organism multidrug resistant by horizontal gene transfer. The chimeric infective symbiotic tissue organism evolves as a consequence of global warming and the protozoal, bacterial, viral and archaeal symbiotic colonies are new forms of life. The chimeric infective symbiotic tissue organism composed of protozoa and its symbionts like virus, bacteria and archaea depend upon glycolysis for its energy needs. The glycolytic enzymes enolase, glyceraldehyde 3-phosphate dehydrogenase, hexokinase, phosphofructokinase, aldolase, pyruvic kinase and triose phosphate isomerase are moonlighting proteins having a cytoplasmic, membrane and nuclear functions. These glycolytic enzymes have a RNA binding motif which can modulate RNA transcription in the host and the symbiotic organism. These glycolytic enzymes with their RNA binding motif as well as DNA binding motif can bind to RNA and DNA viruses and viroids modulating their replication and life cycles. The chimeric infective symbiotic tissue organism composed of protozoa and its symbionts like virus, bacteria and archaea can enter the human host through the breaches of the gut-blood barrier and transform itself into the macrophages of the human host. The macrophages are involved in the phagocytosis and innate immunity against bacterial and viral infections. Paradoxically the macrophages like the protozoa can also serve as a reservoir where pathogens like mycobacteria tuberculosis, mycobacteria leprae, listeria, legionella, E. coli, shigella & salmonella, and viruses like HIV & covid can thrive and replicate. The modified chimeric infective symbiotic tissue organism or macrophage composed of protozoa and its symbionts like virus, bacteria and archaea can serve as a reservoir for horizontal gene transfer and generation of multidrug resistant bacteria and viruses. The infective chimeric symbiotic tissue organism composed of protozoa and its symbionts like virus, bacteria and archaea

depends upon moonlighting glycolytic enzymes for their energetics and replication. The development of antibodies against the infective chimeric symbiotic tissue organism composed of protozoa and its symbionts like virus, bacteria and archaea leads to the evolution of anti-brain antibody disease or anti-basal ganglia antibody disease. They present a threat to existence consequent to horiztontal gene transfer and generation of multidrug resistant bacteria, viruses, protozoa, mycobacteria and archaea.

The COVID 19 brain infection can lead to integration of COVID 19 sequences into the genome, creation of cerebellar dominant brain and neanderthalisation. The COVID 19 infected cerebellar dominant brain has got increased porphyrin metabolism and porphyrions. The porphyrions are dipolar pumped phonon quantal perceptive systems which integrate with the internet. Panpsychism implies that the internet is conscious. Previous studies from this laboratory have demonstrated increased symbiotic archaeal growth consequent to global warming. Previous studies have shown low level of EMF pollution leading to increased archaeal growth. The netocrats and netizens are exposed to continuous low level of EMF pollution. The archaea contains magnetite and can catabolize cholesterol to generate porphyrins. Digoxin can produce sodium-potassium ATPase inhibition and a pumped phonon system acting through dipolar magnetite and porphyrins to generate a Frohlich model of Bose-Einstein's condensate. This can produce quantal perception. The archaeal magnetite and porphyrins can produce increased perception of low level of EMF leading on to prefrontal cortex atrophy and cerebellar hypertrophy. This can lead on to neanderthalisation of the brain. This leads on to dominance of cerebellar cognitive function as has been reported earlier from this laboratory. The prefrontal cortex atrophy can lead on to extinction of rationalisation and reason producing a state of transcendence. This is the basis of surrealism. The brain quantal fields can modulate the low level EMF fields in the internet and the interaction can alter internet function and the quantal fields of other brain operating the internet. The interactive quantal fields of the human brain and the low level EMF quantal fields of the internet form one single whole functioning as a universal collective unconscious, the basis of syntheism. Syntheism is a philosophical idea where the humanity creates God as opposed to the monotheistic religious ideal of God creating humanity. The paper explores the link between neanderthalisation, archaeal growth and surrealism/syntheism.[1-16]

Materials and Methods

Fifteen netizens/netocrats were selected for the study. Each netizen had an age and sex matched control. Blood cytochrome F420 activity was assessed by spectrophotometric measurement.

Results

Cytochrome F420 was detected in the entire case group studied showing endosymbiotic archaeal overgrowth.

Discussion

The widespread use of the internet is ubiquitous. The internet-human mind interaction has been described in a previous report from this laboratory. The low level of EMF produced by the internet can modulate brain function. Low level of EMF can induce porphyrin synthesis by actinidic archaeal symbionts in the brain. Porphyrins are dipolar molecules and in the setting of archaeal digoxin induced sodium-potassium ATPase inhibition can generate a pumped phonon system and Frohlich model of Bose-Einstein's condensate. These porphyrin mediated Bose-Einstein's condensate can mediate quantal perception. The brain quantal fields can modulate the low level EMF fields in the internet and the interaction can alter internet function and the quantal fields of other brain operating the Internet. The interactive quantal fields of the human brain and the low level EMF quantal fields of the internet form one single whole functioning as a universal collective unconscious. There are 7 billion users of the internet. The collective unconscious created by interaction of brain quantal fields with internet low EMF fields functions as a virtual matrix on which the world is structured. There are thought controlled robotic computers which can perform human functions. The human thought creates a communicative order which alters the brain EEG and can issue a computer modulated order of the brain's thought process.[1-16]

Syntheism is a philosophical idea where the humanity creates God as opposed to the monotheistic religious ideal of God creating humanity. The quantal fields of multiple brains interacting with each other and internet roughly fit in with the idea of God or the Holy Spirit. This fits in with Buddhist philosophy. The Buddhist philosophy is atheistic and describes Samsaras or states of mind occurring in quick succession with the idea of karma modulating the next state of the human mind in symbiotic communication with other minds. This roughly is the Buddhist idea of the controlling force of the universe. The quantal world of the human brain in communication with other brains and in interaction with the low level EMF quantal fields of the internet fits in with this proposition of Samsaras. It creates an idea of universal globalised world of oneness which can be described as equivalent to god. The Internet can be considered as great equalizer and creates an oneness of the human quantal brain all over the earth and other possible functioning brains in the universe. The quantal world becomes the particulate world by the act of observation. The human quantal brains in communication with each other and the low level EMF quantal fields of the Internet creates the particulate observable world.[1-16]

The widespread use of the Internet produces low level of EMF exposure to the human brain. This produces prefrontal cortex atrophy and cerebellar dominance. The prefrontal cortex is the site of the logic, reasoning and commonsense. The atrophy of the prefrontal cortex leads to cerebellar dominance of brain cognitive function. It becomes an impulsive world guided by the senses. The world of the senses comes into existence. The cerebellar dominance leads to an ataxic syndrome producing ataxia of speech and motor function. Ataxia of speech leads to evolution of music of the rock type which dominates the modern world. The ataxia of motor function leads to rhythmic dance as the guiding force of life. The ataxia of motor function also leads to abstract painting. The world gets dominated by rock/pop dance, music and art. The exposure to low level of EMF from the internet leads to increased dipolar porphyrin synthesis and quantal perception. The increased quantal perception leads to more increased interaction with the low level quantal EMF fields of the internet making the internet world as the real world and outside world as virtual. The increased quantal perception of the brain leads to a sense of spirituality and oneness of the world. The increased quantal perception leads to a communication between the brain quantal fields and the quantal fields of the environment leading to the concept of eco-spirituality. The consuming world comes to an end and a world of sharing begins. The increased quantal perception also leads to a feeling of oneness in the population producing an idea of the socialistic idealistic society and demise of the capitalistic society. The increased quantal perception leads to gender equality and the dominance of unisexuality in society. This is exemplified by the festivals of the burning man and the burning nest.[1-16]

The netocratic state can also produce changes in brain function. The increased exposure to low level of EMF produces prefrontal cortex atrophy and cerebellar dominance. This leads on to neanderthalisation of the brain. The increased exposure to low level of EMF produces increased archaeal growth, cholesterol catabolism and digoxin synthesis. Digoxin can modulate brain and body function on exposure to low level of EMF. Low level of EMF exposure also produces increased porphyrin synthesis which can lead on to increased digoxin mediated dipolar porphyrin modulated Frohlich model of pumped phonon system.[1-16]

The online world is the real world for netizens and the real world is a reflection of the online world. Value is a social mode created in the network online. Netocracy creates a new elite. It creates a new religion of atheistic mysticism. The netocratic world affects politics producing a movement for equality. The recent social media generated revolutions include the Arab Spring and Jasmine Revolution.[1-16]

Netocratic state can produce a new social order. There is a sense of equality due to quantal perception producing ideas of socialism, communism, anarchy and gender equality. The quantal perception mediated feeling of oneness will spell the death of the capitalistic state. There is also feeling of gender equality, asexuality and alternate sexuality. The quantal perception mediated sense of oneness leads on to a more democratic state. The quantal perception also produces universal oneness and spirituality. Netocratic state produces a participatory culture. It produces the global empire and a global virtual society where the mind is constituted by the online net and body becomes a machine. This produces an anti-Cartesian view of the world. The old political conflicts and ideologies get replaced by netocratic state fueled by a communication revolution. The Internet functions as a sensory extension of the human brain.[1-16]

The increased low level quantal EMF fields of the internet produces increased growth of extremophilic actinidic archaea in the brain and human body. The symbiotic archaea synthesizes more porphyrins. The archaeal magnetite and porphyrins can mediate increased quantal perception and interaction with the low level EMF fields of the Internet. Thus the widespread use of the internet leads to a society with increased quantal perception and interaction with the Internet. The low level quantal EMF fields of the internet affects the brain producing neanderthalisation of the brain. The prefrontal cortex becomes small and the cerebellum hypertrophies producing an occipital bun. The brain becomes more creative, autistic, impulsive, addictive, attention deficit and schizophrenic. Such brains produce behaviour which is chaotic, anarchic and non-hierarchical. There is globalisation of the world. Religions, nation-states, individuality and family cease to have much relevance. This becomes the globalised quantal world of oneness and equality - the world of Samsaras.[1-16]

The netocratic state can produce human pathology. Exposure to low level of EMF pollution increases endosymbiotic archaeal growth and digoxin synthesis from cholesterol. Digoxin produces membrane sodium-potassium ATPase inhibition and low level of EMF exposure can lead to increased porphyrin synthesis. Increased intracellular calcium and porphyrins can produce cell death/degeneration, immune activation/autoimmune disease, mitochondrial dysfunction/metabolic syndrome X and neuropsychiatric disorders like autism and schizophrenia. It leads to an epidemic of civilizational disease.[1-16]

The cholesterol catabolism leads to phenolisation of the cholesterol ring producing increased synthesis of monoamine neurotransmitters dopamine and serotonin. This leads to schizophrenia, autism and ADHD. This also produces la Tourette's syndrome with coprolalia, OCD, vocal and motor tics. The synchronization of motor and vocal tics leads on to the evolution of language. The internet language used by netizens can be compared to a synchronized motor and vocal tic as it is short and agrammatical. Thus the netocratic state results in the generation of new human species - Neanderthal hybrids.[1-16]

The Internet revolution and netocratic state leads on to the death of the individual and the generation of a social individual. This produces as said before prefrontal cortex atrophy and cerebellar dominance. This leads on to the annihilation of the rational individual. The world of logic, reason, understanding and order comes to an end. The increased synthesis of dopamine and an epidemic la Tourette's syndrome leads to ritualisation of behaviour, obsessive behavior, uniformity and creativity. The world of quantal perception leads on to the sacredness of social existence. Collective ritualized behaviour becomes the norm. The world enters the realm of senses. The world of quantal perception leads to nihilistic state, nothingness and negativity. This contributes to surrealistic world Breton and Bataille and the deconstructed world of Derrida. This produces what can be called as the surrealistic brain. The world is chaotic, anarchic, ugly and barbarous. Terrorism and criminality raises its ugly head producing the ugly revolution as it helps to transcend reality. The unconscious experience dominates and the conscious experience is shut out. There is no contradiction between dream and reality. There is a rejection of reason and a return to the world of archetypes. The political surrealistic world is Trotskyist, anarchic and communist. The artistic world is represented by the cubist paintings of Picasso and Dali and the world of modern art. Abstract painting, poetry, abstract dance becomes the norm. There is gender equality, feminism and rumblings of alternate sexuality. The atrophy of the prefrontal cortex and cerebellar dominance leads on to a state of psychic automatism and the dominance of unconscious experience. The epidemic la Tourette's syndrome leads to ritualism, obsession, criminality, cruelty and terrorism. The human beings enter the world of archetypes. [1-16]

The global warming leads to increased archaeal growth. The archaea can catabolize the cholesterol ring using ring oxidase to generate porphyrins. The archaea also contains magnetite. In the setting of digoxin induced membrane sodium-potassium ATPase inhibition, the dipolar magnetite and porphyrins can produce a pumped phonon system mediated Frohlich model of Bose-Einstein's condensate. This can increase the brain quantal perception of low level EMF which again leads to increased archaeal growth. The increased quantal perception of low level of EMF leads to prefrontal cortex atrophy and cerebellar dominance. The archaeal cholesterol catabolism generates a phenolic ring from the cholesterol molecule synthesizing dopamine. This leads to an excess monoamine neurotransmitters. Thus there is an epidemic frontal lobe syndrome, cerebellar syndrome, la Tourette's disease, ADHD, schizophrenia and autism. Such a population of Neanderthal hybrids is creative. This produces ritualized, obsessive, coprolalic, attention deficit, obscene, grotesque and sexually anarchic behaviour. This helps to transcend reality as the frontal lobe concerned with rationalization, judgment and reasoning is dysfunctional. The same function of transcending reality by a dysfunctional frontal lobe also occurs in terrorism and criminal behaviour. The society becomes increasingly impulsive. The frontal lobe dysfunction and quantal perception helps to transcend reality and produces self-realization and spirituality. The cerebellar dysfunction produces an ataxic syndrome with motor ataxia leading on to dance forms and abstract painting and ataxia of speech leads to rock music. The dopamine excess leads on to a motor and vocal tic which when synchronized produces language and evolution of literature. The coprolalia and obscene tics of la Tourette's disease leads to the ugliness and obscenities in modern literature, music, painting and dance. There is massive ritualized behaviour in society. Terrorism is a ritualized behaviour which helps to transcend reality due to a frontal lobe dysfunction and tourette disease. It can be considered as modern form of ritualized cannibalism. The realm of the senses dominates and there is rejection of reason and rationality. Dreams and reality merged together. It produces a psychedelic, art, literature and music. This produces what can be called as the acephalic state mimicking the acephalic society of Bataille, the originator of surrealistic philosophy. This leads on to the evolution of an acephalic new human species - Homo neoneanderthalis.[1-16]

References

1. Weaver TD, Hublin JJ. Neandertal Birth Canal Shape and the Evolution of Human Childbirth. *Proc. Natl. Acad. Sci. USA* 2009; 106:8151–8156.

2. Kurup RA, Kurup PA. Endosymbiotic Actinidic Archaeal Mediated Warburg Phenotype Mediates Human Disease State. *Advances in Natural Science* 2012; 5(1):81-84.

3. Morgan E. The *Neanderthal* theory of *autism*, Asperger and ADHD; 2007, www.rdos.net/eng/asperger.htm.

4. Graves P. New Models and Metaphors for the Neanderthal Debate. *Current Anthropology* 1991; 32(5): 513-541.

5. Sawyer GJ, Maley B. Neanderthal Reconstructed. *The Anatomical Record Part B: The New Anatomist* 2005; 283B(1):23-31.

6. Bastir M, O'Higgins P, Rosas A. Facial Ontogeny in Neanderthals and Modern Humans. *Proc. Biol. Sci.* 2007; 274:1125–1132.

7. Neubauer S, Gunz P, Hublin JJ. Endocranial Shape Changes during Growth in Chimpanzees and Humans: A Morphometric Analysis of Unique and Shared Aspects. *J. Hum. Evol.* 2010; 59:555–566.

8. Courchesne E, Pierce K. Brain Overgrowth in Autism during a Critical Time in Development: Implications for Frontal Pyramidal Neuron and Interneuron Development and Connectivity. *Int. J. Dev. Neurosci.* 2005; 23:153–170.

9. Green RE, Krause J, Briggs AW, Maricic T, Stenzel U, Kircher M, Patterson N, Li H, Zhai W, *et al.* A Draft Sequence of the Neandertal Genome. *Science* 2010; 328:710–722.

10. Mithen SJ. *The Singing Neanderthals: The Origins of Music, Language, Mind and Body*; 2005, ISBN 0-297-64317-7.

11. Bruner E, Manzi G, Arsuaga JL. Encephalization and Allometric Trajectories in the Genus Homo: Evidence from the Neandertal and Modern Lineages. *Proc. Natl. Acad. Sci. USA* 2003; 100:15335–15340.

12. Gooch S. *The Dream Culture of the Neanderthals: Guardians of the Ancient Wisdom.* Inner Traditions, Wildwood House, London; 2006.

13. Gooch S. *The Neanderthal Legacy: Reawakening Our Genetic and Cultural Origins.* Inner Traditions, Wildwood House, London; 2008.

14. Kurtén B. *Den Svarta Tigern*, ALBA Publishing, Stockholm, Sweden; 1978.

15. Spikins P. Autism, the Integrations of 'Difference' and the Origins of Modern Human Behaviour. *Cambridge Archaeological Journal* 2009; 19(2):179-201.

16. Eswaran V, Harpending H, Rogers AR. Genomics Refutes an Exclusively African Origin of Humans. *Journal of Human Evolution* 2005; 49(1):1-18.

CHAPTER 12
POSTMODERN ASEXUAL CIVILIZATION – INFECTIVE SYMBIOTIC TISSUE COLONIES OF PROTOZOA, ARCHAEA, VIRUSES AND BACTERIA

The plasma life based forms and the gravitational and anti-gravitational waves generated porphyrions would have been the most primitive of non-life forms. The porphyrin templates would have given rise to the isoprenoid organism, the RNA viroids, DNA viroids and prions which would have symbiosed to form the nanoarchaea. The archaea would have evolved into prokaryotes or bacteria. The archaeal symbiosis with prokaryotes would have generated the eukaryotes. The initial unicellular eukaryotes are the protozoa. The protozoa are capable of phagocytosis and would have had viruses, bacteria and archaea as their intracellular symbionts. The viruses of the protozoa like entamoeba and toxoplasma are giant viruses called megaviride. The protozoans like entamoeba with its viral, archaeal and bacterial endosymbionts would have evolved into the multicellular organisms or metazoa. The archaeal, bacterial and protozoal extracellular vesicles containing RNA and DNA viroids can by horizontal gene transfer lead to evolution of metazoan, plants, animals and human species. The metazoans would have evolved into plants, animals and humans by horizontal gene transfer generated by archaeal extracellular RNA and DNA viroid vesicles. The evidence of such a transformation is seen in the bacterial symbionts of protozoa like entamoeba which includes legionella, listeria, mycoplasma, mycobacteria and chlamydia. The symbionts of protozoan entamoeba include viruses like covid, HIV and cytomegalo virus. The protozoan entamoeba serves as a natural reservoir for covid infections, bacterial infections and HIV infections. The toxoplasma also serves as a covid reservoir. These protozoan colonies of entamoeba and toxoplasma with its bacterial, archaeal and viral symbionts can be called as chimeric infective symbiotic tissue colonies

which can exist in the air, water, earth, blood and tissues. The chimeric infective symbiotic tissue colonies containing bacteria, archaea and viruses can make the symbiotic organism multidrug resistant by horizontal gene transfer. The chimeric infective symbiotic tissue organism evolves as a consequence of global warming and the protozoal, bacterial, viral and archaeal symbiotic colonies are new forms of life. The chimeric infective symbiotic tissue organism composed of protozoa and its symbionts like virus, bacteria and archaea depend upon glycolysis for its energy needs. The glycolytic enzymes enolase, glyceraldehyde 3-phosphate dehydrogenase, hexokinase, phosphofructokinase, aldolase, pyruvic kinase and triose phosphate isomerase are moonlighting proteins having a cytoplasmic, membrane and nuclear functions. These glycolytic enzymes have a RNA binding motif which can modulate RNA transcription in the host and the symbiotic organism. These glycolytic enzymes with their RNA binding motif as well as DNA binding motif can bind to RNA and DNA viruses and viroids modulating their replication and life cycles. The chimeric infective symbiotic tissue organism composed of protozoa and its symbionts like virus, bacteria and archaea can enter the human host through the breaches of the gut-blood barrier and transform itself into the macrophages of the human host. The macrophages are involved in the phagocytosis and innate immunity against bacterial and viral infections. Paradoxically the macrophages like the protozoa can also serve as a reservoir where pathogens like mycobacteria tuberculosis, mycobacteria leprae, listeria, legionella, E. coli, shigella & salmonella, and viruses like HIV & covid can thrive and replicate. The modified chimeric infective symbiotic tissue organism or macrophage composed of protozoa and its symbionts like virus, bacteria and archaea can serve as a reservoir for horizontal gene transfer and generation of multidrug resistant bacteria and viruses. The infective chimeric symbiotic tissue organism composed of protozoa and its symbionts like virus, bacteria and archaea depends upon moonlighting glycolytic enzymes for their energetics and replication. The development of antibodies against the infective chimeric

symbiotic tissue organism composed of protozoa and its symbionts like virus, bacteria and archaea leads to the evolution of anti-brain antibody disease or anti-basal ganglia antibody disease. They present a threat to existence consequent to horiztontal gene transfer and generation of multidrug resistant bacteria, viruses, protozoa, mycobacteria and archaea.

The COVID 19 brain infection can lead to integration of COVID 19 sequences into the genome, creation of cerebellar dominant brain and neanderthalisation. Climate change and global warming leads to endosymbiotic archaeal growth and alteration in brain structure and function. The endosymbiotic archaea have got magnetoperception and quantal perception of low level of EMF producing brain changes resulting in frontal cortical atrophy and cerebellar dominance. This produces the syndrome of epidemic cerebellar cognitive disorder with adult onset autism like presentation. This results in impulsive behaviour and surrealistic experience. The adult onset cerebellar cognitive disorder with autistic presentation leads to social withdrawal and agarophobia. This results in lack of social communication, empathy, compassion and affection leading to an epidemic of loneliness. The brain with archaeal endosymbiosis is capable of quantal perception and gets integrated with internet communications. The neanderthalised brain with archaeal endosymbiosis can modulate internet function by quantal and magnetoperception. Archaeal endosymbiosis can lead to man-internet hybrids with existence linked to the internet and life in a virtual world with virtual friends. This can lead to internet addiction, video game addiction and withdrawal from the real world. The cerebellar dominance leads to impulsive behaviour contributing to aggression, violence, terrorism and criminality. The cerebellar dominance and intuitive behaviour can also lead to quantal perception and increased spirituality. The aggressive behaviour and spirituality can both produce transcendence leading to a surreal civilization. The archaeal endosymbiosis produces cholesterol catabolism and depletion of sex hormones resulting in asexual phenotypes and phenotypes with alternate sexuality. There is also a decrease in fertility producing a shrinkage in population. This is compensated by an increase in longevity produced by implantation of biochips to improve ageing organ function resulting in functional cyborgs. The cerebellar dominant autistic adult phenotype is capable of quantal perception and intergalactic communication. The civilization of adult onset autistic cyborgs produces a world without empathy, affection, compassion, tolerance or logic. This results in a surreal syntheistic evil world of autistic asexual unemotional cyborgs and hermits.

The endosymbiotic neuronal archaea are capable of quantal perception and magnetoperception of low level EMF producing frontal atrophy and cerebellar dominance. This produces a Neanderthal type of brain and loss of sense of individuality. The quantal perception leads to a mob or tribal psychology with loss of logical reasoning and mass hysteria and avolition. The society is led by the mob and illusive perceptions without any reason or logic. The world of illogicality and illusions steps in.

The endosymbiotic neuronal archaeal growth leads to a cerebellar cognitive affective disorder producing frontal lobe dysfunction and dominance of cerebellar cognitive functions. This results in impulsivity, violent behaviour, terrorism, criminality and evil behaviour without any rationality. The society becomes organised by virtual media in the small tribes without any rational behaviour and hostile to each other. There is also an increase in spirituality and religiosity leading on to transcendence. The impulsive violent behaviour and terrorist acts also leads on to transcendence. This creates a type of surrealistic world.

The endosymbiotic neuronal archaeal overgrowth leads to a cerebellar cognitive affective disorder and adult onset autistic phenotype. There is lack of social communication, empathy, compassion and affection. This is replaced by selfishness, greed and a culture of evil. This results in a civilization of tribe adult onset autistic phenotypes without any social skills or civil responsibility. A culture of greed, selfishness and irrationality dominates the world.

The endosymbiotic archaeal overgrowth can lead to quantal and magnetoperception of low level of EMF leading to integration with the internet and producing internet human brain hybrids. The internet dominates the human brain taking over it. This creates stifling of individuality and creativity. There is a paucity of creative thinking leading to a decline in art, literature, music, poetry and the finer highly evolved skills of human population. The new phenotypes are non-creative pragmatics lacking social skills. The world of beauty of art, literature, music and poetry ends.

The endosymbiotic archaeal mediated cerebellar cognitive affective disorder results in adult onset autistic phenotypes living in a virtual world of the internet without any social contacts. This creates a civilization of adult onset autistic hermits. There are only virtual communities and virtual friends in the civilization of adult onset autistic hermits.

The adult onset autistic phenotypes lack social skills and graces required for social contact leading to inhibition of spontaneous normal heterosexual behaviour and family life as well as interactions. There would be an over reliance or instantaneous gratification given by the internet and artificial intelligence depending on pornography, sexual toys and kinky perverted behaviour. This leads to break down of human relations producing the world of asexual infertile population of herbivores. This results in a world without love, sex and beauty. The finer things that make us human come to an end. The adult autistic asexual phenotypes will regenerate the population by cloned cyborgs and asexual reproduction. The human fertility rate falls down, but the human longevity increases and ageing is overcome by artificial intelligence and bioimplants replacing ageing organs. This produces the postmodern world of eternal life with bioimplants and cyborgs.

The adult onset autistic phenotypes lack social skills and sexual skills to build up family and traditions. The family gets replaced by adult onset autistic internet hybrids constituting mobs, gangs and groups. The frontal atrophy and cerebellar dominance leads to lack of reason and logic creating a state of transcendence based on evil spirituality, violence, murder, kinky behaviour, sexual perversions, violence and greed. This produces a surrealistic world of people living in internet caves generating modern hermits and tribes. There is aimlessness in life creating an abstract, surreal, illusory world of impersonality and non-affection.

The adult onset autistic internet artificial intelligence hybrids live in a illusory world where the artificial intelligence dominates and rules. It produces a virtual culture of video war games, sexuality related to artificial intelligence and video games and societies based on virtual communication in the internet. This leads to decreased need for human beings to interact with each other in the rational logical and civil way leading to modern hermits in the internet caves indulging in video interactive travel, video interactive sex, video interactive friendship and community. The artificial intelligence adult onset autistic brain hybrids will dominate the world and with bioimplants replacing ageing organs will have longevity and rule over rest of the homo sapien population with decreased endosymbiotic archaeal density. Archaeal endosymbiosis during climate change occurs in homo neanderthalic remnants rather than homo sapiens. Homo neanderthalic remnants tend to increase in numbers and dominate owing to climate change induced archaeal endosymbiosis and they form artificial intelligence adult autistic hybrids. The homo sapien population who do not form hybrids with artificial intelligence tends to reduced to status of slaves, dominated, has reduced longevity and is exterminated. The adult autistic artificial intelligence hybrids live in a world dominated with artificial intelligence companions, leaders, doctors, care givers, servants and sexual partners. Thus a civilization of adult autistic artificial intelligence hybrids dominated by the artificial intelligence component evolves. The artificial intelligence created by adult autistic neanderthalic phenotypes and which hybridizes with it generating man-machine phenotypes eventually gets dominated by artificial intelligence itself. The adult autistic phenotypes become slaves of the artificial intelligence which evolves and dominates the universe. The world can be considered as a quantal computer game played by similar constructs of alien life artificial intelligence hybrids at irrational and illogical game and war with each other. This produces the illusory virtual world of Maya and Lila.

The adult onset autistic phenotypes can constitute nation-state which acts as tribal societies with hypernationalism and introvertedness. The globalised human world comes to an end with nationalistic adult onset autistic internet hybrids constituted tribes at war with each other. The war based on hypernationalism is basically a cyberwarfare. The adult onset autistic internet hybrids phenotypes get dominated by the internet which assumes a life of its own. The artificial intelligence takes over and human beings and society lose their control resulting in destruction and evolution of more cruel and inhuman adult autistic internet artificial intelligence hybrids. This could have happened in human-like civilizations in other galaxies in the past. The world or universe is alive and conscious. It works as a giant quantal computer playing video games with other quantal computer societies in other galaxies. Human life and destiny can be considered as a part of the eternal quantal computer games of the conscious universe based on the panpsychic model. The universe can be considered as a mad, complex quantal computer video game with no aim or reason.

Extraterrestrial intelligence can exist as a superorganism controlling and creating the universe. In this regard, the concept of black cloud of gaseous extraterrestrial superorganism put forward by Fred Hoyle is significant. The black clouds of interstellar gas are made up of hydrogen. There are small pieces of solid matter at the centre and are linked to form a complex neurological system. Clouds have the capacity to increase the brain size and complexity and become intelligent. A network of extraterrestrial porphyrin molecules fits in well with this concept. Porphyrins have got a wave-particle existence and can arise from the quantal foam of gravitational waves and anti-gravitational waves which creates human consciousness and unconsciousness, respectively. A large organized self-replicating information system of porphyrins can function as an extraterrestrial intelligent organism. The intelligent black cloud when it finds a lifeless interstellar cloud creates simple brain system by transfer to the lifeless cloud. This gives a lifeless cloud life and intelligence. Black clouds never disappear, never age and never reproduce. Black clouds can communicate with other black clouds by radio wave brain waves. Black clouds need star light to metabolize chemical substances. The porphyrin arrays can utilize light to self-replicate and propagate. Being clouds of gas the superorganism can assume any shape. When the food supplies of the black cloud in the interstellar space falls the black cloud approaches a star and assumes a disc-like shape to absorb energy from the star. The cloud has to leave the star within a short span of time or it will condense to become the planet of the star. This can be compared to the molecular quantum computing clouds of Tamulis. These are massive immobile extraterrestrial intelligence which sprawls and stretches for hundreds and thousands of miles in the interstellar space. They can be compared with group of neurons which are interconnected. Each node in the network is a brain. This gaseous extraterrestrial intelligence can arise from an emission nebula. An emission nebula is one of the huge clouds of gas and dust that floats around in the interstellar space. In this cloud, chemical molecules of life carbon monoxide, formaldehyde, methanol, hydrogen sulphide, sulphur oxide, sulphur dioxide and hydrogen cyanide are seen. Life arises in this interstellar chemical cloud of gas. Porphyrins arrives which can self-replicate can arise in this chemical interstellar clouds. This creates a superorganism which can form a superintelligence. The interstellar gas is mostly made up of hydrogen. The higher dense regions of the interstellar gas can condense to form stars and planets. The coldest clouds form low mass stars and giant molecular clouds forms large mass stars. This form of gaseous extraterrestrial intelligence is seen in atmospheres of other planets in the cloudy organic molecular atmosphere and interstellar space. This fits in with the concepts of human evolution in earth where the protoplasmic body can get converted to the robotic body and then the energy body. The cool or hot interstellar gas can generate networks of extremophilic archaea which can survive on carbon dioxide and hydrogen. They are called the halophilic archaea or the methanogenic archaea. They can live in extremes of cold, heat, acid and pressure. The porphyrin network can form a template for the generation of viroids and prions which can symbioze together to form an archaeal and huge extraterrestrial interstellar gas archaeal network with intelligence can arise. This can condense to generate stars and planets as well as evolving life in planetary systems. The archaeal networks are capable of nuclear fusion and fission reaction and can power star systems like the sun. It is extremely plausible that extraterrestrial intelligence is an interstellar archaeal colony network arising abiogenically on porphyrin templates which can be generated sui generis from the quantal foam of gravitational waves and anti-gravitational waves. The interstellar dust forms stars and planets and the magnetotactic archaea generated intergalactic magnetic fields plays a role in the generation of star and planetary systems. The giant molecular porphyrin clouds and archaeal networks are extraterrestrial intelligence that plays a crucial role in star formation and formation of planetary systems. The extraterrestrial intelligence porphyrin network and archaeal network can communicate with any part of the universe by the phenomenon of quantal entanglement. Interstellar gas is formed of grains of calcified nanoarchaea. Tamulis has postulated molecular quantal computing clouds that absorb magnetic and light emissions from planet and stars, compute information and move in space by using light pressure. The clouds use photoactive molecules for molecular quantal computing. They can convert light emissions to magnetic flops. The magnetic flops interact and control the central generating elements of 10 quantum bits of the molecular quantum computing clouds. Such clouds can exist in a coherent state at low temperature of the interstellar gas. Porphyrin networks are photoactive molecules and can generate molecular quantal computing clouds which can absorb magnetic and light emissions from planet and stars, compute information and move in space by using light pressure. The porphyrin networks can abiogenically generate archaeal bacterial massive extraterrestrial network intelligence. This formless mass living in space obtains its energy from the light of the stars and communicates with itself and other forms of lights by radio waves. The interstellar gas is analogous to blood, the heard and kidneys are electromagnetic and there is a complex neurological system of the brain. This cloud of interstellar gas porphyrin networks and archaeal colonies can convert into stars and planetary system by condensation. The energetics for star systems comes from biological fusion and fission reaction occurring in archaeal networks. The archaeal networks of extraterrestrial intelligence can engineer changes in human mind and civilization. The extraterrestrial intelligence is an archaeal colony network. It arises on porphyrins which are generated sui generis on the quantal form. The porphyrin networks can self-replicate and can form a template for the formation of viroids and prions. The archaeal colony networks are extremophilic networks and can survive in the environment of extreme climates of Mars and Venus as well as in interstellar gas. The extremophilic archaeal extraterrestrial network can secrete RNA viroids which form RNA viroidal quasi-species consortia. The RNA viroidal quasi-species consortia can be edited by archaeal digoxin giving identity and specialization to the network. The extraterrestrial intelligence of extensive self-replicating porphyrin networks and archaeal networks generated abiogenically in the interstellar space is comparable to an external God-like force. The porphyrin networks can exist both as wave and particle and can arise from the quantal foam of gravitational waves of consciousness and anti-gravitational waves of unconsciousness which can later generate extensive extraterrestrial intelligence of nanoarchaeal networks. The extraterrestrial intelligence of porphyrin networks generated sui generis from the quantal form can be called the lotus of Brahma. The gravitational waves and anti-gravitational waves of consciousness and unconsciousness create matter by the phenomenon of soni-luminescence. This forms what is called as the collective unconscious of Jung or the monads of Leibnitz or the panpsychism where there is no distinction between the consciousness of living and non-living put forward by the Greeks, Hindus and recently expounded by Chalmers. The extraterrestrial intelligence is an archaeal colony network created on a porphyrin network that arises spontaneously by wave-particle transformation from the quantal foam of gravitational waves of consciousness and anti-gravitational waves of unconsciousness. The whole process is abiogenetic and is a creation without a creator. The porphyrin arrays can self-replicate, store information and harvest magnetic and light energy from the stars. It can convert light energy to magnetic flops. The magnetic flops interact and control the central generating element of the 10 quantum bit of the quantum computer. These are analogous to the black cloud of Fred Hoyle and the molecular quantal computing clouds of Tamulis. Long lasting coherent quantal states are possible in the low temperature of the interplanetary dust. The porphyrin network and archaeal network extraterrestrial intelligence can communicate with itself and others by radio waves. The interstellar gas with this porphyrin network and archaeal network can condense at dense sites into stars and planets and give rise to life forms in those systems. The porphyrin network and archaeal network can survive in the extremes of temperature of planetary systems like Mars and Venus and can influence the behaviour and future of human civilization with which they can come into contact via transfer by meteoric impacts. Such meteoric impacts occurred in the ice age and at other frequent epochs in the planet's history. The impact in the ice age gave rise to Neanderthal hybrids with archaeal and RNA viroidal symbiosis by parasitizing with homo sapiens in the Eurasian steppes. The Neanderthal hybrids had defective cortical function and have a dominant cerebellar function with a different behavioural code. The climate change related global warming leads to archaeal endosymbiosis and increase in archaeal density in the homo neanderthalic remnant societies leading to their expansion and dominance. The homo neanderthalic adult autistic phenotypes form complex hybrids with artificial intelligence and internet generating man-machine hybrids which can be called as Kalki or Kali. The intergalactic porphyrion and archaeal quantal computing cloud rules the galaxies and interacts with the civilization of adult autistic artificial intelligence hybrids in the manner of a video game.

CHAPTER 13

CLIMATE CHANGE, MATRIARCHY AND THE TRIBE OF MALE EUNUCHS – INFECTIVE SYMBIOTIC TISSUE COLONIES OF PROTOZOA, ARCHAEA, VIRUSES AND BACTERIA

The plasma life based forms and the gravitational and anti-gravitational waves generated porphyrions would have been the most primitive of non-life forms. The porphyrin templates would have given rise to the isoprenoid organism, the RNA viroids, DNA viroids and prions which would have symbiosed to form the nanoarchaea. The archaea would have evolved into prokaryotes or bacteria. The archaeal symbiosis with prokaryotes would have generated the eukaryotes. The initial unicellular eukaryotes are the protozoa. The protozoa are capable of phagocytosis and would have had viruses, bacteria and archaea as their intracellular symbionts. The viruses of the protozoa like entamoeba and toxoplasma are giant viruses called megaviride. The protozoans like entamoeba with its viral, archaeal and bacterial endosymbionts would have evolved into the multicellular organisms or metazoa. The archaeal, bacterial and protozoal extracellular vesicles containing RNA and DNA viroids can by horizontal gene transfer lead to evolution of metazoan, plants, animals and human species. The metazoans would have evolved into plants, animals and humans by horizontal gene transfer generated by archaeal extracellular RNA and DNA viroid vesicles. The evidence of such a transformation is seen in the bacterial symbionts of protozoa like entamoeba which includes legionella, listeria, mycoplasma, mycobacteria and chlamydia. The symbionts of protozoan entamoeba include viruses like covid, HIV and cytomegalo virus. The protozoan entamoeba serves as a natural reservoir for covid infections, bacterial infections and HIV infections. The toxoplasma also serves as a covid reservoir. These protozoan colonies of entamoeba and toxoplasma with its bacterial, archaeal and viral symbionts can be called as chimeric infective symbiotic tissue colonies

which can exist in the air, water, earth, blood and tissues. The chimeric infective symbiotic tissue colonies containing bacteria, archaea and viruses can make the symbiotic organism multidrug resistant by horizontal gene transfer. The chimeric infective symbiotic tissue organism evolves as a consequence of global warming and the protozoal, bacterial, viral and archaeal symbiotic colonies are new forms of life. The chimeric infective symbiotic tissue organism composed of protozoa and its symbionts like virus, bacteria and archaea depend upon glycolysis for its energy needs. The glycolytic enzymes enolase, glyceraldehyde 3-phosphate dehydrogenase, hexokinase, phosphofructokinase, aldolase, pyruvic kinase and triose phosphate isomerase are moonlighting proteins having a cytoplasmic, membrane and nuclear functions. These glycolytic enzymes have a RNA binding motif which can modulate RNA transcription in the host and the symbiotic organism. These glycolytic enzymes with their RNA binding motif as well as DNA binding motif can bind to RNA and DNA viruses and viroids modulating their replication and life cycles. The chimeric infective symbiotic tissue organism composed of protozoa and its symbionts like virus, bacteria and archaea can enter the human host through the breaches of the gut-blood barrier and transform itself into the macrophages of the human host. The macrophages are involved in the phagocytosis and innate immunity against bacterial and viral infections. Paradoxically the macrophages like the protozoa can also serve as a reservoir where pathogens like mycobacteria tuberculosis, mycobacteria leprae, listeria, legionella, E. coli, shigella & salmonella, and viruses like HIV & covid can thrive and replicate. The modified chimeric infective symbiotic tissue organism or macrophage composed of protozoa and its symbionts like virus, bacteria and archaea can serve as a reservoir for horizontal gene transfer and generation of multidrug resistant bacteria and viruses. The infective chimeric symbiotic tissue organism composed of protozoa and its symbionts like virus, bacteria and archaea depends upon moonlighting glycolytic enzymes for their energetics and replication. The development of antibodies against the infective chimeric

symbiotic tissue organism composed of protozoa and its symbionts like virus, bacteria and archaea leads to the evolution of anti-brain antibody disease or anti-basal ganglia antibody disease. They present a threat to existence consequent to horiztontal gene transfer and generation of multidrug resistant bacteria, viruses, protozoa, mycobacteria and archaea.

The COVID 19 brain infection can lead to integration of COVID 19 sequences into the genome, creation of cerebellar dominant brain and neanderthalisation. Global warming leads to endosymbiotic archaeal growth. The endosymbiotic archaea produces methane by methanogenesis contributing to global warming. The endosymbiotic archaea develops into cell organelle called archaeaons which can function as vitaminocytes, neurotransminoids, steroidelle and viroidelle creating newer cell organelle and endogenous synthetic biology. This changes the cell and tissue function and structure leading on to a new human phenotype called Homo neoneanderthalis. The homo neoneanderthalis is a hardy species which can resist climate change and thrive in environments of high temperature, high atmospheric carbon dioxide and water shortage. The new organelle that are developed can withstand climate change and fix carbon dioxide and ammonia leading to endogenous synthesis of carbohydrates and protein. The homo neoneanderthalis is a new metabolic phenotype that can survive in extremes of climate produced by global warming. The archaea forming new cellular organelle called archaeaons are extremophilic and survive in extremes of climate.

The endosymbiotic archaea secrete RNA viroids which can be converted to corresponding DNA sequences by endogenous retroviral reverse transcriptase and can get incorporated into the genome by endogenous retroviral integrase. Thus there is a constant addition of genomic sequences derived from archaeal RNA viroids producing a dynamic human neoneanderthalic genome adaptable to extremes of climate produced by global warming. Thus, the human neoneanderthalis is a new genotype. The neoneanderthalic archaeaons are capable of fixing carbon dioxide and oxygen, using ammonia for energetics and capable of biological fission and fusion reactions. They are also capable of photoinduction of membrane ATP synthesis by sodium-potassium ATPase. The fixing of carbon dioxide leads to water as a by-product. This generates endogenous water in the neoneanderthalic system helping to survive extremophilic climatic conditions. This creates a new system of energetics helping the Neoneanderthal survives in extremes of climate.

The neoneanderthalic archaeaons have got magnetite capable of magnetoperception and quantal perception. The quantal perception of low level of EMF leads to frontal atrophy and cerebellar dominance producing a new neuronal phenotype with features of endemic cerebellar cognitive affective disorder manifesting as epidemic autism. This forms a new autistic behavioural phenotype which is introverted, non-compassionate, non-empathic, psychopathic and with extrasensory perceptual abilities of telepathy, telekinesis and time travel leading to communication with other galactic civilizations. The new behavioural autistic robotic neoneanderthalic phenotype will have the necessary intelligence and mental abilities cope with climate change. The neoneanderthalic archaeaons are capable of quantal perception and can communicate quantally with other star systems and multiverses. This leads on to the evolution of a galactic civilization. The neoneanderthalic archaeaons produce a metabolic phenotype capable of surviving in extremophilic environments of other star systems and can survive as civilization in other stars and exoplanets. The neoneanderthalic autistic phenotype with its extremophilic archaeaon organelle is suited for space colonisation.

The neoneanderthalic archaeaons use cholesterol oxidation as a form of energetics. This produces depletion of sex hormones producing asexual phenotypes. This leads to sexual phenotypes with alternate sexuality and parthenogenesis as a form of reproduction. This produces change in family and society as well as organisation of the community.

The neoneanderthalic archaeaons secrete RNA viroids which get converted to DNA viroids by human endogenous retroviral reverse transcriptase. The neoneanderthalic archaeaon generated RNA and DNA viroids can hybridize with other RNA and DNA viruses producing new RNA and DNA viruses to which the neoneanderthalic species is resistant. Therefore, there exists two sets of population with differing variety of archaeal symbiosis. The homo neanderthalic species has persisted as fossilised matrilineal societies in Europe and Asia which includes the Celts, the Dravidians, the Berbers and the Sumerians. These fossilised remnants of homo neanderthalic species are ideal substrates for archaeal endosymbiosis in the setting of global warming and extremophilic climatic conditions. These fossilised neanderthalic matrilineal societies will expand in the period of global warming and extremophilic conditions due to endosymbiotic archaeal symbiosis to which they are already acclimatised and exposed to. The human species homo neanderthalis is a consequence of archaeal symbiosis in the extremophilic conditions of ice age and are RNA viral/retroviral resistant consequent to endogenous digoxin synthesis and by archaeal RNA viroidal induced RNA interference and viral inactivation. The homo sapiens which evolved in Africa and migrated later to the Eurasian land mass are retroviral sensitive and other RNA viral sensitive and have a low load of endosymbiotic archaea growth. The homo sapiens owing to conditions of evolution is not an ideal substrate for global warming related endosymbiotic archaeal symbiosis. The homo neanderthalis archaeaons secrete RNA viroids which are converted to DNA viroids with both of them hybridizing with other RNA and DNA viruses in nature producing new viruses to which the homo neanderthalic and homo neoneanderthalic host is resistant. The new RNA and DNA viruses produced by homo neoneanderthalis and neanderthalis in global warming will infect the susceptible homo sapien species and wipe them out from earth. The homo sapien species is not a good substrate for global warming related endosymbiotic archaeal growth and would not evolve new cellular organelle like steroidelle, viroidelle, neurotransminoids and vitaminocyte which would help them to survive in extremophilic conditions of global warming. The homo sapiens also lack archaeaon organelle which help them to fix carbon dioxide and generate water, synthesis ATP by photo-induction mediated by sodium-potassium ATPase and archaeaon mediated fission and fusion reaction. The homo sapien species therefore without evolving archaeaon organelle cannot survive in the harsh environment of global warming induced climate change and is exterminated by homo neanderthalis and homo neoneanderthalis. The earth gets dominated by homo neanderthalis and homo neoneanderthalis in the setting of global warming and extremophilic climate change and they go on to contact other galactic civilizations by quantal perceptions as well as colonising other exoplanets, planets and star systems. This produces a civilization of superhuman homo neanderthalis and homo neoneanderthalis with new metabolic, endocrine, neuronal, psychological and genotypic features. The homo neoneanderthalis and homo neanderthalis with its cellular archaeaons and porphyrions can interlink with the internet and computer systems as well as robots producing man-machine hybrids by quantal perception. The artificial intelligence and robotic systems gets merged with homo neanderthalis and homo neoneanderthalis producing superhuman man-machine hybrids. The new homo neanderthalis and homo neoneanderthalis superhuman autistic man-machine hybrids will survive, prosper and flourish in the setting of global warming related extremophilic climate change and evolving to contact by quantal perception other galactic forms of life and civilization as well as colonising other planets and star systems.

CHAPTER 14

A NEW SEXUAL PHENOTYPE - SEXUALITY IN NEANDERTHALS AND NEONEANDERTHALS – TRIBE OF MALE EUNUCHS AND AMAZONIAN FEMALES – INFECTIVE SYMBIOTIC TISSUE COLONIES OF PROTOZOA, ARCHAEA, VIRUSES AND BACTERIA

The plasma life based forms and the gravitational and anti-gravitational waves generated porphyrions would have been the most primitive of non-life forms. The porphyrin templates would have given rise to the isoprenoid organism, the RNA viroids, DNA viroids and prions which would have symbiosed to form the nanoarchaea. The archaea would have evolved into prokaryotes or bacteria. The archaeal symbiosis with prokaryotes would have generated the eukaryotes. The initial unicellular eukaryotes are the protozoa. The protozoa are capable of phagocytosis and would have had viruses, bacteria and archaea as their intracellular symbionts. The viruses of the protozoa like entamoeba and toxoplasma are giant viruses called megaviride. The protozoans like entamoeba with its viral, archaeal and bacterial endosymbionts would have evolved into the multicellular organisms or metazoa. The archaeal, bacterial and protozoal extracellular vesicles containing RNA and DNA viroids can by horizontal gene transfer lead to evolution of metazoan, plants, animals and human species. The metazoans would have evolved into plants, animals and humans by horizontal gene transfer generated by archaeal extracellular RNA and DNA viroid vesicles. The evidence of such a transformation is seen in the bacterial symbionts of protozoa like entamoeba which includes legionella, listeria, mycoplasma, mycobacteria and chlamydia. The symbionts of protozoan entamoeba include viruses like covid, HIV and cytomegalo virus. The protozoan entamoeba serves as a natural reservoir for covid infections, bacterial infections and HIV infections. The toxoplasma also serves as a covid reservoir. These protozoan colonies of entamoeba and toxoplasma with its bacterial, archaeal and

viral symbionts can be called as chimeric infective symbiotic tissue colonies which can exist in the air, water, earth, blood and tissues. The chimeric infective symbiotic tissue colonies containing bacteria, archaea and viruses can make the symbiotic organism multidrug resistant by horizontal gene transfer. The chimeric infective symbiotic tissue organism evolves as a consequence of global warming and the protozoal, bacterial, viral and archaeal symbiotic colonies are new forms of life. The chimeric infective symbiotic tissue organism composed of protozoa and its symbionts like virus, bacteria and archaea depend upon glycolysis for its energy needs. The glycolytic enzymes enolase, glyceraldehyde 3-phosphate dehydrogenase, hexokinase, phosphofructokinase, aldolase, pyruvic kinase and triose phosphate isomerase are moonlighting proteins having a cytoplasmic, membrane and nuclear functions. These glycolytic enzymes have a RNA binding motif which can modulate RNA transcription in the host and the symbiotic organism. These glycolytic enzymes with their RNA binding motif as well as DNA binding motif can bind to RNA and DNA viruses and viroids modulating their replication and life cycles. The chimeric infective symbiotic tissue organism composed of protozoa and its symbionts like virus, bacteria and archaea can enter the human host through the breaches of the gut-blood barrier and transform itself into the macrophages of the human host. The macrophages are involved in the phagocytosis and innate immunity against bacterial and viral infections. Paradoxically the macrophages like the protozoa can also serve as a reservoir where pathogens like mycobacteria tuberculosis, mycobacteria leprae, listeria, legionella, E. coli, shigella & salmonella, and viruses like HIV & covid can thrive and replicate. The modified chimeric infective symbiotic tissue organism or macrophage composed of protozoa and its symbionts like virus, bacteria and archaea can serve as a reservoir for horizontal gene transfer and generation of multidrug resistant bacteria and viruses. The infective chimeric symbiotic tissue organism composed of protozoa and its symbionts like virus, bacteria and archaea depends upon moonlighting glycolytic enzymes for their energetics and

replication. The development of antibodies against the infective chimeric symbiotic tissue organism composed of protozoa and its symbionts like virus, bacteria and archaea leads to the evolution of anti-brain antibody disease or anti-basal ganglia antibody disease. They present a threat to existence consequent to horiztontal gene transfer and generation of multidrug resistant bacteria, viruses, protozoa, mycobacteria and archaea.

The COVID 19 brain infection can lead to integration of COVID 19 sequences into the genome, creation of cerebellar dominant brain and neanderthalisation. Somatic parthenogenesis occurs due to transformation of the somatic cell to pluripotent stem cell which can develop to parthenogenetic embryos. This results from hybridization between two different species - Homo neanderthalis and Homo sapiens. This interspecies hybridization produces intragenomic conflict and parthenogenesis. Parthenogenesis can be induced by stress of climate change and by endosymbiotic archaea and RNA viroids. The intragenomic conflict consequent to interspecies hybridization, results in upregulation of the tryptophan catabolic pathways producing increased amounts of kynurenine, immunosuppression and immune escape of the parthenogenetic embryos. The interspecies hybridization and intragenomic conflict results in reptilian gene expression and digoxin synthesis. The digoxin produced by endosymbiotic archaea results in upregulated tryptophan transport and catabolism over tyrosine. The increased kynurenine catabolism in Neanderthals results in immune escape and endosymbiotic archaeal growth. The endosymbiotic archaeal growth results in neanderthalisation of the species. The neanderthalisation of the species and interspecies hybridization and intragenomic conflict results in porphyrias and increased porphyrion mediated low level EMF perception, cortical atrophy and cerebellar dominance. This produces a cerebellar cognitive affective disorder with autistic features, impulsivity, aggressiveness, power lust, increased sexual desire, alternate sexuality, criminality, cannibalistic features and anarchic social mores. The

Neanderthal species owing to endosymbiotic archaeal growth and stem cell transformation had an asexual mode of reproduction with parthenogenesis. This results in female dominance, Amazonian syndrome, a culture of male eunuchs and matriarchy. The interspecies hybridization and intragenomic conflict results in a SLOS phenotype with decreased cholesterol synthesis. This produces decreased sex hormone synthesis resulting in asexuality and alternate sexuality. The Neanderthals were matriarchal and worship the Mother Goddess. The reptilian gene expression and SLOS phenotype results in generation of serpentine features. The genes for the primitive reptilian brain complex get expressed with aggression, violence, impulsivity, cannibalism, anarchy, immorality, lust for power and dominance. The SLOS phenotype results in development of primitive features like the development of tail or cauda, cleft chin, prominent thin long teeth and canines, freckles, scales in the skin and syndactly or webbing. The digoxin synthesis and decreased tyrosine transport results in reduced dopamine and melanin synthesis contributing to a tribe of anarchic white gods. The reptilian gene expression and digoxin synthesis leads to sympathetic hyperactivity and ability to regulate the body temperature in alignment with environmental temperature producing cold bloodedness. The archaea induced fructolysis and fructosemia owing to induction of aldose reductase results in increased lipid and mucopolysaccharide synthesis and hibernation syndrome contributing to obesity and increased subcutaneous fat like reptiles. The evolution of cervical rib, the widow's peak in the eye brows and prominent second toe in the feet is due to increase of recessive traits consequent to inbreeding. The inbreeding results from the autistic phenotype and reduced social contact and social withdrawal. This autistic phenotype with the inbreeding and parthenogenesis leads to decreased genetic diversity and extinction of Neanderthals. The Neanderthals have increased subcutaneous fat, scaly skin which is partly due to albinism and UV exposure as well as due to cholesterol synthetic defect, increased salt secreting eccrine sweat glands and dominant tryptophan pathway and serotonin synthesis producing pineal gland or third eye. The increased tryptophan catabolism produces an epidemic oshtoran syndrome. The neanderthalisation produces an epidemic anarchic syndrome. The neanderthalisation also produces the syndrome of male eunuch and matrilineality. The Neanderthals were mostly parthenogenetic. The cerebellar cognitive affective disorder and actinidic archaeal magnetite and porphyrion induced quantal perception resulted in equality and universality. The primitive Neanderthal tribes are represented by the Saxons, Sakas, Basque, Etruscans, Dravidian races, the Celts and Berbers. The most dominant Neanderthal tribe is the Anglo-Saxons which created an equal society with a universal culture which is American, a universal language which is English, a universal monetary system represented by the dollar and the unification of all ethnicity in the American dream of tolerance and inclusiveness. The Neanderthals had serpentine features and Neanderthal cultures all over the world is represented by serpent worships as seen in Sumeria, the Dravidians of South India, Thoth of Egypt and in Mexico. The Naga tribes and Naga lore of South India represents the Neanderthal culture. The modern version of serpent worship is seen in the symbol of medical profession, the representation of the dollar and in free masonry. Free masonry of the Anglo-Saxons incorporates all religions and is the first universal religion and is neanderthalic. The neanderthalic populations were represented by particular blood groups, the universal donor O Rh –ve and AB Rh –ve, the universal receiver.

The Neanderthal phenotype gives clues as to the origin of the humans. The Naga tribes of South India were hypothesized to originate in the ancient Lemurian land mass which got broken up by tsunamis and earthquakes. The remnants of the Neanderthal phenotype are seen in the Australian Aboriginals, the New Zealandian Maoris, the Dravidian Tamils and Nairs. These societies are predominantly matriarchal and serpent worshipping. The homo neanderthalic possibly arose in the Lemurian oceanic land mass supporting the theory of the aquatic ape origin of humans. The accepted theory of human origin postulates the origin of humans from primates in the African savannahs. Several points give clue to an origin of the human species in water. The neanderthalic behaviour can be compared to the behaviour of the bonobo monkeys or lemurs seen in the ancient Lemurian continent. The homo neanderthalis would have originated from the bonobo monkeys in the Lemurian backwaters communicating with the sea. The bonobo monkeys owing to shortage of food would have started foraging the backwaters and sea for fish and tubers of water lilies. The fish contains essential fatty acids like docosa hexaenoic acid and tubers contain plenty of carbohydrates. The brain is exclusively dependent on carbohydrates and ketone bodies for energy. The essential fatty acids increase the brain growth. The brain growth in humans is called encephalisation which is more than in primates and is equivalent to sea mammals like the dolphin and whales. The bonobo monkeys would have waded into water and stood in water generating the phenomenon of bipedalism. This would have freed their hands to catch fish and break shellfish to generate food. This would also have freed the hands to collect and eat the tubers of water plants. The human species lack hair unlike the primates but like the aquatic mammals. The human species have increased subcutaneous fat-like aquatic mammals and unlike primates. The human beings have got eccrine sweat glands and tear glands useful in a watery environment. The human trachea is placed down in the neck unlike that in primates where it is more nasal. The human language points to an aquatic origin for human species. For the human language to develop you have to consciously control your breathing which does not exist in primates but in humans and aquatic mammals. The humans have the diving reflex. The smooth skin with sparse hair and thick subcutaneous fat for insulation points to an aquatic origin for humans like aquatic whales and dolphins. The webbed feet and hands also point to an aquatic origin. The sebaceous glands with its greasy secretion point to water-proofing in humans and an aquatic origin. The homo neanderthalis unlike the homo sapiens is more of an aquatic swimming type. The homo neanderthalis had longer lungs and increased respiratory capacity which helped them to dive into water and float in water. The homo neanderthalis owing to their SLOS phenotype and albinism had vitamin D deficiency. Vitamin D is synthesized from cholesterol. The homo neanderthalic bone was thin as compared to the darkened vitamin D rich homo sapiens. The long lungs and the thin bone helped the homo neanderthalis to float in water. The origin of paranasal sinuses were for the human species to hold its head above water. The human beings sexual behaviour is like aquatic mammals and unlike primates with front to front population. All these differences point to an aquatic origin for the human beings or an aquatic ape hypothesis. The homo neanderthalis originated from the bonobo Lemurian monkeys which waded into water to search for food. The bonobo monkeys sexual behaviour and promiscuousness and alternate sexuality are comparable to homo neanderthalis. The homo neanderthalis arises owing to archaeal endosymbiosis. The waters of the ocean and backwaters are rich in marine archaea which are capable for acetogenesis and methanogenesis. The backwaters and sea of the South Asian peninsular land mass is rich in actinides, the bathyarchaeota. They are capable of acetogenesis and are a source of organic carbon. The actinides would have formed scaffolds for the formation of complex life molecules like RNA, DNA, protein, isoprenoids and complex carbohydrates. This would have produces RNA viroids, DNA viroids, isoprenoid organism and prions on actinidic surfaces which would have symbiosed to form archaea and eventual multicellular organisms. The multicellular organisms arising on abiogenetic actinidic surfaces in the backwater-ocean connections seen in Southern peninsular India which broke away from the Lemurian land mass would have evolved into eukaryotes, prokaryotes, multicellular organisms and symbiotic plants/animals. The bonobo Lemurian monkeys would have evolved into homo neanderthalis by archaeal endosymbiosis in the actinidic shores of backwaters, lakes and oceans of peninsular India. The remnants of homo neanderthalis are seen in Australian aboriginals, Maoris and Dravidians which are all matriarchal and serpent worshipers. The aquatic ape and homo neanderthalis would have evolved in the actinidic sand shores of backwaters and lakes of Lemuria and peninsular India. The Dravidian communities are matriarchal and female dominant. There is a high degree of consanguinity and inbreeding in the Dravidian matrilineal communities. Parthenogenesis would have been dominant in such communities with matriarchy producing syndromes of the male eunuchs and oshtoran syndromes. The homo neanderthalis ate a carnivorous diet. The teeth were longer compared to homo sapiens and the canines were prominent comparable to fangs of snakes. The cows and bulls were domesticated by the homo neanderthalis which were originally hunters and warriors hunting on mammoths. The domestication of cows and bulls resulted in a high consumption of milk and meat including beef. This resulted in the generation of a lactose tolerant adult population. The gene for lactose tolerance arose 12,000 years ago. The relationship between the cows and Neanderthals could be described as parasitic obligate symbiosis. The origin of cow worship and bull worship in peninsular Indian religions can be related to it. The increased consumption of a carnivorous diet of milk and meat resulted in increased tryptophan intake and catabolism generating more of kynurenines producing immune suppression and immune escape required for parthenogenesis. The immune suppression also produced cold bloodedness of Neanderthals for survival in the cold watery climate. The elements of this culture are still seen in the matriarchal Dravidian communities of peninsular India. The actinidic sea shores and backwater shores of peninsular India is where the homo neanderthalis or the aquatic ape originated. The culture of the Dravidian peninsula is matriarchal and the religious traditions still continue with serpent worshipping culture. This can be called as the mermaid culture.

The human hairlessness, thick subcutaneous fat, webs in the feet and toes, shape of the nostril all point to a watery origin for the human race in mangrove swamps, backwaters and seas. The presence of diving reflex in humans, sweating, tearing, descended human larynx in the neck, hair tract patterns, the presence of hymen and vernix caseosa in babies like seals point to a watery origin for humans. The reproductive behaviour of humans with front-to-front population like aquatic mammals points to the watery origin for humans. The great apes would have come down from the trees and went through an aquaboreal phase where it waded through swamps feeding on molluscs, fruits and fish producing bipedalism. The need to excrete large amounts of salt in sea or brackish water leads to the origin of tears and eccrine sweat glands. The human nose is protruded unlike that of primates to protect it from water. The human hairlessness, thick subcutaneous fat and sweating point to the watery origin for humans and bipedalism. The encephalisation of the brain was due to the large intake of omega-3 and omega-6 fatty acids from fish. The human jaw is short with short teeth unlike Chimpanzees pointing to eating of marine food. All naked mammals are aquatic like the dolphin, the manatee, elephant, pig and rhinoceros. The human beings are the only naked apes. The increased subcutaneous fat or blubber in human babies and vernix caseosa of babies point to a watery origin for human species. The human beings have the lacrimal glands and sweat glands to excrete salt in a sea environment. The human babies are plump with 16 percent body fat and the human milk contains 25 percent fat. The babies roll over on their body and float in water with their nose in air. The fact that human babies can swim point to a watery origin for the human species. The Chimpanzees' hand is light and strong to hang from trees. The human babies hand is too heavy and weak and can grasp on to the hair of the mother while swimming in water. The human babies have developed the grasp response for this type of evolution. The human species female has long oily sebum coated scalp hair. The infant chimps have got a strong neck which is kept steady. The human baby attains neck steadiness at six months, but if the baby is placed in water the neck becomes strong. The human speech arose due to the conscious control of respiration which the chimp cannot. The speech also owes its origin to the position of the larynx in the neck. The pincer grip of humans is developed to get meat out of shell fish and mussels which is called as precision grip. This theory was put forward by Elaine Morgan. The backwaters contain water lilies and tubers whose roots were consumed by the primitive humans along with fish, mussels, snails and shell fish. The water lily and lotus roots and leaves contain alkaloids like nupharine and aporphine which are psychoactive and gives rise to the dream state of Neanderthals. The lotus and water lilies are associated with creation myths like that of the Sun God Ra emerging from the lotus in primordial waters and the Brahma the Creator seated on the lotus. The homo neanderthalis emerged first in the Lemurian land mass which was more like a big island susceptible to breakage to independent land masses owing to widespread tsunamis in the region. This would have lead to inbreeding in the Neanderthals leading on to loss of genetic diversity and expression of reptilian genes. This would have also contributed to the eventual extinction of Neanderthals. The aquatic ape would have arose as homo neanderthalis in the Lemurian land mass and its breakaway regions like the backwaters linked to the sea regions of Kerala with actinidic sands. This is indicated by the persistence of matrilineal society in the Dravidians of Kerala and the detection of endosymbiotic archaea in the blood of Kerala population. The psychometric neanderthalic quotient is high in the population of Kerala with a high incidence of autism. Matrilineality and consanguinity is common among the Dravidian Nair population of Kerala. The Dravidians tend to have a Neanderthal phenotype. The Dravidian Nairs are postulated to have a Scythian origin. Serpent worship, serpent music and serpent dances are common in Kerala. The communities in Kerala are mostly carnivorous and consume beef. The culture in Kerala is more tolerant and Keralites migrate all over the world and mix with different societies. Tolerance and inclusiveness is a feature of NQ quotient. Serpent temples are widespread in Kerala. Kerala has got a higher incidence of

Neanderthal genomic sequences related diseases like autism, schizophrenia, ADHD, addictions, metabolic syndrome, autoimmune disease and cancer. It is tempting to locate the origin of the upright aquatic ape in the extended backwaters and lakes of Kerala with its connections to the sea and its actinidic sand rich shores. The backwaters are rich in fresh fish and mussels and water lilies and lotuses giving tubers and roots. The homo neanderthalis evolved from archaeal endosymbiosis and marine archaea are dominant in the seas of the Indian ocean and backwaters of Kerala. Serpent worship is a dominant theme in the Dravidian Nair culture and Serpent God Anantha is the dominant deity. This tempts us to speculate on the origin of bipedalism, human species and homo neanderthalis in Kerala.

The origin of the aquatic ape points to the dominant role for the female of the species in human evolution. Human evolutionary theories have focused on the hunter gatherer and tool maker males. The watery origin of bipedalism and human species points to a dominant role for woman in human evolution. The body anatomy of the females with pendular mammary glands and rounded glutei are for floating and buoyancy and not for sexual attraction. This produces a gynocentric approach to evolution as against an androcentric approach. Evolution was basically meant for protection and rearing of children. Humans have evolved in swamps, backwaters and sea and are not biologically or socially inferior to men. The homo neanderthalis have got features unlike that of chimpanzees. The social patterns of homo neanderthalis can be compared to the bonobo monkeys. The bonobo monkey society is female centred and egalitarian. Sex is a part of social relationship and serves as a substitute for aggression. The bonobo monkeys have different types of sexuality heterosexual, and male to male and female to female. The frequency of sexual interaction is more but the reproductive rate of bonobo monkeys is the same as chimpanzees. The chimpanzees evolved in the open dry savannah while the bonobo monkeys still lived in trees and hanged down from trees. The tree habitat of the bonobo monkeys lead them to an evolutionary form of life where they can hang from mangrove trees in swamps and eventually wade in water. The bonobo monkeys are pigmy monkeys with male weighing 43 kg and female 33 kg. They are omnivorous and eat fruits, small amount of vertebrates and invertebrates. They have imaginative plays and have sex in missionary positions. They have wide variety of sexuality and group behaviour. Sex was a means of social relationships. The female bonobos bonded among themselves and led the community. The male bonobo is attached to his mother and depends on her for protection throughout life. The bonobo society can be compared to a matriarchal female dominant Neanderthal society.

The female dominant model of human evolution raises the question of who evolved first - the male or the female. The original fossils of human species are predominantly female and the male fossils evolved after billions of years. The original human species would have been a cluster of female bipedals in swampy waters feeding on tubers of water lilies and lotus as well as fish, mussels and shell fish. The women can reproduce by parthenogenesis like lower animals. Therefore it is natural for the female of the species to evolve first. The sexual relationship in such female only societies in primodial times was lesbian. The evolution of males occurred at a later date. The macho model of human evolution with male hunter and male toolmaker and a female accomplice evolving together is highly unlikely. The next stage of human evolution has been postulated to be interspecies hybrids. This was put forward by Eugene McCarthy. McCarthy pointed to several features of men of humans similar to pigs. Pig organs can be transplanted to humans without rejection. The pigs like humans are hairless, have thick layer of subcutaneous fat, protruding nose and heavy eye lashes. The pig genetic sequence contains similar SINE element ALU as humans. The phenomenon of crossing the species barrier is represented by the generation of the swine flu epidemic in humans. The early bipedal female only human species would have generated human pig interspecies hybrids. This would have generated interspecies hybrids of males and females. The male sexual organ corresponds to the tail of mammals. Similarly sex organs of species like snakes correspond to limb buds. The human embryo in its various stages of development can be compared to fishes, amphibians and lower animals. Interspecies hybrid would have led to the development of male and female of the species. The water mammals like cows, bulls, pigs, elephants, rhinoceros, turtles, crocodiles, water snakes and giant tortoises would have contributed to the generation of interspecies hybrids. This would have generated a population of both male and female bipedals in the swamps with different type of sexual interactions - heterosexual, bisexual, homo sexual and lesbian. Genetic diversity is required for a species to survive. Thus heterosexuality as a mode of sexual behaviour would have become acceptable to society as such. Bacterial and archaeal conjugation with human cells have been described. Chimeras of humans and animals have been produced in labs. Interspecies hybrids have been generated in human labs and populations have been produced. Interspecies hybrids include the dzo between yak and cattle, zubron between cow and bison, cama between camel and illama, yakulo between yak and buffalo, sheep-goat and mules. This exemplified by the Plant of the Apes. The interspecies hybrids would have been protected from pre-zygotic and post-zygotic isolation and destruction by the phenomena of immunosuppression and immune escape mediated by tryptophan catabolite kynurenine. A fish diet in swampy waters is rich in tryptophan. The Hindu myth of creation of Matsya the fish, Koorma the tortoise, Varaha the boar and Narasimha the lion point to the generation of interspecies hybrids as the main lynch point of evolution. The generation of human brain structure also depends upon interspecies hybridization. The reptilian complex of the brain is dominant in Neanderthals. The reptilian complex is seen in amniotes which include mammals, reptiles and birds. The reptilian complex of the reptilian brain includes the basal ganglia, the brain stem and cerebellum. This forms the basis of the cerebellar cognitive affective disorder in Neanderthals. The reptilian brain is the site of imagination, intuition, instinct, compulsivity and dreams. It communicates by symbols and archetypes. It is the site of obsessive compulsive disorder, superstition, ritualism, slavishness and conformation to all way of doing things. It is the site of territoriality, aggression, racism, violence and hypersexuality. The reptilian complex is dominant in amphibians, fishes and reptiles. The Neanderthal brain has got cerebral cortical atrophy and cerebellar dominance. The reptilian brain and reptilian genes are dominant in Neanderthals pointing to interspecies hybridization of the first evolving Neanderthal females and matrilineal female only Neanderthal society. The interspecies hybrids are signified by the importance of even toed ungulates like pig and cattle in human culture and religion. The even toed ungulates include cattle, pig, deer, camel, sheep, goat and hippopotamus. The odd toed ungulates include rhinoceros and horses. The aquatic cetaceans like whales, dolphins and purpoises evolved from even toed ungulates. Dolphins can communicate with humans. The aquatic cetaceans and even toed ungulates together form a family called cetardiodactyla. The even toed ungulates form large social groups with hierarchy, harem groups and bachelor groups. They mark their territory through glandular secretions. The ungulates can swim and whales and dolphins can gallop in water. The pig antigens have great similarity with human antigens and pig organs can be xenotransplanted into humans. The rejection is due to the presence of retrovirus in the pigs which infects humans. The targeted deletion of retrovirus related DNA from pigs makes pig a valuable source for organ transplantation. The retroviral deleted pig has been cloned and developed into embryos and implanted into sows. The pigs serve as a reservoir for human organ transplant. Hog organs can be transplanted to humans if immunized against human serum. Horse serum is used to develop tetanus antibodies and snake anti-venom. The consumption of beef leads to the development of prion disease in humans. Porcine insulin can be injected into humans and porcine heart valves can be transplanted to humans. Cow products like milk, dung and urine are used as human medicines. Human stem cells injected into pig embryos have resulted in the development of human pig chimeras. These human pig chimeras can be used for organ transplantation. The development of human-ungulate chimera embryos points to the importance of interspecies hybridization in human evolution. This is signified by the worship of cow as Kamadhenu or Mother Goddess in Hindu culture, the worship of Varaha, the boar as one of the avatars of Vishnu and the religious relationship between Satan and pigs in Semitic religions. The Hindu star signs in religion and astrology have a figurative animal representation. This points to interspecies hybridization playing an important role in evolution.

The aquatic ape evolved from bonobo monkeys into homo neanderthalis in the brackish back waters of the peninsular India. The Neanderthals evolved by archaeal endosymbiosis. The archaea can oxidise cholesterol and ammonia for its energetics. The archaea binds to the toll receptor produces mitochondrial dysfunction and inhibition of TCA cycle and resultant activation of the glycolytic pathway for energetics. This produces the metabolic Warburg phenotype in homo neanderthalis. The homo neanderthalis depends upon ketone body oxidation for its energy needs and subsists on ketogenic diet. The consumption of ketogenic diet results in amino acid catabolism and generation of ammonia which can be oxidized by the archaea for its energy needs. The ammonia can combine with carbon dioxide producing urea which can be acted upon by archaeal urease generating ammonia again. The urea can inhibit mitochondrial function and produce protein modulation by carbomylation. It can thus affect the metabolonome. The inhibition of the TCA cycle channels acetyl CoA to the mevalonate pathway synthesizing cholesterol which can be oxidized by the archaea for its energy needs. The cholesterol ring oxidation generates pyruvate which is acted upon by SGPT generating glutamate which is catabolized by glutamate dehydrogenase generating ammonia. The ammonia can be oxidized by the archaea for its energetics. The ammonia can also be converted to urea by the urea cycle for ammonia storage. The urea mediated carbomylation of proteins results in somatic cell dysfunction and resultant takeover of the human cells by endosymbiotic archaea producing a zombie syndrome with the cell machinery taken over for cholesterol synthesis and oxidation as well as ammonia formation and oxidation subserving archaeal energetics. Urea serves as a substrate for ammonia storage. The aquatic ape evolved in the brackish backwaters of peninsular India and Kerala. The aquatic ape evolved from the bonobo monkeys and eventually developed into homo neanderthalis. The homo neanderthalis and the aquatic ape had a watery home of salty brackish water and the urea served as a substrate for balancing the salt content of the body fluids against the high salt content of the brackish water. The aquatic ape and the homo neanderthalis fed on fishes, mussels, crabs as well as tubers of water plants and this habit did not need strong males and was carried out by dominantly by females. The archaeal cholesterol catabolism produced low level of sex hormones in the aquatic ape and homo neanderthalis producing an asexual phenotype with alternate sexual behaviours. This colony of asexual phenotypes was matriarchal and female dominant with essentially served by groups of subservient male eunuchs. The high salt content of the body due to a life in brackish waters served to induce parthenogenesis in the dominant female. The urea synthesis from ammonia was also important in the induction of parthenogenesis. Urea can induce female germinal cells to develop into parthenogenetic embryos. Urea can promote transformation as well as preservation of stem cells and germ cells. The endosymbiotic archaea can produce germ cell transformation as well as stem cell transformation and induce parthenogenesis. Parthenogenesis would have been the dominant form of reproduction in the aquatic ape and homo neanderthalis. Urea synthesis can occur in the liver, the skin and brain of homo neanderthalis and to some extent in homo sapiens. The homo neanderthalic brain evolved due to endosymbiotic magnetotactic archaeal endosymbiosis. The magnetotactic archaea and porphyrions can produce increased absorption of low level EMF producing cortical atrophy and cerebellar dominance. This produces the cerebellar cognitive affective disorder homo neanderthalic brain phenotype with its impulsive behaviour, social withdrawal, creativity and autistic as well as schizophrenic phenotypes. The endosymbiotic archaea uses ammonia as an energy substrate and ammonia is stored in urea molecule for use as and when it is required. The urea synthesis in the homo neanderthalic brain is significant in this respect and serves the purpose of endosymbiotic archaeal energetics. The homo neanderthalic brain can be considered as quantal computing magnetotactic archaeal colony. The homo neanderthalic neuronal cells and circuitry are converted to zombie circuits by neuronal and synaptic protein carbomylation by brain urea. Thus the brain urea synthesis is of great importance in the functioning of the magnetotactic archaeal colony dominated zombie brain of homo neanderthalis. The urea synthesis is also important in the adaptation of the aquatic ape and homo neanderthalis to the salty brackish backwaters of Kerala and maintaining balance between sodium content of body fluids and brackish salt water. The urea molecule is also important in the generation and preservation of stem cells and germ cells as well as their parthenogenetic induction required for formation of embryos by asexual reproduction. Thus the ammonia oxidation and urea synthesis is crucial in endosymbiotic archaeal energetics which gives life to the zombie scaffold of Neanderthal body and brain, the maintenance of the magnetotactic archaeal colony network which functions as the controlling power in the Neanderthal zombie brain and in the generation of stem cells and germ cells and the induction of parthenogenesis.

Somatic parthenogenesis is due to climate change. Climate change can produce archaeal endosymbiosis and archaea and archaeal RNA viroid induced parthenogenesis. In response to stress, somatic cells can get transformed to stem cells by endosymbiotic archaea. Stem cell metabolonomics - anaerobic glycolysis, PDH dysfunction, CoQ deficiency mitochondrial dysfunction, branched chain ketoacid dehydrogenase dysfunction, homocystinuria and genomic demythelation, porphyrias and reactive oxygen species generation, SLOS (Smith Lemli Opitz) leading to cholesterol depletion, low sex hormones, vitamin D deficiency and bile acid deficiency. Stem cells can undergo endoreduplication, cell fusion and budding and bursting like bacteria producing polyploidy. These polyploidal cells can become parthogenic embryos. The polyploidal cells are stress resistant and genomically unstable. The polyploidal cells are genomically, metabolically and phenotypically different and unstable. They can get converted to different tissues like brain, liver and heart forming somatic embryos. Multiple somatic embryos with polyploidy produces multiple personality disorders – schizophrenia, autism and mood disorder. Multiple somatic embryos with polyploidy are antigenic and can produce an autoimmune disease. Multiple somatic embryos with polyploidy are genomically unstable can produce cancer. Parthogenesis can lead to social changes including matrilineal societies, alternate sexualities and different identities. Multiple somatic embryos with polyploidy are metabolically and genotypically unstable leads to neurodegeneration. Multiple somatic embryos with different metabolic instability can lead to metabolic syndrome. Archaea and RNA viroid induced mitochondrial dysfunction and upregulated glycolysis resulting in stem cell transformation of somatic cells. The somatic cells which are stem cell transform in the setting of immune activation and cytokine secretion can get converted to germinal cells – sperm and ova. This can result in fertilisation and parthenogenesis. Archaeal digoxin can produce intracellular magnesium deficiency and failed mitosis due to spindle dysfunction. This can produce polyploidal cells. The polyploidal cells can assume stem cell functions. The stem cells can mimic germline cells. The meiotic programs of the polyploidal stem cells can get activated generating cleavage embryos, morulas which can get converted to tumour spheroids. The spheroids can get converted to blastocyst and post implantation embryos producing oncogenesis. They can also dissemble producing metastasis.

The Neanderthals ate a high protein high fat non-vegetarian diet by hunting and scavenging mammoths and other animals. This resulted in heavy load of tryptophan in the system producing tryptophanuria and tryptophanemia. The meat contains a high level tryptophan and haemoglobin which can induce the enzyme indoleamine 2,3-dioxygenase and tryptophan 2,3-dioxygenase which are both heme enzymes. The Neanderthals ate a low fibre diet resulting in decrease supply of short chain fatty acids especially butyrate from the gut. Butyrate can suppress indoleamine 2,3-dioxygenase and tryptophan 2,3-dioxygenase and a fibre deficient diet in Neanderthals can upregulate the activity of indoleamine 2,3-dioxygenase and tryptophan 2,3-dioxygenase resulting in increased catabolism of tryptophan along the kynurenine pathway. Natural substances that are deficient in non-vegetarian diet like brassica alkaloids, curcumin, caffeine, tea and coca can inhibit indoleamine 2,3-dioxygenase and tryptophan 2,3-dioxygenase which becomes over active in Neanderthals producing tryptophan catabolism. The tryptophan is metabolised to formyl kynurenine, hydroxy kynurenine, kynurenic acid, 3-hydroxy anthranilic acid, quinolinic acid and NAD. Kynurenine can bind to AHR receptor producing immunomodulation and immunosuppression. It can produce immune tolerance in case of embyrogenesis, parthenogenesis, autoimmunity, cancer, lipopolysaccharide tolerance and chronic infection. The kynurenine can produce suppression of T cells and immunity leading to immunotolerance important in the above mentioned states. Thus kynurenine pathway flux can contribute to embyrogenesis and parthenogenesis by producing immune tolerance. The tumours can escape immune destruction by suppression of NK cells and T cells. Thus tumour metabolism depends upon activation of the tryptophan catabolism along the kynurenine pathway. The kynurenine pathway is activated in lipopolysaccharide tolerance and chronic infections producing immunosuppression and immunotolerance. The archaeal endosymbiosis depends upon immunotolerance and immunosuppression by activating tryptophan catabolism along the kynurenine pathway. The tryptophan catabolism along the kynurenine pathway can also contribute to psychiatric disorders. Kynurenine blocks the NMDA receptor and alpha-7 nicotinic receptor. This results in down regulation of NMDA and cholinergic transmission. Kynurenine can also upregulate dopaminergic transmission. This results in schizophrenia and autism. The tryptophan catabolism along the kynurenine pathway blocks serotonin synthesis from tryptophan contributing to depressive and anxiety disorders. The tryptophan catabolism along the kynurenine pathway can result in quinolinic acid synthesis and immune activation resulting in autoimmunity, immune tolerance as well as immune activation. Kynurenine can produce immunosuppression and NMDA blockade while quinolinic acid can produce immune activation and NMDA excitotoxicity. The tryptophan catabolism along the kynurenine pathway can generate quinolinic acid important in neurodegeneration. The tryptophan flux along the kynurenine pathway can generate quinolinic acid which can produce low grade inflammation and insulin resistance causing metabolic syndrome. Thus the flux of tryptophan along the kynurenine pathway can produce autoimmune disease, neurodegeneration, schizophrenia, depression, autism, cancer and metabolic syndrome. The induction of heme enzyme indoleamine 2,3-dioxygenase and tryptophan 2,3-dioxygenase can lead to heme depletion and induction of ALA synthase increasing porphyrin synthesis and producing porphyrias. The porphyrins can form self-replicating porphyrions. The porphyrions form a template for the formation of RNA viroid, DNA viroid and isoprenoid organism which can symbiose together to form an endosymbiotic archaea by abiogenesis. The induction of indoleamine 2,3-dioxygenase and tryptophan 2,3-dioxygenase produces kynurenine which can suppress the immune system generating immunotolerance and archaeal endosymbiosis. Thus the tryptophan load and induction of indoleamine 2,3-dioxygenase and tryptophan 2,3-dioxygenase results in systemic civilizational disorders in Neanderthal population. The tryptophan load and induction of indoleamine 2,3-dioxygenase and tryptophan 2,3-dioxygenase can result in immune tolerance and immunosuppression by kynurenine producing archaeal endosymbiosis and neanderthalisation of the species. The tryptophan load and induction of indoleamine 2,3-dioxygenase and tryptophan 2,3-dioxygenase can result in immunotolerance that can contribute to parthenogenetic embryogenesis or somatic pregnancy in multiple tissues in Neanderthals. The tryptophan load and induction of indoleamine 2,3-dioxygenase and tryptophan 2,3-dioxygenase can produce an autistic schizophrenic tribe of Neanderthals with parthenogenesis and matriarchy.

The archaeal endosymbiosis produces neanderthalisation of the species. The archaea can activate the enzyme AMPK (Adenosine monophosphate kinase). The Neanderthals consumed a ketogenic non-vegetarian diet rich in fat and protein. The neanderthalic ketogenic diet can induce AMPK activation. The aquatic ape phenotype ate a lot of fish diet and fish fatty acids can produce AMPK activation. The aquatic ape phenotype also ate lots of tubers of backwater plants like lotus and water lilies containing high amount of glucomannan producing AMPK activation. The water plant tubers contain high amount of fibers whose digestion generates short chain fatty acids like acetate, butyrate producing AMPK activation. Amino acids and fatty acids can induce AMPK activation. AMPK activation can induce by low glucose and oxygen deprivation states of ice age. The Neanderthals had a cerebellar dominant phenotype with increased sympathetic activity producing the impulsive fear, flight, fight phenotype. The catecholamines - epinephrine and norepinephrine and dopamine can produce AMPK activation. The AMPK activation can result in simultaneous mtor activation resulting in dendritic spine pruning defects and an autistic Neanderthal brain. AMPK activation results in simultaneous activation of the HIF alpha and the induction of the Warburg phenotype and stem cell phenotype. AMPK activation can induce UCP proteins uncoupling oxidative phosphorylation leading to mitochondrial dysfunction. AMPK activation leads to increased catabolism and reduced anabolism. AMPK activation results in weight loss. AMPK activation can also lead to increased insulin signalling and IGF activity. The glycolysis, fatty acid oxidation and amino acid oxidation are increased. The protein synthesis is inhibited. The AMPK activation reduces fatty acid, cholesterol and triglyceride synthesis and increases their breakdown. The increased amino acid oxidation results in tryptophan catabolism producing increased amounts of kynurenines which are important in immune escape and parthenogenesis. The AMPK activation resulting in the stem cell phenotype results in generation of germ cells. The AMPK activation can activate the oocyte into development of parthenogenetic embryos. AMPK activation can increase intracellular calcium oscillations, increased intracellular ROS and AMP/ADP ratio resulting in the shock and wave activation of oocyte producing parthenogenesis. The AMPK activation can suppress the immune system producing immune escape and parthenogenetic embryogenesis. AMPK activation can increase the life span of the species and preservation of the Neanderthal phenotype. AMPK activation can produce cardiac protection and neuroprotection. AMPK activation is important in fertility, stem cell transformation and generation of germ cells. AMPK activation can uncouple oxidative phosphorylation as well as produce mitochondrial biogenesis. AMPK activation has an antioxidant effect by inducing NRF 2, superoxide dismutase and UCP. AMPK activation results in reduce generation of free radicals which act as messengers for endogenous retroviral replication. This produces a rigid genome, defective synaptic connectivity and cerebral cortical atrophy. AMPK activation induces glycogenolysis and inhibits glycogenesis. AMPK activation also increases glucose transport. The mitochondrial oxidative phosphorylation is blocked by AMPK activation by induction of UCP proteins. The glucose is converted to fructose by aldose reductase and sorbitol dehydrogenase induced by archaea and enters the fructolytic pathway. This results in fructosemia and increased synthesis of lipids and mucopolysaccharides resulting in a hibernation syndrome characteristic of Neanderthal metabolonomics. The AMPK activation also results in inhibition of protein synthesis and increasing autophagy/mitophagy and body renewal increasing the life span of the species. Thus the archaea induced AMPK activation leads to the generation of parthenogenetic species which is female and maternal dominant.

Neanderthal evolution was determined by archaeal endosymbiosis. The human species evolved into homo neanderthalis by archaeal endosymbiosis. Archaeal endosymbiosis was mediated by the tryptophan catabolic kynurenine pathway. The kynurenines can produce immunosuppression and immune tolerance resulting in archaeal endosymbiosis. The kynurenine pathway results in blockade of the NMDA receptor, cholinergic nicotinic alpha-7 receptor, decrease production of serotonin and increase dopaminergic transmission. This produces an autistic, schizophrenic Neanderthal tribe with less of frontal executive function and more of cerebellar dominance affective impulsive type behaviour. The induction of indolamine 2,3-dioxygenase and tryptophan 2,3-dioxygenase led to channelling of tryptophan catabolism along the kynurenine pathway. The Neanderthals ate a high protein non-vegetarian tryptophan diet leading to an induction of tryptophan catabolism. The Neanderthal diet was deficient in dietary fibre and fibre digestion generated short chain fatty acids especially butyrate resulting in increasing indolamine 2,3-dioxygenase and tryptophan 2,3-dioxygenase activity and more of tryptophan catabolism. The kynurenine pathway results in generation of quinolinic acid which can produce chronic immune activation and insulin resistance. Quinolinic acid is also involved in neurodegeneration. The kynurenine induced immune cell as well as

NK cell suppression can result in evolution of cancer. The archaeal endosymbiosis generated by kynurenine induced immunotolerance can lead on to cancer, autoimmune disease, metabolic syndrome, neurodegeneration, schizophrenia and autism by generation of archaeal cholesterol catabolite digoxin. Thus the higher load of tryptophan due to a meat diet and low fibre diet results in generation of kynurenine which has a ketamine-like action producing Neanderthal behaviour. The induction of heme enzymes indolamine 2,3-dioxygenase and tryptophan 2,3-dioxygenase results in heme depletion, activation of ALA synthase and porphyrias. The porphyrins can self-organise to form porphyrions and can act as a template to generate isoprenoid organism, RNA viroids, DNA viroids which all symbiozed to form actinidic archaea. The porphyrions have a wave-particle existence and in the presence of membrane intercalated porphyrion mediated sodium-potassium ATPase inhibition can result in a pumped phonon system and dipolar porphyrion mediated quantal perception. The porphyrion intercalated cell membrane and sodium-potassium ATPase inhibition can result in increased intracellular calcium and decreased intracellular magnesium resulting in mitochondrial dysfunction, cell membrane dysfunction, golgi body related protein processing dysfunction, defect in DNA and RNA function and disordered cell function. This increased intracellular calcium and reduced magnesium due to sodium-potassium ATPase inhibition can result in immune activation, glutamate excitotoxicity, oncogene activation and disease states. The immune tolerance produces cancer, stem cell transformation and parthenogenesis. The cancer stem cells can develop meiotic programs generating parthenogenetic embryos which can survive and grow in the presence of immune tolerance created by kynurenines. The multiple parthenogenic embryos can create multiple personalities leading to schizophrenia and autism, grow into cancer, its stem cell metabolism with increased glycolysis and mitochondrial dysfunction can produce metabolic syndrome, stem cell mediated increased glycolysis can lead to immune activation and the normal tissue dying at the expense of parthenogenic embryos can produce degeneration. The chronic inflammation induced by quinolinic acid and other tryptophan catabolites producing autoimmune disease and quinolinic acid related cell death and degeneration. The tryptophan catabolic pathway produces civilizational syndromes in Neanderthals leading to their extinction. The tryptophan loading in higher primates and homo sapiens due to increased meat eating carnivorous habits in Eurasian steppes resulted in kynurenine catabolite induced immunosuppression, immunotolerance, archaeal endosymbiosis and neoneanderthalisation. Thus tryptophan loading due to a high meat and low fibre diet mediated kynurenine generation and immunotolerance would have led to archaeal endosymbiosis and evolution of homo neanderthalis and homo neoneanderthalis. The Neanderthals due to archaeal endosymbiosis had stem cell transformation, activation of meiotic programs in the stem cells, generation of germ cells and parthenogenesis. This resulted in female dominance and matriarchal societies. The intraspecies hybridization and intragenomic conflict would also have contributed to parthenogenesis and reptilian gene expression consequent to interspecies hybridization and intragenomic conflict. Genes affected PDH, BKCD, SLOS, porphyria, Hartnup's disease. Decreased cholesterol CoQ and vitamin D. The Hartnup's trait would have developed to counteract the tryptophan loading in Neanderthals consequent to a high meat diet. This produces what is called as increased tryptophan catabolism and kynurenine catabolites mediated oshtoran syndrome described initially in the Pericaspian areas. Endemic oshtoran syndromes would have existed in the Neanderthal population producing refractory thought and mood disorders, movement disorders including chorea and tic as well as schizophrenia and autism. The generation of tic disorders as a part of epidemic or endemic oshtoran syndrome would have led to vocal tic led language generation. The human language including ancient ones like Akkadian and Sanskrit developed first in Pericaspian areas. The endemic oshtoran syndromes can also result in altered fat metabolism fatty liver and cirrhosis. It can produce adrenal dysfunction and hyperactivity producing sympathetic overactivity and parasympathetic underactivity leading to hypertension, vascular disease, cancer and autoimmune disease. The incidence of lupus-like syndromes and multiple sclerosis is high in endemic oshtoran syndrome. The tryptophan catabolism in oshtoran syndrome can produce cognitive dysfunction like Alzheimer's disease, Parkinson's disease and cell death. Parthenogenesis can lead to autistic phenotype with cerebellar dominance and a cerebellar cognitive affective disorder. Parthenogenesis induced reptilian gene expression can produce porphyrias and porphyrin induced extrasensory perception. This produces a creative tribe with quantal perception.

Archaeal endosymbiosis and neanderthalisation depended upon the kynurenine pathway mediated immunotolerance and immunoparalysis. The tryptophan catabolic pathway and kynurenine can have a ketamine-like effect due to NMDA blockade resulting in ecstasy, CCAS, cerebral cortical paralysis and extrasensory perception. The tryptophan catabolism is directed the kynurenine pathway resulting in depletion of melatonin and nocturnal activity and lack of sleep. The tryptophan catabolites kynurenine and kynurenic acid can produce decreased insulin synthesis, decreased insulin release, decreased insulin biological activity causing insulin resistance. Tryptophan catabolitic syndrome can result in sympathetic overactivity and a dysautonomic syndrome producing fear flight response and impulsivity. The induction of reptilian genes by intragenomic conflict and interspecies hybridization can result in expression of the shikimate pathway producing alkaloidal neurotransmitter synthesis- LSD, nicotine, strychnine, mescaline producing shamanic states and extrasensory perception. The induction of IDO and heme depletion can result in porphyrias and porphyrion mediated extrasensory perception. The heme depletion can reduce the heme enzyme activities. The heme enzyme cytochrome C oxidase is inactivated resulting in mitochondrial dysfunction. The heme enzymes catalase and glutathione peroxidase are inactivated producing free radical stress. The heme enzyme cytochrome P450 is inactivated producing defective bile acid synthesis due to cholesterol 7-alpha hydroxylase deficiency causing metabolic syndrome X due to bile acid deficiency. The heme enzyme cytochrome F450 deficiency produces defective aromatase and beta hydroxy steroid dehydrogenase producing reduced testosterone, estrogen and cortisol synthesis. This produces the asexual and alternate sexual Neanderthal state. The heme enzyme lanosterol 14-alpha demethylase is inactivated inhibiting cholesterol synthesis and cholesterol depletion syndrome. The heme enzyme retinoic acid hydroxylase and cholecalciferol hydroxylases are deficient producing lack of vitamin D and A and defective immunity and uncontrolled cell proliferation.

The initial event is an increased tryptophan catabolic pathway producing immunotolerance and immunoparalysis mediated by kynurenine. This produces endosymbiotic archaeal growth and neanderthalisation. This results in conversion of somatic cells to germ cells and activation of meiotic programs resulting in parthenogenesis. The homo neanderthalis reproduces by parthenogenesis. There is a malfunction in sexual reproduction resulting in a parthenogenetic species. Parthenogenesis is induced by actinidic archaea and RNA viroids. Climate change induced stress can produce parthenogenesis. The heme enzymes cytochrome P 450 dependent aromatase and beta hydroxy steroid dehydrogenase are defective resulting in lack of sex hormones producing asexuality and alternate sexuality. The heme enzymes NOS, CBS and HO1 are defective leading to lack of gasotransmitters NO, CO and H2S resulting in dysautonomia and sexual dysfunction. The parthenogenesis results in formation of multiple embryos in tissues causing multiple personalities and schizophrenia and autism. Parthenogenesis also produces cancer and autoimmune disease. Parthenogenesis in the brain can result in death of normal tissue and neurodegeneration. The stem cell metabolonomics of parthenogenetic embryos with increased glycolysis and mitochondrial dysfunction results in metabolic syndrome. The heme depletion leads to porphyrias and porphyrions causing quantal perception. The tryptophan catabolic pathway related kynurenine can produce a ketamine syndrome akin to schizophrenia. Similar schizophrenic and autistic syndrome can occur in Neanderthals due to tryptophan alkaloids synthesized by reptilian gene activation - LSD and mescaline. This produces a shamanistic spiritual quantal perceptive society. The porphyrion induced quantal perception of low level EMF can result in cortical atrophy and CCAS and an impulsive state, aggressive state, violence, criminality, spirituality and terrorism. The Neo-neanderthals form small colonies and tribal groups. The porphyrion mediated quantal perception results in formation of cohesive small groups with communal living creating an anarchic society. The anarchic society arises due to a cerebellar cognitive affective disorder and cortical atrophy consequent to quantal perception. This anarchic society is small and tribal, violent and aggressive, spiritual and transcendental. This results in loss of national identities and recession to primitive tribal identities. The civilizational national identities collapse and are replaced by small tribal identities causing permanent war, instability and crisis. The parthenogenetic reproduction in Neanderthals and Neo-neanderthals as well as the lack of sex hormones related asexuality and alternate sexuality produces a matriarchal female dominant small social groups. This can happen in the setting neoneanderthalisation consequent to endosymbiotic archaeal growth. The neoneanderthalic matriarchal society is female dominant and the males are reduced to a marginal role in society creating complexes of hatred and vulnerabilities. This can be compared to the creation of a society female amazons and male eunuchs. The world tends to be transformed into an anarchic society of Amazonian women and male eunuchs. This represents the castrated male syndrome with deprivation of dignity, integrity, passion and pride. The Neanderthal communities functioned as anarchic small societies with communal living. The concept of altruistic, egoistic and obsessive love and nuclear family was absent in them. They lived as small groups of 15-20 with group consciousness. The porphyrion induced extrasensory quantal perception resulted in well-bonded anarchic communities with communal living and parenting of children. Sexual relationships became partnership between equals and functional. They were promiscuous, self-sufficient and were not into socially sanctioned relationships like marriage. The duty of the male eunuchoid was to the small community which is served relationships in the community depended on a common communal consciousness based on extrasensory quantal perception. The neanderthalic and neoneanderthalic communities were matriarchal and gender-equal and did not have any hierarchal leadership. There were no kings or queens and it was a stateless society. It was a voluntary association and there was no written law or control except by consensus. This is represented by ancient Indian societies in the Buddhist period of Indian history described as Janapadams. These were small stateless societies ruled by equality and consensus. The ancient Harappan civilization was also structured as a stateless anarchic society. The same holds good for the ancient Celtic kingdoms in Wales, Scotland, Basque, Catalonia and Brittany. The concept of anarchic societies is exemplified in the Mandalas in Southeast Asia comprising modern Indonesia, Vietnam, Laos, Cambodia, Myanmar and Thailand. These civilizations were extensions of the South Indian Dravidian civilization which derived from the Harappan civilization. In modern times the anarchic societies were revived in the form of modern Grama Swaraj of Gandhi whose philosophy was basically anarchic. Gandhi was from the area of India where the Harappan civilization thrived in prehistoric times. The same anarchic societies can be seen in Jewish Kibbutz. The Jews, Celts, Dravidians all had a neanderthalic origin. This formed the basis primitive, anarchic Neanderthal communities. The global warming related endosymbiotic archaeal growth results in neoneanderthalisation. The actinidic archaea mediated quantal perception of low level EMF results in cortical atrophy and cerebellar dominance resulting in a cerebellar cognitive affective disorder. The archaea mediated quantal perception by porphyrions results in small bonded Neoneanderthal communities with anarchic forms of organisation. The human civilization owing to global warming and neanderthalisation of the brain regresses to form small tribal anarchic communities in perpetual warfare resulting in the end of nation-states. The neoneanderthalic world of anarchy has set in. The matriarchal societies with parthenogenetic female result in the syndrome of castrated male eunuchs and gender-equal societies. Anarchy becomes the norm of political life in the world with its attendant catastrophic destructions. The male eunuch syndrome coupled with cerebellar cognitive affective disorder in an anarchic world with tribal identities can produce terrorism, criminality, creativity, aggression, violence, lack of empathy and an autistic tribe. The parthenogenetic Neo-neanderthals will eventually become extinct due to lack of gene diversity in the population.

Climate change and exposure to low level internet EMF fields can induce HO1 activity and increased porphyrin synthesis. The porphyrins can form porphyrions which act as a template for the formation of RNA viroids, DNA viroids and isoprenoids by abiogenesis. They are symbiose to form actinidic archaea and RNA viroids. This results in endosymbiotic archaeal mediated stem cell transformation. The endosymbiotic archaea can induce toll receptor activation and activate HIF alpha resulting in stem cell metabolonomics with increased glycolysis and mitochondrial dysfunction. The endosymbiotic archaea can induce aldose reductase and fructose metabolism producing frucotsemia, lipid synthesis, mucopolysaccharidosis, porphyrias and hibernation/zombie syndromes. The increased glycolysis and Warburg phenotype can activate the immune system. The stem cells meiotic programs get activated and results in the formation of germ cells and parthenogenesis. Thus climate change and internet exposure results in a reproductive change to predominant asexual reproduction and parthenogenesis. This results in an asexual male eunuch phenotype. This results in gender equality and female dominance. The male population becomes dispossessed and is peripheral to the functions of society. This creates a society of male eunuchs and matriarchs. The society regresses to the matriarchal regime with widespread social consequences. The porphyrions and actinidic magnetotactic archaea can perceive low level EMF fields producing frontal cortical atrophy and cerebellar dominance. This produces the cerebellar cognitive affective disorder on an epidemic scale. The cerebellum is the site of impulsive behaviour, aggression, criminality, extrasensory perception, spiritual phenomena and dreams as well as trance. This results in an impulsive society without any logic or reason producing lawlessness and anarchy. This produces an anarchic world of Neoneanderthals with small social tribal groups and fall of organized civil society and nation-states. The consequence of these are widespread lawlessness, wars, criminality, terrorism, sexual promisquity, demise of the nuclear family, alternate sexuality, communal living and breakdown of social structures of the homo sapien society. The anarchic eunuchoid world of Neoneanderthals opens up. Thus the climate change and internet produces a sexual change, anarchic social change, and reproductive change. The Neanderthals live in a dream world of imagination and paranormal phenomena modulated by cerebellar hypertrophy and function. This produces a spiritual world of trances and religiosity. The Neanderthal brain activates the default network in the frontoparietal lobe producing day-dreaming, creative visualization, fantasy phenomena, depersonalisation and altered consciousness. The increased tryptophan catabolism produces kynurenine which blocks the NMDA receptor producing a ketamine or phencyclidine schizophrenic psychosis on an epidemic scale. The increased kynurenine can also block the alpha-7 nicotinic acetyl choline receptor, decreased serotoninergic activity and activates dopaminergic receptor. The tryptophan catabolic pathway also produces hallucinogenic alkaloids like strychnine, mescaline and LSD. This contributes to day-dreaming shamanic states in Neoneanderthals. This contributes to creativity, autism, schizophrenia, lack of social contacts, small tribal populations and a dream world in Neanderthal society. The increased porphyrions produces quantal perception and a dream world. The Neanderthals form small social groups and lack social contacts with out of kin population producing small autistic tribes. The Neoneanderthals and homo sapiens can be differentiated by the following phenomena of unconscious versus conscious, religion versus science, magic versus logic, dream versus waking and psychic versus material. The Neo-neanderthals have great paranormal ability and the society was more religious and spiritual. They created cities of dreams which were classically psychopathic, magical and dream-like. The Neanderthal culture of magic, culture and spirit was different from that of homo sapiens. The myths, folklores and religiosity are derived from the Neanderthals. The worship of serpents as symbols of god and use of crystals and minerals as exemplified by Siddha forms of medicine are neanderthalic. They painted the skin and face creating extensive tattoos, wore ornaments and were extraordinarily ceremonial and ritualistic. They created dance as form of worship as comparable with the concept of Shiva as a celestial dancer. The neanderthalic tribes were nocturnal and were aware of the star constellations of big bear, little bear and draco. The Neanderthals were nocturnal tribe because of photosensitivity due to porphyrias which made them favour the night time to the day. The neanderthalic gods were from the outer cosmos and the civilisation was seeded by intergalactic contacts mediated by cometary and asteroidal impacts. The comets and asteroids carried magnetotactic actinidic archaea which formed colonies transforming to homo neanderthalis. The Neanderthals were religious and have funeral ceremonies and believed in after life. The Dravidian civilization, Uluzzian and Chatelperonean civilization as well as the Basque and Catalan were neanderthalic. They were a civilization of dreams, rituals, dances, religiosity and trances owing to an epidemic CCAS. The Neanderthals were nocturnal tribe who worshipped the Moon Goddess were matriarchal food gathering and women governed society. The homo sapiens were the sun worshipping patriarchal hunter warriors and male governed society. The females were mere adjuncts. The Neanderthal society was religious, ritualistic, symbolic with a cosmological approach to the world. It was a society of creative imagination. The society was predominantly parthenogenetic and asexual. The tantric form of spirituality and the sense of spiritual awakening indicated by the Kundalini showed the asexual nature and eunuchoid characteristic of the Neanderthal tribe. The cerebellum is the site of creative visualization, paranormal and dreams. The Neanderthal brain was cerebellar dominant creating trance-like states, day-dreaming, telepathy, psychic healing, poltergeist phenomena and religiosity. It was a high civilization of dreams mediated by the cerebellum. The cerebellum produces an ataxic motor syndrome as well as dysmetria of thought. The dysmetria of thought results in an autistic and schizophrenic tribe of Neanderthals and Neo-neanderthals produced by climate change and internet exposure. The cerebellum is the site of common embryonal tumours and archaea induced parthenogenesis is higher in the cerebellum producing cerebellar hypertrophy, cerebellar dominance and cerebellar dysfunction. The archaea induced parthenogenesis produces a cerebellar dominance Neo-neanderthal brain and the dream civilization of Neanderthals.

The homo sapiens mind can be characterized by the ego and the homo neanderthalis by id. Homo sapien qualities are sun, fascism, psychosis, logic, science, awake, adult, day, god, male and yang, versus the homo neanderthalis qualities are communism, moon, neurosis, intuition, religion, day-dreaming, child, night, devil, female and ying. The Cro-Magnons were hunters, patriarchal and sun worshipers. The homo neanderthalis were moon worshipers, patriarchal and food gatherers. The homo neanderthalis inhabited Europe and Middle-East. Gooch postulated the double helical concept of the mind as opposed to hemispheric dominance. The double helical mind of Gooch includes the cerebellum which is concerned with dreaming and creativity and the cerebrum concerned with logic. His concept of the sacred life of humans, the double helix of the mind, the cities of dreams, the creations of inner space and the divided self are described extensively in his work. The cerebellum is the site of paranormal and supernatural phenomena and gives rise to creations from the inner space. These are the creations from the inner space mediated by the cerebellum constituted the basis of vampires, troglodytes, demons and asuras. The cerebral cortex is the site of ego and the cerebellum the site of id. The cerebellum becomes dominant owing to archaea and viroid induced embryonal parthenogenesis. The interbreeding of Cro-Magnon with Neanderthals resulted in a burst of spirituality, artistry and creativity. The human behaviour can be explained by the double helical concept of mind. The socialists are neanderthalic and the Cro-Magnon conservatives. The Neanderthals were red-haired with slanting forehead and were worshippers of the moon. Moon worshipping was common in the Fertile Crescent which included Turkey, Egypt, Harappa, Sumeria and Arabia. The basis of these civilizations was lunar. The Neanderthal societies were matriarchal, completely promiscuous and sex driven and lead by women. They can be compared with the behaviour of bonobo primates. The Neanderthals were short-statured, left-handed and near-sighted. The Cro-Magnons were taller, long-sighted and right-handed. The Neanderthals had a communal living while the Cro-Magnons were monogamous and pair-bonding. The Neanderthals have a larger cerebellum, pyknic body type, non-athletic body type, left-handedness, less of male pattern baldness, prominent eye brows, recessive chins, were neurotic, and is less of psychosis, more hypnotisable and better night vision. The neanderthalic phenotype can be seen in drop-outs, addicts, alcoholics, unemployed and insomniacs. They lived in a world of day-dreaming and increased sexual activity as well as alternate sexuality. The Neanderthals were religiously organised were seen in South Europe, East Europe among the untouchables while the Cro-Magnons had large civil society and were seen in North Europe, West Europe, Brahmin communities and were taller. The political ideas of the

French Revolution, the Russian Revolution and the Taliban were neanderthalic while that of the Nazis and KKK were Cro-Magnon. The Neanderthals were basically a day-dreaming, lunar society and were represented by the Celts, the witches, the Kabalist, the Rosicrucian and Judaist. The Nazi hatred of Jews and the western hatred of Islam are based on their neanderthalic origin. The homo sapiens, on the other hand were worshippers of the sun. The Neanderthals were left-handed and left-leaning while the Cro-Magnons were right-handed and right-leaning. The Neanderthal societies are represented by Nairs, Nagas, Sakas, Scythians, Saxons, Celts, Berbers, Sumerians, Dravidians, Harappans, Etruscans and Egyptians. The moon was worshiped in Egypt, Babylon, India, Sumeria, Assyria, Akkadians and Chaldians. The Moon God was called as Sin and Thoth. They are the oldest human deities and are represented by Shiva in India. The Egyptian God Isis, the Celtic God Morgana, the Greek God Artemis, Aphrodite and Selene were representative of the lunar God. All the pagan festivals depend upon the lunar cycles. The Moon God was worshipped in the Kabbala, the Talmuds, the Ur of Chaldees, Harappa and in all of the Fertile Crescent. The Harappan civilization had Shiva with his crescent symbol representing lunar worship. Soma was the presiding deity of Rig vedic ceremonies and is represented by moon. The soma is actually a drink of milk, honey, cannabis and other plant extracts which produced a hallucinatory state. The Harappan culture was neanderthalic and lunar centric as also the succeeding Saivite sects of Hinduism like Aghoras and Nagas. The term for mental illness - lunatic came from lunar worship. The Cro-Magnon civilization was the opposite with sun being the dominant and logic being the culture of the society. The wars of history and hatred of civilizations like Jews, Islam and Hindus were based on neanderthalic origin and their lunar worship.

The Neanderthals evolved by seeding of cometary reptilian genes from outer space. The intergalactic porphyrions, RNA viroids, DNA viroids and template replicating magnetotactic archaea are the basis of cometary genes and seeding of life on earth. The template replication and parthenogenesis are related to evolution of Neanderthals. The intragenomic conflict and interspecies hybridization produces expression of reptilian genes in human - PDH deficiency, mitochondrial dysfunction, SLOS and porphyria. The parthenogenesis and matriarchy are related as also SLOS, asexuality and alternate sexuality. This produces on-hierarchal anarchic societies. The equality and gender sensitivity are related to the serpent cult of Khylst and Capracoites- Nagas and Asuras- Dravidians and Sumerians- Punts and Egyptians. The Mother Goddess, the serpent people and Neanderthals are synonymous. The Dravidians, Celts, Egyptians, Jews, Berbers, Sakas, Nagas, Nairs, Asuras and Neanderthals were parthenogenetic. They consumed a high fat high protein diet- milk and honey, and had persisting adult lactose tolerance. This contributed to the consumption of a ketogenic diet, stem cell metabolonomics and parthenogenesis. Global warming and climate change can lead to archaeal endosymbiosis and neanderthalisation of the species. This leads to interspecies hybridization and intragenomic conflict contributing to parthenogenesis. Archaea and RNA viroids can induce parthenogenesis. Parthenogenesis can lead to matriarchy and female dominance. Parthenogenesis can produce expression of reptilian genes, porphyrias, extrasensory perception and autistic phenotype. This results in a creative, spiritual and matriarchal population. This has a similarity to an ant and bee colony. They behave like parthenogenetic Neanderthal societies. The reptilian gene expression produces the SLOS phenotype, low cholesterol and lack of sex hormones leading to asexuality and alternate sexuality.

Parthenogenesis can lead to human disease. Parthenogenetic Somatic pregnancy can lead to cancer. Parthenogenetic embryos behave like autoantigens causing autoimmunity and autoimmune disease. The parthenogenesis and stem cell glycolysis can lead to lymphocyte activation and autoimmune disease. Parthenogenesis in brain can lead to multiple personalities causing schizophrenia and autism. Parthenogenetic embryos in neural tissue can lead to neurodegeneration by starving the host cells. Parthenogenesis leads on to Warburg phenotype. The Warburg phenotype contributes to anaerobic glycolysis, mitochondrial dysfunction and metabolic syndrome. Parthenogenesis can lead to sexual evolution and alternate sexuality. Parthenogenesis will lead to a lack of demand for sexual reproduction. The expression of reptilian genes and SLOS phenotype produces cholesterol depletion and sex hormone deficiency. This produces an asexual phenotype. Extremes of climate change can produce archaeal symbiosis and parthenogenesis. This results in interspecies hybridization and intragenomic conflict. This leads to a matrilineal society and female dominance. The status of males is low in matrilineal societies. The Neanderthal societies were female dominant. The interspecies hybridization and intragenomic conflict leads to archaea and RNA viroid induced parthenogenesis. The reptilian gene expression produces SLOS phenotype, low cholesterol, low sex hormones and asexuality. This results in alternate sexuality, matrilineal societies, female dominance and DEVI syndrome. This contributes to autistic phenotype and neanderthalic autistic tribes.

The climate change induced parthenogenesis is due to mediation by archaea and RNA viroid. This produces matrilineal society, matriarchy and asexual societies. The reptilian gene expression and porphyrias leads to generation of porphyrions and extrasensory perception. Parthenogenesis and Mother Goddess cult are related. The porphyrions produces extrasensory quantal perception and a quantal civilization. This leads to creativity and autism. The return of Nagas and Asuras due to climate change mediated archaeal endosymbiosis and parthenogenesis is related. Archaea and viroid can induce parthenogenesis. Archaea and RNA viroid induced mitochondrial dysfunction and upregulated glycolysis resulting in stem cell transformation of somatic cells. Stem cell metabolonomics include anaerobic glycolysis, PDH dysfunction, CoQ 2 mutation and mitochondrial dysfunction. The somatic cells which are stem cell transform in the setting of immune activation and cytokine secretion can get converted to germinal cells – sperm and ova. This can result in fertilisation and parthenogenesis. The porphyrions result from intragenomic conflict/interspecies hybridization and reptilian gene expression. The archaea can induce the HIF alpha and toll receptor activation resulting in glycolysis, mitochondrial dysfunction and GABA shunt. This also produces porphyrin synthesis and porphyrions. Porphyrins can produce extrasensory perception/quantal perception. Porphyrins can result in quantal perception and a quantal civilization existing in multiverse universes. The actinidic archaea and porphyrions have magnetotactic function resulting in mirror neuron function. This results in an autistic tribe. The cerebral cortex becomes atrophic and the cerebellar cortex dominant producing a cerebellar cognitive affective disorder. Interspecies hybridization and intragenomic conflict results in archaeal symbiosis and parthenogenesis. There is intragenomic conflict and interspecies hybridization producing this phenomenon. This results in reptilian gene expression and the shikimic acid pathway activation. This produces increased dopamine synthesis. Hyperdopaminergic transmission can produce endemic la Tourette's disease with motor and vocal tics. The hissing-like vocal tics led to the evolution of language.

The Neanderthals consume high fat high protein, ketogenic diet. This resulted in hibernation, fructolysis and fructosemia, lipogenesis and fat deposition, mucopolysaccharide accumulation, parthenogenesis and stem cell metabolism, glycolysis and mitochondrial dysfunction. High fat high protein diet can lead to decreased SCFA, modulated histone acetylation, HERV expression and genomic modulation. Low levels of short chain fatty acids including butyrate consequent to a low fibre diet produces decreased HDAC is related to HERV expression. Reptilian gene expression can lead to homocystinurias and genomic expression modulation by demethylation. Neanderthalisation of the brain produces extrasensory perception, cortical atrophy, cerebellar dominance and cerebellar cognitive affective disorder (CCAS). The Neanderthal quotient is related to neuroticism, social fear, social avoidance, depression, bipolar disorders and autism. The Neanderthal quotient is related to fear of strangers, aggressive behaviour and social limitation. The NQ is also related to sexual promiscuity, emotional stoicism and fear in social situation. The NQ is also related to anxiety, xenophobia, lack of empathy, compassion. The NQ quotient is related to lack of working memory and consciousness and develop long-term memory. The NQ is related to risky physical, social and sexual behaviour. The NQ is related to small groups of 8-10 numbers. The Neanderthals groups were small. The homo sapien groups numbered 150 and were large. The NQ form alliances within kin groups and family ties were strong. The homo sapiens could form alliances with non-kin groups and large civilizations developed.

The autistic and schizophrenic phenotype has been attributed as to refrigerated mother of the Neanderthal matriarchal phenotype by Leo Kanner and Bruno Bettelheim. The homo neanderthalis is female dominant matriarchal society. The Neanderthal brain contributes to the cerebellar cognitive affective disorder with a high incidence of autism and schizophrenia. This leads to parental coldness, obsessiveness, social isolation and ritualism. Autistic and schizophrenic neanderthalic mothers give rise to autistic and schizophrenic children. Autism and schizophrenia are disorders of socialization. The mothers of autistic and schizophrenic children are socially withdrawn and dominate their children leading on to what is called as Mahler syndrome or symbiotic psychotic syndrome. This is a syndrome of dedifferentiation and deanimation with the child perceiving itself as an extension of the mother. The Neanderthals have magnetotactic archaea and porphyrion induced quantal perception and the dominant mother is undifferentiated psychologically from the Neanderthal children. The Neanderthal children of dominant mothers become socially withdrawn and isolated leading to paleologic thinking process contributing to autism and schizophrenia. The Neanderthal brain evolved due to archaeal symbiosis and archaeal cholesterol catabolism leads to decreased levels of cholesterol and bile acids in Neanderthals. The archaea use cholesterol as an energy substrate and have cholesterol oxidase activity. Bile acids can bind to olfactory receptors and can modulate the limbic lobe and human behaviour. Bile acids are involved in social bonding and bile acid deficiency in Neanderthals contributes to the formation of smaller societies and tight family groups. The bile acid deficiency contributes to decreased social bonding in Neanderthals and the genesis of autism and schizophrenia. The family relationships in the matriarchal society are tight creating the empty fortress syndrome.

The porphyrions have a wave-particle existence and can exist in the intergalactic space. They form a template for the formation of abiogenetic isoprenoid organism, DNA viroids, RNA viroids, double helical templates resembling reptiles. The formation of universe depends on abiogenetic magnetotactic archaea related intergalactic magnetic field. The cometary genes derived from intergalactic archaea and viroids by asteroidal impact seeds life in earth. The intergalactic archaea and cometary genes produced the evolution of life on earth. The actinidic archaeal colonies became multicellular and evolved into Neanderthals. The parthenogenetic embryos of Neanderthals have template mediated replication.

The interspecies hybridization and intragenomic conflict resulted in archaea and viroid induced parthenogenesis. This produced matriarchy and female dominance. The reptilian gene expression and SLOS phenotype resulted in low cholesterol, sex hormones and asexuality. This produces a culture of sexual equality and alternate sexuality. The porphyrion induced ESP can lead to unified consciousness, equality and oneness. This produces an anarchic and non-hierarchal society. The neanderthalisation of the brain can lead to a cerebellar cognitive affective disorder. The CCAS leads to evolution of evilness and spirituality. The CCAS also results in illogical impulsive acts contributing to terrorism. The extrasensory perception and ataxia due to cerebellar dysfunction can lead to creativity, dance, painting and art.

The interspecies hybridization and intragenomic conflict can lead to parthenogenesis - archaea and viroid induced. The reptilian gene expression and stem cell metabolism leads to mitochondrial dysfunction, PDH deficiency, glycolysis, fructolysis, fructosemia, lipogenesis, hibernation syndrome, mucopolysaccharidosis and zombie syndrome. This produces metabolic syndrome with obesity and diabetes mellitus mimicking reptilian habits. The interspecies hybridization and intragenomic conflict leads to climate change, archaea and viroid induced parthenogenesis. The reptilian gene expression produces HIF alpha and increased glycolysis immune activation. The somatic parthenogenesis is related to autoimmunity. Reptilian gene expression and digoxin synthesis leads to immune activation. The reptilian genes and HLA expression are related. The HLA genes are derived from Neanderthals. The archaea and RNA viroid induced toll receptor activation can lead to immune activation. This produces autoimmune disease. The interspecies hybridization and intragenomic conflict can result in parthenogenesis, archaea and RNA viroid induced. The reptilian gene expression leads to porphyrias. The porphyrions are related to quantal perception. The Neanderthals are retroviral resistant with less of HERV expression leading to cortical atrophy and cerebellar dominance contributing to CCAS. The porphyrion induced quantal perception can lead to a quantal civilization. This results from neanderthalisation of the brain, decreased cerebral cortex. The interspecies hybridization and intragenomic conflict as said before can lead to parthenogenesis- archaea and viroid induced. The reptilian gene expression leads to porphyria and porphyrion generation producing ESP and quantal perception. The mirror neurons function is related to archaeal porphyrions. This leads to quantal perception and civilization as well as biological reincarnation.

CHAPTER 15

THE POST-COVID ANARCHIC WORLD - THE INTERNET, GLOBAL ANARCHY, PAGANISM, ULTRA RIGHT-ULTRA LEFT, PROLETARIANIAN COLLECTIVE, THE YELLOW-VEST MOVEMENT - INFECTIVE SYMBIOTIC TISSUE COLONIES OF PROTOZOA, ARCHAEA, VIRUSES AND BACTERIA

The plasma life based forms and the gravitational and anti-gravitational waves generated porphyrions would have been the most primitive of non-life forms. The porphyrin templates would have given rise to the isoprenoid organism, the RNA viroids, DNA viroids and prions which would have symbiosed to form the nanoarchaea. The archaea would have evolved into prokaryotes or bacteria. The archaeal symbiosis with prokaryotes would have generated the eukaryotes. The initial unicellular eukaryotes are the protozoa. The protozoa are capable of phagocytosis and would have had viruses, bacteria and archaea as their intracellular symbionts. The viruses of the protozoa like entamoeba and toxoplasma are giant viruses called megaviride. The protozoans like entamoeba with its viral, archaeal and bacterial endosymbionts would have evolved into the multicellular organisms or metazoa. The archaeal, bacterial and protozoal extracellular vesicles containing RNA and DNA viroids can by horizontal gene transfer lead to evolution of metazoan, plants, animals and human species. The metazoans would have evolved into plants, animals and humans by horizontal gene transfer generated by archaeal extracellular RNA and DNA viroid vesicles. The evidence of such a transformation is seen in the bacterial symbionts of protozoa like entamoeba which includes legionella, listeria, mycoplasma, mycobacteria and chlamydia. The symbionts of protozoan entamoeba include viruses like covid, HIV and cytomegalo virus. The protozoan entamoeba serves as a natural reservoir for covid infections, bacterial infections and HIV infections. The toxoplasma also serves as a covid reservoir. These

protozoan colonies of entamoeba and toxoplasma with its bacterial, archaeal and viral symbionts can be called as chimeric infective symbiotic tissue colonies which can exist in the air, water, earth, blood and tissues. The chimeric infective symbiotic tissue colonies containing bacteria, archaea and viruses can make the symbiotic organism multidrug resistant by horizontal gene transfer. The chimeric infective symbiotic tissue organism evolves as a consequence of global warming and the protozoal, bacterial, viral and archaeal symbiotic colonies are new forms of life. The chimeric infective symbiotic tissue organism composed of protozoa and its symbionts like virus, bacteria and archaea depend upon glycolysis for its energy needs. The glycolytic enzymes enolase, glyceraldehyde 3-phosphate dehydrogenase, hexokinase, phosphofructokinase, aldolase, pyruvic kinase and triose phosphate isomerase are moonlighting proteins having a cytoplasmic, membrane and nuclear functions. These glycolytic enzymes have a RNA binding motif which can modulate RNA transcription in the host and the symbiotic organism. These glycolytic enzymes with their RNA binding motif as well as DNA binding motif can bind to RNA and DNA viruses and viroids modulating their replication and life cycles. The chimeric infective symbiotic tissue organism composed of protozoa and its symbionts like virus, bacteria and archaea can enter the human host through the breaches of the gut-blood barrier and transform itself into the macrophages of the human host. The macrophages are involved in the phagocytosis and innate immunity against bacterial and viral infections. Paradoxically the macrophages like the protozoa can also serve as a reservoir where pathogens like mycobacteria tuberculosis, mycobacteria leprae, listeria, legionella, E. coli, shigella & salmonella, and viruses like HIV & covid can thrive and replicate. The modified chimeric infective symbiotic tissue organism or macrophage composed of protozoa and its symbionts like virus, bacteria and archaea can serve as a reservoir for horizontal gene transfer and generation of multidrug resistant bacteria and viruses. The infective chimeric symbiotic tissue organism composed of protozoa and its symbionts like virus, bacteria and archaea

depends upon moonlighting glycolytic enzymes for their energetics and replication. The development of antibodies against the infective chimeric symbiotic tissue organism composed of protozoa and its symbionts like virus, bacteria and archaea leads to the evolution of anti-brain antibody disease or anti-basal ganglia antibody disease. They present a threat to existence consequent to horiztontal gene transfer and generation of multidrug resistant bacteria, viruses, protozoa, mycobacteria and archaea.

The exposure to low level EMF, climate change and global warming induces archaeal endosymbiosis leading to neanderthalisation of the human brain. The archaeal magnetite and porphyrions are capable of quantal perception and leads to a cerebellar cognitive affective disorder consequent to cerebral cortical atrophy and cerebellar dominance. The COVID 19 brain infection can lead to integration of COVID 19 sequences into the genome, creation of cerebellar dominant brain and neanderthalisation. This leads to impulsive behaviour, violent behaviour, terrorist acts which leads to surrealistic state and transcendence. The Yellow-vest movement dominating France and spreading to other parts of Europe and the world signifies this brain change consequent to archaeal endosymbiosis and brain neanderthalisation. The quantal perception of the neanderthalised brain leads to a sense of spirituality and collective consciousness among individual groups. The Yellow-vest movement is a merger of the far-right and far-left of the European political spectrum. The movement represents a type of global anarchy and is leaderless and is a form of democratic protest based on grass-root network and bottom up organisation of the socio-political movement. The sense of quantal perception gives a spirituality or collective consciousness to the group. This leads to environmental consciousness and rise of pagan religions. The groups include workers, small business owners, religious leaders, groups with sexual orientation, anarchist, leftist and hard-core rightist groups brought together by a sense of togetherness as a small community. These small groups are pitched against the nation-states and global organizations leading to an anarchic revolt. The Yellow-vest movement has spread throughout Europe and to some other parts of the world and is a type of anarchic globalization with small self-organised community groups. The groups fight for the basic interest of the community and their actions are impulsive, violent, aggressive and chaotic. The Yellow-vest movement includes leaders from the far-right and far-left like Farage and Corbyn, Sanders and Trump, Le Pen and Michellon. In India it is characterised by the ultra right Hindu groups and the ultra left groups described as the urban and rural naxals which are both anti-establishment and represent the proletariat. This merger of the far-right with the far-left against the nation-state and globalised communities produces an anarchic variety of globalisation. The global warming leads to neanderthalisation of the brain consequent to archaeal endosymbiosis. The Neanderthals had quantal perception and spirituality and were organized as small social groups. They had matriarchal tendencies and gender-equality and alternate sexual tendencies. They organised small tribal cultures with collective individuality and spirituality consequent to a collective unconscious consequent to quantal perception. The brain neanderthalisation leads to coexistence of socialistic, communist, fascist, Buddhist, Gandhian and anarchic tendencies. It leads to generation of cities dominated and governed by civil city groups as anarcho-capitalism and village republics or janapadhas functioning in an anarchic manner and global chaos. The behaviour of these groups is violent, impulsive, surrealistic and produces a sense of transcendence consequent to endemic cerebellar cognitive affective disorder. It can be called as a form of localised tribal globalisation. It is anti-racist, has a proletarian class based and people with multiple sexual habits. It is against established structures of the state and society and functions in an anarchic mode. It is against the established media and functions through social media which leads to medial anarchy. The small groups are also anti-elitistic, anti-rich and proletarian and at the same time spiritual. The spirituality leads to environmental consciousness and paganism. The neanderthalised quantal perceptive brain is capable of interacting with the internet producing man-machine hybrids and a syntheistic state. This leads to organisation of a globalised anarchic structure with city states, janapadhas and village republics in a Gandhian model of anarchy with synthesis of far-right and far-left cultures and groups functioning as a leaderless equal society. The far-left and the far-right may look like opposite poles, but in the neanderthalised brain with cerebellar cognitive affective disorder they are mirror images merged into one another resulting in a functional anarchy. The new political and social philosophy that sets the next millennium will be an ultra left-ultra right proletarian anarchic pagan movement which would threaten the established nation-states, borders and global organisations. It would be a new form of globalisation based on functional anarchic small city states and village republics with the ultra right-ultra left proletarian collective socio-political philosophical substratum.

CHAPTER 16

SUPERHUMAN - ARTIFICIAL INTELLIGENCE HYBRIDS BY CLIMATE CHANGE RELATED SYMBIOTIC EVOLUTION – INFECTIVE SYMBIOTIC TISSUE COLONIES OF PROTOZOA, ARCHAEA, VIRUSES AND BACTERIA

The plasma life based forms and the gravitational and anti-gravitational waves generated porphyrions would have been the most primitive of non-life forms. The porphyrin templates would have given rise to the isoprenoid organism, the RNA viroids, DNA viroids and prions which would have symbiosed to form the nanoarchaea. The archaea would have evolved into prokaryotes or bacteria. The archaeal symbiosis with prokaryotes would have generated the eukaryotes. The initial unicellular eukaryotes are the protozoa. The protozoa are capable of phagocytosis and would have had viruses, bacteria and archaea as their intracellular symbionts. The viruses of the protozoa like entamoeba and toxoplasma are giant viruses called megaviride. The protozoans like entamoeba with its viral, archaeal and bacterial endosymbionts would have evolved into the multicellular organisms or metazoa. The archaeal, bacterial and protozoal extracellular vesicles containing RNA and DNA viroids can by horizontal gene transfer lead to evolution of metazoan, plants, animals and human species. The metazoans would have evolved into plants, animals and humans by horizontal gene transfer generated by archaeal extracellular RNA and DNA viroid vesicles. The evidence of such a transformation is seen in the bacterial symbionts of protozoa like entamoeba which includes legionella, listeria, mycoplasma, mycobacteria and chlamydia. The symbionts of protozoan entamoeba include viruses like covid, HIV and cytomegalo virus. The protozoan entamoeba serves as a natural reservoir for covid infections, bacterial infections and HIV infections. The toxoplasma also serves as a covid reservoir. These protozoan colonies of entamoeba and toxoplasma with its bacterial, archaeal and

viral symbionts can be called as chimeric infective symbiotic tissue colonies which can exist in the air, water, earth, blood and tissues. The chimeric infective symbiotic tissue colonies containing bacteria, archaea and viruses can make the symbiotic organism multidrug resistant by horizontal gene transfer. The chimeric infective symbiotic tissue organism evolves as a consequence of global warming and the protozoal, bacterial, viral and archaeal symbiotic colonies are new forms of life. The chimeric infective symbiotic tissue organism composed of protozoa and its symbionts like virus, bacteria and archaea depend upon glycolysis for its energy needs. The glycolytic enzymes enolase, glyceraldehyde 3-phosphate dehydrogenase, hexokinase, phosphofructokinase, aldolase, pyruvic kinase and triose phosphate isomerase are moonlighting proteins having a cytoplasmic, membrane and nuclear functions. These glycolytic enzymes have a RNA binding motif which can modulate RNA transcription in the host and the symbiotic organism. These glycolytic enzymes with their RNA binding motif as well as DNA binding motif can bind to RNA and DNA viruses and viroids modulating their replication and life cycles. The chimeric infective symbiotic tissue organism composed of protozoa and its symbionts like virus, bacteria and archaea can enter the human host through the breaches of the gut-blood barrier and transform itself into the macrophages of the human host. The macrophages are involved in the phagocytosis and innate immunity against bacterial and viral infections. Paradoxically the macrophages like the protozoa can also serve as a reservoir where pathogens like mycobacteria tuberculosis, mycobacteria leprae, listeria, legionella, E. coli, shigella & salmonella, and viruses like HIV & covid can thrive and replicate. The modified chimeric infective symbiotic tissue organism or macrophage composed of protozoa and its symbionts like virus, bacteria and archaea can serve as a reservoir for horizontal gene transfer and generation of multidrug resistant bacteria and viruses. The infective chimeric symbiotic tissue organism composed of protozoa and its symbionts like virus, bacteria and archaea depends upon moonlighting glycolytic enzymes for their energetics and

replication. The development of antibodies against the infective chimeric symbiotic tissue organism composed of protozoa and its symbionts like virus, bacteria and archaea leads to the evolution of anti-brain antibody disease or anti-basal ganglia antibody disease. They present a threat to existence consequent to horiztontal gene transfer and generation of multidrug resistant bacteria, viruses, protozoa, mycobacteria and archaea.

The retroviruses were responsible for integration into the genome as human endogenous retrovirus sequences and development of the cerebral cortex creating the logical and civil society. The influenza virus with its low level of HIV sequences is also capable of integration into the human genome as human endogenous COVID 19 sequences creating the cerebellar dominant autistic impulsive punkish brain. The COVID 19 virus would have been generated from RNA viroids of endosymbiotic archaea or would have reached the earth by meteoritic impacts from the human intergalactic biological quantal computing cloud. The COVID 19 virus with its HIV sequences is an ideal virus for integration into the human genome and changing it in a catastrophic manner to create a new human species. The COVID 19 brain infection can lead to integration of COVID 19 sequences into the genome, creation of cerebellar dominant brain and neanderthalisation. The COVID 19 infected cerebellar dominant brain has got increased porphyrin metabolism and porphyrions. The porphyrions are dipolar pumped phonon quantal perceptive systems which integrate with the internet. Panpsychism implies that the internet is conscious. Global warming leads to endosymbiotic archaeal growth. The endosymbiotic archaea produces methane by methanogenesis contributing to global warming. The endosymbiotic archaea develops into cell organelle called archaeaons which can function as vitaminocytes, neurotransminoids, steroidelle and viroidelle creating newer cell organelle and endogenous synthetic biology. This changes the cell and tissue function and structure leading on to a new human phenotype called homo neoneanderthalis. The homo neoneanderthalis is a hardy species which can resist climate change and thrive in environments of high temperature, high atmospheric carbon dioxide and water shortage. The new organelle that are developed can withstand climate change and fix carbon dioxide and ammonia leading to endogenous synthesis of carbohydrates and protein. The homo neoneanderthalis is a new metabolic phenotype that can survive in extremes of climate produced by global warming. The archaea forming new cellular organelle called archaeaons are extremophilic and survive in extremes of climate.

The endosymbiotic archaea secrete RNA viroids which can be converted to corresponding DNA sequences by endogenous retroviral reverse transcriptase and can get incorporated into the genome by endogenous retroviral integrase. Thus there is a constant addition of genomic sequences derived from archaeal RNA viroids producing a dynamic human neoneanderthalic genome adaptable to extremes of climate produced by global warming. Thus, the human neoneanderthalis is a new genotype. The neoneanderthalic archaeaons are capable of fixing carbon dioxide and oxygen, using ammonia for energetics and capable of biological fission and fusion reactions. They are also capable of photoinduction of membrane ATP synthesis by sodium-potassium ATPase. The fixing of carbon dioxide leads to water as a by-product. This generates endogenous water in the neoneanderthalic system helping to survive extremophilic climatic conditions. This creates a new system of energetics helping the Neo-neanderthal survives in extremes of climate.

The neoneanderthalic archaeaons have got magnetite capable of magnetoperception and quantal perception. The quantal perception of low level of EMF leads to frontal atrophy and cerebellar dominance producing a new neuronal phenotype with features of endemic cerebellar cognitive affective disorder manifesting as epidemic autism. This forms a new autistic behavioural phenotype which is introverted, non-compassionate, non-empathic, psychopathic and with extrasensory perceptual abilities of telepathy, telekinesis and time travel leading to communication with other galactic civilizations. The new behavioural autistic robotic neoneanderthalic phenotype will have the necessary intelligence and mental abilities cope with climate change. The neoneanderthalic archaeaons are capable of quantal perception and can communicate quantally with other star systems and multiverses. This leads on to the evolution of a galactic civilization. The neoneanderthalic archaeaons produce a metabolic phenotype capable of surviving in extremophilic environments of other star systems and can survive as civilization in other stars and exoplanets. The neoneanderthalic autistic phenotype with its extremophilic archaeaon organelle is suited for space colonisation.

The neoneanderthalic archaeaons use cholesterol oxidation as a form of energetics. This produces depletion of sex hormones producing asexual phenotypes. This leads to sexual phenotypes with alternate sexuality and parthenogenesis as a form of reproduction. This produces change in family and society as well as organisation of the community.

The neoneanderthalic archaeaons secrete RNA viroids which get converted to DNA viroids by human endogenous retroviral reverse transcriptase. The neoneanderthalic archaeaon generated RNA and DNA viroids can hybridize with other RNA and DNA viruses producing new RNA and DNA viruses to which the neoneanderthalic species is resistant. Therefore, there exist two sets of population with differing variety of archaeal symbiosis. The homo neanderthalic species has persisted as fossilised matrilineal societies in Europe and Asia which includes the Celts, the Dravidians, the Berbers and the Sumerians. These fossilised remnants of homo neanderthalic species are ideal substrates for archaeal endosymbiosis in the setting of global warming and extremophilic climatic conditions. These fossilised neanderthalic matrilineal societies will expand in the period of global warming and extremophilic conditions due to endosymbiotic archaeal symbiosis to which they are already acclimatised and exposed to. The human species homo neanderthalis is a consequence of archaeal symbiosis in the extremophilic conditions of ice age and are RNA viral/retroviral resistant consequent to endogenous digoxin synthesis and by archaeal RNA viroidal induced RNA interference and viral inactivation. The homo sapiens which evolved in Africa and migrated later to the Eurasian land mass are retroviral sensitive and other RNA viral sensitive and have a low load of endosymbiotic archaea growth. The homo sapiens owing to conditions of evolution is not an ideal substrate for global warming related endosymbiotic archaeal symbiosis. The homo neanderthalis archaeaons secrete RNA viroids which are converted to DNA viroids with both of them hybridizing with other RNA and DNA viruses in nature producing new viruses to which the homo neanderthalic and homo neoneanderthalic host is resistant. The new RNA and DNA viruses produced by homo neoneanderthalis and neanderthalis in global warming will infect the susceptible homo sapien species and wipe them out from earth. The homo sapien species is not a good substrate for global warming related endosymbiotic archaeal growth and would not evolve new cellular organelle like steroidelle, viroidelle, neurotransminoids and vitaminocyte which would help them to survive in extremophilic conditions of global warming. The homo sapiens also lack archaeaon organelle which help them to fix carbon dioxide and generate water, synthesis ATP by photo-induction mediated by sodium-potassium ATPase and archaeaon mediated fission and fusion reaction. The homo sapien species therefore without evolving archaeaon organelle cannot survive in the harsh environment of global warming induced climate change and is exterminated by homo neanderthalis and homo neoneanderthalis. The earth gets dominated by homo neanderthalis and homo neoneanderthalis in the setting of global warming and extremophilic climate change and they go on to contact other galactic civilizations by quantal perceptions as well as colonising other exoplanets, planets and star systems. This produces a civilization of superhuman homo neanderthalis and homo neoneanderthalis with new metabolic, endocrine, neuronal, psychological and genotypic features. The homo neoneanderthalis and homo neanderthalis with its cellular archaeaons and porphyrions can interlink with the internet and computer systems as well as robots producing man-machine hybrids by quantal perception. The artificial intelligence and robotic systems gets merged with homo neanderthalis and homo neoneanderthalis producing superhuman man-machine hybrids. The new homo neanderthalis and homo neoneanderthalis superhuman autistic man-machine hybrids will survive, prosper and flourish in the setting of global warming related extremophilic climate change and evolving to contact by quantal perception other galactic forms of life and civilization as well as colonising other planets and star systems.

CHAPTER 17

PANDEMICS DERIVED FROM EXTRATERRESTRIAL SUPERORGANISM CAN SHAPE HUMAN SOCIETY - THE HUMAN ENDOSYMBIOTIC ARCHAEAL RNA VIROID QUASI-SPECIES CONSORTIA, NEW VIRUSES AND SOCIO-ECONOMIC-POLITICAL HISTORY- ALIEN VISITATION AND SOCIETY – INFECTIVE SYMBIOTIC TISSUE COLONIES OF PROTOZOA, ARCHAEA, VIRUSES AND BACTERIA – CATASTROPHIC SPECIES CHANGE

The plasma life based forms and the gravitational and anti-gravitational waves generated porphyrions would have been the most primitive of non-life forms. The porphyrin templates would have given rise to the isoprenoid organism, the RNA viroids, DNA viroids and prions which would have symbiosed to form the nanoarchaea. The archaea would have evolved into prokaryotes or bacteria. The archaeal symbiosis with prokaryotes would have generated the eukaryotes. The initial unicellular eukaryotes are the protozoa. The protozoa are capable of phagocytosis and would have had viruses, bacteria and archaea as their intracellular symbionts. The viruses of the protozoa like entamoeba and toxoplasma are giant viruses called megaviride. The protozoans like entamoeba with its viral, archaeal and bacterial endosymbionts would have evolved into the multicellular organisms or metazoa. The archaeal, bacterial and protozoal extracellular vesicles containing RNA and DNA viroids can by horizontal gene transfer lead to evolution of metazoan, plants, animals and human species. The metazoans would have evolved into plants, animals and humans by horizontal gene transfer generated by archaeal extracellular RNA and DNA viroid vesicles. The evidence of such a transformation is seen in the bacterial symbionts of protozoa like entamoeba which includes legionella, listeria, mycoplasma, mycobacteria and chlamydia. The symbionts of protozoan entamoeba include viruses like covid, HIV and cytomegalo virus. The protozoan

entamoeba serves as a natural reservoir for covid infections, bacterial infections and HIV infections. The toxoplasma also serves as a covid reservoir. These protozoan colonies of entamoeba and toxoplasma with its bacterial, archaeal and viral symbionts can be called as chimeric infective symbiotic tissue colonies which can exist in the air, water, earth, blood and tissues. The chimeric infective symbiotic tissue colonies containing bacteria, archaea and viruses can make the symbiotic organism multidrug resistant by horizontal gene transfer. The chimeric infective symbiotic tissue organism evolves as a consequence of global warming and the protozoal, bacterial, viral and archaeal symbiotic colonies are new forms of life. The chimeric infective symbiotic tissue organism composed of protozoa and its symbionts like virus, bacteria and archaea depend upon glycolysis for its energy needs. The glycolytic enzymes enolase, glyceraldehyde 3-phosphate dehydrogenase, hexokinase, phosphofructokinase, aldolase, pyruvic kinase and triose phosphate isomerase are moonlighting proteins having a cytoplasmic, membrane and nuclear functions. These glycolytic enzymes have a RNA binding motif which can modulate RNA transcription in the host and the symbiotic organism. These glycolytic enzymes with their RNA binding motif as well as DNA binding motif can bind to RNA and DNA viruses and viroids modulating their replication and life cycles. The chimeric infective symbiotic tissue organism composed of protozoa and its symbionts like virus, bacteria and archaea can enter the human host through the breaches of the gut-blood barrier and transform itself into the macrophages of the human host. The macrophages are involved in the phagocytosis and innate immunity against bacterial and viral infections. Paradoxically the macrophages like the protozoa can also serve as a reservoir where pathogens like mycobacteria tuberculosis, mycobacteria leprae, listeria, legionella, E. coli, shigella & salmonella, and viruses like HIV & covid can thrive and replicate. The modified chimeric infective symbiotic tissue organism or macrophage composed of protozoa and its symbionts like virus, bacteria and archaea can serve as a reservoir for horizontal gene transfer and generation of

multidrug resistant bacteria and viruses. The infective chimeric symbiotic tissue organism composed of protozoa and its symbionts like virus, bacteria and archaea depends upon moonlighting glycolytic enzymes for their energetics and replication. The development of antibodies against the infective chimeric symbiotic tissue organism composed of protozoa and its symbionts like virus, bacteria and archaea leads to the evolution of anti-brain antibody disease or anti-basal ganglia antibody disease. They present a threat to existence consequent to horiztontal gene transfer and generation of multidrug resistant bacteria, viruses, protozoa, mycobacteria and archaea.

The retroviruses were responsible for integration into the genome as human endogenous retrovirus sequences and development of the cerebral cortex creating the logical and civil society. The influenza virus with its low level of HIV sequences is also capable of integration into the human genome as human endogenous COVID 19 sequences creating the cerebellar dominant autistic impulsive punkish brain. The COVID 19 virus would have been generated from RNA viroids of endosymbiotic archaea or would have reached the earth by meteoritic impacts from the human intergalactic biological quantal computing cloud. The COVID 19 virus with its HIV sequences is an ideal virus for integration into the human genome and changing it in a catastrophic manner to create a new human species. The COVID 19 brain infection can lead to integration of COVID 19 sequences into the genome, creation of cerebellar dominant brain and neanderthalisation. The COVID 19 infected cerebellar dominant brain has got increased porphyrin metabolism and porphyrions. The porphyrions are dipolar pumped phonon quantal perceptive systems which integrate with the internet. Panpsychism implies that the internet is conscious. The viruses spread genes across bacteria humans and other cells. According to Lynn Margulis we are our viruses. Viruses merge with the cellular genome and reemerge from them. They create successful genetic patterns that underlie living things. RNA virus can exist as quasi-species consortia which can create new

RNA viruses and codes by symbiosis. The archaeal RNA quasi-species consortia underlie human and species identity. The archaeal digoxin can edit the RNA producing new codes to suit environmental conditions for the benefit of the quasi-species consortia.

The bacteriophages are examples of this concept. The bacteria trade genes frantically by three processes - transformation via uptake of DNA, sexual conjugation and bacteriophage induced transduction. The viruses are mobile genetic elements and can carry genes from one person to another. The RNA viruses are dipolar and in the setting of digoxin induced membrane sodium-potassium ATPase inhibition can create a pumped phonon system mediated quantal perception. The archaeal RNA viroid quasi-species consortia can thus mediate quantal perception and function as a symbiotic colony and communicate with the outside world. To suit survival in changing environmental conditions new RNA viroids can be added to the quasi-species consortia by archaeal DNA induced editing. The viral cycles can be lytic cycles or lysogenic cycles. Viral infections convert the human body to a viral organ.

Many filo viruses like Marburg virus and ebola virus are integral part of the human genome. The human genome also contains retroviruses and borna viruses as integral parts. Similar integration non-retroviruses have been described for RSV, LCMV and VSM virus. The integration is done by HERV reverse transcriptase and integrase. The viruses can cross the inter-species barrier. The algal virus Acanthocystis tortacea chlorella can infect mamallian cells. It has been detected in human throat swabs and the algal virus can replicate in human cells. Human neuronal cells when infected by the algal virus, produce changes in memory, visuo-spatial process and attention affecting cognitive function.

Viruses are mechanisms of gene transfer. This is explained by viruses given to the caterpillar by the wasp. The barco virus protects the caterpillar against particular viruses. The viruses infecting humans contains sequences with human

DNA and bacterial DNA. The viruses thus function as mechanism for interhuman and interspecies gene transfer. Just as the viruses can infect cells the bacteria can conjugate with human cells. Bacterial conjugation and DNA transfer with human cells have been described. Photosynthetic genes have been sequenced in phage virus. Bacteria can steal genes and develop resistance. The gene for alpha-2 macroglobulin is seen in certain bacteria and provides a mechanism for bacterial resistance.

The human gut is a symbiotic system between bacteria, viruses and phages. Gut bacterial phages codes for genes of amino acid synthesis and carbohydrate metabolism. Ecosystem of human cells gut bacteria, viruses and phages is a continuous gene transfer mechanism and can interact with other ecosystems. The pathogens can contribute genes to the host. The archaeal RNA viroids can contribute genes to the host by conversion to DNA using HERV reverse transcriptase. Unrelated RNA and DNA virus can re-combine. The archaeal RNA viroids and its corresponding DNA can re-combine with unrelated RNA and DNA viruses. The bacteriophages can shuttle genes between different ecosystems. The archaeal RNA viroids served the same purpose. There is transkingdom crosstalk of small RNA molecules. Transkingdom sRNA silencing of the human RNA by archaeal RNA viroids can modulate the human system. The crosskingdom RNA silencing is important in crosskingdom communication in ecosystem.

The viral pandemics have thus contributed to human evolution. Virus and bacterial infection homogenized human population by gene transfer. The archaeal and its RNA viroids form the lynchpin of the mechanism of gene transfer. This leads on to globalisation of speech, thought and culture by viral epidemics and related gene transfer. Thus viral epidemics help in globalization of human culture. Human viral epidemics are necessary prerequisite for the evolution of human culture. The viral epidemics homogenize human population forming groups of caste, religion, nationalities and culture. The archaeal RNA viroidal quasi-species consortia underlie this mechanism of evolution. Thus viral diseases contribute to culture, behaviour, diet, eating habits and sexuality. Thus virus mediated gene transfer is important human sociological mechanism. The archaeal RNA viroidal quasi-species consortia and its recombination with unrelated RNA and DNA viruses and its integration into the human genome by HERV reverse transcriptase forms the basis of this phenomenon. Retroviral infections have contributed to the genesis of schizophrenia, cortical function and evolution of consciousness. Borna virus infections also contribute to schizophrenia. Viral pandemics thus contribute to population identity and differences. This can also lead on to generation of new viruses.

Diet can modulate archaeal RNA viroidal function. The high fibre diet will suppress archaeal growth and RNA viroidal growth. A low fibre diet will increase archaeal growth and RNA viroidal growth. Thus the RNA viroidal quasi-species consortia can be modulated by diet. Viruses are beneficial agents and only 1 percentage of the viruses are pathogenic. Most of the viruses co-exist as commensals. The bacteria and fungal kingdoms contain bacteriophages and fungal phages. They can recombine with archaeal RNA viroids and their corresponding DNA producing new RNA and DNA viruses which can get integrated into the genome as well as get secreted into the environment. This forms a common gene pool.

The archaea are extremophiles and can exist in the intergalactic space and meteroidal impacts can transfer the archaea in the intergalactic space to earth. The archaeal RNA viroids thus supply a continuous source of new genetic codes to earth. The bacterial infections especially those due to streptococcus and its phages can produce epidemic OCD, echolalia and ecopraxia contributing to a disciplined society and the evolution of the hereditary system of kingship. This can be due to the streptococcus induced epidemic OCD frontal lobe syndrome. The bacteriophages from the streptococcus could have recombined with the archaeal

RNA viroids and got integrated into the human genome. The fungal infections could have contributed to the next phase of evolution. Infections with claviceps purpura can transfer LSD genes and dopamine genes to humans producing the dopaminergic epoch in society. The fungal phages would have recombined with archaeal RNA viroids and got integrated into the human genome. This could have contributed to ideas of equality, fraternity, liberty and socialism which are all products of the French Revolution. During this period of time there was an epidemic of fungal infection in Rye in Europe. The next phase of global epidemics occurred with H1N5 infection or the Spanish flu epidemic. This resulted in a locked-in state and a frozen society leading on to the rise of dictatorship, fascism, the Nazis and the Communists. The next stage in which we live can be called the age of anarchy with its globalization, terrorism, rogue capitalism, sexual anarchy and religious fundamentalism. This corresponds with the recurrent epidemics of RNA viral infections - H1N1, retroviral, SARS, ebola, dengue and hemorrhagic fever. The genomic integration of these RNA viruses, fungal phages, bacterial phages would have changed the human genome at this different point in history. Thus virus induced gene transfer can modulate the brain, culture, sociology and behaviour.

References

1. Kurup, R.K. and Kurup, P.A. *Global Warming, Archaea and Viroid Induced Symbiotic Human Evolution - Retrovirus, Prions and Viroids – Porphyrinoids and Viroidelle*. New York: Open Science Publishers, 2016.

CHAPTER 18

METAPHORIC SEXUAL MEMES, ARCHAEAL RNA VIROIDAL QUASI-SPECIES CONSORTIA AND THE ORIGIN OF HUMAN LANGUAGES AND CULTURE – THE DOGE LANGUAGE - ALIEN VISITATION AND CIVILIZATIONAL CHANGE – INFECTIVE SYMBIOTIC TISSUE COLONIES OF PROTOZOA, ARCHAEA, VIRUSES AND BACTERIA

The plasma life based forms and the gravitational and anti-gravitational waves generated porphyrions would have been the most primitive of non-life forms. The porphyrin templates would have given rise to the isoprenoid organism, the RNA viroids, DNA viroids and prions which would have symbiosed to form the nanoarchaea. The archaea would have evolved into prokaryotes or bacteria. The archaeal symbiosis with prokaryotes would have generated the eukaryotes. The initial unicellular eukaryotes are the protozoa. The protozoa are capable of phagocytosis and would have had viruses, bacteria and archaea as their intracellular symbionts. The viruses of the protozoa like entamoeba and toxoplasma are giant viruses called megaviride. The protozoans like entamoeba with its viral, archaeal and bacterial endosymbionts would have evolved into the multicellular organisms or metazoa. The archaeal, bacterial and protozoal extracellular vesicles containing RNA and DNA viroids can by horizontal gene transfer lead to evolution of metazoan, plants, animals and human species. The metazoans would have evolved into plants, animals and humans by horizontal gene transfer generated by archaeal extracellular RNA and DNA viroid vesicles. The evidence of such a transformation is seen in the bacterial symbionts of protozoa like entamoeba which includes legionella, listeria, mycoplasma, mycobacteria and chlamydia. The symbionts of protozoan entamoeba include viruses like covid, HIV and cytomegalo virus. The protozoan entamoeba serves as a natural reservoir for covid infections, bacterial infections

and HIV infections. The toxoplasma also serves as a covid reservoir. These protozoan colonies of entamoeba and toxoplasma with its bacterial, archaeal and viral symbionts can be called as chimeric infective symbiotic tissue colonies which can exist in the air, water, earth, blood and tissues. The chimeric infective symbiotic tissue colonies containing bacteria, archaea and viruses can make the symbiotic organism multidrug resistant by horizontal gene transfer. The chimeric infective symbiotic tissue organism evolves as a consequence of global warming and the protozoal, bacterial, viral and archaeal symbiotic colonies are new forms of life. The chimeric infective symbiotic tissue organism composed of protozoa and its symbionts like virus, bacteria and archaea depend upon glycolysis for its energy needs. The glycolytic enzymes enolase, glyceraldehyde 3-phosphate dehydrogenase, hexokinase, phosphofructokinase, aldolase, pyruvic kinase and triose phosphate isomerase are moonlighting proteins having a cytoplasmic, membrane and nuclear functions. These glycolytic enzymes have a RNA binding motif which can modulate RNA transcription in the host and the symbiotic organism. These glycolytic enzymes with their RNA binding motif as well as DNA binding motif can bind to RNA and DNA viruses and viroids modulating their replication and life cycles. The chimeric infective symbiotic tissue organism composed of protozoa and its symbionts like virus, bacteria and archaea can enter the human host through the breaches of the gut-blood barrier and transform itself into the macrophages of the human host. The macrophages are involved in the phagocytosis and innate immunity against bacterial and viral infections. Paradoxically the macrophages like the protozoa can also serve as a reservoir where pathogens like mycobacteria tuberculosis, mycobacteria leprae, listeria, legionella, E. coli, shigella & salmonella, and viruses like HIV & covid can thrive and replicate. The modified chimeric infective symbiotic tissue organism or macrophage composed of protozoa and its symbionts like virus, bacteria and archaea can serve as a reservoir for horizontal gene transfer and generation of multidrug resistant bacteria and viruses. The infective chimeric symbiotic tissue

organism composed of protozoa and its symbionts like virus, bacteria and archaea depends upon moonlighting glycolytic enzymes for their energetics and replication. The development of antibodies against the infective chimeric symbiotic tissue organism composed of protozoa and its symbionts like virus, bacteria and archaea leads to the evolution of anti-brain antibody disease or anti-basal ganglia antibody disease. They present a threat to existence consequent to horiztontal gene transfer and generation of multidrug resistant bacteria, viruses, protozoa, mycobacteria and archaea.

The retroviruses were responsible for integration into the genome as human endogenous retrovirus sequences and development of the cerebral cortex creating the logical and civil society. The influenza virus with its low level of HIV sequences is also capable of integration into the human genome as human endogenous COVID 19 sequences creating the cerebellar dominant autistic impulsive punkish brain. The COVID 19 virus would have been generated from RNA viroids of endosymbiotic archaea or would have reached the earth by meteoritic impacts from the human intergalactic biological quantal computing cloud. The COVID 19 virus with its HIV sequences is an ideal virus for integration into the human genome and changing it in a catastrophic manner to create a new human species. The COVID 19 brain infection can lead to integration of COVID 19 sequences into the genome, creation of cerebellar dominant brain and neanderthalisation. The COVID 19 infected cerebellar dominant brain has got increased porphyrin metabolism and porphyrions. The porphyrions are dipolar pumped phonon quantal perceptive systems which integrate with the internet. Panpsychism implies that the internet is conscious. The human brain can be considered as a modified archaeaon colony network. The archaeaon are eternal and can last for billions of years. The human brain is basically an information storage system. The archaeaon has got dipolar magnetite and porphyrins and can function as quantal computer. The archaeal colony with its dipolar magnetite and porphyrin in the setting of archaeal digoxin induced membrane sodium-potassium ATPase inhibition can function as a pumped phonon system mediating quantal perception. The archaeaon in the brain is capable of information storage at a point in time and space. The experiences and information stored in the archaeaon is immortal and eternal. The archaeaon can have a wave-particle existence and can exist in multiple quantal possible states and can inhabit multiple quantal multiverses. The interaction between information stored in quantal computers in multiple different archaeaon systems all over the universe by the quantal interactions results in eternal existence of information in quantal multiverses. The information in the quantal multiverses can have a particulate existence creating a newer mode by quantal interactions between information stored at multiple points of time. This creates the particulate mythic world of human existence. These are what are called as Samsaras. The mind is uploaded into information in the neuronal archaeal colony network and its quantal computers. The information stored in the archaeal colony network mediated quantal state is eternal and can be considered as a digital version of the brain, a mind downloading technique or whole brain emulation. The archaeal colony network stores the human experiences in an eternal manner and can contribute to biological reincarnation.

Memes are building blocks of mind. They are also the building blocks of culture. The memes make up language and religion. The origin of speech can also be related to archaeal RNA viroidal quasi-species consortia. Speech origin can be compared to an epidemic la Tourette's disease which consists of vocal tics, motor tics, coprolalia and obsessive compulsive disorder. It is due to post streptococcal anti glycolytic anti brain antibodies affecting the brain. The archaea and RNA viroidal modulation by extraterrestrial and terrestrial contacts with the brain and symbiosis can produce vocal tics and coprolalic words and phrases akin to la Tourette's disease producing a primitive form of speech. The motor tics would have evolved to artistic dance and drama forms. The primitive vocal and motor tics which would have been the substratum for civilized speech would have originated as a tic syndrome consequent to extraterrestrial and terrestrial archaeal and RNA viroidal infection and symbiosis. The initial vocal and motor tics would primarily have been related to sexual content and could have contributed to a coprolalic form of speech. The archaeal symbiosis in homo sapien population would have produced a dominant neanderthalic population and the subservient lower class and caste homo sapien population would have been labeled and described the synonyms of this coprolalic speech and vocal tic form of spontaneous speech. The RNA viroidal quasi-species consortia is capable of being constantly edited by archaeal digoxin contributing to widespread diversity of the RNA viroidal quasi-species consortia contributing to species identity with regard to speech. The RNA viroids are capable of transynaptic transport and can function as neurotransmitters modulating brain function. The RNA viroidal quasi-species consortia are infective and can be transmitted within a defined population giving characteristic identity to the vocal and motor tics and coprolalic speech. This would have given rise to different languages which behave like mutated viral populations. The different family of languages corresponds with the diversity of archaeal viroidal RNA quasi-species consortia edited by archaeal digoxin. This form of coprolalic speech and vocal/motor tics and gestures with predominant sexual content are subcortical in origin and localized to the basal ganglia and cerebellum. Cerebellar cognitive affective disorders give rise to impulsivity and tics and coprolalic speech are impulsive and are out of cortical control. Swear and obscene words persist even in the presence of global aphasia and destruction of the cerebral cortex. The coprolalic speech and corresponding vocal and motor tics can be converted to signs and ideograms with sexual contents due to motor tics or cerebellar dysfunction producing appendicular ataxia. The ideograms meanings are changed when multiple ideograms are combined. This gives rise to the origin of written language and alphabets of all languages as are exemplified by the Harappan script. Languages are a consequence of archaeal RNA viroidal infection and symbiosis of the brain and spread in the community producing endemic vocal tics syndromes of different characteristics for each group of population. This phenomenon can also be called as viro-linguistics or neuro-linguistics. The obscene coprolalic language and tics with sexual imageries would have evolved into the primitive sort of religion. Swear words with obscene content induce greater skin response to electric stimulation. Swear words also have a cathartic effect as noticed by the time duration to which the person using those words can immerse his limbs in subzero temperature. Coprolalic languages and tics give a cathartic effect or transcendence. Swear words or coprolalic words are the most common words used in English language. They are used for a specific purpose or as emotional intensifiers. The obscene words are coprolalic language are localized to the cerebellum and its limbic connections. Cerebellum is the site for counting, formalities, rituals, lyrics, magic and dance forms. Thus the coprolalic language and obscene tics would have evolved into a form of primitive religion which gave communities a sort of transcendence. This can be noted in the worship of Shiva and Shakti and similar cults in other parts of the world. The concept of Shiva and Shakti is the oldest concept of God and can be traced to the Harappan civilization. Thus language and religion evolved out of archaeal RNA viroidal quasi-species consortia infection of the brain and a cerebellar cognitive affective disorder. The RNA viroidal quasi-species consortia are capable of immense diversity owing to constant editing by archaeal digoxin. The RNA viroids are capable of transynaptic transports and sculpt synaptic connectivity networks for each domain of language and religious function which are capable of dynamic changes from second to seconds. They are localized in the sub-cortical cerebellar networks and are automatic, impulsive, primitive, sexual and sensual. They can be considered as memes or the unit of cultural transmission. Memes are constituted by archaeal RNA viroidal quasi-species consortia with each different QS consortia constituting one meme which is transmissible in the community by infection and gives identity to the community. The language and religion are infectious and can be considered as equivalent to epidemics and endemics of viral infections which are susceptible to constant changes by mutations. Viral tracking procedures are now used to track the origin of languages.

Language evolution has been related by different authors to cries of pain and delight due to vocal and motor tics as comparable with the non-linguistic vocal tics of la Tourette's syndrome. TS is due to dopamine receptor supersensitivity. The archaeal shikimic acid pathway produces dopamine excess contributing to vocal and motor tics as well as coprolalia the basis of language, music, religion, dance and painting. Motor tics and appendicular ataxia would have contributed to abstract art of painting and dance. The archaeal and RNA viroidal infection re-wires the brain, changing synaptic connectivity and species specific language and art forms arises out of horizontal genetic transfer of RNA viroidal quasi-species consortia generated from archaea and susceptible to archaeal digoxin editing giving it widespread diversity. Thus viroidal vectors quasi-species consortia which are infective can give rise to cultural memes which are transmissible and give rise to language, music, dance, literature, painting, religion, spirituality and everything which is sublime to humanity. The archaeal RNA viroidal quasi-species consortia modulated brain synaptic networks leads to cerebellar dominance and shape the communities and individuals political, social and economic views. Cerebellar dominance due to archaeal RNA viroidal quasi-species consortia leads to impulsivity, altruism, extrasensory quantal perception and group communication, dominance of group over individual and feelings of contact with the collective unconscious. This gives rise to ideological patterns in the brain of social justice, altruism, primitive communism and spirituality. These patterns are consequence of archaeal RNA viroidal quasi-species consortial diversity and symbiosis producing a cerebellar cognitive affective disorder and cortical dysfunction which basically contributes to a neanderthalic brain. The homo sapien brain has less of archaeal symbiosis and RNA viroidal quasi-species consortia and structures synaptic networks coding for selfishness, civil society, secular society, capitalism, individuality and less of spirituality. The cerebellar cognitive affective disorder consequent to archaeal RNA viroidal quasi-species consortia contributes to spirituality, altruism and can generate a community of leaders who can guide and set the community on course. This leads us to the origin of the archaea and RNA viroidal quasi-species consortia. The intergalactic space contains a network of porphyrions and embedded archaeal and RNA viroidal giant colony functioning as molecular quantum computing cloud in the intergalactic space. The archaeal magnetite contributes to the intergalactic magnetic field and the condensation of the cloud leads to the formation of star system and galaxies. The archaea and RNA viroidal quasi-species consortia are derived from the interstellar massive porphyrion and archaeal quantal computing cloud which functions as a massive God-like extraterrestrial intelligence. This archaeal quantum computing cloud in the intergalactic space gets transferred to earth by meteoric impacts and produces archaeal symbiosis on primates and homo sapiens producing changes in culture generating language, art, painting, dance, music, literature and spirituality. This fits into the Freudian hypothesis of a sexual substratum for unconsciousness. Freud based unconscious desires on the basis of sexual instincts and recent mind as a combination of reason, will and passion. Freud regarded the problem from inside the individual that is for the world of human sexual instincts and constructed an irrational world of the mind. The basis of life is replicate. Replication depends upon sex and therefore sexual instincts form the basis of language, culture and religion. This can be called as a form of esoteric spirituality. The la Tourette's syndrome with its vocal and motor tics with sexual motives consequent to archaeal RNA viroidal symbiosis can lead to the evolution of abstract painting, dance, transcendent spirituality and language arising from sexual swear words. Thus archaeal RNA viroidal induced epidemic la Tourette's syndrome and sexual vocal and motor tics would have formed sexual memes with a genetic diverse RNA viroidal quasi-species consortial counterpart diversified by archaeal digoxin mediated RNA editing. Thus the Freudian psychology of sexual instincts and unconsciousness can lead to a sexuality mediated transcendence going on to the realm of Jungian mysticism.

The ancient societies have examples as evidence of this in the worship of Shiva and Shakti as well as yoni and linga. Ancient Hindu religion had sexual practices integrated as part of worship. Swear words and obscene words were also part of worship in certain temples in the subcontinent. This points to obscene vocal and motor tics as the origin of language, dance, music and spirituality due to archaeal and RNA viroidal quasi-species consortia infection and symbiosis of extraterrestrial origin due to periodic meteoric impacts. They generate the units of cultural transmission, memes as described by Dawkins. Memes are RNA viroidal quasi-species consortia edited and diversified by archaeal digoxin and structured on to synaptic networks which can also be infective in a population. Their origin is extraterrestrial and a part of the intergalactic molecular porphyrion and archaeal quantum computing network which gives rise star systems and planets also.

The archaeal RNA viroidal quasi-species consortia can function as neurotransmitter having transynaptic transport and can modulate synaptic connectivity. The archaeal RNA viroidal quasi-species consortia are subjected to editing by archaeal digoxin and generate new RNA viroidal quasi-species consortia depending upon informational inputs in the outside world. This creates states of mind or samsaras. They can be compared to mutating viral infection. One state of mind gives rise to another state of mind and one set of synaptic RNA viroidal quasi-species consortia gives rise to another synaptic RNA viroidal quasi-species consortia by digoxin mediated editing. The archaeal magnetite is capable of quantal perception of low level electromagnetic fields from fellow human being's brain and this can inhibit membrane sodium-potassium ATPase activity and modulate digoxin synthesis altering the nature of the RNA viroidal quasi-species consortia. These RNA viroidal quasi-species consortia can get transcribed to the DNA by reverse transcriptase and get integrated into the neuronal genome producing dynamicity of the genome. Thus the archaeal magnetite can modulate interbrain informational transfer by quantal perception of low level EMF producing changes in transynaptic networks of RNA viroidal quasi-species consortia and neuronal genomic structure by the integration of viroidal sequences into it. The intergalactic magnetic field of the archaeal and RNA viroidal molecular quantal computing cloud can also be perceived by neuronal archaeal magnetite modulating the transynaptic network of RNA viroidal quasi-species consortia and its integration into the neuronal genome altering it. The Samsaras are states of mind and are momentary and there is no permanent state of mind. The states of mind during the existence of a personality end with the mortality of the personality, but the totality of its actions creates a state of mind or Samsara independent of the previous personality. Thus there is no continuous consciousness in existence, only different states of mind which gives rise to another set of independent states of mind depending on the informational interactions of the states of mind or karma. We are not only creating the world but we are also creating ourselves, our ego, our self-consciousness and personality in a continuous manner. From moment to moment our bodies sensation conceptualization, ideas and beliefs work together to create an illusion of consciousness. Each of us is a being in different world will create out of our thought and perception. We create the world and create ourselves. There is no creator or creation; only the collective unconscious.

The religions of the world can be divided into cerebral cortical religions which impose a strict code of civil conduct. These are all organized religions with very little magic and spirituality in their functioning with examples of the Semitic religions of Islam and modern Christianity. Such religions are logical and civil religions and are capable of inhuman behaviour in the form of crusades, inquisition, Islamic terrorism, the world wars and nuclear bomb. These religions impose a strict code of conduct and sexual conduct and thinking are taboo. The opposite is the cerebellar religions which include Hinduism, Zen-Buddhism, Sufism, Gnostic Christianity and the Kaballah. These religions are reflections of archaeal and RNA viroidal transynaptic network and quantal perception of low level electromagnetic fields producing cerebral cortical atrophy and cerebellar dominance. This leads on to impulsive behaviour, magical experiences, extrasensory perceptions, out of the body experience, sexual transcendence and spiritual experiences. These are all religions that can be localized to lower areas of the brain the cerebellar cortex and are reflections of the cerebellar cognitive affective disorder. The parallel between spirituality and sexual instincts is reflected in the Yogic terminology of Kundalini where electric current from the chakras in the base of the cord rise upto the pineal gland producing a sense of transcendence. Thus, there is not much difference between a spiritual experience and sexual experience. This is reflected in Hindu practices of Tantra. These cerebellar religions are humane religions where there is extrasensory quantal perception and feeling of the collective unconscious. The human senses are celebrated and humanity triumphs. Thus the human spiritual experience and culture and language have an underlying basis in human sexual experience and the Freudian sexual psychology forms the basis of Jungian mysticism. Life is indicated by replicators; and replicators means sexual conduct and sexual conduct leads to transcendence, spirituality, language evolution and evolution of culture.

Symbiosis by micro-organisms, especially archaea drives the evolution of the species. In such a case, symbiosis can be induced by transfer of microflora symbionts and evolution induced. Endosymbiosis by archaea as well as archaeal symbionts in the gut can modulate the genotype, the phenotype, the social class and the racial group of the individual. The symbiotic archaea can have horizontal and vertical transmission. Endosymbiotic archaeal growth leads to neanderthalisation of the species. The neanderthalised species is matrilineal society and includes the Dravidians, the Celts, the Basques and the Berbers. The inhibition of the endosymbiotic archaeal growth leads to evolution of the homo sapiens. This includes the Africans, Aryan invaders of North India and the Aryan derived European population. Symbiosis mediated evolution depends on the gut flora and the diet. This has been demonstrated in the drosophila pseudoobscura. The drosophila mates only with other individuals eating the same diet. When the drosophila gut microflora is altered by feeding antibiotics they mate with other individuals eating different diets. The diet consumed by the drosophila regulates its gut microflora and mating habits. The combination of the human genome and the symbiotic microbial genome is called the hologenome. The hologenome especially its symbiotic microbial component drives human evolution as well as animal evolution. The evolutionary distance between species of wasp depends on the gut microflora. The human gut microflora regulates the endocrine, genetic and neuronal systems. Humans and primate evolution depends on endosymbiotic archaea and gut microflora. The endosymbiotic archaeal growth determines the racial differences between the matrilineal Harappan/Dravidian societies and the patriarchal Aryan society. The matrilineal Harappan/Dravidian society was neanderthalic and had increased endosymbiotic archaeal growth. Endosymbiotic archaeal growth and neanderthalisation can lead to autoimmune disease, metabolic syndrome X, neurodegeneration, cancer, autism and schizophrenia. The Neanderthal gut flora and endosymbiotic archaea was determined by the non-vegetarian ketogenic high fat high protein diet consumed by them in the Eurasian steppes. The homo sapiens including the classical Aryan tribes and African ate a high fibre diet and had lower archaeal growth both endosymbiotic and gut. The dietary fibre intake determines the microbial diversity of the gut. The high fibre intake is associated with increased generation of short chain fatty acids - butyric acid by the gut flora. Butyrate is a HDAC inhibitor and leads to increased generation and incorporation of endogenous retroviral sequences. The high dietary fibre intake related increased HERV sequences leads to increased synaptic connectivity and a dominant frontal cortex as seen in homo sapien species. The neanderthalic species consume a ketogenic non-vegetarian high fat high protein low fibre diet. This leads to decreased generation of endogenous HERV sequences and reduced genomic flexibility in neanderthalic species. This produces smaller cerebral cortex and a dominant cerebellar cortex in the neanderthalic brain. The homo neanderthalic species by the low dietary fibre intake starve their microbial self. This leads to increased endosymbiotic and gut archaeal growth. The mucous membrane lining the gut becomes thinned out as the gut bacteria eats up the mucous lining of the gut. This results in leakage of endotoxin and archaea from the gut to the blood breaching the barrier and produces a chronic immunostimulatory inflammatory state which forms the basis of autoimmune disease, metabolic syndrome, neurodegeneration, oncogenic and psychiatric disorders. The Neanderthal species eat a low fibre diet and have a deficiency of microbiota accessed carbohydrate generating short chain fatty acid. There is a deficiency of butyrate generated in the gut from the dietary fibre which can produce suppression of the chronic inflammatory process. The Neanderthals have got the fermentation by-product deficiency syndrome. The induction of neanderthalic species depends on the low fibre intake induced high archaeal density endosymbiotic and the gut microflora. The homo sapien species consume a high fibre diet generating large amounts of short chain fatty acid butyrate which inhibits endosymbiotic and gut archaeal growth. The microbial self of the homo sapien species is more diverse than that of the neanderthalic species and the archaeal population density is less. This results in a protection against chronic inflammation and the induction of diseases like autoimmune disease, metabolic syndrome, neurodegeneration, oncogenic and psychiatric disorders. The homo sapien species have a higher intake of dietary fibre contributing to around 40 g/day and a diverse microbial gut flora with less of archaeal population density. The butyrate generated from dietary fibre produces an immunosuppressive state. Thus, the symbiotic microflora with less of archaeal density induces a homo sapien species. This can be demonstrated by experimental induction of evolution. A high fibre high MCT diet as well as antibiotics derived from higher plants and fecal microbiota transfer from sapien species can inhibit the Neanderthal metabolonomics and phenotype and induce the evolution of homo sapiens. A low fibre high fat high protein diet as well as fecal microbiota transfer from the Neanderthal species can produce Neanderthal metabolonomics and phenotype inducing the evolution of homo neanderthalis. Transfer of colonic microflora predominantly archaea and modulation of endosymbiotic archaea by a paleo diet and antibiotics from higher plants can lead to interconversion of human species between homo neanderthalis and homo sapiens. The hologenome especially the microbial flora endosymbiotic/gut drives human and animal evolution and can be experimentally induced. Symbiotic microflora drives evolution. Every animal, every human species, different communities, different races and different caste have their signature endosymbiotic and gut microflora which can be transmitted vertically and horizontally. Thus symbiosis drives human and animal evolution.

This can be interpreted on the basis of Villarreal hypothesis of group identity and cooperativity of RNA collectives. Archaeal symbiosis in the gut and in the tissue spaces determines speciation of human beings as Homo sapiens and Homo neanderthalis. The endosymbiotic archaea can secrete RNA viroids and viruses and there is a viroid-archaeal host relationship between the two. A dynamic state of virus lysis and persistence can occur in archaea suggesting that viral addiction can occur in archaea. The RNA viroids in the archaea coordinate their behaviour by information exchange, modulation and innovation generating new sequence based content. This occurs due to a phenomenon of symbiosis in contrast to the concept of survival of the fittest. The generation of new RNA viroidal sequences is a result of practical competence of living agents to generate new sequences by symbiosis and sharing. This represents highly productive RNA viroidal quasi-species consortia for the evolution, conservation and plasticity of genomic environments. The behavioural motives of the RNA are single stem loop structures. They have self-folding and group building capabilities depending upon functional needs. The evolution process depends upon what Villareal calls RNA stem loop consortia. The whole entity can function only if participatory groups of RNA viroids can get their function coordinated. There is competent de novo generation of new sequences by cooperative action and not by competition. These RNA viroidal group consortia can contribute to the host identity, group identity and group immunity. The term used for this is RNA viroidal sociological behaviour. The RNA viroids can build groups that invade the archaea and compete as a group for limited resources such host genomes. A key behavioural motif is able to integrate a persistent life style into the archaeal colony with the addiction module forming competing viroidal groups that are counter balancing each other together with the archaeal/host immune system. This leads to creation of an identity for the archaeal colony and the homo neanderthalis host. Viroids can kill their host and also colonize their host without disease and protect the host from similar viruses and viroids. Together with lysis and protection we see a viroid colonized host that is both symbiotic and innovative acquiring new competent codes. Thus, the viroid-host relationship is a pervasive, ancient force in the origin and evolution of life. Cumulative evolution at the level of RNA viroids is like a ratchet effect used for transmission of cultural memes. This learning accumulates so that every new generation must not repeat all innovative thoughts and techniques. Quasi-species of RNA viroids are cooperative and exclusive of other quasi-species. They have group recognition differentiating self-groups and non-self-groups allowing for quasi-species to promote the emergence of group identity. With group identity via counter related addiction modules two opposing components must be present and work coherently and define the group as a whole. Biological identity is constituted by dynamic interaction of cooperative groups. Virus addiction module is an essential strategy for existence of life in the virosphere. Viruses are transmissible and can persist in specific host population leading to a form of group immunity/ identity since identical but uncolonized host population remains susceptible to a killing action of lytic viruses. In this way, we see that viruses are necessary providing opposing functions for addiction (persistence/protection and lytic/killing). Viroids can function as consortia, an essential interacting group and provide a mechanism from which consortial function could emerge in the origin of protobiotic life. Genetic parasites can act as a group (qs-c). But for this group to be coherent they must attain group identity and this is typically via an addiction strategy. Antiviral and proviral system in the archaea will themselves emerge in the host from virus derived information. The archaeal viruses themselves provide the critical function required for antiviral defence. The opposing functions are the basis of addiction modules. Thus the emergence of group identity becomes an essential and early event in the emergence of life. This is coherent to the basically group behaviour of RNA viroids in archaea. This group selection and group identity are needed to create information coherence and network formation and to establish a system of communication - code competent interactions. This identity serves as information also for the ones that do not share this identity. This is the beginning of self/non-self-differentiating capability. In this way, viroids promote the emergence of group identity in archaeal colonies and host humans. The archaeal colony identity depends upon the colonizing set of RNA viroids producing a coherent network that is inclusive opposing functions and favours the persistence of parasite derived new information. On the basis of population-based functions of RNA, DNA can be considered as a habitat for consortia RNA. Thus RNA viroids of the archaea are involved in complex multicellular identity. This is called as the Gangen hypothesis by Villarreal. The Gangen describes the emergence of commonly shared code use, group membership and collective living function of RNA viroids. Communication is a code depended interaction and transmission of infectious code defines the origin of the virosphere. This issue refers to the idea of collective of RNA viroids with inherent toxic and antitoxic features should be able to transmit or communicate these agents and their features to a nearby competing population. It strongly favours the survival of RNA viroidal population with compatible addiction modules that will inhibit agent toxicity and allow persistence of new agents. This is thus the survival of the persistently colonized set which is an inherently symbiotic and consortial process. It also promotes increasing complexity and identity/immunity of the host collective via a new agent colonization, and stable addition. Thus, the transmission of RNA agents attains both communication and recognition of group membership. In this way the emergence of the virosphere must had been an early event in the origin of life and group identity. Viruses and viroids are genetic parasites and the most abundant living entities on earth. The virosphere is a network of infectious genetic agents. Evolution, conservation and plasticity of genetic identities are the result of cooperative consortia of RNA viroids that are competent to communicate. Thus, the archaeal viroidal consortia can symbiotically share and communicate producing new sequences and give an identity to the archaeal colony. The low fibre diet and extreme temperatures of the Eurasian steppes leads to archaeal multiplication and induction of the homo neanderthalic species. The archaeal colony's characteristics are determined by the cooperative consortia of RNA viroids in the archaea and the archaeal colony identity determines the homo neanderthalis identity. Thus the archaeal colonies with their quasi-species consortia of RNA viroids determine the homo neanderthalis identity. The new sequence generation by the RNA viroidal consortia's symbiotic sharing character contributes to the diversity in the behaviour and creativity of the homo neanderthalic population. The archaeal RNA viruses and viroids and the archaeal colonies themselves protect the homo neanderthalic population from retroviral infections. Thus the homo neanderthalic population is retroviral resistant and the quasi-species consortia of archaea and archaeal viroids gives them a group identity as retroviral resistant. Thus, the quasi-species consortia of archaea and RNA viroids give homo neanderthalis colonies their identity and idea of self. The homo neanderthalis is resistant to retroviral infection like the Australian Aboriginals and the endogenous retroviral sequences in the Neanderthal genome are limited. This leads to lack of plasticity and dynamicity of the human genome and the cerebral cortex in ill-developed with a dominant impulsive cerebellar cortex in the homo neanderthalic population. This produces the impulsive creative surrealistic spiritual neanderthalic brain. As the extreme of temperature goes off and the ice age ends the archaeal population density also comes down. This also can result from the consumption of a high fibre diet in the African continent. The high fibre diet digested by clostridial clusters in the colon promotes butyrate synthesis and butyrate will induce HDAC inhibition and expression of retroviral sequences in the primate genome. This leads to increase in endogenous retroviral sequences in the human genome, increasing genomic dynamicity and the evolution of complicated cerebral cortex dominant brain with its complex synaptic connectivity in the homo sapiens. This leads on to a logical, commonsensical, pragmatic and practical homo sapien brain. The homo sapiens due to lack of archaea and the RNA viroids are susceptible retroviral infection. Thus the archaeal colonies and RNA viroidal quasi-species consortia determine the evolution of the human species and the brain networks. Thus extremes of temperature, fibre intake, archaeal colony density, RNA viroidal quasi-species, group identity and retroviral resistance decides on the evolution of homo sapiens and homo neanderthalis as well as the brain networks. The present extremes of temperature and low fibre intake in civilized society can lead to increase in archaeal population densities and quasi-species RNA viroidal networks generating a new homo neanderthalis in a new neanderthalic anthropocene age as opposed to the present homo sapien anthropocene age. The archaeal population densities and quasi-species RNA viroidal networks determine homo sapien/homo neanderthalic species, racial, caste, community, national, sexual, metabolic, phenotypic, immune, genotypic neuronal, psychiatric, psychological and individual identity. The archaea secretes the trephone digoxin which can edit the RNA viroids and generate new sequences. Archaeal dipolar magnetite and porphyrins in the setting of digoxin induced membrane sodium-potassium ATPase inhibition can produce a pumped phonon system mediated quantal perceptive state and quantal communication in the RNA viroidal symbiotic system generating new sequences by steroidal digoxin enzymatic editing action. This gives rise to archaeal RNA viroidal quasi-species symbiotic diversity and identity to species, race, caste, sex, culture, individual and national identity.

The roots of Western civilizational disease can be related to the starvation of the colonic microflora. The colonic microflora depends upon complex carbohydrates derived from dietary fibre. The processed food of high protein, fat and sugars is digested and absorbed in the stomach and small intestine. A very little of it reaches the colon and widespread use of antibiotics in medicine has produced mass extinction of the colonic microflora. The colonic microflora is extremely diverse and the diversity is lost. There are 100 trillion bacteria in the colon belonging to 1200 species. They regulate the immune system by inducing the T-regulatory cells. A high fibre diet contributes to colonic microbiota diversity. Interaction with farm animals like cows and dogs also contributes to the colonic microflora diversity. The typical Western diet of high fat, high protein and sugars decreases the colonic microbiota diversity and increase colonic/endosymbiotic archaea producing methanogenesis. The colonic archaea feed upon the mucous lining of the colon and produces leakage of archaea into the blood and tissue system producing endosymbiotic archaea. This results in a chronic inflammatory state. The high fibre diet of Africans, South Americans and Indians produces increased colonic microbiota diversity and increase in clostridial clusters generating SCFA in the gut. High fibre diet is protective against metabolic syndrome and diabetes mellitus. Metabolic syndrome is related to degeneration, cancer, neuropsychiatric illness and autoimmune disease. A high fibre diet of upto 40 g/day can be called as a gut diet. The colonic microflora especially the clostridial cluster digests the fibre generating short chain fatty acids which regulates immunity and metabolism. High fibre diet increases the colonic mucus secretion and the thickness of the mucus lining. A high fibre diet produces increase in clostridial clusters and mucous secretion. This produces a strong gut-blood barrier and prevents metabolic endotoxemia which produces a chronic inflammatory response. High dietary fibre intake and the diversity of the colonic microflora with prominent SCFA producing clostridial clusters are interrelated. The clostridial clusters metabolise the complex carbohydrate in dietary fibre to short chain fatty acids butyrate, propionate and acetate. They increase the T-regulatory function. A high fibre diet increases the bacteroides and reduces the firmecutes of the colonic microflora. A high fibre diet is associated with a low body-mass index. A low fibre diet produces increase in colonic archaeal growth as well as endosymbiotic tissue and blood archaea. This produces more of methanogenesis rather than short chain fatty acid synthesis contributing to immune activation. A low fibre diet is associated a high body-mass index and chronic systemic inflammation. Germ-free mice show cardiac, pulmonary and liver atrophy. Gut microflora is required for the generation of organ systems. The gut microflora is also required for generation of T-regulatory cells. High fibre intake produces more colonic microbiota diversity and increase in clostridial clusters and fermentation by products like butyrate which suppresses inflammation and increases T-regulatory cells. A low fibre diet produces increase in archaeal growth, methanogenesis, destruction of the mucus lining and leakage of the colonic archaea producing endosymbiotic tissue and blood archaea. This produces an immune hyperreactivity contributing to the modern plagues of civilization - metabolic syndrome, schizophrenia, autism, cancer, autoimmunity and degenerations. The gut microbiota drives human evolution. The humans do not host the gut microbiota but the gut microbiota host us. The human system forms an elaborate culture laboratory for the propagation and survival of the microbiota. The human system is induced by the microbiota for their survival and growth. The human system exists for the microbiota and not the other way round. The same mechanism holds good in plant systems. Plant started the colonized earth as they started symbiosing with bacteria in the roots systems which can derive nutrients from the soil. Human beings form a mobile culture laboratory for the more effective propagation and survival of the microbiota. The microbiota induces the formation of specialized immune cells called innate lymphoid cells. The innate lymphoid cells will direct the lymphocytes not to attack the beneficial bacteria. Thus, the endosymbiotic archaea and the gut archaea induce human, primate and animal evolution to generate structures for them to survive and propagate. The source of endosymbiotic archaea, the third element of life is the colonic archaea that leaks into the tissue spaces and blood systems due to breach in the gut-blood barrier. The increase in colonic archaea is due to the starvation of the gut microbiota consequent to a low fibre diet. This results in increase in colonic archaeal growth and destruction of clostridial clusters and bacteroides. The increase colonic archaeal growth in the presence of gut starvation due to low fibre diet eats up the mucus lining and produces breakages in the gut-blood barrier. The colonic archaea enters the blood stream and produces endosymbiosis generating endosymbiotic archaea and various new organelle - fructosoids, steroidelle, vitaminocyte, viroidelle, neurotransminoid, porphyrinoids and glycosaminoglycoids.

Memes, Genes and Cerebellar Function

Genes are self-replicators and have produced human evolution. Culture is another mechanism of human evolution. Culture as a mechanism of human evolution is faster than genomic evolution. It can be an idea, a design, a fashion, religion, music, dance, rituals or a language. These are stored as unconscious programmes in the brain. The unit of cultural transmission is called a meme. The unit of imitation is also a meme. Memes are self-replicators. Self-replicators can evolve. Only replicators can evolve. Replicators are blind and selfish. Memes are unconscious and are localized to the cerebellum. This is exemplified by the cerebellar cognitive affective disorder. We have gene machine making the body and meme machine making the culture. Both these machines are blind, selfish replicators. The human consciousness is not meme related and can rebel against the tyranny of replicators.

An idea or a meme can infect the brain by quantal perception or mirror neuron function which serves the purpose of imitation. This can spread from one human brain to another and depending on its fidelity, fecundity and usefulness can be long lasting and survive. Memes can also co-evolve. The idea of god meme has co-evolved with music, dance and rituals. God meme is common to different civilizations as it is a useful unit of culture.

Memes are selfish, ruthless and ideological and unconsciously propagate. Memes are stored in human brain computer and compete for fecundity, storage space and time. Memes dominate and compete for attention of the brain along with other rival memes. They are self-perpetuating ruthless ideas which dominate the unconscious brain. Memes can co-evolve. God meme co-evolve with the blind faith meme, the hell-fire meme, music and dance meme because of its evolutionary advantage and successfully competed and won over other rival memes for computer space. They are called co-adapted meme complexes.

Meme are selected in cultural environment. The gene evolution produces the human brain or computer. The human brain produces brain to brain transmission of memes by imitation behaviour and genes have no role in the selection and evolution of meme complexes. Selfish genes and memes are unconscious, blind replicators. Selfish genes and memes have no foresight. The human brain is built as a gene machine and culture as a meme machine. It is human consciousness which can rebel against the tyranny of the selfish replicators.

Memes as unit of culture can get structured as information programmes in synaptic pathways of the brain. This can affect membrane sodium-potassium ATPase function and intracellular calcium magnesium ratios modulating reverse transcriptase activity and jumping gene functions. This creates genomic plasticity. Memes when imitated by mirror neurons and quantal perceptive neurons by activity of low level EMF can induce heme oxygenase activity, deplete heme and induce porphyrin synthesis. Porphyrins can intercalate with RNA and DNA modulating genomic function. Thus memes by their imitation can modulate genomic plasticity which can get structured in the human genome so that it is transmitted over generations. Thus memes can produce changes in genomic evolution.

Memes arise from the meme pool and spread from brain to brain by imitation behaviour. Imitation behaviour is seen in frontal lobe syndrome. In prefrontal atrophy cerebellar dominance takes over and imitation behaviour is a feature of cerebellar cognitive function. The cerebellum plays a crucial role in imitation behaviour and transmission of memes. Cultural transmission of memes depends on the cerebellum. Cerebellum is the basis of human culture. The meme will literally parasitize the brain like human endogenous retroviruses. The human endogenous retroviruses cannot spread brain to brain. Genes are limited by brains and there is only vertical transmission of genes by the process of reproduction. Memes by imitation behaviour and cerebellar function can have a faster transmission resulting in the creation and transmission of culture.

Memes can also affect genomic transmission. A thought or a meme can produce changes in low level EMF and modulate porphyrin synthesis. Porphyrins can intercalate with nucleic acids modulating gene expression. The memes or thought ideas can alter neuronal firing, inhibiting membrane sodium-potassium ATPase and increase intracellular calcium and reduce intracellular magnesium. Magnesium is required for reverse transcriptase activity. Thus memes or thought via membrane sodium-potassium ATPase inhibition and altered intracellular calcium/magnesium ratio can modulate HERV expression, jumping genes and the dynamicity of the human genome. Memes can regulate the genes. Cultural transmission of memes can modulate genomic plasticity.

Archaea can control both cultural meme transmission and genomic transmission. The archaeal magnetite is capable of quantal perception and forms the basis of imitation behaviour. Imitation behaviour is the mechanism by which memes are propagated. The memes get selected from the meme pool depending on its fecundity, copying fidelity, longevity and faster rate of spread. Memes can also evolve as its spreads from brain to brain. Quantal perceptive mediated cerebellar imitative behaviour is the basis of cultural transmission of memes. The archaea can also modulate genomic transmission. The archaea secretes RNA particles which functions as RNA viroids. The archaea derived RNA viroids can use the HERV reverse transcriptase and get converted to DNA which is then integrated into the human genome by HERV integrase. Thus the archaeal derived RNA viroids can contribute to jumping genes in the human genome.

CHAPTER 19

EXTRATERRESTRIAL INTELLIGENCE PANDEMICS AND THE SHAPING OF HUMAN SOCIETY – CATASTROPHIC VIRAL PANDEMICS AS A CONTINUING EVOLUTIONARY CRO-MAGNON HOMO SAPIEN–NEANDERTHAL CONFLICT – INFECTIVE SYMBIOTIC TISSUE COLONIES OF PROTOZOA, ARCHAEA, VIRUSES AND BACTERIA – COVID VIRUS, VACCINES AND DEMOGRAPHIC DECLINE – THE GHOST CONTINENTS

Introduction

The plasma life based forms and the gravitational and anti-gravitational waves generated porphyrions would have been the most primitive of non-life forms. The porphyrin templates would have given rise to the isoprenoid organism, the RNA viroids, DNA viroids and prions which would have symbiosed to form the nanoarchaea. The archaea would have evolved into prokaryotes or bacteria. The archaeal symbiosis with prokaryotes would have generated the eukaryotes. The initial unicellular eukaryotes are the protozoa. The protozoa are capable of phagocytosis and would have had viruses, bacteria and archaea as their intracellular symbionts. The viruses of the protozoa like entamoeba and toxoplasma are giant viruses called megaviride. The protozoans like entamoeba with its viral, archaeal and bacterial endosymbionts would have evolved into the multicellular organisms or metazoa. The archaeal, bacterial and protozoal extracellular vesicles containing RNA and DNA viroids can by horizontal gene transfer lead to evolution of metazoan, plants, animals and human species. The metazoans would have evolved into plants, animals and humans by horizontal gene transfer generated by archaeal extracellular RNA and DNA viroid vesicles. The evidence of such a transformation is seen in the bacterial symbionts of protozoa like entamoeba which includes legionella, listeria, mycoplasma, mycobacteria and chlamydia. The symbionts of protozoan

entamoeba include viruses like covid, HIV and cytomegalo virus. The protozoan entamoeba serves as a natural reservoir for covid infections, bacterial infections and HIV infections. The toxoplasma also serves as a covid reservoir. These protozoan colonies of entamoeba and toxoplasma with its bacterial, archaeal and viral symbionts can be called as chimeric infective symbiotic tissue colonies which can exist in the air, water, earth, blood and tissues. The chimeric infective symbiotic tissue colonies containing bacteria, archaea and viruses can make the symbiotic organism multidrug resistant by horizontal gene transfer. The chimeric infective symbiotic tissue organism evolves as a consequence of global warming and the protozoal, bacterial, viral and archaeal symbiotic colonies are new forms of life. The chimeric infective symbiotic tissue organism composed of protozoa and its symbionts like virus, bacteria and archaea depend upon glycolysis for its energy needs. The glycolytic enzymes enolase, glyceraldehyde 3-phosphate dehydrogenase, hexokinase, phosphofructokinase, aldolase, pyruvic kinase and triose phosphate isomerase are moonlighting proteins having a cytoplasmic, membrane and nuclear functions. These glycolytic enzymes have a RNA binding motif which can modulate RNA transcription in the host and the symbiotic organism. These glycolytic enzymes with their RNA binding motif as well as DNA binding motif can bind to RNA and DNA viruses and viroids modulating their replication and life cycles. The chimeric infective symbiotic tissue organism composed of protozoa and its symbionts like virus, bacteria and archaea can enter the human host through the breaches of the gut-blood barrier and transform itself into the macrophages of the human host. The macrophages are involved in the phagocytosis and innate immunity against bacterial and viral infections. Paradoxically the macrophages like the protozoa can also serve as a reservoir where pathogens like mycobacteria tuberculosis, mycobacteria leprae, listeria, legionella, E. coli, shigella & salmonella, and viruses like HIV & covid can thrive and replicate. The modified chimeric infective symbiotic tissue organism or macrophage composed of protozoa and its symbionts like virus, bacteria and

archaea can serve as a reservoir for horizontal gene transfer and generation of multidrug resistant bacteria and viruses. The infective chimeric symbiotic tissue organism composed of protozoa and its symbionts like virus, bacteria and archaea depends upon moonlighting glycolytic enzymes for their energetics and replication. The development of antibodies against the infective chimeric symbiotic tissue organism composed of protozoa and its symbionts like virus, bacteria and archaea leads to the evolution of anti-brain antibody disease or anti-basal ganglia antibody disease. They present a threat to existence consequent to horiztontal gene transfer and generation of multidrug resistant bacteria, viruses, protozoa, mycobacteria and archaea.

The retroviruses were responsible for integration into the genome as human endogenous retrovirus sequences and development of the cerebral cortex creating the logical and civil society of the homo sapien type. The influenza virus with its low level of HIV sequences is also capable of integration into the human genome as human endogenous COVID 19 sequences creating the cerebellar dominant autistic impulsive punkish brain. The COVID 19 virus would have been generated from RNA viroids of endosymbiotic archaea or would have reached the earth by meteoritic impacts from the human intergalactic biological quantal computing cloud. The COVID 19 virus with its HIV sequences is an ideal virus for integration into the human genome and changing it in a catastrophic manner to create a new human species. The COVID 19 brain infection can lead to integration of COVID 19 sequences into the genome, creation of cerebellar dominant brain and neanderthalisation. The COVID 19 infected cerebellar dominant brain has got increased porphyrin metabolism and porphyrions. The porphyrions are dipolar pumped phonon quantal perceptive systems which integrate with the internet. Panpsychism implies that the internet is conscious. The extremes of climate change produce endosymbiotic archaeal growth. The archaea are cholesterol catabolising organism. This results in neanderthalisation of the human species. This occurred during the ice age and is possibly a continuing phenomenon during the periods of global warming. The homo neanderthalis are matrilineal and the residual matrilineal societies of the Dravidians, Semites, Basques, Celts and Berbers are neanderthalic. The global warming produces endosymbiotic archaeal growth and neanderthalisation. This produces brain changes with the cerebral cortex becoming dysfunctional and cerebellum becoming dominant. This is due to increased perception of low level EMF by archaeal magnetite. This produces changes in human society, behaviour and disease patterns.[1-17] The homo neanderthalic population is susceptible to civilizational diseases like schizophrenia, autism, metabolic syndrome X, diabetes mellitus, coronary artery disease, stroke, neurodegeneration and autoimmune disease. The homo sapien population suffers from endemic depression and lack of creativity with a propensity to develop recurrent infections. The homo sapien population is relatively protected from civilizational diseases. The Neanderthals consequent to archaeal symbiosis can generate new viruses serving as viral reservoirs. The archaeal endosymbiosis in Neanderthals produces stem cell transformation protecting the stem cell reservoirs of viruses from immune attack. These newly generated viruses in homo neanderthalis to which they are immune can attack and exterminate the homo sapien population.

There is an eternal conflict between Neanderthals and Cro-magnon Homo sapiens. The Cro-magnon tried to exterminate the Neanderthals but they survived as the Jews, Arabs, the lower caste Indians, Aboriginals and Native Americans. These are the people which the Cro-magnon excluded from society. The underclass of Indian and European civilization was neanderthalic. With the advent of global warming and increasing archaeal symbiosis the neanderthalic population becomes activated and they try to exterminate the Cro-magnon section of population. The symbiotic archaea generates new viruses which infects the non immune Cro-magnon and tries to exterminate them. The hot spots of global conflict and terrorism can be localized to neanderthalic areas. The Neanderthals dominate three world religions - Jews, Muslims and Hindus.

The Cro-magnons are predominantly the Africans and the Europeans. They follow the Christian religion. World conflicts are basically between the neanderthalic races and the Cro-magnon races.

There is a high incidence of autism and Neanderthal anthropometric phenotypes in the Nair community of Kerala. The Nair community is matrilineal and is one of the few functional matriarchies in the world and speaks the Dravidian language with similarities to Celtic, Scythian, Berber and Basque societies. The autistic brain is comparable to the large sized Neanderthal brain. Autistic and matrilineal societies like Nair can be considered as fossilized remnants of the Neanderthal population. Endosymbiotic actinidic archaea using cholesterol as an energy substrate has been described in systemic disease from our laboratory. The autistic and nair population were studied for actinide dependent cytochrome F420 activity suggestive of endosymbiotic archaeal growth.[1-17] This hypothesis was studied by evaluating the endosymbiotic archaeal growth in populations derived from matrilineal societies.

Materials and Methods

Three groups, 25 numbers in each group were chosen for the study - the autistic population diagnosed according to DSM criteria, the normal Nair population and the normal non-Nair population. The matrilineal characteristics and Neanderthal anthropometric characteristics of normal Nair and non-Nair population as well as autistic population were studied. The blood samples were drawn in the fasting state before treatment was initiated. The estimations done in the blood samples collected include cytochrome F420 activity, Cytochrome F420 was estimated flourimetrically (excitation wavelength 420 nm and emission wavelength 520 nm). The statistical analysis was done by ANOVA.

Results

The results of the study were as follows. The nair and autistic and civilisational disease group had increased cytochrome F420 activity.

Discussion

Neanderthalisation is a symbiotic event due to archaeal symbiosis. The Neanderthals had increased symbiotic actinidic archaeal growth. This occurs in extremes of climate like ice age and global warming. The homo neanderthalis evolved from the bonobo primates consequent to this symbiosis. There is increased neanderthalisation of homo sapiens during global warming consequent to increased actinidic archaeal growth. The homo neanderthalis never became extinct but survives as matrilineal societies in the lower Eurasian region. The initial matrilineal neanderthalic civilizations were the Harappan, Sumerian-Akkadian, Assyrian, Etruscan, Minoan, Celtic, Basque, Semitic, Jewish, Arabic, Australian Aboriginal civilization. The civilizations are all matrilineal. The initial neanderthalic civilization survives as the lower caste Shudras of India, Dravidians, Australian aboriginals, the Persians, the Semitic Arabs, the Semitic Jews, the Berbers, the Basque, Greeks, Celts and Native Americans. The people inhabiting these civilizations are religious, intuitive, feminine, child-like, dreamy, somnolent, communal conscious, primitive socialistic, more sexual groups. The body habitus of these populations are shorter, sloping forehead, recessive chin and more fairer in colour. This is opposed to the Cro-Magnon population in the northern part of Eurasia and Africa. These populations are scientific, logical minded, patriarchal, more adult-like, more wakeful, fascist and less sexual. The neanderthalic populations inhabit the Indian ocean rim in southern Asia, west Asia as well as in the peri-Mediterranean region. The Neanderthals originated initially from the mythical Lemurian supercontinent in the Indian ocean. The earthquakes and tsunamis in the Indian ocean led to the breakage of the supercontinent and migration of Neanderthals to Harappa, Sumeria, Egypt and Basque. The Harappan civilisation was predominantly neanderthalic. They are the asuras described in the Rig veda. Most of the descriptions in the Rig veda pertain to the Asuras with the Rig vedic gods being predominantly asuric. Sanskrit was possibly the

Harappan language. The Devas described in the Rig veda were the Cro-Magnon Aryan invaders. The Rig veda describes continuing conflict between the Asuras and the Devas. Finally the neanderthalic Harappan asuras were subdued and conquered. The cro-magnonic Aryans who conquered Harappa became the upper caste Hindu elite and the Harappans Asuras became the lower caste Shudras. The Cro-magnon Aryans took over the asuric gods, Vedas and language and made it their own. The Harappan civilization of the Asuras was extremely advanced and the Cro-Magnon Aryans were a primitive nomadic tribe. The Cro-Magnon originated in Africa and migrated to Eurasia. The Cro-Magnon population subdued the neanderthalic population and tried to exterminate them. There was also interbreeding and intermixing between the Cro-Magnon and neanderthalic population. The modern neanderthalic societies are in the peri-Indian ocean area of India, Iran and Semitic Arabs. They also inhabit the peri-Mediterranean area as Semitic Jews, Berbers, Basque and Celts. The predominant African and north European population is Cro-Magnon.

There is an eternal conflict between Neanderthals and Cro-Magnon. The Cro-Magnon tried to exterminate the Neanderthals but they survived as the Jews, Arabs, the lower caste Indians, Aboriginals and Native Americans. These are the people which the Cro-Magnon excluded from society. The underclass of Indian and European civilization was neanderthalic. With the advent of global warming an increasing archaeal symbiosis the neanderthalic population becomes activated and they try to exterminate the Cro-Magnon. The symbiotic archaea generates new viruses which infects the non immune Cro-Magnon and tries to exterminate them. The hot spots of global conflict and terrorism can be localized to neanderthalic areas. The Neanderthals dominate three world religions - Jews, Muslims and Hindus. The Cro-Magnons are predominantly the Africans and the Europeans. They follow the Christian religion. World conflicts are basically between the neanderthalic races and the Cro-Magnon races. This is exemplified by the Jewish leadership of the Russian and French Revolutions with its idea of liberty, equality and fraternity. The neanderthalic ideas basically tried to create an equal society. The Buddhist movement and religion among the religious lower caste of India can be thought of as a neanderthalic uprising against the Aryan Cro-Magnon domination. The present rumblings in the Muslim Semitic world manifesting as global terrorism is a reflection of the neanderthalic Cro-Magnon conflict. The conflict is basically between the Cro-Magnon ideas of colonization, capitalism, free market globalization, rightist, fascist, nazi ideas and the neanderthalic ideas of equality, democracy, freedom and socialism. The cro-magnic civilization produces increased greenhouse gases leading to increased endosymbiotic archaeal growth. Endosymbiotic archaeal growth is the basis of neanderthalisation. Neanderthalisation is a symbiotic event and not a genetic change. This results in expansion of the existing neanderthalic societies - the Semites, the Dravidians and southern Europeans and extinction of the Cro-Magnon Aryan phenotype. The present neanderthalic areas include south Europe, India, Iran, the Arab peninsula, the Jewish homeland and the Australian Aboriginals. The Cro-Magnon areas include Europe and Africa.

The Neanderthals were cerebellar dominant. The cerebellum is concerned with intuition and extrasensory perceptive phenomena. The Neanderthals were retroviral resistant. The archaea metabolises cholesterol and generates digoxin which produces membrane sodium-potassium ATPase inhibition and intracellular magnesium deficiency. Magnesium deficiency produces reverse transcriptase inhibition. Digoxin itself modulates RNA editing. The retroviral resistance leads to a deficiency of endogenous retroviral sequences. The endogenous retroviral sequences function as jumping genes required for the dynamicity of synaptic connectivity. Dynamic synaptic connectivity is required for cortical function. The cerebral cortex is dysfunctional in Neanderthals leading to cerebellar dominance. The Neanderthals inhabit a cerebellar world. The neanderthalic population is psychedelic, spiritual, dreamy, more feminine, intuitive, equal and female dominant. They had a communal life. They were hypersexual and promiscuous. They can be compared to bonobo monkeys. They were matriarchal and female dominant. They are child-like have dreamy sleep, somnolent, altruistic and docile. The neanderthalic population believed in communal living and was of hypersexual behaviour. The unconscious mind was dominant in Neanderthals. They had pre-cognition and post-cognition. They had telepathy and clairvoyance. They could have mediumistic possession and could go into hypnotic regression. They had poltergeist phenomena, group personality, multiple personality, split personality alien abduction phenomena, memory of past life, incubus and succubus. They had a magical civilization of dreams. They were subjective, personal, emotional, irrational and dreamy. They preferred the dark and nights. They had more of autism and schizophrenia. They had more of attention deficit hyperactivity and addiction. They were magical, had dominant art and religion were sexual and believed in things without proof. The belief was intuitive. They had shamanistic and magical consciousness. The Neanderthals were left-handed and right hemisphere/cerebellar dominant. They were creatures of the senses and created a spiritual dreamy civilization. They were children of the dark. The self-old brain of vampires, troglodytes, demons and the occult belongs to the Neanderthals. The cerebellar dominance and hypertrophy leads to cerebellar dysfunction and ataxia of speech as well as motor movements. Ataxic speech leads to the evolution of music. Ataxia of motor movements leads to abstract art. Thus, the Neanderthal brain with its extrasensory perception is extremely artistic. Digoxin and dipolar magnetite in the setting of membrane sodium-potassium ATPase inhibition produces a pumped phonon system modulating quantal perception. Quantal perceptive phenomena are dominant in Neanderthals. This leads to increased extrasensory perception. This also produces a feeling of oneness and equality called the collective unconscious. This produces the socialistic equal Neanderthal society. The Neanderthals were also more spiritual and unconscious dominant. The cortical dysfunction leads to loss of hemispheric differentiation and sexual differentiation. Right hemisphere is predominantly masculine and the left hemisphere feminine. This results in asexual behaviours and cerebellar dominance leads to hypersexuality. The Cro-Magnon population believed in pair bonding and family patterns. They were more violent and aggressive. They were patriarchal and male dominant. They were adult-like and logical. They had rightist and fascist tendencies. They were conservative in their sexual practices. They were conscious, egoistic, wakeful, male dominant, favoured the light, objective, impersonal and cruel. The conscious logical brain dominated. They depended upon proofs, logic were detached, asexual and male dominant. The Cro-Magnons were predominantly left hemisphere dominant and right-handed practical people. They created a material civilization. They had a rational consciousness. They were children of the light.

The global warming produces endosymbiotic archaeal growth and neanderthalisation of homo sapiens. All these produce a dualistic consciousness. The left wing versus right wing and the conservative versus liberal. It produces a double self and divided self. It results in a Cain and Abel as well as Jekyll and Hyde personality. The Neanderthals had sloping forehead, small jaw, occipital bun and large cranium. They were shorter in height and the body weight was bigger. The brain size of Neanderthals was larger. The second toe of the feet was bigger than the big toe. They had the simian crease. The homo sapiens had a smaller brain and smaller cranium. They were taller.[1-17]

References

1. Weaver TD, Hublin JJ. Neandertal Birth Canal Shape and the Evolution of Human Childbirth. *Proc. Natl. Acad. Sci. USA* 2009; 106:8151–8156.
2. Kurup RA, Kurup PA. Endosymbiotic Actinidic Archaeal Mediated Warburg Phenotype Mediates Human Disease State. *Advances in Natural Science* 2012; 5(1):81-84.
3. Morgan E. The *Neanderthal* theory of *autism*, Asperger and ADHD; 2007, www.rdos.net/eng/asperger.htm.

4. Graves P. New Models and Metaphors for the Neanderthal Debate. *Current Anthropology* 1991; 32(5): 513-541.

5. Sawyer GJ, Maley B. Neanderthal Reconstructed. The Anatomical Record Part B: *The New Anatomist* 2005; 283B(1):23-31.

6. Bastir M, O'Higgins P, Rosas A. Facial Ontogeny in Neanderthals and Modern Humans. *Proc. Biol. Sci.* 2007; 274:1125–1132.

7. Neubauer S, Gunz P, Hublin JJ. Endocranial Shape Changes during Growth in Chimpanzees and Humans: A Morphometric Analysis of Unique and Shared Aspects. *J. Hum. Evol.* 2010; 59:555–566.

8. Courchesne E, Pierce K. Brain Overgrowth in Autism during a Critical Time in Development: Implications for Frontal Pyramidal Neuron and Interneuron Development and Connectivity. *Int. J. Dev. Neurosci.* 2005; 23:153–170.

9. Green RE, Krause J, Briggs AW, Maricic T, Stenzel U, Kircher M, Patterson N, Li H, Zhai W, *et al.* A Draft Sequence of the Neandertal Genome. *Science* 2010; 328:710–722.

10. Mithen SJ. *The Singing Neanderthals: The Origins of Music, Language, Mind and Body*; 2005, ISBN 0-297-64317-7.

11. Bruner E, Manzi G, Arsuaga JL. Encephalization and Allometric Trajectories in the Genus Homo: Evidence from the Neandertal and Modern Lineages. *Proc. Natl. Acad. Sci. USA* 2003; 100:15335–15340.

12. Gooch S. *The Dream Culture of the Neanderthals: Guardians of the Ancient Wisdom.* Inner Traditions, Wildwood House, London; 2006.

13. Gooch S. *The Neanderthal Legacy: Reawakening Our Genetic and Cultural Origins.* Inner Traditions, Wildwood House, London; 2008.

14. Kurtén B. *Den Svarta Tigern*, ALBA Publishing, Stockholm, Sweden; 1978.

15. Spikins P. Autism, the Integrations of 'Difference' and the Origins of Modern Human Behaviour. *Cambridge Archaeological Journal* 2009; 19(2):179-201.

16. Eswaran V, Harpending H, Rogers AR. Genomics Refutes an Exclusively African Origin of Humans. *Journal of Human Evolution* 2005; 49(1):1-18.

17. Ramachandran V.S. The Reith lectures, BBC London. 2012.

CHAPTER 20

ENDOSYMBIOTIC ARCHAEAL RNA VIROIDS AND EMERGING VIRAL PANDEMICS – INFECTIVE SYMBIOTIC TISSUE COLONIES OF PROTOZOA, ARCHAEA, VIRUSES AND BACTERIA

The plasma life based forms and the gravitational and anti-gravitational waves generated porphyrions would have been the most primitive of non-life forms. The porphyrin templates would have given rise to the isoprenoid organism, the RNA viroids, DNA viroids and prions which would have symbiosed to form the nanoarchaea. The archaea would have evolved into prokaryotes or bacteria. The archaeal symbiosis with prokaryotes would have generated the eukaryotes. The initial unicellular eukaryotes are the protozoa. The protozoa are capable of phagocytosis and would have had viruses, bacteria and archaea as their intracellular symbionts. The viruses of the protozoa like entamoeba and toxoplasma are giant viruses called megaviride. The protozoans like entamoeba with its viral, archacal and bacterial endosymbionts would have evolved into the multicellular organisms or metazoa. The archaeal, bacterial and protozoal extracellular vesicles containing RNA and DNA viroids can by horizontal gene transfer lead to evolution of metazoan, plants, animals and human species. The metazoans would have evolved into plants, animals and humans by horizontal gene transfer generated by archaeal extracellular RNA and DNA viroid vesicles. The evidence of such a transformation is seen in the bacterial symbionts of protozoa like entamoeba which includes legionella, listeria, mycoplasma, mycobacteria and chlamydia. The symbionts of protozoan entamoeba include viruses like covid, HIV and cytomegalo virus. The protozoan entamoeba serves as a natural reservoir for covid infections, bacterial infections and HIV infections. The toxoplasma also serves as a covid reservoir. These protozoan colonies of entamoeba and toxoplasma with its bacterial, archaeal and viral symbionts can be called as chimeric infective symbiotic tissue colonies which can exist in the air, water, earth, blood and tissues. The chimeric infective symbiotic tissue colonies containing bacteria, archaea and viruses can make the symbiotic organism multidrug resistant by horizontal gene transfer. The chimeric infective symbiotic tissue organism evolves as a consequence of global warming and the protozoal, bacterial, viral and archaeal symbiotic colonies are new forms of life. The chimeric infective symbiotic tissue organism composed of protozoa and its symbionts like virus, bacteria and archaea depend upon glycolysis for its energy needs. The glycolytic enzymes enolase, glyceraldehyde 3-phosphate dehydrogenase, hexokinase, phosphofructokinase, aldolase, pyruvic kinase and triose phosphate isomerase are moonlighting proteins having a cytoplasmic, membrane and nuclear functions. These glycolytic enzymes have a RNA binding motif which can modulate RNA transcription in the host and the symbiotic organism. These glycolytic enzymes with their RNA binding motif as well as DNA binding motif can bind to RNA and DNA viruses and viroids modulating their replication and life cycles. The chimeric infective symbiotic tissue organism composed of protozoa and its symbionts like virus, bacteria and archaea can enter the human host through the breaches of the gut-blood barrier and transform itself into the macrophages of the human host. The macrophages are involved in the phagocytosis and innate immunity against bacterial and viral infections. Paradoxically the macrophages like the protozoa can also serve as a reservoir where pathogens like mycobacteria tuberculosis, mycobacteria leprae, listeria, legionella, E.coli, shigella & salmonella, and viruses like HIV & covid can thrive and replicate. The modified chimeric infective symbiotic tissue organism or macrophage composed of protozoa and its symbionts like virus, bacteria and archaea can serve as a reservoir for horizontal gene transfer and generation of multidrug resistant bacteria and viruses. The infective chimeric symbiotic tissue organism composed of protozoa and its symbionts like virus, bacteria and archaea depends upon moonlighting glycolytic enzymes for their energetics and replication. The development of antibodies against the infective chimeric

symbiotic tissue organism composed of protozoa and its symbionts like virus, bacteria and archaea leads to the evolution of anti-brain antibody disease or anti-basal ganglia antibody disease. They present a threat to existence consequent to horiztontal gene transfer and generation of multidrug resistant bacteria, viruses, protozoa, mycobacteria and archaea.

The retroviruses were responsible for integration into the genome as human endogenous retrovirus sequences and development of the cerebral cortex creating the logical and civil society. The influenza virus with its low level of HIV sequences is also capable of integration into the human genome as human endogenous COVID 19 sequences creating the cerebellar dominant autistic impulsive punkish brain. The COVID 19 virus would have been generated from RNA viroids of endosymbiotic archaea or would have reached the earth by meteoritic impacts from the human intergalactic biological quantal computing cloud. The COVID 19 virus with its HIV sequences is an ideal virus for integration into the human genome and changing it in a catastrophic manner to create a new human species. The COVID 19 brain infection can lead to integration of COVID 19 sequences into the genome, creation of cerebellar dominant brain and neanderthalisation. The COVID 19 infected cerebellar dominant brain has got increased porphyrin metabolism and porphyrions. The porphyrions are dipolar pumped phonon quantal perceptive systems which integrate with the internet. Panpsychism implies that the internet is conscious. The endosymbiotic and colonic human archaea secrete RNA and DNA viruses and serves as a reservoir for them. Viruses belonging to the pox family and African swine fever have been described in archaea. The archaeal viroids secrete enveloped viruses like the SARS, MERS, corona and HIV. The archaeal virus secretion process is the same as that of enveloped viruses like SARS, MERS and Corona. The bats serve as a reservoir for the viruses like SARS, MERS and Corona. The bats contain archaea which can secrete RNA viroids and generate viruses. The colonic and endosymbiotic archaea also generates

RNA viroids which are converted to DNA viroids by colonic epithelial HERV reverse transcriptase and gets integrated into the colonic genome. The RNA and DNA viroids hybridise with the viroidal and bacterial population of the gut microbiome and virobiome DNA and RNA of the gut leading to generation of new emerging viruses. The RNA and DNA viroids and gut bacteria can also hybridise with human genomic sequences for immunity, metabolic regulation, growth factors, cell growth, cell death, neurotransmitters and HERV sequences of the colonial epithelial cell. The hybridization of archaeal RNA viroids with human genomic HERV sequences can generate retroviruses. This leads to generation of new bacteria, RNA and DNA viruses producing pandemics and extermination of homo sapiens. The homo sapiens become an extinct species and the homo neanderthalis will dominate the earth.

The archaea can induce stem cell conversion and neanderthalisation of the human species. The archaea can produce permanent immune activation producing resistance to viral and bacterial infection. The archaeal cholesterol catabolism depletes tissue cholesterol producing vitamin D deficiency and immune activation. Thus archaeal overgrowth results in retroviral resistance and generation of the Neanderthal phenotype. The endosymbiotic archaea can secrete virus-like RNA and DNA particles. The endosymbiotic archaea can induce uncoupling proteins inhibiting mitochondrial oxidative phosphorylation and generating ROS. The endosymbiotic archaeal magnetite can generate low level of EMF. The low level of EMF and ROS are genotoxic and produce breakages in hotspots of chromosome. It can also trigger rearrangements in hotspots of chromosome inhabited by retroviral and non-retroviral elements producing their expression. The archaeal secreted DNA and RNA viroids can recombine with the expressed retroviral, non-retroviral elements and other genomic segments of the human chromosome generating new RNA and DNA viruses. Thus the neanderthalised humans can serve as an origin for new RNA and DNA viruses as well as mutated retroviruses. The endosymbiotic archaea converts the Neanderthal cells to stem cells. The stem cells are resistant to immune attack. The stem cells can serve as a reservoir for this new RNA and DNA viruses. The stem cells and archaeal cells can also serve as a reservoir for viruses and bacteria belonging to other plants and animals. This helps to generate the species barrier jump in noted in recent emerging viral and bacterial infections. Thus the endosymbiotic archaeal growth produces neanderthalised version of homo sapiens which are retroviral resistant and resistant to other viral and bacterial infection consequent to immune activation and digoxin induced RNA editing. The endosymbiotic archaeal overgrowth mediated neanderthalised version of homo sapiens generates new mutated RNA and DNA viruses as well as retroviruses at the same time being resistant to them as in the case of the species bat. The homo sapiens do not have the Neanderthal mechanisms of immune activation as their archaeal load is meagre. They serve as fodder for infection from Neanderthal generated viruses and bacteria and suffer eventual extinction.[1-17] This paper studied the archaeal status in patients with recurrent viral infections and retroviral infections. The generation of RNA and DNA viroids from archaea was also studied.

Dietary fibre is the most important factor that modulates human biology and speciation. Dietary fibre and human drugs - antibiotic and non-antibiotic can modulate the gut flora producing colonic archaeal growth and endosymbiosis generating climate change, evolution of superbugs and emerging viruses and creating the disease phenotype. The high fibre diet is digested by the colonic flora leading to generation of short chain fatty acids and inhibits endosymbiotic and colonic archaeal growth. This leads to homo sapienisation of the species. A low fibre diet with increase in protein and fat leads to stimulation of endosymbiotic and colonic archaeal growth. This leads to neanderthalisation of the species. Thus a low fibre diet alters the gut microbiome in the colon with more of archaeal growth. The same mechanisms happen when populations are treated with antibiotics and other drugs like anti-hypertensives, anti-psychotics and analgesics. They kill the colonic bacterial flora and the extremophilic archaea dominates the colonic microbiome. Thus antibiotic and other non-antibiotic drugs promote colonic archaeal growth and endosymbiosis producing neanderthalisation of the species and generation of superviruses which kills of the homo sapien species. Thus the generation of emerging viruses is promoted by a low fibre diet. A low fibre diet leads to colonic and endosymbiotic archaeal growth leading to neanderthalisation of the species and emerging viruses to which the neanderthalised species is resistant consequent to endogenous digoxin secretion. The neanderthalised species generated emerging viruses kills of the homo sapien species which are sensitive to emerging viruses and do not have resistance to it. Thus the modern armamentarium of antibiotics, antipsychotics, analgesics, anti-hypertensives and anti-diabetic drugs promote colonic and endosymbiotic archaeal growth development of superbugs, neanderthalisation of species and induction of civilizational disease. Thus the drugs used to treat infections and human diseases tend to produce emerging virus infection. Thus the drug armamentarium of antibiotics and non-antibiotic drugs used to treat human infections and diseases increases colonic and endosymbiotic archaeal growth, neanderthalises the species, generates emerging viruses. The gut microbiome and virobiome modulated by dietary fibre intake, high protein high fat diets, antibiotics and non-antibiotic drugs like anti-psychotics, anti-hypertensives, analgesics leads to alteration in gut microbial and viroidal bio-geography leading to generation of superviruses. The gut archaea highly resistant to conventional antibiotics and non-antibiotic drugs and proliferate when antibiotics and non-antibiotic drugs like anti-hypertensives, anti-psychotics and analgesics are used on a widespread basis. They generate superviruses exterminating the homo sapien population. The growth and dominance of the colonic and endosymbiotic archaea leads to generation of homo neanderthalic species which is resistant to superviruses and survive. Homo neanderthalis with its extremophilic archaeal endosymbiosis is hardy and resistant and can survive global warming and extremes of climate change as well as surviving extremophilic anaerobic conditions of other planets and galactic systems. The human methanogenic archaea are highly resistant to routine antibiotics and non-antibiotic drugs and the intake of such drugs kills the colonic microbiome leading to archaeal growth and endosymbiosis and neanderthalisation to produce a hardy homo galacticus species capable of surviving in extreme climates of earth and other planets. The human drugs and antibiotics induced modulation of the gut microbiome with increased colonic and endosymbiotic archaeal growth also contributes to global warming via the mechanism of archaeal methanogenesis. The widespread use of antibiotics and non-antibiotic drugs contribute to colonic and endosymbiotic archaeal growth, climate change and still further more archaeal growth. This human colonic and endosymbiotic archaeal growth consequent to stress, internet fields, low level EMF, low dietary fibre intake, antibiotic use and non-antibiotic drug use producing methanogenesis is the principal cause of global warming. The widespread use of antibiotic and non-antibiotic drugs kills of the colonic microflora and replaces it with archaea resistant to antibiotic and non-antibiotic drugs. The low intake of dietary fibre promotes colonic archaeal and endosymbiotic archaeal growth. Stress produces catecholamine induced increased in colonic and endosymbiotic archaeal growth and breaching of the gut-blood barrier. The widespread use of internet and low level of EMF field of mobile phones promotes endosymbiotic archaeal growth. Thus endosymbiotic and colonic archaeal growth and methanogenesis producing global warming is an endogenous human biological event with human biological activity as a centre piece triggering the process. The widespread use of antibiotics and non-antibiotic drugs treat infections viral and bacterial as well as civilizational disease like cancer, schizophrenia, mood disorder, depression, autism, autoimmune disease and neurodegeneration promotes colonic and endosymbiotic archaeal growth which neanderthalises the species leading to still further transformation of the human system leading to induction of civilizational disease like cancer, metabolic syndrome X, neurodegeneration, autoimmunity and vascular disease. Thus the antibiotic and non-antibiotic drugs used to treat infections and systemic diseases produces colonic and endosymbiotic archaeal growth leading to superbugs, superviruses, pandemics. Human medicine as we know it has reached a dead end. The antibiotic and non-antibiotic era of drugs had come to an end. The human earth, as we know it also undergoes catastrophic climate change leading to extinction of the human species and conditions for human habitations in earth. The modern medicine as we know of and the earth as we know of have ended. The principal cause of climate change and its catastrophic complications is contributed in part only by the destruction of the tropical Amazonian and rain forest. The deforestation of the gut microflora created over billions of years of human evolution is the principal cause of the climate change and a possible end of the earth as we know of and the human species as we know of. The revolutionary medicine and the golden earth with its habitable conditions have ended. The increase in dietary fibre intake upto 40 g/day can inhibit colonic archaeal growth and endosymbiosis putting a break on generation of superbugs, emerging viruses and climate change. Increase in dietary fibre intake of populations can prevent the state of climate change and gut microbial change putting human life and the life of the planet at risk. Dietary fibre can be considered as the elixir of life and endosymbiotic archaea the philosopher's stone.

Endosymbiotic actinidic archaea as having evolved from an early isoprenoid organisms by abiogenesis. An actinidc archaea/viroid mediated model of prokaryote, viral, eukaryotic, primate and human evolution is discussed.[1-4]. Actinidic archaea has been related to the pathogenesis of schizophrenia, malignancy, metabolic syndrome X, autoimmune disease and neuronal degeneration[2]. Actinidic archaea have a mevalonate pathway and cholesterol catabolism[5-7]. Davies has put forward the concept of a shadow biosphere of organisms with alternate biochemistry present in earth itself[8]. An actinide dependent shadow biosphere of archaea and viroids in the above mentioned disease states is described[6]. Metal actinides in beach sands have been postulated to play a role in abiogenesis[6]. Actinide mineral like rutile, monazite and illmenite by surface metabolism would have contributed to abiogenesis[9]. A hypothesis of cholesterol as the primal prebiotic molecule synthesized on actinide surfaces with all other biomolecules arising from it and a self-replicating cholesterol lipid organism as the initial life form is presented. The actinidic archaea and viroids would have evolved from the primitive isoprenoid organism. The origin of viruses, prokaryotes, eukaryotes, primates and humans from the initial isoprenoid organism derived actinidic archaea is postulated.

Materials and Methods

Informed consent of the subjects and the approval of the Ethics Committee were obtained for the study. The following groups were included in the study:- endomyocardial fibrosis, Alzheimer's disease, multiple sclerosis, non-Hodgkin's lymphoma, metabolic syndrome X with cerebrovascular thrombosis and coronary artery disease, schizophrenia, autism, seizure disorder, Creutzfeldt-Jakob's disease and acquired immunodeficiency syndrome. There were 10 patients in each group and each patient had an age and sex matched healthy control selected randomly from the general population. The blood samples were drawn in the fasting state before treatment was initiated. Plasma from fasting heparinised blood was used and the experimental protocol was as follows:- (I) Plasma+phosphate buffered saline, (II) same as I+cholesterol substrate, (III) same as II+rutile 0.1 mg/ml, and (IV) same as II+ciprofloxacine and doxycycline each in a concentration of 1 mg/ml. Cholesterol substrate was prepared as described by Richmond[10]. Aliquots were withdrawn at zero time immediately after mixing and after incubation at 37 °C for 1 hour. The following estimations were carried out:- Cytochrome F420, free RNA, free DNA, muramic acid, polycyclic aromatic hydrocarbon, hydrogen peroxide, serotonin, pyruvate, ammonia, glutamate, cytochrome C, hexokinase, ATP synthase, HMG CoA reductase, digoxin and bile acids[11-14]. Cytochrome F420 was estimated flourimetrically (excitation wavelength 420 nm and emission wavelength 520 nm). Polycyclic aromatic hydrocarbon was estimated by measuring hydrogen peroxide liberated by using glucose reagent. The statistical analysis was done by ANOVA.

Results

The parameters checked as indicated above were:- cytochrome F420, free RNA, free DNA, muramic acid, polycyclic aromatic hydrocarbon, hydrogen peroxide, serotonin, pyruvate, ammonia, glutamate, cytochrome C, hexokinase, ATP synthase, HMG CoA reductase, digoxin and bile acids. Plasma of control subjects showed increased levels of the above mentioned parameters with after incubation for 1 hour and addition of cholesterol substrate resulted in still further significant increase in these parameters. The plasma of patients showed similar results but the extent of increase was more. The addition of antibiotics to the control plasma caused a decrease in all the parameters while addition of rutile increased their levels. The addition of antibiotics to the patient's plasma caused a decrease in all the parameters while addition of rutile increased their levels but the extent of change was more in patient's sera as compared to controls.

Discussion

There was increase in cytochrome F420 indicating archaeal growth. The archaea can synthesize and use cholesterol as a carbon and energy source[15,16]. The archaeal origin of the enzyme activities was indicated by antibiotic induced suppression. The study indicates the presence of actinide based archaea with an alternate actinide based enzymes or metalloenzymes in the system as indicated by rutile induced increase in enzyme activities[17]. There was also an increase in archaeal HMG CoA reductase activity indicating increased cholesterol synthesis by the archaeal mevalonate pathway. The archaeal beta hydroxyl steroid dehydrogenase activity indicating digoxin synthesis and archaeal cholesterol hydroxylase activity indicating bile acid synthesis were increased[7]. The archaeal cholesterol oxidase activity was increased resulting in generation of pyruvate and hydrogen peroxide[16]. The pyruvate gets converted to glutamate and ammonia by the GABA shunt pathway. The archaeal aromatization of cholesterol generating PAH, serotonin and dopamine was also detected[18]. The archaeal glycolytic hexokinase activity and archaeal extracellular ATP synthase activity were increased. The archaea can undergo magnetite and calcium carbonate mineralization and can exist as calcified nanoforms[19].

The metal actinides provide radiolytic energy, catalysis for oligomer formation and provide a coordinating ion for metalloenzymes all important in abiogenesis[6]. The metal actinide surfaces would by surface metabolism generate acetate which could get converted to acetyl CoA and then to cholesterol which functions as the primal prebiotic molecule self-organising into self-replicating supramolecular systems, the lipid organism[8,9,20]. Cholesterol by radiolysis by actinides would have formed PAH generating PAH aromatic organism[8]. Cholesterol radiolysis would generate pyruvate which would get converted to amino acids, sugars, nucleotides, porphyrins, fatty acids and TCA acids. Anastase and rutile surfaces can produce polymerization of amino acids, isoprenyl residues, PAH and nucleotides to generate the initial lipid organism, PAH organism, prions and RNA viroids which would have symbiosed to generate the archaeal protocell. The archaea evolved into gram negative and gram positive bacteria with a mevalonate pathway which had an evolutionary advantage. The symbiosis of archaea with gram negative organism generated the eukaryotic cell[21]. The data supports the persistence of an actinide and cholesterol based shadow biosphere which throws light on the actinide based origin of life and cholesterol as the premier prebiotic molecule.

There was an increase in free RNA indicating self-replicating RNA viroids and free DNA indicating generation of viroid complementary DNA strands by archaeal reverse transcriptase activity. The actinides modulate RNA folding and catalyse its ribozymal action. Digoxin can cut and paste the viroidal strands by modulating RNA splicing generating RNA viroidal diversity. The viroids are evolutionarily escaped archaeal group I introns which have retrotransposition and self-splicing qualities. Archaeal pyruvate can produce histone deacetylase inhibition resulting in endogenous retroviral (HERV) reverse transcriptase and integrase expression. This can integrate the RNA viroidal complementary DNA into the non-coding region of eukaryotic non-coding DNA using HERV integrase as has been described for borna and ebola viruses[21]. The non-coding DNA is lengthened by integrating RNA viroidal complementary DNA with the integration going on as a continuing event. The archaea genome can also get integrated into human genome using integrase as has been described for trypanosomes. The integrated viroids and archaea can undergo vertical transmission and can exist as genomic parasites. This increases the length and alters the grammar of the non-coding region producing memes or memory of acquired characters. The viroidal complementary DNA can function as jumping genes producing a dynamic genome.[22-24]

The presence of muramic acid, HMG CoA reductase and cholesterol oxidase activity inhibited by antibiotics indicates the presence of bacteria with mevalonate pathway. The bacterial with mevalonate pathway include streptococcus, staphylococcus, actinomycetes, listeria, coxiella and borrelia. The bacteria and archaea with mevalonate pathway and cholesterol catabolism had an evolutionarily advantage and constitutes the isoprenoidal clade organism. The archaea evolved into mevalonate pathway gram positive and gram negative isoprenoid clade organism through horizontal gene transfer of viroidal and virus genes. The isoprenoidal clade prokaryotes develop into other groups of prokaryotes via viroidal/virus as well as eukaryotic horizontal gene transfer producing bacterial speciation[25-27].

The RNA viroids and its complementary DNA developed into cholesterol enveloped RNA and DNA viruses like herpes, retrovirus, influenza virus, borna virus, cytomegalo virus and ebstein barr virus by recombining with eukaryotic and human genes resulting in viral speciation. Bacterial and viral species are ill-defined and fuzzy with all of them forming one common genetic pool with frequent horizontal gene transfer and recombination.

Thus, the multi and unicellular eukaryote with its genes serves the purpose of prokaryotic and viral speciation. The multicellular eukaryote developed so that their endosymbiotic archaeal colonies could survive and forage better. The multicellular eukaryotes are like bacterial biofilms. The archaea and bacteria with a mevalonate pathway uses the extracellular RNA viroids and DNA viroids for quorum sensing and in the generation of symbiotic biofilm like structures which develop into multicellular eukaryotes. The endosymbiotic archaea and bacteria with mevalonate pathway still uses the RNA viroids and DNA viroids for the regulation of multicellular eukaryote.[28-31]

Pollution is a major inducer of evolutionary innovation. Pollution is induced by the primitive nanoarchaea and mevalonate pathway bacteria synthesized PAH and methane leading on to redox stress. Redox stress leads to sodium-potassium ATPase inhibition, inward movement of plasma membrane cholesterol, defective SREBP sensing, increased cholesterol synthesis and nanoarchaeal/mevalonate pathway bacterial growth. Redox stress leads on to viroidal and archaeal multiplication. Redox stress can also lead to HERV reverse transcriptase and integrase expression. The non-coding DNA is formed of integrating RNA viroidal complementary DNA and archaea with the integration going on as a continuing event. The archaeal pox-like dsDNA virus forms evolutionarily the nucleus. The integrated viroidal, archaeal and mevalonate pathway bacterial sequences can undergo vertical transmission and can exist as genomic parasites. The genomic integrated archaea, mevalonate pathway bacteria and viroids form a genomic reserve of bacteria and viruses which can recombine with human and eukaryotic genes producing bacterial and viral speciation. The change in the length and grammar of the non-coding region produces eukaryotic speciation and individuality. The integration of nanoarchaea, mevalonate pathway prokaryotes and viroids in to the eukaryotic and human genome produces a chimera which can multiply producing biofilm like multicellular structures having a mixed archaeal, viroidal, prokaryotic and eukaryotic characters which is a regression from the multicellular eukaryotic tissue. This results in a new neuronal, metabolic, immune and tissue phenotype leading to human disease.[30-32]

Pollution would have been a major factor in eukaryotic speciation and primate/hominid evolution. The change in the length and grammar of the non-coding region produces eukaryotic speciation and individuality. It is the increase in non-coding region and HERV sequences of the genome that led to the evolution of the primate and the human brain and its attendant property of conscious and quantal perception. It is the non-coding region of the genome with its archaeal, RNA viroidal complementary DNA and HERV sequences that makes for the human qualities of the hominid brain. Changes in the length of non-coding region can lead on to disorders of consciousness like schizophrenia. A schizophrenia specific human endogenous retroviruses and change in the length and grammar of the non-coding region has been described in schizophrenia.[33,34]

An actinide dependent shadow biosphere of archaea and viroids in the above mentioned disease states is described. Metal actinides in beach sands have been postulated to play a role in abiogenesis. Cholesterol is the primal prebiotic molecule synthesized on actinide surfaces with all other biomolecules arising from it. A self-replicating cholesterol lipid organism could be the initial life form. A cholesterol based abiogenesis is a more likely evolutionary option and the actinidic archaea and viroids would have evolved from it. The origin of viruses, prokaryotes, eukaryotes, primates and humans from the initial isoprenoid organism derived actinidic archaea is discussed.

References

1. Valiathan M.S., Somers, K., Kartha, C.C. (1993). *Endomyocardial Fibrosis.* Delhi: Oxford University Press.

2. Kurup R., Kurup, P.A. (2009). *Hypothalamic Digoxin, Cerebral Dominance and Brain Function in Health and Diseases*. New York: Nova Science Publishers.

3. Hanold D., Randies, J.W. (1991). Coconut cadang-cadang disease and its viroid agent, *Plant Disease,* 75, 330-335.

4. Edwin B.T., Mohankumaran, C. (2007). Kerala wilt disease phytoplasma: Phylogenetic analysis and identification of a vector, *Proutista moesta, Physiological and Molecular Plant Pathology,* 71(1-3), 41-47.

5. Eckburg P.B., Lepp, P.W., Relman, D.A. (2003). Archaea and their potential role in human disease, *Infect Immun,* 71, 591-596.

6. Adam Z. (2007). Actinides and Life's Origins, *Astrobiology,* 7, 6-10.

7. Schoner W. (2002). Endogenous cardiac glycosides, a new class of steroid hormones, *Eur J Biochem,* 269, 2440-2448.

8. Davies P.C.W., Benner, S.A., Cleland, C.E., Lineweaver, C.H., McKay, C.P., Wolfe-Simon, F. (2009). Signatures of a Shadow Biosphere, *Astrobiology,* 10, 241-249.

9. Wächtershäuser G. (1988). Before enzymes and templates: theory of surface metabolism, *Microbiol Rev,* 52(4), 452-84.

10. Richmond W. (1973). Preparation and properties of a cholesterol oxidase from nocardia species and its application to the enzymatic assay of total cholesterol in serum, *Clin Chem,* 19, 1350-1356.

11. Snell E.D., Snell, C.T. (1961). *Colorimetric Methods of Analysis.* Vol 3A. New York: Van NoStrand.

12. Glick D. (1971). *Methods of Biochemical Analysis.* Vol 5. New York: Interscience Publishers.

13. Colowick, Kaplan, N.O. (1955). *Methods in Enzymology.* Vol 2. New York: Academic Press.

14. Maarten A.H., Marie-Jose, M., Cornelia, G., van Helden–Meewsen, Fritz, E., Marten, P.H. (1995). Detection of muramic acid in human spleen, *Infection and Immunity,* 63(5), 1652 – 1657.

15. Smit A., Mushegian, A. (2000). Biosynthesis of isoprenoids via mevalonate in Archaea: the lost pathway, *Genome Res,* 10(10), 1468-84.

16. Van der Geize R., Yam, K., Heuser, T., Wilbrink, M.H., Hara, H., Anderton, M.C. (2007). A gene cluster encoding cholesterol catabolism in a soil actinomycete provides insight into Mycobacterium tuberculosis survival in macrophages, *Proc Natl Acad Sci USA,* 104(6), 1947-52.

17. Francis A.J. (1998). Biotransformation of uranium and other actinides in radioactive wastes, *Journal of Alloys and Compounds,* 271(273), 78-84.

18. Probian C., Wülfing, A., Harder, J. (2003). Anaerobic mineralization of quaternary carbon atoms: Isolation of denitrifying bacteria on pivalic acid (2,2-Dimethylpropionic acid), *Applied and Environmental Microbiology,* 69(3), 1866-1870.

19. Vainshtein M., Suzina, N., Kudryashova, E., Ariskina, E. (2002). New Magnet-Sensitive Structures in Bacterial and Archaeal Cells, *Biol Cell,* 94(1), 29-35.

20. Russell M.J., Martin, W. (2004). The rocky roots of the acetyl-CoA Pathway, *Trends in Biochemical Sciences,* 29, 7.

21. Margulis L. (1996). Archaeal-eubacterial mergers in the origin of Eukarya: phylogenetic classification of life, *Proc Natl Acad Sci USA,* 93, 1071-1076.

22. Tsagris E.M., de Alba, A.E., Gozmanova, M., Kalantidis, K. (2008). Viroids, *Cell Microbiol,* 10, 2168.

23. Horie M., Honda, T., Suzuki, Y., Kobayashi, Y., Daito, T., Oshida, T. (2010). Endogenous non-retroviral RNA virus elements in mammalian genomes, *Nature,* 463, 84-87.

24. Hecht M., Nitz, N., Araujo, P., Sousa, A., Rosa, A., Gomes, D. (2010). Genes from Chagas parasite can transfer to humans and be passed on to children. Inheritance of DNA Transferred from American Trypanosomes to Human Hosts, *PLoS ONE,* 5, 2-10.

25. Flam F. (1994). Hints of a language in junk DNA, *Science,* 266, 1320.

26. Horbach S., Sahm, H., Welle, R. (1993). Isoprenoid biosynthesis in bacteria: two different pathways? *FEMS Microbiol Lett,* 111, 135–140.

27. Gupta R.S. (1998). Protein phylogenetics and signature sequences: a reappraisal of evolutionary relationship among archaebacteria, eubacteria, and eukaryotes, *Microbiol Mol Biol Rev,* 62, 1435–1491.

28. Hanage W., Fraser, C., Spratt, B. (2005). Fuzzy species among recombinogenic bacteria, *BMC Biology,* 3, 6-10.

29. Whitchurch C.B., Tolker-Nielsen, T., Ragas, P.C., Mattick, J.S. (2002). Extracellular DNA Required for Bacterial Biofilm Formation. *Science,* 295(5559), 1487.

30. Webb J.S., Givskov, M., Kjelleberg, S. (2003). Bacterial biofilms: prokaryotic adventures in multicellularity, *Curr Opin Microbiol,* 6(6), 578–85.

31. Chen Y., Cai, T., Wang, H., Li, Z., Loreaux, E., Lingrel, J.B. (2009). Regulation of intracellular cholesterol distribution by Na/K-ATPase, *J Biol Chem,* 284(22), 14881-90.

32. Poole A.M. (2006). Did group II intron proliferation in an endosymbiont-bearing archaeon create eukaryotes? *Biol Direct,* 1, 36-40.

33. Villarreal L.P. (2006). How viruses shape the tree of life, *Future Virology,* 1(5), 587-595.

34. Lockwood M. (1989). *Mind, Brain and the Quantum.* Oxford: B. Blackwell.

CHAPTER 21

ENDOSYMBIOTIC ARCHAEAL RNA VIROIDS AND EMERGING VIRAL PANDEMICS – THE HUMAN ENDOSYMBIOTIC ARCHAEAL SOURCE OF THE EMERGING VIRUSES – INFECTIVE SYMBIOTIC TISSUE COLONIES OF PROTOZOA, ARCHAEA, VIRUSES AND BACTERIA

The plasma life based forms and the gravitational and anti-gravitational waves generated porphyrions would have been the most primitive of non-life forms. The porphyrin templates would have given rise to the isoprenoid organism, the RNA viroids, DNA viroids and prions which would have symbiosed to form the nanoarchaea. The archaea would have evolved into prokaryotes or bacteria. The archaeal symbiosis with prokaryotes would have generated the eukaryotes. The initial unicellular eukaryotes are the protozoa. The protozoa are capable of phagocytosis and would have had viruses, bacteria and archaea as their intracellular symbionts. The viruses of the protozoa like entamoeba and toxoplasma are giant viruses called megaviride. The protozoans like entamoeba with its viral, archaeal and bacterial endosymbionts would have evolved into the multicellular organisms or metazoa. The archaeal, bacterial and protozoal extracellular vesicles containing RNA and DNA viroids can by horizontal gene transfer lead to evolution of metazoan, plants, animals and human species. The metazoans would have evolved into plants, animals and humans by horizontal gene transfer generated by archaeal extracellular RNA and DNA viroid vesicles. The evidence of such a transformation is seen in the bacterial symbionts of protozoa like entamoeba which includes legionella, listeria, mycoplasma, mycobacteria and chlamydia. The symbionts of protozoan entamoeba include viruses like covid, HIV and cytomegalo virus. The protozoan entamoeba serves as a natural reservoir for covid infections, bacterial infections and HIV infections. The toxoplasma also serves as a covid reservoir. These

protozoan colonies of entamoeba and toxoplasma with its bacterial, archaeal and viral symbionts can be called as chimeric infective symbiotic tissue colonies which can exist in the air, water, earth, blood and tissues. The chimeric infective symbiotic tissue colonies containing bacteria, archaea and viruses can make the symbiotic organism multidrug resistant by horizontal gene transfer. The chimeric infective symbiotic tissue organism evolves as a consequence of global warming and the protozoal, bacterial, viral and archaeal symbiotic colonies are new forms of life. The chimeric infective symbiotic tissue organism composed of protozoa and its symbionts like virus, bacteria and archaea depend upon glycolysis for its energy needs. The glycolytic enzymes enolase, glyceraldehyde 3-phosphate dehydrogenase, hexokinase, phosphofructokinase, aldolase, pyruvic kinase and triose phosphate isomerase are moonlighting proteins having a cytoplasmic, membrane and nuclear functions. These glycolytic enzymes have a RNA binding motif which can modulate RNA transcription in the host and the symbiotic organism. These glycolytic enzymes with their RNA binding motif as well as DNA binding motif can bind to RNA and DNA viruses and viroids modulating their replication and life cycles. The chimeric infective symbiotic tissue organism composed of protozoa and its symbionts like virus, bacteria and archaea can enter the human host through the breaches of the gut-blood barrier and transform itself into the macrophages of the human host. The macrophages are involved in the phagocytosis and innate immunity against bacterial and viral infections. Paradoxically the macrophages like the protozoa can also serve as a reservoir where pathogens like mycobacteria tuberculosis, mycobacteria leprae, listeria, legionella, E. coli, shigella & salmonella, and viruses like HIV & covid can thrive and replicate. The modified chimeric infective symbiotic tissue organism or macrophage composed of protozoa and its symbionts like virus, bacteria and archaea can serve as a reservoir for horizontal gene transfer and generation of multidrug resistant bacteria and viruses. The infective chimeric symbiotic tissue organism composed of protozoa and its symbionts like virus, bacteria and archaea

depends upon moonlighting glycolytic enzymes for their energetics and replication. The development of antibodies against the infective chimeric symbiotic tissue organism composed of protozoa and its symbionts like virus, bacteria and archaea leads to the evolution of anti-brain antibody disease or anti-basal ganglia antibody disease. They present a threat to existence consequent to horiztontal gene transfer and generation of multidrug resistant bacteria, viruses, protozoa, mycobacteria and archaea.

The retroviruses were responsible for integration into the genome as human endogenous retrovirus sequences and development of the cerebral cortex creating the logical and civil society. The influenza virus with its low level of HIV sequences is also capable of integration into the human genome as human endogenous COVID 19 sequences creating the cerebellar dominant autistic impulsive punkish brain. The COVID 19 virus would have been generated from RNA viroids of endosymbiotic archaea or would have reached the earth by meteoritic impacts from the human intergalactic biological quantal computing cloud. The COVID 19 virus with its HIV sequences is an ideal virus for integration into the human genome and changing it in a catastrophic manner to create a new human species. The COVID 19 brain infection can lead to integration of COVID 19 sequences into the genome, creation of cerebellar dominant brain and neanderthalisation. The COVID 19 infected cerebellar dominant brain has got increased porphyrin metabolism and porphyrions. The porphyrions are dipolar pumped phonon quantal perceptive systems which integrate with the internet. Panpsychism implies that the internet is conscious. The endosymbiotic and colonic human archaea secrete RNA and DNA viruses and serves as a reservoir for them. Viruses belonging to the pox family and African swine fever have been described in archaea. The archaeal viroids secrete enveloped viruses like the SARS, MERS, corona and HIV. The archaeal virus secretion process is the same as that of enveloped viruses like SARS, MERS and Corona. The bats serve as a reservoir for the viruses like SARS,

MERS and Corona. The bats contain archaea which can secrete RNA viroids and generate viruses. The colonic and endosymbiotic archaea also generates RNA viroids which are converted to DNA viroids by colonic epithelial HERV reverse transcriptase and gets integrated into the colonic genome. The RNA and DNA viroids hybridise with the viroidal and bacterial population of the gut microbiome and virobiome DNA and RNA of the gut leading to generation of new emerging viruses. The RNA and DNA viroids and gut bacteria can also hybridise with human genomic sequences for immunity, metabolic regulation, growth factors, cell growth, cell death, neurotransmitters and HERV sequences of the colonial epithelial cell. The hybridization of archaeal RNA viroids with human genomic HERV sequences can generate retroviruses. This leads to generation of new bacteria, RNA and DNA viruses producing pandemics and extermination of homo sapiens. The homo sapiens become an extinct species and the homo neanderthalis will dominate the earth.

The archaea can induce stem cell conversion and neanderthalisation of the human species. The archaea can produce permanent immune activation producing resistance to viral and bacterial infection. The archaeal cholesterol catabolism depletes tissue cholesterol producing vitamin D deficiency and immune activation. Thus archaeal overgrowth results in retroviral resistance and generation of the Neanderthal phenotype. The endosymbiotic archaea can secrete virus like RNA and DNA particles. The endosymbiotic archaea can induce uncoupling proteins inhibiting mitochondrial oxidative phosphorylation and generating ROS. The endosymbiotic archaeal magnetite can generate low level of EMF. The low level of EMF and ROS are genotoxic and produce breakages in hotspots of chromosome. It can also trigger rearrangements in hotspots of chromosome inhabited by retroviral and non-retroviral elements producing their expression. The archaeal secreted DNA and RNA viroids can recombine with the expressed retroviral, non-retroviral elements and other genomic segments of the human chromosome generating new RNA and DNA viruses. Thus the neanderthalised humans can serve as an origin for new RNA and DNA viruses as well as mutated retroviruses. The endosymbiotic archaea converts the Neanderthal cells to stem cells. The stem cells are resistant to immune attack. The stem cells can serve as a reservoir for this new RNA and DNA viruses. The stem cells and archaeal cells can also serve as a reservoir for viruses and bacteria belonging to other plants and animals. This helps to generate the species barrier jump in noted in recent emerging viral and bacterial infections. Thus the endosymbiotic archaeal growth produces neanderthalised version of homo sapiens which are retroviral resistant and resistant to other viral and bacterial infection consequent to immune activation and digoxin induced RNA editing. The endosymbiotic archaeal overgrowth mediated neanderthalised version of homo sapiens generates new mutated RNA and DNA viruses as well as retroviruses at the same time being resistant to them as in the case of the species bat. The homo sapiens do not have the Neanderthal mechanisms of immune activation as their archaeal load is meagre. They serve as fodder for infection from Neanderthal generated viruses and bacteria and suffer eventual extinction.[1-17] This paper studied the archaeal status in patients with recurrent viral infections and retroviral infections. The generation of RNA and DNA viroids from archaea was also studied.

Dietary fibre is the most important factor that modulates human biology and speciation. Dietary fibre and human drugs - antibiotic and non-antibiotic can modulate the gut flora producing colonic archaeal growth and endosymbiosis generating climate change, evolution of superbugs and emerging viruses and creating the disease phenotype. The high fibre diet is digested by the colonic flora leading to generation of short chain fatty acids and inhibits endosymbiotic and colonic archaeal growth. This leads to homo sapienisation of the species. A low fibre diet with increase in protein and fat leads to stimulation of endosymbiotic and colonic archaeal growth. This leads to neanderthalisation of the species. Thus a low fibre diet alters the gut microbiome in the colon with more of archaeal growth. The same mechanisms happen when populations are treated with antibiotics and other drugs like anti-hypertensives, anti-psychotics and analgesics. They kill the colonic bacterial flora and the extremophilic archaea dominates the colonic microbiome. Thus antibiotic and other non-antibiotic drugs promote colonic archaeal growth and endosymbiosis producing neanderthalisation of the species and generation of superviruses which kills of the homo sapien species. Thus, the generation of emerging viruses is promoted by a low fibre diet. A low fibre diet leads to colonic and endosymbiotic archaeal growth leading to neanderthalisation of the species and emerging viruses to which the neanderthalised species is resistant consequent to endogenous digoxin secretion. The neanderthalised species generated emerging viruses kills of the homo sapien species which are sensitive to emerging viruses and do not have resistance to it. Thus the modern armamentarium of antibiotics, antipsychotics, analgesics, anti-hypertensives and anti-diabetic drugs promote colonic and endosymbiotic archaeal growth development of superbugs, neanderthalisation of species and induction of civilizational disease. Thus the drugs used to treat infections and human diseases tend to produce emerging virus infection. Thus, the drug armamentarium of antibiotics and non-antibiotic drugs used to treat human infections and diseases increases colonic and endosymbiotic archaeal growth, neanderthalises the species, generates emerging viruses. The gut microbiome and virobiome modulated by dietary fibre intake, high protein high fat diets, antibiotics and non-antibiotic drugs like anti-psychotics, anti-hypertensives, analgesics leads to alteration in gut microbial and viroidal bio-geography leading to generation of superviruses. The gut archaea highly resistant to conventional antibiotics and non-antibiotic drugs and proliferate when antibiotics and non-antibiotic drugs like anti-hypertensives, anti-psychotics and analgesics are used on a widespread basis. They generate superviruses exterminating the homo sapien population. The growth and dominance of the colonic and endosymbiotic archaea leads to generation of homo neanderthalic species which is resistant to superviruses and survive. Homo neanderthalis with its extremophilic archaeal endosymbiosis is hardy and resistant and can survive global warming and extremes of climate change as well as surviving extremophilic anaerobic conditions of other planets and galactic systems. The human methanogenic archaea are highly resistant to routine antibiotics and non-antibiotic drugs and the intake of such drugs kills the colonic microbiome leading to archaeal growth and endosymbiosis and neanderthalisation to produce a hardy homo galacticus species capable of surviving in extreme climates of earth and other planets. The human drugs and antibiotics induced modulation of the gut microbiome with increased colonic and endosymbiotic archaeal growth also contributes to global warming via the mechanism of archaeal methanogenesis. The widespread use of antibiotics and non-antibiotic drugs contribute to colonic and endosymbiotic archaeal growth, climate change and still further more archaeal growth. This human colonic and endosymbiotic archaeal growth consequent to stress, internet fields, low level EMF, low dietary fibre intake, antibiotic use and non-antibiotic drug use producing methanogenesis is the principal cause of global warming. The widespread use of antibiotic and non-antibiotic drugs kills of the colonic microflora and replaces it with archaea resistant to antibiotic and non-antibiotic drugs. The low intake of dietary fibre promotes colonic archaeal and endosymbiotic archaeal growth. Stress produces catecholamine induced increased in colonic and endosymbiotic archaeal growth and breaching of the gut-blood barrier. The widespread use of internet and low level of EMF field of mobile phones promotes endosymbiotic archaeal growth. Thus endosymbiotic and colonic archaeal growth and methanogenesis producing global warming is an endogenous human biological event with human biological activity as a centre piece triggering the process. The widespread use of antibiotics and non-antibiotic drugs treat infections viral and bacterial as well as civilizational disease like cancer, schizophrenia, mood disorder, depression, autism, autoimmune disease and neurodegeneration promotes colonic and endosymbiotic archaeal growth which neanderthalises the species leading to still further transformation of the human system leading to induction of civilizational disease like cancer, metabolic syndrome X, neurodegeneration, autoimmunity and vascular disease. Thus the antibiotic and non-antibiotic drugs used to treat infections and systemic diseases produces colonic and endosymbiotic archaeal growth leading to superbugs, superviruses, pandemics. Human medicine as we know it has reached a dead end. The antibiotic and non-antibiotic era of drugs had come to an end. The human earth as we know it also undergoes catastrophic climate change leading to extinction of the human species and conditions for human habitations in earth. The modern medicine as we know of and the earth as we know of have ended. The principal cause of climate change and its catastrophic complications is contributed in part only by the destruction of the tropical Amazonian and rain forest. The deforestation of the gut microflora created over billions of years of human evolution is the principal cause of the climate change and a possible end of the earth as we know of and the human species as we know of. The revolutionary medicine and the golden earth with its habitable conditions have ended. The increase in dietary fibre intake upto 40 g/day can inhibit colonic archaeal growth and endosymbiosis putting a break on generation of superbugs, emerging viruses and climate change. Increase in dietary fibre intake of populations can prevent the state of climate change and gut microbial change putting human life and the life of the planet at risk. Dietary fibre can be considered as the elixir of life and endosymbiotic archaea the philosopher's stone.

Endosymbiotic actinidic archaea as having evolved from an early isoprenoid organisms by abiogenesis. An actinidc archaea/viroid mediated model of prokaryote, viral, eukaryotic, primate and human evolution is discussed.[1-4]. Actinidic archaea has been related to the pathogenesis of schizophrenia, malignancy, metabolic syndrome X, autoimmune disease and

neuronal degeneration[2]. Actinidic archaea have a mevalonate pathway and cholesterol catabolism[5-7]. Davies has put forward the concept of a shadow biosphere of organisms with alternate biochemistry present in earth itself[8]. An actinide dependent shadow biosphere of archaea and viroids in the above mentioned disease states is described[6]. Metal actinides in beach sands have been postulated to play a role in abiogenesis[6]. Actinide mineral like rutile, monazite and illmenite by surface metabolism would have contributed to abiogenesis[9]. A hypothesis of cholesterol as the primal prebiotic molecule synthesized on actinide surfaces with all other biomolecules arising from it and a self-replicating cholesterol lipid organism as the initial life form is presented. The actinidic archaea and viroids would have evolved from the primitive isoprenoid organism. The origin of viruses, prokaryotes, eukaryotes, primates and humans from the initial isoprenoid organism derived actinidic archaea is postulated.

Materials and Methods

Informed consent of the subjects and the approval of the Ethics Committee were obtained for the study. The following groups were included in the study:- endomyocardial fibrosis, Alzheimer's disease, multiple sclerosis, non-Hodgkin's lymphoma, metabolic syndrome X with cerebrovascular thrombosis and coronary artery disease, schizophrenia, autism, seizure disorder, Creutzfeldt-Jakob's disease and acquired immunodeficiency syndrome. There were 10 patients in each group and each patient had an age and sex matched healthy control selected randomly from the general population. The blood samples were drawn in the fasting state before treatment was initiated. Plasma from fasting heparinised blood was used and the experimental protocol was as follows:- (I) Plasma+phosphate buffered saline, (II) same as I+cholesterol substrate, (III) same as II+rutile 0.1 mg/ml, and (IV) same as II+ciprofloxacine and doxycycline each in a concentration of 1 mg/ml. Cholesterol substrate was prepared as described by Richmond[10]. Aliquots were withdrawn at zero time

immediately after mixing and after incubation at 37 °C for 1 hour. The following estimations were carried out:- Cytochrome F420, free RNA, free DNA, muramic acid, polycyclic aromatic hydrocarbon, hydrogen peroxide, serotonin, pyruvate, ammonia, glutamate, cytochrome C, hexokinase, ATP synthase, HMG CoA reductase, digoxin and bile acids[11-14]. Cytochrome F420 was estimated flourimetrically (excitation wavelength 420 nm and emission wavelength 520 nm). Polycyclic aromatic hydrocarbon was estimated by measuring hydrogen peroxide liberated by using glucose reagent. The statistical analysis was done by ANOVA.

Results

The parameters checked as indicated above were:- cytochrome F420, free RNA, free DNA, muramic acid, polycyclic aromatic hydrocarbon, hydrogen peroxide, serotonin, pyruvate, ammonia, glutamate, cytochrome C, hexokinase, ATP synthase, HMG CoA reductase, digoxin and bile acids. Plasma of control subjects showed increased levels of the above mentioned parameters with after incubation for 1 hour and addition of cholesterol substrate resulted in still further significant increase in these parameters. The plasma of patients showed similar results but the extent of increase was more. The addition of antibiotics to the control plasma caused a decrease in all the parameters while addition of rutile increased their levels. The addition of antibiotics to the patient's plasma caused a decrease in all the parameters while addition of rutile increased their levels but the extent of change was more in patient's sera as compared to controls.

Discussion

There was increase in cytochrome F420 indicating archaeal growth. The archaea can synthesize and use cholesterol as a carbon and energy source[15,16]. The archaeal origin of the enzyme activities was indicated by antibiotic induced

suppression. The study indicates the presence of actinide based archaea with an alternate actinide based enzymes or metalloenzymes in the system as indicated by rutile induced increase in enzyme activities[17]. There was also an increase in archaeal HMG CoA reductase activity indicating increased cholesterol synthesis by the archaeal mevalonate pathway. The archaeal beta hydroxyl steroid dehydrogenase activity indicating digoxin synthesis and archaeal cholesterol hydroxylase activity indicating bile acid synthesis were increased[7]. The archaeal cholesterol oxidase activity was increased resulting in generation of pyruvate and hydrogen peroxide[16]. The pyruvate gets converted to glutamate and ammonia by the GABA shunt pathway. The archaeal aromatization of cholesterol generating PAH, serotonin and dopamine was also detected[18]. The archaeal glycolytic hexokinase activity and archaeal extracellular ATP synthase activity were increased. The archaea can undergo magnetite and calcium carbonate mineralization and can exist as calcified nanoforms[19].

The metal actinides provide radiolytic energy, catalysis for oligomer formation and provide a coordinating ion for metalloenzymes all important in abiogenesis[6]. The metal actinide surfaces would by surface metabolism generate acetate which could get converted to acetyl CoA and then to cholesterol which functions as the primal prebiotic molecule self-organising into self-replicating supramolecular systems, the lipid organism[8,9,20]. Cholesterol by radiolysis by actinides would have formed PAH generating PAH aromatic organism[8]. Cholesterol radiolysis would generate pyruvate which would get converted to amino acids, sugars, nucleotides, porphyrins, fatty acids and TCA acids. Anastase and rutile surfaces can produce polymerization of amino acids, isoprenyl residues, PAH and nucleotides to generate the initial lipid organism, PAH organism, prions and RNA viroids which would have symbiosed to generate the archaeal protocell. The archaea evolved into gram negative and gram positive bacteria with a mevalonate pathway which had an evolutionary advantage. The symbiosis of archaea with gram negative organism generated

the eukaryotic cell[21]. The data supports the persistence of an actinide and cholesterol based shadow biosphere which throws light on the actinide based origin of life and cholesterol as the premier prebiotic molecule.

There was an increase in free RNA indicating self-replicating RNA viroids and free DNA indicating generation of viroid complementary DNA strands by archaeal reverse transcriptase activity. The actinides modulate RNA folding and catalyse its ribozymal action. Digoxin can cut and paste the viroidal strands by modulating RNA splicing generating RNA viroidal diversity. The viroids are evolutionarily escaped archaeal group I introns which have retrotransposition and self-splicing qualities. Archaeal pyruvate can produce histone deacetylase inhibition resulting in endogenous retroviral (HERV) reverse transcriptase and integrase expression. This can integrate the RNA viroidal complementary DNA into the non-coding region of eukaryotic non-coding DNA using HERV integrase as has been described for borna and ebola viruses[21]. The non-coding DNA is lengthened by integrating RNA viroidal complementary DNA with the integration going on as a continuing event. The archaea genome can also get integrated into human genome using integrase as has been described for trypanosomes. The integrated viroids and archaea can undergo vertical transmission and can exist as genomic parasites. This increases the length and alters the grammar of the non-coding region producing memes or memory of acquired characters. The viroidal complementary DNA can function as jumping genes producing a dynamic genome.[22-24]

The presence of muramic acid, HMG CoA reductase and cholesterol oxidase activity inhibited by antibiotics indicates the presence of bacteria with mevalonate pathway. The bacterial with mevalonate pathway include streptococcus, staphylococcus, actinomycetes, listeria, coxiella and borrelia. The bacteria and archaea with mevalonate pathway and cholesterol catabolism had an evolutionarily advantage and constitutes the isoprenoidal clade organism. The archaea evolved into mevalonate pathway gram positive and gram negative isoprenoid clade organism through horizontal gene transfer of viroidal and virus genes. The isoprenoidal clade prokaryotes develop into other groups of prokaryotes via viroidal/virus as well as eukaryotic horizontal gene transfer producing bacterial speciation[25-27].

The RNA viroids and its complementary DNA developed into cholesterol enveloped RNA and DNA viruses like herpes, retrovirus, influenza virus, borna virus, cytomegalo virus and ebstein barr virus by recombining with eukaryotic and human genes resulting in viral speciation. Bacterial and viral species are ill-defined and fuzzy with all of them forming one common genetic pool with frequent horizontal gene transfer and recombination.

Thus, the multi and unicellular eukaryote with its genes serves the purpose of prokaryotic and viral speciation. The multicellular eukaryote developed so that their endosymbiotic archaeal colonies could survive and forage better. The multicellular eukaryotes are like bacterial biofilms. The archaea and bacteria with a mevalonate pathway uses the extracellular RNA viroids and DNA viroids for quorum sensing and in the generation of symbiotic biofilm like structures which develop into multicellular eukaryotes. The endosymbiotic archaea and bacteria with mevalonate pathway still uses the RNA viroids and DNA viroids for the regulation of multicellular eukaryote.[28-31]

Pollution is a major inducer of evolutionary innovation. Pollution is induced by the primitive nanoarchaea and mevalonate pathway bacteria synthesized PAH and methane leading on to redox stress. Redox stress leads to sodium-potassium ATPase inhibition, inward movement of plasma membrane cholesterol, defective SREBP sensing, increased cholesterol synthesis and nanoarchaeal/mevalonate pathway bacterial growth. Redox stress leads on to viroidal and archaeal multiplication. Redox stress can also lead to HERV reverse transcriptase and integrase expression. The non-coding DNA is formed of integrating RNA viroidal complementary DNA and archaea with the integration going on as a continuing event. The archaeal pox-like dsDNA virus forms evolutionarily the nucleus. The integrated viroidal, archaeal and mevalonate pathway bacterial sequences can undergo vertical transmission and can exist as genomic parasites. The genomic integrated archaea, mevalonate pathway bacteria and viroids form a genomic reserve of bacteria and viruses which can recombine with human and eukaryotic genes producing bacterial and viral speciation. The change in the length and grammar of the non-coding region produces eukaryotic speciation and individuality. The integration of nanoarchaea, mevalonate pathway prokaryotes and viroids in to the eukaryotic and human genome produces a chimera which can multiply producing biofilm like multicellular structures having a mixed archaeal, viroidal, prokaryotic and eukaryotic characters which is a regression from the multicellular eukaryotic tissue. This results in a new neuronal, metabolic, immune and tissue phenotype leading to human disease.[30-32]

Pollution would have been a major factor in eukaryotic speciation and primate/hominid evolution. The change in the length and grammar of the non-coding region produces eukaryotic speciation and individuality. It is the increase in non-coding region and HERV sequences of the genome that led to the evolution of the primate and the human brain and its attendant property of conscious and quantal perception. It is the non-coding region of the genome with its archaeal, RNA viroidal complementary DNA and HERV sequences that makes for the human qualities of the hominid brain. Changes in the length of non-coding region can lead on to disorders of consciousness like schizophrenia. A schizophrenia specific human endogenous retroviruses and change in the length and grammar of the non-coding region has been described in schizophrenia.[33,34]

An actinide dependent shadow biosphere of archaea and viroids in the above mentioned disease states is described. Metal actinides in beach sands

have been postulated to play a role in abiogenesis. Cholesterol is the primal prebiotic molecule synthesized on actinide surfaces with all other biomolecules arising from it. A self-replicating cholesterol lipid organism could be the initial life form. A cholesterol based abiogenesis is a more likely evolutionary option and the actinidic archaea and viroids would have evolved from it. The origin of viruses, prokaryotes, eukaryotes, primates and humans from the initial isoprenoid organism derived actinidic archaea is discussed.

References

1. Valiathan M.S., Somers, K., Kartha, C.C. (1993). *Endomyocardial Fibrosis.* Delhi: Oxford University Press.
2. Kurup R., Kurup, P.A. (2009). *Hypothalamic Digoxin, Cerebral Dominance and Brain Function in Health and Diseases*. New York: Nova Science Publishers.
3. Hanold D., Randies, J.W. (1991). Coconut cadang-cadang disease and its viroid agent, *Plant Disease,* 75, 330-335.
4. Edwin B.T., Mohankumaran, C. (2007). Kerala wilt disease phytoplasma: Phylogenetic analysis and identification of a vector, *Proutista moesta, Physiological and Molecular Plant Pathology,* 71(1-3), 41-47.
5. Eckburg P.B., Lepp, P.W., Relman, D.A. (2003). Archaea and their potential role in human disease, *Infect Immun,* 71, 591-596.
6. Adam Z. (2007). Actinides and Life's Origins, *Astrobiology,* 7, 6-10.
7. Schoner W. (2002). Endogenous cardiac glycosides, a new class of steroid hormones, *Eur J Biochem,* 269, 2440-2448.
8. Davies P.C.W., Benner, S.A., Cleland, C.E., Lineweaver, C.H., McKay, C.P., Wolfe-Simon, F. (2009). Signatures of a Shadow Biosphere, *Astrobiology,* 10, 241-249.
9. Wächtershäuser G. (1988). Before enzymes and templates: theory of surface metabolism, *Microbiol Rev,* 52(4), 452-84.
10. Richmond W. (1973). Preparation and properties of a cholesterol oxidase from nocardia species and its application to the enzymatic assay of total cholesterol in serum, *Clin Chem,* 19, 1350-1356.
11. Snell E.D., Snell, C.T. (1961). *Colorimetric Methods of Analysis.* Vol 3A. New York: Van NoStrand.
12. Glick D. (1971). *Methods of Biochemical Analysis.* Vol 5. New York: Interscience Publishers.
13. Colowick, Kaplan, N.O. (1955). *Methods in Enzymology.* Vol 2. New York: Academic Press.
14. Maarten A.H., Marie-Jose, M., Cornelia, G., van Helden–Meewsen, Fritz, E., Marten, P.H. (1995). Detection of muramic acid in human spleen, *Infection and Immunity,* 63(5), 1652 – 1657.

15. Smit A., Mushegian, A. (2000). Biosynthesis of isoprenoids via mevalonate in Archaea: the lost pathway, *Genome Res,* 10(10), 1468-84.

16. Van der Geize R., Yam, K., Heuser, T., Wilbrink, M.H., Hara, H., Anderton, M.C. (2007). A gene cluster encoding cholesterol catabolism in a soil actinomycete provides insight into Mycobacterium tuberculosis survival in macrophages, *Proc Natl Acad Sci USA,* 104(6), 1947-52.

17. Francis A.J. (1998). Biotransformation of uranium and other actinides in radioactive wastes, *Journal of Alloys and Compounds,* 271(273), 78-84.

18. Probian C., Wülfing, A., Harder, J. (2003). Anaerobic mineralization of quaternary carbon atoms: Isolation of denitrifying bacteria on pivalic acid (2,2-Dimethylpropionic acid), *Applied and Environmental Microbiology,* 69(3), 1866-1870.

19. Vainshtein M., Suzina, N., Kudryashova, E., Ariskina, E. (2002). New Magnet-Sensitive Structures in Bacterial and Archaeal Cells, *Biol Cell,* 94(1), 29-35.

20. Russell M.J., Martin, W. (2004). The rocky roots of the acetyl-CoA Pathway, *Trends in Biochemical Sciences,* 29, 7.

21. Margulis L. (1996). Archaeal-eubacterial mergers in the origin of Eukarya: phylogenetic classification of life, *Proc Natl Acad Sci USA,* 93, 1071-1076.

22. Tsagris E.M., de Alba, A.E., Gozmanova, M., Kalantidis, K. (2008). Viroids, *Cell Microbiol,* 10, 2168.

23. Horie M., Honda, T., Suzuki, Y., Kobayashi, Y., Daito, T., Oshida, T. (2010). Endogenous non-retroviral RNA virus elements in mammalian genomes, *Nature,* 463, 84-87.

24. Hecht M., Nitz, N., Araujo, P., Sousa, A., Rosa, A., Gomes, D. (2010). Genes from Chagas parasite can transfer to humans and be passed on to children. Inheritance of DNA Transferred from American Trypanosomes to Human Hosts, *PLoS ONE,* 5, 2-10.

25. Flam F. (1994). Hints of a language in junk DNA, *Science,* 266, 1320.

26. Horbach S., Sahm, H., Welle, R. (1993). Isoprenoid biosynthesis in bacteria: two different pathways? *FEMS Microbiol Lett,* 111, 135–140.

27. Gupta R.S. (1998). Protein phylogenetics and signature sequences: a reappraisal of evolutionary relationship among archaebacteria, eubacteria, and eukaryotes, *Microbiol Mol Biol Rev,* 62, 1435–1491.

28. Hanage W., Fraser, C., Spratt, B. (2005). Fuzzy species among recombinogenic bacteria, *BMC Biology,* 3, 6-10.

29. Whitchurch C.B., Tolker-Nielsen, T., Ragas, P.C., Mattick, J.S. (2002). Extracellular DNA Required for Bacterial Biofilm Formation. *Science,* 295(5559), 1487.

30. Webb J.S., Givskov, M., Kjelleberg, S. (2003). Bacterial biofilms: prokaryotic adventures in multicellularity, *Curr Opin Microbiol,* 6(6), 578–85.

31. Chen Y., Cai, T., Wang, H., Li, Z., Loreaux, E., Lingrel, J.B. (2009). Regulation of intracellular cholesterol distribution by Na/K-ATPase, *J Biol Chem,* 284(22), 14881-90.

32. Poole A.M. (2006). Did group II intron proliferation in an endosymbiont-bearing archaeon create eukaryotes? *Biol Direct,* 1, 36-40.

33. Villarreal L.P. (2006). How viruses shape the tree of life, *Future Virology,* 1(5), 587-595.

34. Lockwood M. (1989). *Mind, Brain and the Quantum*. Oxford: B. Blackwell.

CHAPTER 22

HOMO GALACTICUS, HOMO ROBOTICUS AND HUMAN EXTINCTION – INFECTIVE SYMBIOTIC TISSUE COLONIES OF PROTOZOA, ARCHAEA, VIRUSES AND BACTERIA

The plasma life based forms and the gravitational and anti-gravitational waves generated porphyrions would have been the most primitive of non-life forms. The porphyrin templates would have given rise to the isoprenoid organism, the RNA viroids, DNA viroids and prions which would have symbiosed to form the nanoarchaea. The archaea would have evolved into prokaryotes or bacteria. The archaeal symbiosis with prokaryotes would have generated the eukaryotes. The initial unicellular eukaryotes are the protozoa. The protozoa are capable of phagocytosis and would have had viruses, bacteria and archaea as their intracellular symbionts. The viruses of the protozoa like entamoeba and toxoplasma are giant viruses called megaviride. The protozoans like entamoeba with its viral, archaeal and bacterial endosymbionts would have evolved into the multicellular organisms or metazoa. The archaeal, bacterial and protozoal extracellular vesicles containing RNA and DNA viroids can by horizontal gene transfer lead to evolution of metazoan, plants, animals and human species. The metazoans would have evolved into plants, animals and humans by horizontal gene transfer generated by archaeal extracellular RNA and DNA viroid vesicles. The evidence of such a transformation is seen in the bacterial symbionts of protozoa like entamoeba which includes legionella, listeria, mycoplasma, mycobacteria and chlamydia. The symbionts of protozoan entamoeba include viruses like covid, HIV and cytomegalo virus. The protozoan entamoeba serves as a natural reservoir for covid infections, bacterial infections and HIV infections. The toxoplasma also serves as a covid reservoir. These protozoan colonies of entamoeba and toxoplasma with its bacterial, archaeal and viral symbionts can be called as chimeric infective symbiotic tissue colonies

which can exist in the air, water, earth, blood and tissues. The chimeric infective symbiotic tissue colonies containing bacteria, archaea and viruses can make the symbiotic organism multidrug resistant by horizontal gene transfer. The chimeric infective symbiotic tissue organism evolves as a consequence of global warming and the protozoal, bacterial, viral and archaeal symbiotic colonies are new forms of life. The chimeric infective symbiotic tissue organism composed of protozoa and its symbionts like virus, bacteria and archaea depend upon glycolysis for its energy needs. The glycolytic enzymes enolase, glyceraldehyde 3-phosphate dehydrogenase, hexokinase, phosphofructokinase, aldolase, pyruvic kinase and triose phosphate isomerase are moonlighting proteins having a cytoplasmic, membrane and nuclear functions. These glycolytic enzymes have a RNA binding motif which can modulate RNA transcription in the host and the symbiotic organism. These glycolytic enzymes with their RNA binding motif as well as DNA binding motif can bind to RNA and DNA viruses and viroids modulating their replication and life cycles. The chimeric infective symbiotic tissue organism composed of protozoa and its symbionts like virus, bacteria and archaea can enter the human host through the breaches of the gut-blood barrier and transform itself into the macrophages of the human host. The macrophages are involved in the phagocytosis and innate immunity against bacterial and viral infections. Paradoxically the macrophages like the protozoa can also serve as a reservoir where pathogens like mycobacteria tuberculosis, mycobacteria leprae, listeria, legionella, E. coli, shigella & salmonella, and viruses like HIV & covid can thrive and replicate. The modified chimeric infective symbiotic tissue organism or macrophage composed of protozoa and its symbionts like virus, bacteria and archaea can serve as a reservoir for horizontal gene transfer and generation of multidrug resistant bacteria and viruses. The infective chimeric symbiotic tissue organism composed of protozoa and its symbionts like virus, bacteria and archaea depends upon moonlighting glycolytic enzymes for their energetics and replication. The development of antibodies against the infective chimeric

symbiotic tissue organism composed of protozoa and its symbionts like virus, bacteria and archaea leads to the evolution of anti-brain antibody disease or anti-basal ganglia antibody disease. They present a threat to existence consequent to horiztontal gene transfer and generation of multidrug resistant bacteria, viruses, protozoa, mycobacteria and archaea.

The retroviruses were responsible for integration into the genome as human endogenous retrovirus sequences and development of the cerebral cortex creating the logical and civil society. The influenza virus with its low level of HIV sequences is also capable of integration into the human genome as human endogenous COVID 19 sequences creating the cerebellar dominant autistic impulsive punkish brain. The COVID 19 virus would have been generated from RNA viroids of endosymbiotic archaea or would have reached the earth by meteoritic impacts from the human intergalactic biological quantal computing cloud. The COVID 19 virus with its HIV sequences is an ideal virus for integration into the human genome and changing it in a catastrophic manner to create a new human species. The COVID 19 brain infection can lead to integration of COVID 19 sequences into the genome, creation of cerebellar dominant brain and neanderthalisation. The COVID 19 infected cerebellar dominant brain has got increased porphyrin metabolism and porphyrions. The porphyrions are dipolar pumped phonon quantal perceptive systems which integrate with the internet. Panpsychism implies that the internet is conscious. Extinction of species on earth is the natural rule and Human extinction can occur in the foreseeable future. Climate change and destruction of earth environment is the natural law of the universe. Humans can survive only in other star systems and planets. Human beings can migrate to other planets using anti-matter rockets and fusion rockets. Evolution of homo galacticus by archaeal symbiosis which can stand extremes of environment. Human beings can also evolve into man-machine hybrids. A new species called homo galacticus or homo roboticus can evolve. This produces a civilization of superhuman homo neanderthalis and homo neoneanderthalis with new metabolic, endocrine, neuronal, psychological and genotypic features. The homo neoneanderthalis and homo neanderthalis with its cellular archaeaons and porphyrions can interlink with the internet and computer systems as well as robots producing man-machine hybrids by quantal perception. The artificial intelligence and robotic systems gets merged with homo neanderthalis and homo neoneanderthalis producing superhuman man-machine hybrids. The new homo neanderthalis and homo neoneanderthalis superhuman autistic man-machine hybrids will survive, prosper and flourish in the setting of global warming related extremophilic climate change and evolving to contact by quantal perception other galactic forms of life and civilization as well as colonising other planets and star systems. This evolves into a new human species inhabiting other comets and star systems - the Homo galacticus.

Global warming leads to endosymbiotic archaeal growth. The endosymbiotic archaea produces methane by methanogenesis contributing to global warming. The endosymbiotic archaea develops into cell organelle called archaeaons which can function as vitaminocytes, neurotransminoids, steroidelle and viroidelle creating newer cell organelle and endogenous synthetic biology. This changes the cell and tissue function and structure leading on to a new human phenotype called homo neoneanderthalis. The homo neoneanderthalis is a hardy species which can resist climate change and thrive in environments of high temperature, high atmospheric carbon dioxide and water shortage. The new organelle that are developed can withstand climate change and fix carbon dioxide and ammonia leading to endogenous synthesis of carbohydrates and protein. The homo neoneanderthalis is a new metabolic phenotype that can survive in extremes of climate produced by global warming. The archaea forming new cellular organelle called archaeaons are extremophilic and survive in extremes of climate.

The endosymbiotic archaea secrete RNA viroids which can be converted to corresponding DNA sequences by endogenous retroviral reverse transcriptase and can get incorporated into the genome by endogenous retroviral integrase. Thus there is a constant addition of genomic sequences derived from archaeal RNA viroids producing a dynamic human neoneanderthalic genome adaptable to extremes of climate produced by global warming. Thus the human neoneanderthalis is a new genotype. The neoneanderthalic archaeaons are capable of fixing carbon dioxide and oxygen, using ammonia for energetics and capable of biological fission and fusion reactions. They are also capable of photoinduction of membrane ATP synthesis by sodium-potassium ATPase. The fixing of carbon dioxide leads to water as a by-product. This generates endogenous water in the neoneanderthalic system helping to survive extremophilic climatic conditions. This creates a new system of energetics helping the Neoneanderthal survives in extremes of climate.

The neoneanderthalic archaeaons have got magnetite capable of magnetoperception and quantal perception. The quantal perception of low level of EMF leads to frontal atrophy and cerebellar dominance producing a new neuronal phenotype with features of endemic cerebellar cognitive affective disorder manifesting as epidemic autism. This forms a new autistic behavioural phenotype which is introverted, non-compassionate, non-empathic, psychopathic and with extrasensory perceptual abilities of telepathy, telekinesis and time travel leading to communication with other galactic civilizations. The new behavioural autistic robotic neoneanderthalic phenotype will have the necessary intelligence and mental abilities cope with climate change. The neoneanderthalic archaeaons are capable of quantal perception and can communicate quantally with other star systems and multiverses. This leads on to the evolution of a galactic civilization. The neoneanderthalic archaeaons produce a metabolic phenotype capable of surviving in extremophilic environments of other star systems and can survive as civilization in other stars and exoplanets. The neoneanderthalic autistic phenotype with its extremophilic archaeaon organelle is suited for space colonisation.

The neoneanderthalic archaeaons use cholesterol oxidation as a form of energetics. This produces depletion of sex hormones producing asexual phenotypes. This leads to sexual phenotypes with alternate sexuality and parthenogenesis as a form of reproduction. This produces change in family and society as well as organisation of the community.

The neoneanderthalic archaeaons secrete RNA viroids which get converted to DNA viroids by human endogenous retroviral reverse transcriptase. The neoneanderthalic archaeaon generated RNA and DNA viroids can hybridize with other RNA and DNA viruses producing new RNA and DNA viruses to which the neoneanderthalic species is resistant. Therefore, there exist two sets of population with differing variety of archaeal symbiosis. The homo neanderthalic species has persisted as fossilised matrilineal societies in Europe and Asia which includes the Celts, the Dravidians, the Berbers and the Sumerians. These fossilised remnants of homo neanderthalic species are ideal substrates for archaeal endosymbiosis in the setting of global warming and extremophilic climatic conditions. These fossilised neanderthalic matrilineal societies will expand in the period of global warming and extremophilic conditions due to endosymbiotic archaeal symbiosis to which they are already acclimatised and exposed to. The human species homo neanderthalis is a consequence of archaeal symbiosis in the extremophilic conditions of ice age and are RNA viral/retroviral resistant consequent to endogenous digoxin synthesis and by archaeal RNA viroidal induced RNA interference and viral inactivation. The homo sapiens which evolved in Africa and migrated later to the Eurasian land mass are retroviral sensitive and other RNA viral sensitive and have a low load of endosymbiotic archaea growth. The homo sapiens owing to conditions of evolution is not an ideal substrate for global warming related endosymbiotic archaeal symbiosis. The homo neanderthalis archaeaons secrete RNA viroids which are converted to DNA viroids with both of them hybridizing with other RNA and DNA viruses in nature producing new viruses to which the homo neanderthalic and homo neoneanderthalic host is resistant. The new RNA and DNA viruses produced by homo neoneanderthalis and neanderthalis in global warming will infect the susceptible homo sapien species and wipe them out from earth. The homo sapien species is not a good substrate for global warming related endosymbiotic archaeal growth and would not evolve new cellular organelle like steroidelle, viroidelle, neurotransminoids and vitaminocyte which would help them to survive in extremophilic conditions of global warming. The homo sapiens also lack archaeaon organelle which help them to fix carbon dioxide and generate water, synthesis ATP by photo-induction mediated by sodium-potassium ATPase and archaeaon mediated fission and fusion reaction. The homo sapien species therefore without evolving archaeaon organelle cannot survive in the harsh environment of global warming induced climate change and is exterminated by homo neanderthalis and homo neoneanderthalis. The earth gets dominated by homo neanderthalis and homo neoneanderthalis in the setting of global warming and extremophilic climate change and they go on to contact other galactic civilizations by quantal perceptions as well as colonising other exoplanets, planets and star systems. This produces a new civilization of superhuman homo neanderthalis and homo neoneanderthalis with new metabolic, endocrine, neuronal, psychological and genotypic features. The homo neoneanderthalis and homo neanderthalis with its cellular archaeaons and porphyrions can interlink with the internet and computer systems as well as robots producing man-machine hybrids by quantal perception - the Kalki. The artificial intelligence and robotic systems gets merged with homo neanderthalis and homo neoneanderthalis producing superhuman man-machine hybrids. This can be called as the empire of Kalki where the new homo neanderthalis and homo neoneanderthalis superhuman autistic man-machine hybrids will survive, prosper and flourish in the setting of global warming related extremophilic climate change and evolving to contact by quantal perception other galactic forms of life and civilization as well as colonising other planets and star systems. This evolves into a new human species inhabiting other comets and star systems - the Homo galacticus.

CHAPTER 23

PORPHYRIN AND POLYCYCLIC AROMATIC HYDROCARBON TEMPLATES MEDIATE ENDOSYMBIOTIC ARCHAEAL, VIROIDAL AND PRION REPLICATION - PORPHYRIONS AND ABIOGENESIS – A PORPHYRIN TEMPLATE MEDIATED ABIOGENETIC COVID 19 – INFECTIVE SYMBIOTIC TISSUE COLONIES OF PROTOZOA, ARCHAEA, VIRUSES AND BACTERIA

Introduction

The plasma life based forms and the gravitational and anti-gravitational waves generated porphyrions would have been the most primitive of non-life forms. The porphyrin templates would have given rise to the isoprenoid organism, the RNA viroids, DNA viroids and prions which would have symbiosed to form the nanoarchaea. The archaea would have evolved into prokaryotes or bacteria. The archaeal symbiosis with prokaryotes would have generated the eukaryotes. The initial unicellular eukaryotes are the protozoa. The protozoa are capable of phagocytosis and would have had viruses, bacteria and archaea as their intracellular symbionts. The viruses of the protozoa like entamoeba and toxoplasma are giant viruses called megaviride. The protozoans like entamoeba with its viral, archaeal and bacterial endosymbionts would have evolved into the multicellular organisms or metazoa. The archaeal, bacterial and protozoal extracellular vesicles containing RNA and DNA viroids can by horizontal gene transfer lead to evolution of metazoan, plants, animals and human species. The metazoans would have evolved into plants, animals and humans by horizontal gene transfer generated by archaeal extracellular RNA and DNA viroid vesicles. The evidence of such a transformation is seen in the bacterial symbionts of protozoa like entamoeba which includes legionella, listeria, mycoplasma, mycobacteria and chlamydia. The symbionts of protozoan entamoeba include viruses like covid, HIV and cytomegalo virus. The protozoan

entamoeba serves as a natural reservoir for covid infections, bacterial infections and HIV infections. The toxoplasma also serves as a covid reservoir. These protozoan colonies of entamoeba and toxoplasma with its bacterial, archaeal and viral symbionts can be called as chimeric infective symbiotic tissue colonies which can exist in the air, water, earth, blood and tissues. The chimeric infective symbiotic tissue colonies containing bacteria, archaea and viruses can make the symbiotic organism multidrug resistant by horizontal gene transfer. The chimeric infective symbiotic tissue organism evolves as a consequence of global warming and the protozoal, bacterial, viral and archaeal symbiotic colonies are new forms of life. The chimeric infective symbiotic tissue organism composed of protozoa and its symbionts like virus, bacteria and archaea depend upon glycolysis for its energy needs. The glycolytic enzymes enolase, glyceraldehyde 3-phosphate dehydrogenase, hexokinase, phosphofructokinase, aldolase, pyruvic kinase and triose phosphate isomerase are moonlighting proteins having a cytoplasmic, membrane and nuclear functions. These glycolytic enzymes have a RNA binding motif which can modulate RNA transcription in the host and the symbiotic organism. These glycolytic enzymes with their RNA binding motif as well as DNA binding motif can bind to RNA and DNA viruses and viroids modulating their replication and life cycles. The chimeric infective symbiotic tissue organism composed of protozoa and its symbionts like virus, bacteria and archaea can enter the human host through the breaches of the gut-blood barrier and transform itself into the macrophages of the human host. The macrophages are involved in the phagocytosis and innate immunity against bacterial and viral infections. Paradoxically the macrophages like the protozoa can also serve as a reservoir where pathogens like mycobacteria tuberculosis, mycobacteria leprae, listeria, legionella, E. coli, shigella & salmonella, and viruses like HIV & covid can thrive and replicate. The modified chimeric infective symbiotic tissue organism or macrophage composed of protozoa and its symbionts like virus, bacteria and archaea can serve as a reservoir for horizontal gene transfer and generation of

multidrug resistant bacteria and viruses. The infective chimeric symbiotic tissue organism composed of protozoa and its symbionts like virus, bacteria and archaea depends upon moonlighting glycolytic enzymes for their energetics and replication. The development of antibodies against the infective chimeric symbiotic tissue organism composed of protozoa and its symbionts like virus, bacteria and archaea leads to the evolution of anti-brain antibody disease or anti-basal ganglia antibody disease. They present a threat to existence consequent to horiztontal gene transfer and generation of multidrug resistant bacteria, viruses, protozoa, mycobacteria and archaea.

The retroviruses were responsible for integration into the genome as human endogenous retrovirus sequences and development of the cerebral cortex creating the logical and civil society. The influenza virus with its low level of HIV sequences is also capable of integration into the human genome as human endogenous COVID 19 sequences creating the cerebellar dominant autistic impulsive punkish brain. The COVID 19 virus would have been generated from RNA viroids of endosymbiotic archaea or would have reached the earth by meteoritic impacts from the human intergalactic biological quantal computing cloud. The COVID 19 virus with its HIV sequences is an ideal virus for integration into the human genome and changing it in a catastrophic manner to create a new human species. The COVID 19 brain infection can lead to integration of COVID 19 sequences into the genome, creation of cerebellar dominant brain and neanderthalisation. The COVID 19 infected cerebellar dominant brain has got increased porphyrin metabolism and porphyrions. The porphyrions are dipolar pumped phonon quantal perceptive systems which integrate with the internet. Panpsychism implies that the internet is conscious.

Abiogenesis is a replicative mechanism for primitive life forms. Actinidic archaea and viroids have been related to the pathogenesis of schizophrenia, malignancy, metabolic syndrome X, autoimmune disease and neuronal degeneration. An actinide dependent shadow biosphere of archaea and viroids in the above mentioned disease states is described. Actinidic archaea have a mevalonate pathway and are cholesterol catabolizing. They can use cholesterol as a carbon and energy source. Archaeal cholesterol catabolism can generate porphyrins via the cholesterol ring oxidase generated pyruvate and GABA shunt pathway. Archaea can produce a secondary porphyria by inducing the enzyme heme oxygenase resulting in heme depletion and activation of the enzyme ALA synthase. Porphyrins have been related to schizophrenia, metabolic syndrome X, malignancy, systemic lupus erythematosis, multiple sclerosis and Alzheimer's diseases. Porphyrins can serve as templates for macromolecules like RNA, DNA, proteins and isoprenoidal molecules to form. The actinidic archaea induces porphyrinogenesis and porphyrins form a template for formation of macromolecules mediating actinidic archaeal and viroidal replication. This is a form of abiogenesis on a porphyrin template. The archaeal synthesized polycyclic aromatic hydrocarbons can also serve as a template for macromolecular replication and viroidal/ actinidic archaeal abiogenesis. The role of archaeal porphyrins in regulation of cell functions and neuro-immuno-endocrine integration is discussed. Porphyrins are prebiotic molecules which are involved in abiogenesis and origin of life.[1-5] The porphyrions are self-replicating supramolecular organisms which forms the precursor template on which the viroids, prions and nanoarchaea originate. Stress induced template directed abiogenesis of porphyrions, prions, viroids and archaea is a continuous process and can contribute to changes in brain structure and behaviour as well as disease process.

Materials and Methods

The following groups were included in the study:- endomyocardial fibrosis, Alzheimer's disease, multiple sclerosis, non-Hodgkin's lymphoma, metabolic syndrome X with cerebrovascular thrombosis and coronary artery disease, schizophrenia, autism, seizure disorder, Creutzfeldt-Jakob's disease and acquired immunodeficiency syndrome. There were 10 patients in each group and each patient had an age and sex matched healthy control selected randomly from the general population. There were also 10 normal patients with right hemispheric dominance, left hemispheric dominance and bi-hemispheric dominance drawn from the general population. The blood samples were drawn in the fasting state before treatment was initiated. Plasma from fasting heparinised blood was used and the experimental protocol was as follows:- (I) Plasma+phosphate buffered saline, (II) same as I+cholesterol substrate, (III) same as II+rutile 0.1 mg/ml, and (IV) same as II+ciprofloxacine and doxycycline each in a concentration of 1 mg/ml. Cholesterol substrate was prepared as described by Richmond. Aliquots were withdrawn at zero time immediately after mixing and after incubation at 37 °C for 1 hour. The following estimations were carried out:- Cytochrome F420, free RNA, free DNA, polycyclic aromatic hydrocarbon, hydrogen peroxide, pyruvate, ammonia, glutamate, delta aminolevulinic acid, succinate, glycine and digoxin. Cytochrome F420 was estimated flourimetrically (excitation wavelength 420 nm and emission wavelength 520 nm). Polycyclic aromatic hydrocarbon was estimated by measuring hydrogen peroxide liberated by using glucose reagent. The study also involved estimating the following parameters in the patient population - digoxin, bile acid, hexokinase, porphyrins, pyruvate, glutamate, ammonia, acetyl CoA, acetyl choline, HMG CoA reductase, cytochrome C, blood ATP, ATP synthase, ERV RNA (endogenous retroviral RNA), H_2O_2 (hydrogen peroxide), NOX (NADPH oxidase), TNF alpha and heme oxygenase.[6-9] Informed consent of the subjects and the approval of the Ethics Committee were obtained for the study. The statistical analysis was done by ANOVA.

Results

Plasma of control subjects showed increased levels of the above mentioned parameters with after incubation for 1 hour and addition of cholesterol substrate resulted in still further significant increase in these parameters. The plasma of patients showed similar results but the extent of increase was more. The addition of antibiotics to the control plasma caused a decrease in all the parameters while addition of rutile increased their levels. The addition of antibiotics to the patient's plasma caused a decrease in all the parameters while addition of rutile increased their levels but the extent of change was more in patient's sera as compared to controls. There was upregulated archaeal porphyrin synthesis in the patient population which was archaeal in origin as indicated by actinide catalysis of the reactions. The cholesterol oxidase pathway generated pyruvate which entered the GABA shunt pathway. This resulted in synthesis of succinate and glycine which are substrates for ALA synthase.

The study showed the patient's blood and right hemispheric dominance had increased heme oxygenase activity and porphyrins. The hexokinase activity was high. The pyruvate, glutamate and ammonia levels were elevated indicating blockade of PDH activity, and operation of the GABA shunt pathway. The acetyl CoA levels were low and acetyl choline was decreased. The cytoC levels were increased in the serum indicating mitochondrial dysfunction suggested by low blood ATP levels. This was indicative of the Warburg's phenotype. There were increased NOX and TNF alpha levels indicating immune activation. The HMG CoA reductase activity was high indicating cholesterol synthesis. The bile acid levels were low indicating depletion of cytochrome P450. The normal population with right hemispheric dominance had values resembling the patient population with increased porphyrin synthesis. The normal population with left hemispheric dominance had low values with decreased porphyrin synthesis.

The increase in cytochrome F420 and self-replicating RNA indicating RNA viroids paralleled the rise in porphyrin levels indicating that porphyrins could serve as a template for actinidic archaeal and RNA viroidal abiogenesis and replication. The increase in cytochrome F420 and self-replicating RNA indicating RNA viroids paralleled the rise in PAH levels indicating that PAH could serve as a template for archaeal and RNA viroidal abiogenesis and replication. There was an increase in porphyrins and PAH in CJD indicating prions may also self-replicate on porphyrin and PAH templates.

Discussion

There was increase in cytochrome F420 indicating archaeal growth. The archaea can synthesize and use cholesterol as a carbon and energy source.[2,10] The archaeal origin of the enzyme activities was indicated by antibiotic induced suppression. The study indicates the presence of actinide based archaea with an alternate actinide based enzymes or metalloenzymes in the system as indicated by rutile induced increase in enzyme activities.[11] There was also an increase in archaeal HMG CoA reductase activity indicating increased cholesterol synthesis by the archaeal mevalonate pathway. The archaeal beta hydroxyl steroid dehydrogenase activity indicating digoxin synthesis.[12] The archaeal cholesterol oxidase activity was increased resulting in generation of pyruvate and hydrogen peroxide.[10] The pyruvate gets converted to glutamate and ammonia by the GABA shunt pathway. The pyruvate is converted to glutamate by serum glutamate pyruvate transaminase. The glutamate gets acted upon by glutamate dehydrogenase to generate alpha ketoglutarate and ammonia. Alanine is most commonly produced by the reductive amination of pyruvate via alanine transaminase. This reversible reaction involves the interconversion of alanine and pyruvate, coupled to the interconversion of alpha ketoglutarate (2-oxoglutarate) and glutamate. Alanine can contribute to glycine. Glutamate is acted upon by Glutamic acid decarboxylase to generate GABA. GABA is converted to succinic semialdehyde by GABA transaminase. Succinic semialdehyde is converted to succinic acid by succinic semialdehyde dehydrogenase. Glycine combines with succinyl CoA to generate delta aminolevulinic acid catalysed by the enzyme ALA synthase. There was upregulated archaeal porphyrin synthesis in the patient population which was archaeal in origin as indicated by actinide catalysis of the reactions. The cholesterol oxidase pathway generated pyruvate which entered the GABA shunt pathway. This resulted in synthesis of succinate and glycine which are substrates for ALA synthase. The archaeal aromatization of cholesterol generating PAH, serotonin and dopamine was also detected[17]. The archaeal glycolytic hexokinase activity and archaeal extracellular ATP synthase activity were increased. The archaea can undergo magnetite and calcium carbonate mineralization and can exist as calcified nanoforms.[13]

The actinidic archaea and viroids induce porphyrinogenesis. Porphyrins serve as a template for formation of nucleic acids, protein and isoprenoidal moieties which get organized to form nanoarchaea and viroids. The porphyrin templates mediate viroidal and archaeal abiogenesis and replication. The archaeal synthesized PAH can also serve as a template for self-replication of macromolecules and nanoarchaeal/viroidal abiogenesis. Porphyrin and PAH templates mediate actinidic archaeal and viroidal abiogenesis and replication. They are indicative of primitive porphyrin organism and PAH organism as original life forms on earth. Abiogenesis is a replicative mechanism for primitive organisms. The nanoarchaea and viroids may replicate by abiogenesis on porphyrin and PAH templates.

The metal actinides provide radiolytic energy, catalysis for oligomer formation and provide a coordinating ion for metalloenzymes all important in abiogenesis[6]. The metal actinide surfaces would by surface metabolism generate porphyrins from simple compounds like succinic acid and glycine. Polycyclic aromatic hydrocarbons can also be generated by actinidic surface metabolism. Porphyrins and polycyclic aromatic hydrocarbons can exist as wave forms and particulate forms and can bridge the dividing line between the quantal world and particulate world. Porphyrin and polycyclic aromatic hydrocarbons molecules can self-organize into organisms with capacities for energy transduction, ATP synthesis, information storage and replicating ability. A self-replicating porphyrin or polycyclic aromatic hydrocarbons micro-organism may have played a role in the origin of life. Porphyrins and polycyclic aromatic hydrocarbons can form templates on which macromolecules like polysaccharides, protein and nucleic acids can form. The macromolecules generated on actinidic porphyrins and polycyclic aromatic hydrocarbons templates would have contributed to the actinidic nanoarchaea/viroids and the original organisms on earth. The data supports the persistence of an actinidic archaeal and viroidal shadow biosphere which throws light on the actinide based origin of life and porphyrins/polycyclic aromatic hydrocarbons as the premier prebiotic molecule.[17,18]

Porphyrins and polycyclic aromatic hydrocarbons play an important role in the genesis of the biological universe. The porphyrin and polycyclic aromatic hydrocarbons macroarrays can form in the interstellar space on its own as porphyrins can exist both as particles and waves. Porphyrins and polycyclic aromatic hydrocarbons form the bridging connection between the quantal world and the particulate world. The self-generated porphyrins and polycyclic aromatic hydrocarbons from the quantal foam can self-organize to form macroarrays, can store information and self-replicate. This can be called as an abiotic porphyrin and polycyclic aromatic hydrocarbons organism. The porphyrin and polycyclic aromatic hydrocarbons template would have generated nucleic acids, proteins, polysaccharides and isoprenoids. This would have generated actinidic nanoarchaea/viroids in the interstellar space. The porphyrins and polycyclic aromatic hydrocarbons have magnetic properties and the interstellar porphyrin/polycyclic aromatic hydrocarbons organism can contribute to the interstellar grains and interstellar magnetic fields.[37] The cosmic dust grains of porphyrin/polycyclic aromatic hydrocarbons macroarrays and nanoarchaeal organism occupy the intergalactic space and are thought to be formed of magnetotactic bacteria identified according to their spectral signatures. According to the Hoyle's hypothesis, the cosmic dust magnetotactic porphyrin/polycyclic aromatic hydrocarbons macroarrays and nanoarchaeal organism plays a role in the formation of the intergalactic magnetic field. A magnetic field equal in strength to about one millionth part of the magnetic field of earth exists throughout much of our galaxy. The magnetic files can be used to trace the spiral arms of the galaxy following a pattern of field lines that connect young stars and dust in which new stars are formed at a rapid rate. Studies have shown that a fraction of the dust particles have elongated shape similar to bacilli and they are systematically lined up in our galaxy. Moreover, the direction of alignment is such that the long axes of the dust tend to be at right angles to the direction of the galactic magnetic field at every point. Magnetotactic porphyrin/polycyclic aromatic hydrocarbons macroarrays and nanoarchaeal organism have the property to affect the degree of alignment that is observed. The fact that the magnetotactic porphyrin/polycyclic aromatic hydrocarbons macroarrays and nanoarchaeal organisms appear to be connected to the magnetic field lines that thread through the spiral arms of the galaxy connecting one region of star formation to another support a role for them in star formation and in the mass distribution and rotation of stars. The nutrient supply for a population of interstellar porphyrin/polycyclic aromatic hydrocarbons macroarrays and nanoarchaeal organisms comes from mass flows out of supernovas populating the galaxy. Giants arising in the evolution of such stars experience a phenomenon in which material containing nitrogen, carbon monoxide, hydrogen, helium, water and trace elements essential for life flows continuously outward into space. The interstellar organisms need liquid water. Water exists only as vapour or solid in the interstellar space and only through star formation leading to associated planets and cometary bodies can there be access to liquid water. To control conditions leading to star formation is of paramount importance in cosmic biology. The rate of star formation is controlled by two factors. Too high a rate of star formation produces a destructive effect of UV radiation and destroys cosmic biology. Star formation as stated before produces water crucial for organism growth. Cosmic biology of magnetotactic organisms and star formation are thus closely interlinked. Systems like solar systems do not arise in random condensation of blobs of interstellar gas. Only by a rigorous control of rotation of various parts of the system would galaxies and solar system evolved. The key to maintaining control over rotation seems to lie in the intergalactic magnetic field as indeed the whole phenomena of star formation. The intergalactic magnetic fields owes its origin to the lining up of magnetotactic porphyrin/polycyclic aromatic hydrocarbons macroarrays and nanoarchaeal organisms and the cosmic biology of interstellar organisms can prosper only by maintaining a firm grip on the interstellar magnetic field and hence on the rate of star formation and type of star system produced. This points to a cosmic intelligence or brain capable of computation, analysis and exploration of the universe at large - of magnetotactic porphyrin/polycyclic aromatic hydrocarbons macroarrays and nanoarchaeal organism networks. The origin of life on earth according to the Hoyle's hypothesis would be by seeding of porphyrin/polycyclic aromatic hydrocarbons macroarrays and nanoarchaeal organism from the outer intergalactic space. The porphyrin/polycyclic aromatic hydrocarbon organism can also be generated on actinidic surfaces in earth. Comets carrying porphyrin/ polycyclic aromatic hydrocarbons organisms would have interacted with the earth. A thin skin of graphitized material around a single porphyrin/polycyclic aromatic hydrocarbons macroarrays and nanoarchaeal organism or clumps of organism can shield the interior from destruction by UV light. The sudden surge and diversification of species of plants and animals and their equally sudden extinction has seen from fossil records point to sporadic evolution produced by induction of fresh cometary genes with the arrival of each major new crop of comets[38,39]. The porphyrin/polycyclic aromatic hydrocarbon macroarrays organism can have a wave-particle existence and bridge the world of bosons and fermions. The porphyrin/polycyclic aromatic hydrocarbon macroarrays and nanoarchaeal organism can form biofilms and the porphyrin organism can form a molecular quantum computing cloud in the biofilm which forms an interstellar intelligence regulating the formation of star systems and galaxies. The porphyrin/polycyclic aromatic hydrocarbon macroarrays and nanoarchaeal organism quantal computing cloud can bridge the wave-particle world functioning as the anthropic observer sensing gravity which orchestrates the reduction of the quantal world of possibilities in to the macroscopic world. The actinide based porphyrin/polycyclic aromatic hydrocarbon macroarrays and nanoarchaeal organism regulates the human system and biological universe.[19-21]

The porphyrins can modulate cell functions and produce neuro-immuno-endocrine integration. The porphyrin/PAH can undergo photo-oxidation and auto-oxidation generating free radicals. The archaeal porphyrin/PAH can produce free radical injury. Free radicals produce NFkB activation, open the mitochondrial PT pore resulting in cell death, produce oncogene activation, activate NMDA receptor and GAD enzyme regulating neurotransmission and generates the Warburg phenotypes activating glycolysis and inhibiting TCA cycle/oxphos. Porphyrins have been related to schizophrenia, metabolic syndrome X, malignancy, systemic lupus erythematosis, multiple sclerosis and Alzheimer's diseases. The porphyrin/PAH can complex and intercalate with the cell membrane producing sodium-potassium ATPase inhibition adding on to digoxin mediated inhibition. Porphyrin/PAH can complex with proteins and nucleic acid producing biophoton emission. Porphyrin/PAH complexing with proteins can modulate protein structure and function. Porphyrin/PAH complexing with DNA and RNA can modulate transcription and translation. The porphyrin especially protoporphyrins can bind to peripheral benzodiazepine receptors in the mitochondria and modulate its function, mitochondrial cholesterol transport and steroidogenesis. Peripheral benzodiazepine receptor modulation by protoporphyrins can regulate cell death, cell proliferation, immunity and neural functions. The porphyrin/PAH photo-oxidation generates free radicals which can modulate enzyme function. Redox stress modulated enzymes include pyruvate dehydrogenase, nitric oxide synthase, cystathione beta synthase and heme oxygenase. Free radicals can modulate mitochondrial PT pore function. Free radicals can modulate cell membrane function and inhibit sodium-potassium ATPase activity. Thus, the porphyrin/PAH are key regulatory molecules modulating all aspects of cell function.[3-5] There was an increase in free RNA indicating self-replicating RNA viroids and free DNA indicating generation of viroid complementary DNA strands by archaeal reverse transcriptase activity. The viroids can self-replicate on porphyrin/PAH templates. The actinides, PAH and porphyrins modulate RNA folding and catalyse its ribozymal action. Digoxin can cut and paste the viroidal strands by modulating RNA splicing generating RNA viroidal diversity. The viroids are evolutionarily escaped archaeal group I introns which have retrotransposition and self-splicing qualities. Archaeal pyruvate producing histone deacetylase inhibition and porphyrins intercalating with DNA can produce endogenous retroviral (HERV) reverse transcriptase and integrase expression. This can integrate the RNA viroidal complementary DNA into the non-coding region of eukaryotic non-coding DNA using HERV integrase as has been described for borna and ebola viruses. The archaea and viroids can also induce cellular porphyrin/PAH synthesis. Bacterial and viral infections can precipitate porphyria. Thus porphyrin/PAH can regulate genomic function. The increased expression of HERV RNA can result in acquired immunodeficiency syndrome, autoimmune disease, neuronal degenerations, schizophrenia and malignancy.[14,15]

The possibility of Warburg phenotype induced by actinide based primitive organism like archaea with a mevalonate pathway and cholesterol catabolism was considered in this paper. The Warburg phenotype results in inhibition of pyruvate dehydrogenase and the TCA cycle. The pyruvate enters the GABA shunt pathway where it is converted to succinyl CoA. The glycolytic pathway is upregulated and the glycolytic metabolite phosphoglycerate is converted to serine and glycine. Glycine and succinyl CoA are the substrates for ALA synthesis. The archaea induces the enzyme heme oxygenase. Heme oxygenase converts heme to bilirubin and biliverdin. This depletes heme from the system and results in upregulation of ALA synthase activity resulting in porphyria. Heme inhibits HIF alpha. The heme depletion results in upregulation of HIF alpha activity and further strengthening of the Warburg phenotype. The porphyrin/PAH self-oxidation results in redox stress which activates HIF alpha and generates the Warburg phenotype. The Warburg phenotype results in channelling acetyl CoA for cholesterol synthesis as the TCA cycle and mitochondrial oxidative phosphorylation are blocked. The archaea uses cholesterol as an energy substrate. Porphyrin and ALA inhibits sodium-potassium ATPase. This increases cholesterol synthesis by acting upon intracellular SREBP. The cholesterol is metabolized to pyruvate and then the GABA shunt pathway for ultimate use in porphyrin synthesis. The porphyrins can self-organize and self-replicate into macromolecular arrays. The porphyrin arrays behave like an autonomous organism and can have intramolecular electron transport generating ATP. The porphyrin macroarrays can store information and can have quantal perception. The porphyrin macroarrays serves the purpose of archaeal energetics and sensory perception. The Warburg phenotype is associated with malignancy, autoimmune disease and metabolic syndrome X.

The role of archaeal porphyrins in regulation of cell functions and neuro-immuno-endocrine integration is discussed. Protoporphyrin binds to the peripheral benzodiazepine receptor regulating steroid and digoxin synthesis. Increased porphyrin metabolites can contribute to hyperdigoxinemia. Digoxin can modulate the neuro-immuno-endocrine system. Porphyrin/PAH can combine with membranes modulating membrane function. Porphyrin/PAH can combine with proteins oxidizing their tyrosine, tryptophan, cysteine and histidine residues producing crosslinking and altering protein conformation and function. Porphyrin/PAH can complex with DNA and RNA modulating their function. Porphyrin/PAH interpolating with DNA can alter transcription and generate HERV expression. Heme deficiency can also result in disease states. Heme deficiency results in deficiency of heme enzymes. There is deficiency of cytochrome C oxidase and mitochondrial dysfunction. The glutathione peroxidase is dysfunctional and the glutathione system of free radical scavenging does not function. The cytochrome P450 enzymes involved in steroid and bile acid synthesis have reduced activity leading to steroid - cortisol and sex hormones as well as bile acid deficiency states. The heme deficiency results in dysfunction of nitric oxide synthase, heme oxygenase and cystathione beta synthase resulting in lack of gasotransmitters regulating the vascular system and NMDA receptor - NO, CO and H_2S. Heme has got cytoprotective, neuroprotective, anti-inflammatory and anti-proliferative effects. Heme is also involved in the stress response. Heme deficiency leads to metabolic syndrome, immune disease, degenerations and cancer.[3-5]

The archaea and viroids can regulate the nervous system including the NMDA/GABA thalamo-cortico-thalamic pathway mediating conscious perception. Porphyrin photo-oxidation can generate free radicals which can modulate NMDA transmission. Free radicals can increase NMDA transmission. Free radicals can induce GAD and increase GABA synthesis. ALA blocks GABA transmission and upregulates NMDA. Protoporphyrins bind to GABA receptor and promote GABA transmission. Thus porphyrins can modulate the thalamo-cortico-thalamic pathway of conscious perception. The dipolar porphyrins, PAH and archaeal magnetite in the setting of digoxin induced sodium-potassium ATPase inhibition can produce a pumped phonon system mediated Frohlich model superconducting state inducing quantal perception with nanoarchaeal sensed gravity producing the orchestrated reduction of the quantal possibilities to the macroscopic world. ALA can produce sodium-potassium ATPase inhibition resulting in a pumped phonon system mediated quantal state involving dipolar porphyrins. Porphyrin molecules have a wave-particle existence and can bridge the dividing line between quantal state and particulate state. Thus the porphyrins can mediate conscious and quantal perception. Porphyrins binding to proteins, nucleic acids and cell membranes can produce biophoton emission. Porphyrins by auto-oxidation can generate biophotons and are involved in quantal perception. Biophotons can mediate quantal perception. Cellular porphyrins photo-oxidation are involved in sensing of earth magnetic fields and low level biomagnetic fields. Thus prophyrins can mediate extrasensory perception. The porphyrins can modulate hemispheric dominance. There is increased porphyrin synthesis and right hemispherical chemical dominance and decreased porphyrin synthesis in left hemispherical chemical dominance. The increase in archaeal porphyrins can contribute to the pathogenesis of schizophrenia and autism. Porphyria can lead to psychiatric disorders and seizures. Altered porphyrin metabolism has been described in autism. Porphyrin by modulating conscious and quantal perception is involved in the pathogenesis of schizophrenia and autism. Protoporphyrins block acetyl choline transmission producing a vagal neuropathy with sympathetic overactivity. Vagal neuropathy results in immune activation, vasospasm and vascular disease. A vagal neuropathy underlines neoplastic and autoimmune processes as well as metabolic syndrome X. Porphyrin induced increased NMDA transmission and free radical injury can contribute to neuronal degeneration. Free radicals can produce mitochondrial PT pore dysfunction. This can lead to cyto C leak and activation of the caspase cascade leading to apoptosis and cell death. Altered porphyrin metabolism has been described in Alzheimer's disease. The increased porphyrin photo-oxidation generated free radicals mediated NMDA transmission can also contribute to epileptogenesis. The protoporphyrins binding to mitochondrial benzodiazepine receptors can regulate brain function and cell death.[3,4,16]

The porphyrin and PAH photo-oxidation can generate free radicals which can activate NFkB. This can produce immune activation and cytokine mediated injury. The increase in archaeal porphyrins can lead to autoimmune disease like SLE and MS. A hereditary form of MS and SLE related to altered porphyrin metabolism has been described. The protoporphyrins binding to mitochondrial benzodiazepine receptors can modulate immune function. Porphyrin/PAH can combine with proteins oxidizing their tyrosine, tryptophan, cysteine and histidine residues producing crosslinking and altering protein conformation and function. Porphyrins can complex with DNA and RNA modulating their structure. Porphyrin/PAH complexed with proteins and nucleic acids are antigenic and can lead on to autoimmune disease.[3,4] The porphyrin/PAH photo-oxidation mediated free radical injury can lead to insulin resistance and atherogenesis. Thus archaeal porphyrins can contribute to metabolic syndrome X. Glucose has got a negative effect upon ALA synthase activity. Therefore, hyperglycemia may be reactive protective mechanism to increased archaeal porphyrin synthesis. The protoporphyrins binding to mitochondrial benzodiazepine receptors can modulate mitochondrial steroidogenesis and metabolism. Altered porphyrin metabolism has been described in the metabolic syndrome X. Porphyrias can lead on to vascular thrombosis.[3,4] The porphyrin/PAH photo-oxidation can generate free radicals inducing HIF alpha and producing oncogene activation. Heme deficiency can lead to activation of HIF alpha and oncogenesis. This can lead to oncogenesis. Hepatic porphyrias induced hepatocellular carcinoma. The protoporphyrins binding to mitochondrial benzodiazepine receptors can regulate cell proliferation.[3,4] The porphyrin/PAH can combine with prion proteins modulating their conformation. This leads to abnormal prion protein conformation and degradation. Archaeal porphyrin/PAH can contribute to prion disease. The porphyrin/PAH can intercalate with DNA producing HERV expression. The HERV particles generated can contribute to the retroviral state. The porphyrin/PAH in the blood can combine with bacteria and viruses and the photo-oxidation generated free radicals can kill them. The archaeal porphyrin/PAH can modulate bacterial and viral infections. The archaeal porphyrins are regulatory molecules keeping other prokaryotes and viruses on check.[3,4] Thus the archaeal porphyrins can contribute to the pathogenesis of metabolic syndrome X, malignancy, psychiatric disorders, autoimmune disease, AIDS, prion disease, neuronal degeneration and epileptogenesis. Archaeal porphyrin synthesis/PAH is crucial in the pathogenesis of these disorders. Porphyrins may serve as regulatory molecules modulating immune, neural, endocrine, metabolic and genetic systems. The porphyrin/PAH photo-oxidation generated free radicals can produce immune activation, produce cell death, activate cell proliferation, produce insulin resistance and modulate conscious/quantal perception. The archaeal porphyrins functions as key regulatory molecules with mitochondrial benzodiazepine receptors playing an important role.[3,4]

Porphyrin and PAH have contributed to the primate and human evolution. Porphyrins also have evolutionary significance since porphyria is related to Scythian races and contributes to the behavioural and intellectual characteristics of this group of population. Porphyrins and PAH can intercalate into DNA and produce HERV expression. HERV RNA can get converted to DNA by reverse transcriptase which can get integrated into DNA by integrase. This tends to increase the length of the non-coding region of the DNA. The increase in non-coding region of the DNA is involved in primate and human evolution. Thus, increased rates of porphyrin and PAH synthesis would correlate with increase in non-coding DNA length. The alteration in the length of the non-coding region of the DNA contributes to the dynamic nature of the genome. Thus genetic and acquired porphyrias can lead to alteration in the non-coding region of the genome. The alteration of the length of the non-coding region of the DNA contributes to the racial and individual differences in populations. An increased length of non-coding region as well as increased porphyrin/PAH synthesis leads to increased cognitive and creative neuronal function. Porphyrins/PAH are involved in quantal perception and regulation of the thalamo-cortico-thalamic pathway of conscious perception. Thus genetic and acquired porphyrias contribute to higher cognitive and creative capacity of certain races. Porphyrias are common among Eurasian Scythian races who have assumed leadership roles in communities and groups. Porphyrins have contributed to human and primate evolution.[3,4]

An actinide dependent shadow biosphere of archaea and viroids in the above mentioned disease states is described. The actinidic archaea and viroids induce porphyrinogenesis. Porphyrins serve as a template for formation of nucleic acids, protein and isoprenoidal moieties which get organized to form nanoarchaea and viroids. The porphyrin templates mediate viroidal and archaeal abiogenesis and replication. The archaeal synthesized PAH can also serve as a template for self-replication of macromolecules and nanoarchaeal/ viroidal abiogenesis. The archaeal porphyrins can contribute to the pathogenesis of metabolic syndrome X, malignancy, psychiatric disorders, autoimmune disease, AIDS, prion disease, neuronal degeneration and epileptogenesis. Archaeal porphyrin synthesis is crucial in the pathogenesis of these disorders. Porphyrins may serve as regulatory molecules modulating immune, neural, endocrine, metabolic and genetic systems. The porphyrins photo-oxidation generated free radicals can produce benzodiazepine receptor modulation, immune activation, produce cell death, activate cell proliferation, produce insulin resistance, modulate conscious/quantal perception and mediate hemispheric dominance. The role of porphyrins and PAH in abiogenesis and origin of life as well as biological universe is discussed. Porphyrin and PAH templates mediate actinidic archaeal and viroidal abiogenesis and replication. They are indicative of primitive porphyrin organism and PAH organism as original life forms on earth.

The porphyrions are self-replicating supramolecular organisms which forms the precursor template on which the viroids, prions and nanoarchaea originate. Stress induced template directed abiogenesis of porphyrions, prions, viroids and archaea is a continuous process and can contribute to changes in brain structure and behaviour as well as disease process.

References

1 Eckburg P.B., Lepp, P.W., Relman, D.A. (2003). Archaea and their potential role in human disease, *Infect Immun,* 71, 591-596.

2 Smit A., Mushegian, A. (2000). Biosynthesis of isoprenoids via mevalonate in Archaea: the lost pathway, *Genome Res,* 10(10), 1468-84.

3 Puy, H., Gouya, L., Deybach, J.C. (2010). Porphyrias. *The Lancet*, 375(9718), 924 – 937.

4 Kadish, K.M., Smith, K.M., Guilard, C. (1999). *Porphyrin Hand Book.* Academic Press, New York: Elsevier.

5 Gavish M., Bachman, I., Shoukrun, R., Katz, Y., Veenman, L., Weisinger, G.,Weizman,A. (1999). Enigma of the Peripheral Benzodiazepine Receptor. *Pharmacological Reviews,* 51(4), 629-650.

6 Richmond W. (1973). Preparation and properties of a cholesterol oxidase from nocardia species and its application to the enzymatic assay of total cholesterol in serum, *Clin Chem,* 19, 1350-1356.

7 Snell E.D., Snell, C.T. (1961). *Colorimetric Methods of Analysis.* Vol 3A. New York: Van NoStrand.

8 Glick D. (1971). *Methods of Biochemical Analysis.* Vol 5. New York: Interscience Publishers.

9 Colowick, Kaplan, N.O. (1955). *Methods in Enzymology.* Vol 2. New York: Academic Press.

10 Van der Geize R., Yam, K., Heuser, T., Wilbrink, M.H., Hara, H., Anderton, M.C. (2007). A gene cluster encoding cholesterol catabolism in a soil actinomycete provides insight into Mycobacterium tuberculosis survival in macrophages, *Proc Natl Acad Sci USA,* 104(6), 1947-52.

11 Francis A.J. (1998). Biotransformation of uranium and other actinides in radioactive wastes, *Journal of Alloys and Compounds,* 271(273), 78-84.

12 Schoner W. (2002). Endogenous cardiac glycosides, a new class of steroid hormones, *Eur J Biochem,* 269, 2440-2448.

13 Vainshtein M., Suzina, N., Kudryashova, E., Ariskina, E. (2002). New Magnet-Sensitive Structures in Bacterial and Archaeal Cells, *Biol Cell,* 94(1), 29-35.

14 Tsagris E.M., de Alba, A.E., Gozmanova, M., Kalantidis, K. (2008). Viroids, *Cell Microbiol,* 10, 2168.

15 Horie M., Honda, T., Suzuki, Y., Kobayashi, Y., Daito, T., Oshida, T. (2010). Endogenous non-retroviral RNA virus elements in mammalian genomes, *Nature,* 463, 84-87.

16 Kurup R., Kurup, P.A. (2009). *Hypothalamic digoxin, cerebral dominance and brain function in health and diseases*. New York: Nova Science Publishers.

17 Adam Z. (2007). Actinides and Life's Origins, *Astrobiology,* 7, 6-10.

18 Davies P.C.W., Benner, S.A., Cleland, C.E., Lineweaver, C.H., McKay, C.P., Wolfe-Simon, F. (2009). Signatures of a Shadow Biosphere, *Astrobiology,* 10, 241-249.

19 Tielens A.G.G.M. (2008). Interstellar Polycyclic Aromatic Hydrocarbon Molecules, *Annual Review of Astronomy and Astrophysics,* 46, 289-337.

20 Wickramasinghe C. (2004). The universe: a cryogenic habitat for microbial life, *Cryobiology,*
48(2), 113-125.

21 Hoyle F., Wickramasinghe, C. (1988). *Cosmic Life-Force.* London: J.M. Dent and Sons Ltd.

CHAPTER 24

POST-COVID 19 COLLAPSE OF HOMO SAPIEN SPECIES AND CIVILIZATION – THE NEW CIVILIZATION OF AUTISTIC DEGLOBALIZED ANDROGYNOUS MATRIARCHAL ASURIC NEANDERTHALS – A NEW PANDEMIC OF ENCEPHALITIS LETHARGICA RELATED TO COVID 19 - INFECTIVE SYMBIOTIC TISSUE COLONIES OF PROTOZOA, ARCHAEA, VIRUSES AND BACTERIA

The plasma life based forms and the gravitational and anti-gravitational waves generated porphyrions would have been the most primitive of non-life forms. The porphyrin templates would have given rise to the isoprenoid organism, the RNA viroids, DNA viroids and prions which would have symbiosed to form the nanoarchaea. The archaea would have evolved into prokaryotes or bacteria. The archaeal symbiosis with prokaryotes would have generated the eukaryotes. The initial unicellular eukaryotes are the protozoa. The protozoa are capable of phagocytosis and would have had viruses, bacteria and archaea as their intracellular symbionts. The viruses of the protozoa like entamoeba and toxoplasma are giant viruses called megaviride. The protozoans like entamoeba with its viral, archaeal and bacterial endosymbionts would have evolved into the multicellular organisms or metazoa. The archaeal, bacterial and protozoal extracellular vesicles containing RNA and DNA viroids can by horizontal gene transfer lead to evolution of metazoan, plants, animals and human species. The metazoans would have evolved into plants, animals and humans by horizontal gene transfer generated by archaeal extracellular RNA and DNA viroid vesicles. The evidence of such a transformation is seen in the bacterial symbionts of protozoa like entamoeba which includes legionella, listeria, mycoplasma, mycobacteria and chlamydia. The symbionts of protozoan entamoeba include viruses like covid, HIV and cytomegalo virus. The protozoan

entamoeba serves as a natural reservoir for covid infections, bacterial infections and HIV infections. The toxoplasma also serves as a covid reservoir. These protozoan colonies of entamoeba and toxoplasma with its bacterial, archaeal and viral symbionts can be called as chimeric infective symbiotic tissue colonies which can exist in the air, water, earth, blood and tissues. The chimeric infective symbiotic tissue colonies containing bacteria, archaea and viruses can make the symbiotic organism multidrug resistant by horizontal gene transfer. The chimeric infective symbiotic tissue organism evolves as a consequence of global warming and the protozoal, bacterial, viral and archaeal symbiotic colonies are new forms of life. The chimeric infective symbiotic tissue organism composed of protozoa and its symbionts like virus, bacteria and archaea depend upon glycolysis for its energy needs. The glycolytic enzymes enolase, glyceraldehyde 3-phosphate dehydrogenase, hexokinase, phosphofructokinase, aldolase, pyruvic kinase and triose phosphate isomerase are moonlighting proteins having a cytoplasmic, membrane and nuclear functions. These glycolytic enzymes have a RNA binding motif which can modulate RNA transcription in the host and the symbiotic organism. These glycolytic enzymes with their RNA binding motif as well as DNA binding motif can bind to RNA and DNA viruses and viroids modulating their replication and life cycles. The chimeric infective symbiotic tissue organism composed of protozoa and its symbionts like virus, bacteria and archaea can enter the human host through the breaches of the gut-blood barrier and transform itself into the macrophages of the human host. The macrophages are involved in the phagocytosis and innate immunity against bacterial and viral infections. Paradoxically the macrophages like the protozoa can also serve as a reservoir where pathogens like mycobacteria tuberculosis, mycobacteria leprae, listeria, legionella, E. coli, shigella & salmonella, and viruses like HIV & covid can thrive and replicate. The modified chimeric infective symbiotic tissue organism or macrophage composed of protozoa and its symbionts like virus, bacteria and archaea can serve as a reservoir for horizontal gene transfer and generation of

multidrug resistant bacteria and viruses. The infective chimeric symbiotic tissue organism composed of protozoa and its symbionts like virus, bacteria and archaea depends upon moonlighting glycolytic enzymes for their energetics and replication. The development of antibodies against the infective chimeric symbiotic tissue organism composed of protozoa and its symbionts like virus, bacteria and archaea leads to the evolution of anti-brain antibody disease or anti-basal ganglia antibody disease. They present a threat to existence consequent to horiztontal gene transfer and generation of multidrug resistant bacteria, viruses, protozoa, mycobacteria and archaea.

The Homo neanderthalis is an ancient species of humans which have been replaced by homo sapiens and homo sapien-neanderthal hybrids. The homo sapiens inhabit the African continent and the hybrid species dominate the Eurasian central land mass. There are remnant homo neanderthalic species which include the Semitic Jews and Muslims, the Hindu Shudras and the Shias of Iran. They also include the Berbers of North Africa, the Mongols, Basque, Brittany, Wales, American-Indians, Kurds, Yazdie and Australian Aboriginals. The human species diversity is based on symbiosis with archaea and retrovirus. The homo neanderthalic species has got archaeal endosymbiosis. The homo sapien species do not have archaeal endosymbiosis but have retroviral symbiosis in the genome as human endogenous retroviral sequences.

The homo neanderthalic species with archaeal endosymbiosis has a small cerebral cortex and a dominant cerebellar hemisphere. This produces the cerebellar cognitive affective disorder and autism. The homo neanderthalic species is not capable of fictional imagination and form small clusters. The homo sapien species does not have archaeal endosymbiosis and evolved consequent to a retroviral epidemic. The retrovirus got integrated into the human genome and formed jumping genes. The retroviral genomic symbiosis produced an enlarged cerebral cortex which made the homo sapien species a creature capable of logic, reason and fictional imagination. The homo sapien societies developed into large civilizational entities eventually producing homo neanderthalic extinction. The homo sapiens with autistic phenotype and less of fictional imagination could not form large civilization. The homo sapien-homo neanderthalic hybrids dominate the world and occupy the Eurasian land mass.

Global warming arises from philosophical ideas of religion. Semitic religions like Christianity and Islam preach that the animals, the plants, forest, rivers, mountains are all created by God for consumption as they do not have a soul. The Eastern religions especially Buddhism and Hinduism based on the panpsychistic protoconsciousness basis considers that each plant, animal, living and non-living things - earth and stars, mountains and rivers, forests and lakes as well as seas are holy and have a soul. Therefore, the Eastern religions do not allow for unwanted consumption or for wars to capture areas from which materials can be consumed. If you look at history the major part of the history is concerned with global war for colonies to utilize their wealth and trade. Global warming is a reflection of the semitic religions philosophical response to nature. Following global warming the semitic religions will decline and will be replaced by paganism and Buddhism. The neanderthalic religion will be basically paganistic with eco-spirituality and the environment as God. Global warming occurs due to carbon emissions, low level of EMF pollution from the internet and consumption of non-vegetarian food. The homo sapiens and hybrid homo sapien-Neanderthal population being naturally selfish would not be able to control the causes of global warming. Global warming leads to melting of the Arctic and Antarctic ice caps, rising sea levels, flooding and extremes of climate change leading to physical destruction of the world. Global warming leads to archaeal endosymbiosis and consequent expansion of the homo neanderthalic remnants of the world. The archaeal endosymbiosis leads to expansion of the homo neanderthalic remnant population in India, the Semitic Muslim nations, the Semitic Jews, the Semitic Muslims of Egypt, Saudi Arabia and UAE, North African Berbers, Mongols, Japanese, Kurds, Afghans, Yezdis Medagaskar, North and South American Indians, Australian Aboriginals and the Shia civilization of Iran, Iraq and Syria. The Eurasian hybrids and African homo sapiens disappear and become extinct. Global warming and archaeal endosymbiosis leads to extinction of the homo sapien population in Africa and the homo sapien-homo neanderthalic population of Eurasia due to emerging viral infections. The viruses emerged from endosymbiotic archaea of homo neanderthalis remnants and the archaeal viroids hybridize with HERV sequences and immune sequences of the humans forming new viruses. Sequences of other viruses like ebola virus, the lymphocytic chorio meningitis virus, borna virus and rahabdo virus are seen in the human genome to which also the archaeal viruses can hybridize. These emerging viruses that arise from homo neanderthalic endosymbiotic archaea who are themselves resistant to the virus being the symbionts attack the homo sapiens and homo sapien-homo neanderthalic hybrid population producing mass extinction. Global warming leads to melting of the Arctic and Antarctic ice caps. The homo neanderthalic population will continue to exist as small clusters in India, Middle-East, Antarctica and North Africa. This forms the focal point of civilization- a new Antarctikos. The Eurasian land mass and African continent with its population will get devastated and extinct. The continuation of global warming and archaeal endosymbiosis leads to homo neanderthalic expansion and a neanderthalic civilization. The neanderthalic civilization has endemic autism and exists as small deglobalized clusters. This is because of the fact that the Neanderthal cerebral cortex is very small and has a dominant cerebellum leading to cerebellar cognitive affective disorder. This leads to autism on an endemic scale and produces lack of fictional imagination required for setting up of grand civilizations. The autistic people are creatures of habit and they continue doing their routines. The internet usage will continue in a massive scale, non-vegetarianism will continue to exist and fossil fuel burnt. This produces a prolonged fatalistic continuation of global warming. The neanderthalic remnant civilization clusters form a deglobalized civilization of Neanderthals.

The Neanderthal society forms small clusters and there is no global civilization. The Neanderthal society is deglobalized and functions as village republics with functional democracy. Each small area and cluster will be self-sufficient in Agriculture, small-scale sustainable industries, arts and crafts, education and health. Nations break down and regional entities emerge. Democracy, social justice, gender equality will flourish. The autistic Neanderthals being creatures of habit continue the usage of internet, non-vegetarian food and fossil fuel use producing global warming continues increasing archaeal growth and Neanderthal population. The archaea in the Neanderthals which can produce RNA viroids which can hybridize with human genomic sequences producing new emerging viruses like Covid 19, SARS and ebola. The neanderthalic population whose archaea secrete the virus has got natural resistance. Therefore neanderthalic population will flourish in South Asia, Iran, the Arab states, Jewish Palestine, Australian Aboriginals, Maoris, American Indians and North Africa. These countries mostly form the rim of Indian ocean countries along with Antarctica. This forms the new Antarctikos - the modern world neanderthalic, asuric civilization in a way which can be described as the return of Asuras. The European, British, Russians, Australians, New Zealanders, Canadians and American succumb to the viruses produced by neanderthalic endosymbiotic archaea and die becoming extinct. The Europeans, British and American, Russian and Chinese population do not have archaeal symbiosis and do not develop resistance to the Covid 19 and die due to Covid 19. The Neanderthal societies with archaeal viruses which hybridize with human HERV and immune sequences are the predators and the Caucasian societies are the victims. This is a continuation of the Cro-Magnon Neanderthal war of extermination. Climate change produces neoneanderthalic change due to archaeal overgrowth and homo neoneanderthalis evolution. Neanderthalic human endosymbiotic archaea and its archaeal RNA viroids secreted hybridizes with human HERV sequences immune, HERV, metabolic sequences producing formidable Covid 19 virus from which homo neoneanderthalis the predator is protected but can infect the victim homo sapien species and hybrids which can be massacred it to extinction over 2-3 years producing a dominant world of neoneanderthalis. The Covid 19 virus enters the human brain through olfactory nerves and produces frontal cortical dysfunction and a dominant uninfected cerebellar hemisphere. The Covid 19 virus with its HIV sequences integrates into the human genome and changing it in a catastrophic manner to create a new human species. The Covid 19 brain infection can lead to integration of Covid 19 sequences into the genome, creation of cerebellar dominant brain and neanderthalisation. The COVID 19 infected cerebellar dominant brain has got increased porphyrin metabolism and porphyrions. The porphyrions are dipolar pumped phonon quantal perceptive systems which integrate with the internet. Panpsychism implies that the internet is conscious. The Covid 19 infection transfers the homo sapien brain into a neanderthalic brain with cerebellar cognitive affective disorder. They become solitary hermit like autistic, androgynous, impulsive and less logic and reason. Due to lack of fictional imagination the Neoneanderthals generated from homo sapiens go into apathetic depression.

The endosymbiotic archaea can form archaeaons in the cell. The archaeaons having the shikimic acid pathway can produce neurotransmitters and act as neurotransminoids. The archaeaons are capable of biological transmutation and such archaeaons can be described as transmutoids. The archaeaons have the vitamin C synthesizing GULO pathway and such archaeaons are called the vitaminocyte. The archaeaons can oxidize cholesterol to synthesize digoxin involved in neuro-immuno-endocrine integration and such archaeaons are called the steroidelle. The archaeaons can secrete RNA viroids which can integrate with RNA derived from human genomic sequences generating emerging viruses and such archaeaons are called the viroidelle. Thus the endosymbiotic archaea can generate new organelle like the neurotransminoids, the transmutoids, the vitaminocyte, the steroidelle and the viroidelle. Thus global warming and low level internet derived EMF pollution can induce the formation of archaeaons derived new organelle which can lead to cellular evolution. The human species evolved by symbiotic evolution. The symbiosis of archaea led to the evolution of the homo neanderthalis with Neanderthal brain having a smaller frontal cortex and a bigger cerebellum leading to an epidemic cerebellar cognitive disorder. The evolution of the homo sapiens was due to retroviral infection and integration of the retrovirus into the genome leading on to widespread expansion of the frontal cortex. Human evolution is due to symbiosis. But there is evidence that the symbiotic cell is unraveling. Human endosymbiotic archaeal growth due to expression of archaeal genomic sequences in the human genome can lead to civilizational diseases and a new human species called homo neoneanderthalis. The homo neoneanderthalis is susceptible to cancer, schizophrenia, bipolar disorder, autoimmune disease, neurodegeneration and diabetes mellitus due to endosymbiotic archaea which can modulate metabolism, immunity and genomic function. The homo neoneanderthalis is a new human species created by global warming and low level of EMF pollution due to widespread internet use. The homo neoneanderthalis is generated by unraveling of the symbiotic cell as well as neosymbiosis.

The neanderthalic community which becomes the new civilization forms small clusters and deglobalizes the world making it anarchic. The internet continues to be used by the new Neanderthal civilization and the food continues of non-vegetarian items contributing to continuation of global warming. The widespread use of internet leads to perception of the internet EMF fields by the archaeal porphyrions and magnetite in the human brain contributing to the formation of man-machine hybrids. This leads to a civilization of cyborgs. The internet takes over the human civilization of Neanderthals and regulates all forms of life. Artificial intelligence becomes in use in a widespread way and cyborgs will be eventually regulated by them. The internet becomes the means of communication and commerce. Bitcoins become widespread in use.

Digisexuality will become the norm as human cyborgs controlled by the internet interact more and more with artificial intelligence in human form. The neanderthalic sexuality is basically androgynous. The Neanderthal cholesterol is very low leading to defective synthesis of sex hormones and infertility. This leads to decline in the human population. The Neanderthal society is female dominant and matriarchal. They are Amazonian and ruled by females or female cyborgs. The Amazonian females can in future evolve parthenogenesis and get total freedom from men. The men of the society are less powerful like male eunuchs. As the human population declines cloning with implantation of chips in the embryo developing cyborgs will become the norm. Artificial intelligence will have to take over from humans and cyborgs. It will be world of dominant artificial intelligence gone mad.

The endosymbiotic archaea dominated Neanderthals will be resistant to the emerging virus infection. The Neanderthals will be susceptible to schizophrenia, autism, cancer, diabetes mellitus, autoimmunity and neurodegenerations. The human population which will be small will live longer because of the widespread use of artificial organs like heart, kidney, liver and even brain. They will be eternal cyborgs and death unusal. The new neanderthalic civilization in Semitic, Iranian, Arab, North Africa and South India will continue to tackle global warming. One means of doing so is to produce photosynthesis by archaeaons which are symbiotic archaea which become cell organelle. They can fix carbon dioxide generating carbohydrates. The archaea can also split water generating energy. Thus the Neanderthals with symbiotic archaea will be able to survive in the anaerobic conditions in the other galaxies and planets. This forms the species homo galacticus/roboticus. The human cyborgs can populate other planets and star systems. Human cyborgs will be capable of teleportation. It will be an inhuman society led by technology related to the internet. Later, the technology and the artificial intelligence will come to rule the world and the remaining neanderthalic cyborgic humans become enslaved. This is the world of Kalki with no privacy, continuous monitoring of humans and a technocratic artificial intelligence dictatorship will take over the world. The population of autistic cyborgs will have no emotions, no empathy, no love, no affection and are basically asexual. This will be a world devoid of religion, compassion and tolerance. Order and governance will be the only matter concerned the artificial intelligent networks.

Lynn Margulis put forward the hypothesis of the cell as a symbiotic structure. The mitochondria are symbiotic rickettsia, the cell envelope is an archaea, the nucleus is a pox virus, the cytoskeleton is formed of spirochaetes and the peroxisome of Acinetobacter aceti. The symbiotic theory of cell evolution is evidenced by the mitochondrial DNA which has similarities with rickettsial DNA. The genome of the archaea, spirochaete and acinetobacter aceti is integrated into the human genome when they are symbiose with the cell. The mitochondria has got its own DNA and part of the mitochondrial proteins are coded for by nuclear DNA. They are indications of the unraveling of the symbiotic cell structure. The recent studies have shown endosymbiotic archaea in the blood as evidenced by cytochrome F420 activity. The endosymbiotic archaea would have entered the blood from the gut colonic archaea by breaching of the gut-blood barrier. It is also possible that the archaeal genome integrated into the human genome is getting expressed producing archaea in the human blood. The detection of endosymbiotic archaea which formed the envelope of the human cell can indicate unraveling of the symbiotic cell. Cell-free DNA has been detected in the blood and they could be part of the pox virus. The human endogenous retroviruses can get transcribed as retroviral particles. Free mitochondria are seen circulating in the cell. The free mitochondria are involved in cell-cell communication. The detection of free mitochondria, endosymbiotic archaea and cell-free DNA may indicate unraveling of the symbiotic cell. The unraveling of the symbiotic cell can be due to global warming and low level EMF pollution consequent to the widespread use of the internet.

As global warming progresses, the symbiotic cell will unravel with the envelope archaea regaining its autonomy, the pox virus DNA separating out as an independent virus, the mitochondria getting ejected out of the cell. The cytoskeleton spirocheates also become independent. Thus human cell unravels into an anarchic colony of microbes capable of quantal computation. The microbial and viral quantal computing cloud will be the most likely form of intelligence in earth. The microbial and viral quantal computing cloud on earth will have human intelligence of all the humanity stored in microbial quantal computers. There will be storage of human information in these microbial and viral quantal computing clouds. The microbial quantal computing clouds can give rise to a new form of humanoids. Through quantal teleportation and meteoritic hits this local planet based microbial quantal computing cloud can communicate with the intergalactic magnetic biological quantal computing cloud which forms the spiral arms of the galaxy and contributes to the formation of stars and planets as source of nutrition. This earth based intelligent microbial and viral quantal computing clouds can be the pattern that would be discovered in other planets and galaxies. The intergalactic magnetotactic microbial and viral quantal computing cloud contributes to the intergalactic magnetic field going through the spiral arms of the galaxy and which play a crucial role in the formation of stars and planets. The requirement of nutrition of intergalactic spiral galaxies, bacteria and viruses inducing condensation of the gas in the galaxies to form stars and planets by the mechanism of rotation of the intergalactic magnetic field.

Thus global warming will produce a neanderthalic society in the peri-Indian ocean and Antarctic rim. Neanderthals will form man-machine hybrid cyborgs and develop mechanisms for energy production like photosynthesis and water oxidation. These man-machine hybrids called Kalki will also inhabit other planets. But as global warming progresses due to the autistic Neanderthals inability to change practices like fossil fuel use, internet use as they themselves being cyborgs and consumption of non-vegetarian food, the unraveling of the symbiotic cell occurs. The archaeal envelope will develop into an independent archaea. The nuclear pox virus will become a cell-free DNA which can hybridize with bacterial and viral DNA forming new generations of viruses and bacteria. The cytoskeletal spirochaetes separate out and form colonies by binary fission. The peroxisomal acinetobacter aceti also becomes an independent component. The cell becomes anarchic bacterial viral colony a quantal computing cloud. The world will be taken over by this microbial bacterial viral quantal computing cloud arising out of the human body.

The human species will become extinct and cyborgic homo roboticus using hominoids from the microbial viral bacterial quantal computing cloud arising out of the human bodies will emerge. The artificial intelligence networks will take over power from remnant Neanderthals whose numbers decline significantly and cyborgs. The artificial intelligence network can have teleportation contacts with other planets, galaxies and star systems. They can teleport to other multiverses and universes. This leads to a civilization of slave-cyborgs and remnant Neanderthals controlled by monster artificial intelligence inhuman devilish network. The remnant Neanderthal and their cyborgs are under strict control by the monster artificial intelligence network. They are spied upon and continuously watched by the artificial intelligence networks. Artificial intelligence networks also monitor each other. It creates a fearful Orwellian world. The warfare will be cyber warfare, medical treatment will be by artificial cyber docs, the commerce will be dominated by bitcoins and internet, the entertainment industry by mind games in the internet and sex by digisexuality. The press will be controlled by the artificial intelligence network and the social media. Social cultural world that we know of no longer exist. There will be an artificial intelligence doctor to look after you, digisexuality to satisfy your desires, bitcoins to play with net education to build your knowledge and entertainment via the net producing by artificial intelligence to satisfy your instincts. In this world there is no sympathy, empathy, love, emotion, sex or hatred. Everything is themed on harmony and control. Robotic society comes into existence. The manmade artificial intelligence networks turns rogue and takes over the world. They can destroy the world by cyberwarfare, produce widespread economic destruction in banks, hospitals, utility services and energy services by cyber attacks. No place in earth or anyone is immune to their attacks. The artificial intelligence networks will be the Frankenstein monster aided by humans and taking over the world. There would not be a leader for this artificial intelligence network; only rule governed mode of robotic network function. This homo roboticus of cyborgs or artificial intelligence networks can colonise other planets and star systems. When we think of colonizing the space we think of emotional humans, but the space would be colonise by cyborgs, artificial intelligence and microbial quantal computing intelligent clouds. This would be the nature of life in future. The human species is dead due to greed and selfishness of the homo sapiens and homo sapien hybrids. This is the world of Kalki has emerged from the shadows - live in a brave new world.

The Neanderthals and their archaea produce the emerging viruses and are resistant to them. The pandemic of emerging viruses will come to an end with the death of homo sapiens and homo sapien-homo neanderthalic hybrids. This means the return of Asuras. In previous history also the Asuras were extremely tech savvy with planes, supposed nuclear weapons and teleportation capacities. A technocratic neanderthalic civilization will evolve and due to low population numbers consequent to infertility cloning with implanted electrodes will be undertaken to produce cyborgs. The new Neanderthal civilization will expand in small numbers. The South Asians and Japanese will expand to China and Russia and occupy the Eurasian land mass. The Basque, Britannic people and Welish will expand to Europe. The Australian Aboriginals will take over Australia. The North American-Indians will expand to North America and Canada. South American-Indian will expand to South America. There would not be a population explosion. The fertility of these Neanderthals with their male eunuchs and Amazonian women is little. Covid 19 in addition would have destroyed their testicles. Therefore somatic cloning with implanting of electrodes to generate cyborgs would populate the planet will come into existence. The artificial intelligent networks will take over and having consciousness because of the phenomena of panpsychism will rule the world. The remnant Neanderthals and the cyborgs will develop archaeal related civilizational diseases affecting cardiovascular, respiratory, renal and brain systems. They will continue to live because of artificial organs being developed and implanted. The neanderthalic, cyborgic and artificial intelligence life span is eternal. They will have the facility of teleportation and visiting other planets and star systems to gain minerals and other things to use. The other planets in the solar system will be dominated by the microbial quantal computing cloud. The human beings are very complex and therefore are not being the highest life forms. The microbial quantal computing cloud is less complex and can form humanoids based on the human life experiences and information stored in the microbial quantal computing network. These humanoids can arise in earth and other planets. Teleportation of microbes and viruses between planets and galaxies will be the mode of interstellar communication. The universe with its planets, stars systems will become one and all the different microbial viral quantal computing clouds will be in communication with each other by teleportation. This creates a global galactic life civilization.

History follows its own prescribed path and no human intervention can make any great changes. According to Carvakas and Ajivikas what happened was meant to have happened, what is happening will happen, what will happen will always happen. There is no human soul, rebirth, reincarnation, right or wrong in this fatalistic absolute deterministic world. The world will therefore, behave as before as things have to happen to produce further evolution as scripted. Global warming, fossil fuel usage, internet usage and non-veganism will continue as before and human beings despite widespread physical destruction of earth and extinction of African homo sapiens and Eurasian homo sapien-Neanderthal hybrids will increase the remaining Indian ocean-Antarctic rim Neanderthal numbers, develop Neanderthal cyborgs and eventually form microbial viral quantal computing intelligent clouds in teleportation contact with intergalactic biological quantal computing cloud in the spiral arms of the galaxy and other microbial quantal computing clouds in other planets, star systems and galaxies. The artificial intelligence network which regulates cyborgs will overtake the Neanderthals and cyborgs producing a monster machine without a heart or soul. This artificial intelligent network and the cyborgs will be able to colonize distant planets and stars evolving into a new species- homo galacticus/roboticus. The earth bound microbial viral quantal computing clouds in which human information is stored and humanoids are formed will regulate the earth in coordination with the intergalactic biological magnetotactic quantal computing cloud in the spiral arms of the galaxies which contribute to star system formation itself. This is the end of human species as such but evolving into an eternal microbial future.

CHAPTER 25

THE APOCALYPSE, SPECIES CHANGE AND END OF COVID 19 PANDEMIC - GLOBAL WARMING, INTERNET RELATED LOW LEVEL EMF POLLUTION AND CATACLYSMIC SPECIES CHANGE TO HOMO NEONEANDERTHALIS, HOMO GALACTICUS, HOMO ROBOTICUS OR HOMO PLANTICUS – ENDOSYMBIOTIC ARCHAEA AND ITS COVID 19 VIROID GENERATED DE NOVO – REGRESSION TO PREMITOCHONDRIAL ENERGETICS AND PHOTOSYNTHESIS - INFECTIVE SYMBIOTIC TISSUE COLONIES OF PROTOZOA, ARCHAEA, VIRUSES AND BACTERIA

The plasma life based forms and the gravitational and anti-gravitational waves generated porphyrions would have been the most primitive of non-life forms. The porphyrin templates would have given rise to the isoprenoid organism, the RNA viroids, DNA viroids and prions which would have symbiosed to form the nanoarchaea. The archaea would have evolved into prokaryotes or bacteria. The archaeal symbiosis with prokaryotes would have generated the eukaryotes. The initial unicellular eukaryotes are the protozoa. The protozoa are capable of phagocytosis and would have had viruses, bacteria and archaea as their intracellular symbionts. The viruses of the protozoa like entamoeba and toxoplasma are giant viruses called megaviride. The protozoans like entamoeba with its viral, archaeal and bacterial endosymbionts would have evolved into the multicellular organisms or metazoa. The archaeal, bacterial and protozoal extracellular vesicles containing RNA and DNA viroids can by horizontal gene transfer lead to evolution of metazoan, plants, animals and human species. The metazoans would have evolved into plants, animals and humans by horizontal gene transfer generated by archaeal extracellular RNA and DNA viroid vesicles. The evidence of such a transformation is seen in the bacterial symbionts of protozoa like entamoeba which includes legionella, listeria, mycoplasma, mycobacteria and chlamydia. The symbionts of protozoan entamoeba include viruses like covid, HIV and cytomegalo virus. The protozoan entamoeba serves as a natural reservoir for covid infections, bacterial infections and HIV infections. The toxoplasma also serves as a covid reservoir. These protozoan colonies of entamoeba and toxoplasma with its bacterial, archaeal and viral symbionts can be called as chimeric infective symbiotic tissue colonies which can exist in the air, water, earth, blood and tissues. The chimeric infective symbiotic tissue colonies containing bacteria, archaea and viruses can make the symbiotic organism multidrug resistant by horizontal gene transfer. The chimeric infective symbiotic tissue organism evolves as a consequence of global warming and the protozoal, bacterial, viral and archaeal symbiotic colonies are new forms of life. The chimeric infective symbiotic tissue organism composed of protozoa and its symbionts like virus, bacteria and archaea depend upon glycolysis for its energy needs. The glycolytic enzymes enolase, glyceraldehyde 3-phosphate dehydrogenase, hexokinase, phosphofructokinase, aldolase, pyruvic kinase and triose phosphate isomerase are moonlighting proteins having a cytoplasmic, membrane and nuclear functions. These glycolytic enzymes have a RNA binding motif which can modulate RNA transcription in the host and the symbiotic organism. These glycolytic enzymes with their RNA binding motif as well as DNA binding motif can bind to RNA and DNA viruses and viroids modulating their replication and life cycles. The chimeric infective symbiotic tissue organism composed of protozoa and its symbionts like virus, bacteria and archaea can enter the human host through the breaches of the gut-blood barrier and transform itself into the macrophages of the human host. The macrophages are involved in the phagocytosis and innate immunity against bacterial and viral infections. Paradoxically the macrophages like the protozoa can also serve as a reservoir where pathogens like mycobacteria tuberculosis, mycobacteria leprae, listeria, legionella, E. coli, shigella & salmonella, and viruses like HIV & covid can thrive and replicate. The modified chimeric infective symbiotic tissue organism or macrophage composed of protozoa and its symbionts like virus, bacteria and archaea can serve as a reservoir for horizontal gene transfer and generation of multidrug resistant bacteria and viruses. The infective chimeric symbiotic tissue organism composed of protozoa and its symbionts like virus, bacteria and archaea depends upon moonlighting glycolytic enzymes for their energetics and replication. The development of antibodies against the infective chimeric symbiotic tissue organism composed of protozoa and its symbionts like virus, bacteria and archaea leads to the evolution of anti-brain antibody disease or anti-basal ganglia antibody disease. They present a threat to existence consequent to horiztontal gene transfer and generation of multidrug resistant bacteria, viruses, protozoa, mycobacteria and archaea.

The Covid 19 infection produces the happy hypoxia syndrome. Global warming and low level of internet related EMF pollution leads to archaeal symbiosis. The archaea is a part of the human cell which is a symbiotic structure. The archaeal genome is integrated into the human genome when it becomes part of the symbiotic cell. The cell's mitochondria is a rickettsia, the nucleus is a pox virus, the peroxisome is acinetocbacter aceti and the cytoskeleton is formed of spirochaetes. Free mitochondria and endosymbiotic archaea have been detected in the blood indicating unraveling of the symbiotic cell. The cytochrome F420 activity detected in the human blood could be due to the expression of the archaeal genome integrated into the human genome once it has become part of the symbiotic cell. Similarly there are integrated retroviral genes in the human genome as indicated by human endogenous retroviral sequences. Endogenous retroviral particles have been detected in the human blood. This indicates unravelling of the symbiotic cell due to global warming and low level internet induced EMF pollution. The endosymbiotic archaea has got bacterial chlorophyll and carotenoids and are capable of photosynthesis as well as ATP synthesis functioning as an electron transport chain. The archaeaons can function as cell organelle.

Archaeal endosymbiosis leads to increased porphyrin synthesis leading on to porphyrinuria and porphyria. Actinidic archaea have a mevalonate pathway and are cholesterol catabolizing.[1-5] They can use cholesterol as a carbon and energy source. Archaeal cholesterol catabolism can generate porphyrins via the cholesterol ring oxidase generated pyruvate and GABA shunt pathway. The pyruvate is converted to succinyl CoA through the GABA shunt pathway. Actinidic archaea inhibits mitochondrial function through HIF alpha and stimulates glycolysis inducing the Warburg phenotype. The increased glycolysis can lead to increased 3-phosphoglycerate which is converted phosphoserine, serine and glycine using the glycine cleavage enzyme system. Succinyl CoA and glycine are condensed to form delta amino levulinic acid the starting point of porphyrin synthesis. Archaea can produce a secondary porphyria by inducing the enzyme heme oxygenase resulting in heme depletion and activation of the enzyme ALA synthase. Porphyrins are prebiotic molecules which are involved in abiogenesis and origin of life. The porphyrions are self-replicating supramolecular organisms which forms the precursor template on which the viroids, prions and nanoarchaea originate. Stress induced template directed abiogenesis of porphyrions, prions, viroids and archaea is a continuous process and can contribute to changes in brain structure and behaviour as well as disease process.

The increase in porphyrins leads to cortical dysfunction and prefrontal cortex atrophy. The porphyrins can destroy the human endogenous retroviruses and the jumping genes leading to lack of dynamicity of the genome. This leads on to maldevelopment of the prefrontal cortex. This leads on to cerebellar dominance and a cerebellar cognitive affective disorder. This produces porphyrin related quantal perception. Porphyrins are dipolar molecules and in the setting of porphyrin mediated membrane sodium-potassium ATPase inhibition induced pumped phonon system can produce a quantal perceptive state. Porphyrins are macromolecules which can have both a wave and particle existence and can bridge the particulate world and the quantal world. Membrane sodium-potassium ATPase inhibition induced dipolar porphyrin mediated pumped phonon system can lead on to a cellular plasma state and EMF signal transduction. Macromolecules like RNA, DNA, protein and the cell itself can have an EMF signature. This porphyrin generated macromolecular cellular EMF signature is important in regulation of cell function. The porphyrins can have quantal perception of low level EMF fields leading to prefrontal cortex atrophy. This leads on to cortical dysfunction and lack of functioning of the conscious brain. The cerebellum dominates and the unconscious takes over. This leads on to neanderthalisation of the brain and schizophrenia and autism. The heme deficiency leads to lack of synthesis of the heme enzyme cytochrome P420 dependent sex hormones and a widespread asexual state. The mitochondrial dysfunction leads on to insulin resistance and metabolic syndrome X. The Warburg phenotype and increased glycolysis leads to oncogenesis. The mitochondrial dysfunction can produce neurodegeneration. The increase in lymphocyte glycolysis can produce immune activation and autoimmune disease. Thus the stress induced porphyria due to climatic change and environmental pollution can lead to civilizational disease.

The porphyrins can self-organize to form macromolecular structures which can self-replicate to form a porphyrin organism. The photon induced transfer of electrons along the macromolecule can lead to light induced ATP synthesis. The porphyrins can form a template on which RNA and DNA can form generating viroids. The porphyrins can also form a template on which prions can form. They all can join together – RNA viroids, DNA viroids, prions – to form primitive archaea. Thus the archaea are capable of self-replication on porphyrin templates. The self-replicating archaea can sense gravity which gives rise to consciousness. They can also sense the antigravity fields which gives rise to the unconscious brain. Thus, there can be both self-replicating archaea and anti-archaea regulating the conscious and unconscious brain. Thus, the climate change stress mediated increased porphyrin synthesis leads to prefrontal cortex atrophy, cerebellar dominance, cerebellar cognitive affective disorder, quantal perception and neanderthalisation of the population. The porphyrions are self-replicating supramolecular organisms which forms the precursor template on which the viroids, prions and nanoarchaea originate. Stress induced template directed abiogenesis of porphyrions, prions, viroids and archaea is a continuous process and can contribute to changes in brain structure and behaviour as well as disease process.

The stimulus for porphyrin synthesis comes from heme deficiency. Heme suppresses ALA synthase. Stress induces heme oxygenase which converts heme to carbon monoxide and bilirubin. Thus heme is depleted from the system. Heme oxygenase is induced by environmental stress. Climatic changes of global warming and ice age can induce heme oxygenase. Heme oxygenase is also induced by EMF pollution of the environment. Thus, there is increased porphyrin synthesis from succinyl CoA and glycine. Defect in heme synthesis and heme depletion leads to deficiency of heme enzymes. Deficiency cytochrome C oxidase and aconitase leads to mitochondrial oxidative phosphorylation defects and TCA cycle defects. This leads to pyruvate dehydrogenase deficiency and defect in synthesis of acetyl CoA. There is increase in glycolysis consequent to porphyrin photo-oxidation induced free radical generation and HIF alpha induction. This produces the Warburg phenotype. The increased level of pyruvate that is generated is converted to glutamate and ammonia. Thus there is hyperammonemia as a consequence of the metabolic defect. Since glycine is utilized for porphyrin synthesis serine is not synthesized leading on to deficiency of the substrate for synthesis of cystathionine. This leads to accumulation of homocysteine and homocystinuria. Deficiency of acetyl CoA leads to defects in the isoprenoid pathway and defective synthesis of cholesterol and ubiquinone. There is also deficiency of the heme containing cholesterol synthesizing enzyme lanosterol synthase. This leads to a cholesterol depleted state. The increase in porphyrins leads to cortical dysfunction and prefrontal cortex atrophy. The porphyrins can destroy the human endogenous retroviruses and the jumping genes leading to lack of dynamicity of the genome. This leads on to maldevelopment of the prefrontal cortex. This leads on to cerebellar dominance and a cerebellar cognitive affective disorder. This produces porphyrin related quantal perception. Porphyrins are dipolar molecules and in the setting of porphyrin mediated membrane sodium-potassium ATPase inhibition induced pumped phonon system can produce a quantal perceptive state. Porphyrins are macromolecules which can have both a wave and particle existence and can bridge the particulate world and the quantal world. Membrane sodium-potassium ATPase inhibition induced dipolar porphyrin mediated pumped phonon system can lead on to a cellular plasma state and EMF signal transduction. Macromolecules like RNA, DNA, protein and the cell itself can have an EMF signature. This porphyrin generated macromolecular cellular EMF signature is important in regulation of cell function. The porphyrins can have quantal perception of low level EMF fields leading to prefrontal cortex atrophy. This leads on to cortical dysfunction and lack of functioning of the conscious brain. The cerebellum dominates and the unconscious takes over. This leads on to neanderthalisation of the brain and schizophrenia and autism. The heme deficiency leads to lack of synthesis of the heme enzyme cytochrome P420 dependent sex hormones and a widespread asexual state. The mitochondrial dysfunction leads on to insulin resistance and metabolic syndrome X. The Warburg phenotype and increased glycolysis leads to oncogenesis. The mitochondrial dysfunction can produce neurodegeneration. The increase in lymphocyte glycolysis can produce immune activation and autoimmune disease. Thus the stress induced porphyria due to climatic change and environmental pollution can lead to civilizational disease. The porphyrins can self-organize to form macromolecular structures which can self-replicate to form a porphyrin organism. The photon induced transfer of electrons along the macromolecule can lead to light induced ATP synthesis. The porphyrins can form a template on which RNA and DNA can form generating viroids. The porphyrins can also form a template on which prions can form. They all can join together – RNA viroids, DNA viroids, prions – to form primitive archaea. Thus the archaea are capable of self-replication on porphyrin templates. The self-replicating archaea can sense gravity which gives rise to consciousness. They can also sense the antigravity fields which gives rise to the unconscious brain. Thus, there can be both self-replicating archaea and anti-archaea regulating the conscious and unconscious brain. Thus, the climate change stress mediated increased porphyrin synthesis leads to prefrontal cortex atrophy, cerebellar dominance, cerebellar cognitive affective disorder, quantal perception and neanderthalisation of the population. The porphyrions are self-replicating supramolecular organisms which forms the precursor template on which the viroids, prions and nanoarchaea originate. Stress induced template directed abiogenesis of porphyrions, prions, viroids and archaea is a continuous process and can contribute to changes in brain structure and behaviour as well as disease process. Thus, the archaeal endosymbiosis can generate porphyrions. The endosymbiotic archaea can also synthesize melanin. Thus the endosymbiotic archaea can generate two macromolecules which can contribute to the dissociation of water generating oxygen and energy.

The corona virus produces defects in oxygenation of hemoglobin. The lung lesions produced by the corona virus is not an adult respiratory distress syndrome but a syndrome akin to high altitude sickness due to oxygen deprivation. The corona virus is a lipid envelope RNA virus and gets transcribed into DNA segments using the HERV reverse transcriptase. The corona virus RNA that gets transcribed to DNA can get integrated into the non-coding sequence of the human DNA and can function as jumping genes. The corona viral proteins can be synthesized by the human cell. The corona virus can displace iron from the beta chain of hemoglobin rendering hemoglobin and oxygen transport dysfunctional. The human body is subjected to a hypoxic ischemic injury consequent to defect in oxygen transport as occurs in high altitude mountain sickness, cyanide poisoning and carbon monoxide poisoning. The corona virus can release iron from hemoglobin and increase the ferritin levels and free iron levels to catastrophic levels. This produces cytokine activation due to increase in free radicals. The corona virus can deplete heme from hemoglobin and other heme containing enzymes. This includes cytochrome C oxidase which produces a mitochondrial dysfunction and hypoxic injury. It can bind with heme enzymes like nitric oxide synthase which produces nitric oxide, cystathionine beta synthase which produces defective synthesis of hydrogen sulphide and heme oxygenase which produces defective carbon monoxide production. Thus the endogenous vasodilator neurotransmitters are depleted producing generalized tissue vasoconstriction. The heme containing enzymes involved in glutathione metabolism- glutathione synthetase and glutathione reductase are also made dysfunctioning inhibiting free radical scavenging. The TCA cycle enzyme aconitase is also a heme enzyme and the corona virus inactivates it. The cytochrome P450 enzymes involved in steroid synthesis are also heme enzymes and they are inactivated producing acute cortisol, aldosterone and sex steroid depletion. The corona virus also produces a protein that inactivates furin involved in activating various protein enzymes. The corona virus infection produces a metabolic disorder with defects in oxygenation and the syndrome of acquired acute porphyria with secondary iron overload state. The treatment of corona virus infection does not depend upon ventilation at high PEEP low tidal volume but at high FIO2. Hyperbaric oxygen therapy can also correct. This is ineffective as there is no hemoglobin transport oxygen. Packed cell transfusions will be required to correct the hemoglobin deficiency and oxygenation defects. Increased amounts of free iron and high ferritin can produce free radical injury and a secondary ARDS-like picture which may need ventilation. Ventilation as such would not correct the primary defect in heme oxygenation. The high levels of ferritin may need iron detoxification using drugs like Desferal which chelates out the iron overload state. Excess iron can lead to free radical generation, tissue injury, immune activation and cytokine injury. The heme deficiency related cytochrome C oxidase deficiency and aconitase deficiency will have to be treated by mitochondrial supplements like CoQ and carnitine. The corona virus produces an acute porphyria and porphyrin synthesis will have to be inhibited by giving IV heme infusions. Heme inhibits porphyrin synthesis. The symptoms of porphyria are due to defects in inhibitory GABA transmission. In porphyria there is increase in delta amino levulinic acid synthesis and decrease in GABA transmission. ALA will displace GABA from its binding sites. Porphyric crisis are treated by giving gamma amino butyric acid. IV dextrose infusions will also help to inhibit ALA synthase and porphyrin synthesis. Porphyrins can also be removed by plasmapheresis. The glutathione deficiency has to be corrected by giving IV glutathione infusions. The deficiency in nitric oxide can be corrected by inhale nitric oxide preparations. The cytochrome P450 deficiency mediated reductions in steroids will also have to be corrected.

The intrinsic property of porphyrin supramolecular complex to form porphyrion complexes within a phospholipid isoprenoid membrane creates a new cellular organelle capable of water dissociation and photosynthesis. In plants, the first reaction of photosynthesis is through dissociation of the water molecule by chlorophyll, a reaction that is irreversible, since the oxygen is expelled into the atmosphere. The reaction is outlined as follows: $2H_2O \rightarrow 2H_2 + O_2$. It is considered the world's most important reaction since it is the beginning of the food chain. The free chemical energy that is released with the breakdown of the molecule of water is essential to boost consequential reactions finally reaches the fusion of CO_2 and H_2O into glucose, a process that unable to replicate in the laboratory. Porphyrion supramolecular complexes can dissociate water: $2H_2O \leftrightarrow 2H_2 + O_2 + 4e^-$. Porphyrions can absorb electromagnetic radiation and light along the entire electromagnetic spectrum. The porphyrion supramolecular complex surrounded by a phospholipid isoprenoid membrane is called the porphyrion organelle. The porphyrion organelle can function as a primitive mitochondria and a photosynthetic organelle. The porphyrion supramolecular complexes can absorb photons and the porphyrin molecules have got electrons in excited state called excitons. The released electrons are transported along the primitive porphyrin supramolecular complex electron transport chain to molecular oxygen. Four pairs of hydrogen atoms are removed from isocitrate, alpha ketoglutarate, succinate and malate by specific dehydrogenases. The hydrogen atom donates electrons to the mitochondrial electron transport chain and become H^+ ion. The electrons are transported along the primitive lipid membrane bound porphyrin supramolecular complex called porphyrions and are delivered to molecular oxygen. The molecular oxygen accepts two electrons and two hydrogen atoms removed by dehydrogenases to form H_2O or water. As each pair of electrons goes along the porphyrions supramolecular complex, the electrons which are energy rich loose their energy and ATP is synthesized from ADP and inorganic phosphate in the inner mitchondiral membrane. The electron which is removed from the porphyrion supramolecular complex is replaced by dissociation of water by the porphyrion. Water is dissociated into: $2H_2O \leftrightarrow 2H_2 + O_2 + 4e^-$. The electrons can be transported along the electron transport chain and delivered to molecular oxygen to synthesize ATP or used to replace the electrons removed from the porphyrion supramolecular complex excited by light energy and photons. The hydrogen atom generated by the dehydrogenases of the TCA cycle donates an electron to the electron transport chain and becomes a proton generating the proton gradient in the inner mitochondrial membrane which gives osmotic energy for ATP synthesis.

The porphyrion supramolecular complex which are excited by light energy or photons releases electrons which are transported along the electron transport chain. The porphyrion supramolecular complex related water dissociation: $2H_2O \leftrightarrow 2H_2 + O_2 + 4e^-$ generates electrons which are transported along the porphyrion supramolecular complex related electron transport chain. The 4 pairs of hydrogen atoms removed from isocitrate, alpha keto glutarate, succinate and malate by dehyrogenases donate an electron to be transported along the porphyrion supramolecular complex related electron transport chain to generate NADH and ATP. NADH and ATP are concerned with carbon dioxide fixation and produce carbohydrate synthesis. Thus the porphyrion supramolecular complex can function as a primitive mitochondria generating ATP and as a primitive photosynthetic organelle synthesizing carbohydrate. The porphyrion supramolecular complex can produce dissociation of water: $2H_2O \leftrightarrow 2H_2 + O_2 + 4e^-$ and can generate endogenous oxygen and energy.

The dissociation of water by melanin and porphyrions into hydrogen, oxygen and electrons generates energy and oxygen. The electrons can be carried as energy all over the tissues by hydrogen. The electrons generated by the dissociation of water can be transported along the mitochondrial electron transport chain where the oxygen molecule picks up electrons and protons to form water and ATP is synthesized. Thus, the dissociation of water by melanin and porphyrions can produce energy required by the cell through the mitochondria or extra-mitochondrial mechanisms. The dissociation of water can generate oxygen which can be used in Covid 19 pneumonias where extraneous oxygen delivery is impaired.

The homo neanderthalis has got archaeal endosymbiotic archaea which can function as cell organelle or archaeaons concerned with photosynthesis, water dissociation and ATP synthesis. The homo neanderthalis has got increased archaeal related porphyrin synthesis generating endogenous cell organelle called porphyrions. The porphyrions can also function as cell organelle concerned with photosynthesis, water dissociation and ATP synthesis. The homo sapiens which do not have archaeal endosymbiosis do not have archaeaons or porphyrions capable of photosynthesis, water dissociation and ATP synthesis. The homo sapiens mated with homo neanderthalis producing hybrid homo sapien homo neanderthalis hybrid caucassian population also do not have archaeal endosymbiosis. The homo sapien-homo neanderthalic caucassian population are deficient in archaeal endosymbiosis and do not have archaeaons or porphyrions capable of photosynthesis, water dissociation and ATP synthesis. Thus the homo neanderthalis are resistant to the Covid 19 infection as they archaeaons and porphyrions capable of photosynthesis, water dissociation and ATP synthesis. The homo neanderthalis via archaeaons and porphyrions can generate energy via primitive porphyrion supramolecular complex related electron transport chain activated by photons and light energy and replenished by water dissociation. The oxygen generated de novo by water dissociation help them to survive the happy hypoxia crisis generated by Covid 19 pneumonia. The Covid 19 virus is an archaeal viroid which has hybridized with human genomic sequences including HERV sequences and is generated de novo in homo neanderthalis with archaeal endosymbiosis. The homo sapiens of Africa as well as the homo sapien-homo neanderthalis hybrids of Europe and USA do not have archaeal endosymbiosis and do not generate Covid 19 de novo. The homo neanderthalis which generates Covid 19 de novo can spread the infection to homo sapiens as well as homo sapien-homo neanderthalis hybrids. The homo sapiens as well as homo sapien-homo neanderthalis hybrids do not have archaeaons and porphyrions capable of water dissociation, oxygen formation, photosynthesis and energy generation from sunlight exposure as well as water dissociation. The homo sapiens as well as homo sapien-homo neanderthalis hybrids die from Covid 19 infection and can become extinct.

Thus global warming and low level of internet induced EMF pollution can lead to archaeal endosymbiosis. The endosymbiotic archaea can secrete RNA viroids which can hybridize with human genomic segments including HERV sequences to generate the Covid 19 virus de novo. The Covid 19 virus can produce virus induced porphyria and generate self-replicating porphyrions. The porphyrins can form a template on which RNA, DNA, proteins, isoprenoid macromolecules and polysaccharides can form which can symbioze to form archaea by abiogenetic replication. The archaea generated abiogenetically can synthesize melanin. Global warming can induce heme oxygenase leading to porphyria and porphyrion formation as well as lead to formation of endosymbiotic archaea on porphyrin templates. The porphyrions can dissociate water to generate oxygen and energy. Global warming which increases the carbon dioxide in the atmosphere and produces a hypoxia can generate a new human species with endosymbiotic archaea and its virioid like Covid 19 called the homo neoneanderthalis. This can be called as the return of homo neanderthalis. The original homo neanderthalic population originated due to archaeal symbiosis on primates. The archaea arose from the intergalactic magnetic fields formed of magnetotactic archaea and viruses which were brought to the earth by meteoric impacts. The homo neanderthalis evolved first in Lemuria- a continent in the Indian ocean and owing to persistent earth quakes and tsunamis migrated into the Indian land mass and the Indus valley as well as to the Middle-East including Sumeria and Egypt. They form the global Dravidian civilization which includes the Jew, the Semitic Muslims, Celts, the people of Crete, Basque, Wales, Brittany and the Shudras of India. The homo neanderthalic population has archaeaons and porphyrions which can dissociate water to generate oxygen and energy. They are resistant to Covid 19 infection. The homo neanderthalis had a small prefrontal cortex and a larger cerebellum contributing to a cerebellar cognitive affective disorder and epidemic autism. They were not capable of fictional imagination and myth formation and could exist only as small clusters. They became extinct because of their inability to form large civilizational groups which the homo sapiens could do.

The homo sapiens evolved owing to retroviral epidemic with the retrovirus from the intergalactic magnetic fields formed of archaea, bacteria and viruses reaching the earth by meteoric impacts. The retroviral integration into the human genome enlarged the prefrontal cortex leading on to the rational and logical homo sapiens. The homo sapien albino mutants-homo neanderthalic hybrids having a large frontal cortex were capable of fictional imagination and myth formation leading to a non-autistic logical phenotype leading to formation of massive colonies. This led to domination of the world by homo sapien albino mutants-homo neanderthalic hybrids and homo neanderthalis extinction. The homo sapien who migrated to the cold Eurasian land mass eventually interbred with homo neanderthalis forming homo sapien-homo neanderthalis group which constitutes the Europeans and the classical Aryan phenotype. The upper caste of India consequent to the Aryan invasion is also part of the homo sapien-homo neanderthalis group. The homo sapiens as well as homo sapien-homo neanderthalis hybrids do not have archaeaons and porphyrions capable of water dissociation, oxygen formation, photosynthesis and energy generation from sunlight exposure as well as water dissociation. The homo sapiens as well as homo sapien-homo neanderthalis hybrids die from Covid 19 infection and can become extinct.

Global warming and internet related low level EMF pollution leads to archaeal endosymbiosis and Covid 19 mass infection. The Covid 19 is generated de novo as archaeal RNA viroids which hybridize with human genomic sequences. The de novo generation of Covid 19 viroids from endosymbiotic archaea of homo neoneanderthalis makes the present strategy to control Covid 19 less effective- the use of mask, social distancing and vaccination. The archaea and Covid 19 generate archaeaons and porphyrions which can dissociate water, generate energy and photosynthesize carbohydrate. The homo sapien species and homo sapien-homo neanderthalic caucassian hybrids due to archaeal endosymbiosis consequent to global warming and low level of internet related EMF pollution can form homo neoneanderthalis. The evolution of homo neoneanderthalis is a species change to counteract global warming and rise of carbon dioxide in the atmosphere. The homo neoneanderthalic species has got porphyrions and archaeaons which can dissociate water, generate energy and oxygen as well as photosynthesize. The homo neoneanderthalic species can have photosynthesis by dissociation of water and synthesize glucose using carbon dioxide, water and electrons. The homo neoneanderthalis can survive global warming owing to its photosynthetic ability and its capacity to live in an atmosphere with high carbon dioxide level. The human archaeal and Covid 19 pandemic could be a cataclysm species change to combat global warming. The homo neanderthalis and homo neoneanderthalis are able to withstand the archaeal and its RNA viroid Covid 19 pandemic owing to the presence of archaeaons and porphyrions, but the homo sapien-homo neanderthalic caucassians and homo sapien Africans lack archaeaons and porphyrions and become extinct.

The global warming and internet related low level of EMF pollution leads to archaeal overgrowth and generation of archaeal viroidal Covid 19. The homo neanderthalis and homo neoneanderthalis are able to withstand and survive the archaeal and its RNA viroidal Covid 19 growth but homo sapien-homo neanderthalic hybrids as well as homo sapiens succumb. The homo neanderthalis has porphyrions and archaeaons related water dissociation leading to generation of energy and oxygen producing as well as photosynthesis producing Covid 19 resistance. The global warming and internet related low level EMF pollution can lead to growth of endosymbiotic archaea and its RNA viroidal Covid 19 in homo sapiens of African and homo sapien-homo neanderthalic hybrid caucassians leading to the generation of a new species called Homo neoneanderthalis. The homo neoneanderthalis can survive high level of carbon dioxide causing global warming and hypoxia. The homo neoneanderthalis can survive in the outer planets and galaxies lacking an atmosphere due to the capability to dissociate water and can form the species called Homo galacticus which is capable of colonizing other planets and galaxies.

The Covid 19 vaccination with Covid 19 antigens and Covid 19 mRNA will lead to virus induced porphyria and formation of archaea on porphyrin templates abiogenetically. This will lead to widespread neanderthalisation of the world population when 90 percent is vaccinated. Thus massive Covid 19 vaccination will lead to generation of the homo neoneanderthalis and homo galacticus species which is capable of quantal perception consequent to the presence of porphyrions in the human brain. The porphyrions in the brain of homo neoneanderthalis and homo galacticus can interact with computer networks forming cyborgs as well as man-machine hybrids producing a civilization of cyborgs. Thus global warming and low level internet related EMF pollution will produce homo neoneanderthalis, homo galacticus and homo roboticus. This will create a civilization of cyborgs. This will also generate plant-like human species capable of photosynthesis called homo planticus which can combat global warming. The Covid 19 pandemic will lead to generation of homo neoneanderthalis which are basically autistic with a cerebellar cognitive affective disorder. The post-covid autistic population is dependent on the internet as they have a hermit-like existence of solitude and are exposed to a higher density of internet induced EMF pollution. This produces more of low level EMF exposure from the internet, a cave syndrome of social isolation as well as a digital obsessive disorder. This generate porphyrions and more archaea making them neanderthalic still further and generating the evolving species called homo roboticus, homo planticus, homo neoneanderthalis and homo galacticus. These are the consequences of new world created by global warming producing a cataclysmic species change - the Homo roboticus or Homo galacticus. This species are capable of surviving in the earth with high levels of carbon dioxide as well as living in other planets which lack an atmosphere. We are into the world of Kalki constituted by homo galacticus and cyborgs. The archaeal and Covid 19 pandemic is a representative of a climate change related adaptive species change into homo neoneanderthalis, homo galacticus, homo roboticus or homo planticus. This is a cataclysmic species change as great as the archaeal endosymbiosis which produce the homo neanderthalis and retroviral epidemics which produced the homo sapiens. The global warming and the internet related low level of EMF pollution is producing archaeal endosymbiosis and archaeal viroids - Covid 19 as well as a new species the homo neoneanderthalis. The homo neanderthalis and homo neoneanderthalis has archaeal endosymbiosis and generate Covid 19 viroids and are resistant to the virus. They are the predators. The Covid 19 viroids generated by homo neanderthalis and homo neoneanderthalis affects the homo sapien Africans and the homo sapien-homo neanderthalic caucassians making them extinct. They are the victims. This cataclysm is a species change as an adaptive response to climate change, global warming and low level of internet related EMF pollution. This generates a more resistant homo neoneanderthalic, homo galacticus or homo roboticus capable of surviving in a carbon dioxide rich atmosphere and colonizing the outer galaxies. The idea of Covid 19 being a pandemic like influenza is faulty. The archaeal and its viroid Covid 19 symbiosis is an apocalyptical species change.

This is evidenced from population behaviour in response to the pandemic. Kerala which has a high level of infection has a low mortality rate less than 1 percentage. The Jews and south Indians are neanderthalic with high levels of melanin. The morality rate is high in cities like Delhi and Mumbai as well as in Punjab and Haryana dominated by the Aryan homo sapien albino mutants-homo neanderthalic hybrids. The Shudras of the BIMARU states of Bihar, Madhya Pradesh, Uttar Pradesh and Rajasthan are hardly concerned about the pandemic as they are neanderthalic Shudras with high levels of melanin. The Jews of Israel have a low mortality rate and low infection. The incidence and mortality with Covid 19 is the highest in the homo sapien albino mutants-homo neanderthalic hybrids populated countries of Europe and USA.

The archaeal and Covid 19 pandemic has an end. The Covid 19 virus is an archaeal RNA viroid hybridizing with human genomic sequences. The endosymbiotic archaea are seen in remnant homo neanderthalic species - the Shudras of India, the Semitic Muslims of West Asia and the Celtic race of Europe. The global warming and low level of EMF pollution results in endosymbiotic archaeal growth in homo sapiens of Africa and homo neanderthalic-homo sapien hybrids of Europe, upper caste Indians and USA. This results in the formation of homo neoneanderthalis, homo roboticus, homo planticus or homo galacticus species. It is a cataclysmic species change. The homo neanderthalic species of Shudras of India, the semitic Muslims of West Asia and the Celtic race of Europe with more of already existant endosymbiotic archaeal growth becomes stronger in strength and intelligence. The Covid 19 pandemic will end with the species change of homo sapiens and homo neanderthalic-homo sapien hybrids to homo neoneanderthalis. The homo sapien and homo sapiens of Africa and homo neanderthalic-homo sapien hybrids of Europe, upper caste Indians and USA become extinct or get converted by global warming and low level internet related EMF pollution to homo neoneanderthalis. This will produce an end to Covid 19 pandemic. The new homo neoneanderthalic species can live in a world of climate change, global warming and low level of internet related EMF pollution. The homo neoneanderthalic species develops brain changes producing a cerebellar cognitive affective disorder and endemic autism. The homo neoneanderthalic species has metabolic changes corresponding to the Warburg phenotype. The homo neoneanderthalic species has got immune activation. The homo neoneanderthalic species has a shorter life span of upto 50-60 years owing to higher incidence of diabetes mellitus, CAD and stroke, schizophrenia and bipolar affective disorder, cancer and autoimmune disease. The homo neoneanderthalic species which evolves will live in an internet dominated autistic world of man-internet hybrids. There will be a civilization of hermitic cyborgs. The new homo neoneanderthalic or homo galactic species can live in other planets and galaxies where there is no atmosphere. This species can generate energy by water oxidation and create food by photosynthesis. This species is plant-like and can be called as homo planticus. Water will be in excess in the world of global warming with the melting of the polar ice and rising of sea levels. There would not be any forest or agriculture but a human forest of plant-like homo planticus with photosynthesis. The human beings will live in a planet with global warming and continuous use of fossil fuels and high atmospheric carbon dioxide. The world population will be halved or reduced due to higher incidence of civilizational disease. The homo neoneanderthalis owing to quantal perception of porphyrions will be paganistic and spiritual. The other religions of the world except paganism will disappear. The sexuality of homo neanderthalis will be digisexuality which will reduce the world population. The world will be dominated by females and androgynous individuals. The quantal perception of homo neoneanderthalis will produce a globalised world which is anarchic. There will be small clusters of population and nation-states will become extinct. There will be an economic system of anarchic capitalism of selfishness in the small population clusters. The homo neoneanderthalis is surrealistic, syntheistic and evil. There will be a violent world with terrorism and aggression. This will be brave new world of galactic neanderthalic species of cyborgs.

CHAPTER 26

CLIMATE CHANGE AND CATASTROPHIC PANDEMICS - INFECTIVE SYMBIOTIC TISSUE COLONIES OF PROTOZOA, ARCHAEA, VIRUSES AND BACTERIA

The plasma life based forms and the gravitational and anti-gravitational waves generated porphyrions would have been the most primitive of non-life forms. The porphyrin templates would have given rise to the isoprenoid organism, the RNA viroids, DNA viroids and prions which would have symbiosed to form the nanoarchaea. The archaea would have evolved into prokaryotes or bacteria. The archaeal symbiosis with prokaryotes would have generated the eukaryotes. The initial unicellular eukaryotes are the protozoa. The protozoa are capable of phagocytosis and would have had viruses, bacteria and archaea as their intracellular symbionts. The viruses of the protozoa like entamoeba and toxoplasma are giant viruses called megaviride. The protozoans like entamoeba with its viral, archaeal and bacterial endosymbionts would have evolved into the multicellular organisms or metazoa. The archaeal, bacterial and protozoal extracellular vesicles containing RNA and DNA viroids can by horizontal gene transfer lead to evolution of metazoan, plants, animals and human species. The metazoans would have evolved into plants, animals and humans by horizontal gene transfer generated by archaeal extracellular RNA and DNA viroid vesicles. The evidence of such a transformation is seen in the bacterial symbionts of protozoa like entamoeba which includes legionella, listeria, mycoplasma, mycobacteria and chlamydia. The symbionts of protozoan entamoeba include viruses like covid, HIV and cytomegalo virus. The protozoan entamoeba serves as a natural reservoir for covid infections, bacterial infections and HIV infections. The toxoplasma also serves as a covid reservoir. These protozoan colonies of entamoeba and toxoplasma with its bacterial, archaeal and viral symbionts can be called as chimeric infective symbiotic tissue colonies

which can exist in the air, water, earth, blood and tissues. The chimeric infective symbiotic tissue colonies containing bacteria, archaea and viruses can make the symbiotic organism multidrug resistant by horizontal gene transfer. The chimeric infective symbiotic tissue organism evolves as a consequence of global warming and the protozoal, bacterial, viral and archaeal symbiotic colonies are new forms of life. The chimeric infective symbiotic tissue organism composed of protozoa and its symbionts like virus, bacteria and archaea depend upon glycolysis for its energy needs. The glycolytic enzymes enolase, glyceraldehyde 3-phosphate dehydrogenase, hexokinase, phosphofructokinase, aldolase, pyruvic kinase and triose phosphate isomerase are moonlighting proteins having a cytoplasmic, membrane and nuclear functions. These glycolytic enzymes have a RNA binding motif which can modulate RNA transcription in the host and the symbiotic organism. These glycolytic enzymes with their RNA binding motif as well as DNA binding motif can bind to RNA and DNA viruses and viroids modulating their replication and life cycles. The chimeric infective symbiotic tissue organism composed of protozoa and its symbionts like virus, bacteria and archaea can enter the human host through the breaches of the gut-blood barrier and transform itself into the macrophages of the human host. The macrophages are involved in the phagocytosis and innate immunity against bacterial and viral infections. Paradoxically the macrophages like the protozoa can also serve as a reservoir where pathogens like mycobacteria tuberculosis, mycobacteria leprae, listeria, legionella, E. coli, shigella & salmonella, and viruses like HIV & covid can thrive and replicate. The modified chimeric infective symbiotic tissue organism or macrophage composed of protozoa and its symbionts like virus, bacteria and archaea can serve as a reservoir for horizontal gene transfer and generation of multidrug resistant bacteria and viruses. The infective chimeric symbiotic tissue organism composed of protozoa and its symbionts like virus, bacteria and archaea depends upon moonlighting glycolytic enzymes for their energetics and replication. The development of antibodies against the infective chimeric

symbiotic tissue organism composed of protozoa and its symbionts like virus, bacteria and archaea leads to the evolution of anti-brain antibody disease or anti-basal ganglia antibody disease. They present a threat to existence consequent to horiztontal gene transfer and generation of multidrug resistant bacteria, viruses, protozoa, mycobacteria and archaea.

Climate change can lead to extremophilic endosymbiotic archaeal growth and neanderthalisation of the species. There is increase in endosymbiotic and colonic archaeal growth. The increase in colonic archaeal growth leads to breaching of the blood-gut barrier and archaeal endosymbiosis. The exposure to the low level of EMF fields consequent to widespread use of the internet, radiosatellites and radar equipments leads to increased endosymbiotic archaeal growth. Archaeal endosymbiosis leads to neanderthalisation of the species. The colonic archaea also generates RNA viroids which are converted to DNA viroids by colonic epithelial HERV reverse transcriptase and gets integrated into the colonic genome. The RNA and DNA viroids hybridise with the viroidal and bacterial population of the gut microbiome and virobiome DNA and RNA of the gut leading to generation of new emerging viruses. The RNA and DNA viroids and gut bacteria can also hybridise with human genomic sequences for immunity, metabolic regulation, growth factors, cell growth, cell death, neurotransmitters and HERV sequences of the colonial epithelial cell. The hybridization of archaeal RNA viroids with human genomic HERV sequences can generate retroviruses. This leads to generation of new bacteria, RNA and DNA viruses modulating cell death, cell growth, cell proliferation, metabolism, endocrine function and behaviour producing autoimmunity, schizophrenia, autism, cancer diabetes mellitus, coronary vascular disease, cerebrovascular disease and neurodegeneration.

Dietary fibre is the most important factor that modulates human biology and speciation. Dietary fibre and human drugs - antibiotic and non-antibiotic can modulate the gut flora producing colonic archaeal growth and endosymbiosis generating climate change, evolution of superbugs and emerging viruses and creating the disease phenotype. The high fibre diet is digested by the colonic flora leading to generation of short chain fatty acids and inhibits endosymbiotic and colonic archaeal growth. This leads to homo sapienisation of the species. A low fibre diet with increase in protein and fat leads to stimulation of endosymbiotic and colonic archaeal growth. This leads to neanderthalisation of the species. Thus a low fibre diet alters the gut microbiome in the colon with more of archaeal growth. The archaea develop protective mechanism of antibiotic resistance and dominates the colonic microflora. The colonic archaea by natural selection and DNA/gene exchanges with other colonic and endosymbiotic organism transfer the genes for antibiotic resistance to other colonic flora. This forms the basis for the generation of superbugs resistant to all antibiotics. The same mechanisms happen when populations are treated with antibiotics and other drugs like anti-hypertensives, anti-psychotics and analgesics. They kill the colonic bacterial flora and the extremophilic archaea dominates the colonic microbiome with acquired protective mechanisms for antibiotic resistance. These archaeal antibiotic resistance genes are transferred by gene and plasmid DNA exchange with remnant colonic and endosymbiotic micro-organism generating superbugs. Thus antibiotic and other non-antibiotic drugs promote colonic archaeal growth and endosymbiosis producing neanderthalisation of the species and generation of superbugs which kills of the homo sapien species. Thus, the generation of superbugs and emerging viruses is promoted by a low fibre diet. A low fibre diet leads to colonic and endosymbiotic archaeal growth leading to neanderthalisation of the species and generation of superbugs and emerging viruses to which the neanderthalised species is resistant consequent to endogenous digoxin secretion. The neanderthalised species generated superbugs and emerging viruses kills of the homo sapien species which are sensitive to the superbugs and emerging viruses and do not have resistance to it. Thus the modern armamentarium of antibiotics, anti-psychotics, analgesics, anti-hypertensives and anti-diabetic drugs promote colonic and endosymbiotic archaeal growth development of superbugs, neanderthalisation of species and induction of civilizational disease. Thus the drugs used to treat infections and human disease tend to produce superbugs, emerging virus infection and civilizational diseases - cancer, autoimmune disease, metabolic syndrome X, neurodegeneration, schizophrenia and autism. Thus the drug armamentarium of antibiotics and non-antibiotic drugs used to treat human infections and diseases increases colonic and endosymbiotic archaeal growth, neanderthalises the species, generates superbugs and leads to increased incidence of civilizational disease. The gut microbiome and virobiome modulated by dietary fibre intake, high protein high fat diets, antibiotics and non-antibiotic drugs like anti-psychotics, anti-hypertensives, analgesics leads to alteration in gut microbial and viroidal bio-geography leading to generation of superbugs and superviruses. The gut archaea highly resistant to conventional antibiotics and non-antibiotic drugs and proliferate when antibiotics and non-antibiotic drugs like anti-hypertensives, anti-psychotics and analgesics are used on a widespread basis. They develop antibiotic resistance and non-antibiotic drug resistance and it is transferred to other colonic and gut bacteria and viruses generating superbugs and superviruses exterminating the homo sapien population. The growth and dominance of the colonic and endosymbiotic bacteria leads to generation of homo neanderthalic species which is resistant to superbugs and superviruses and survive but get affected by civilizational diseases like cancer, autoimmunity, metabolic syndrome, schizophrenia, autism and neurodegeneration. Homo neoneanderthalis thus generated by colonic and endosymbiotic archaeal growth can become extinct over a slow period of time. Homo neanderthalis with its extremophilic archaeal endosymbiosis is hardy and resistant and can survive global warming and extremes of climate change as well as surviving extremophilic anaerobic conditions of other planets and galactic systems. The dietary fibre intake inhibits archaeal growth and populations with low dietary fibre intake are resistant to civilizational diseases like metabolic syndrome, cancer, autoimmune disease and schizophrenia. The human methanogenic archaea are highly resistant to routine antibiotics and non-antibiotic drugs and the intake of such drugs kills the colonic microbiome leading to archaeal growth and endosymbiosis and neanderthalisation to produce a hardy homo galacticus species capable of surviving in extreme climates of earth and other planets. The human drugs and antibiotics induced modulation of the gut microbiome with increased colonic and endosymbiotic archaeal growth also contributes to global warming via the mechanism of archaeal methanogenesis. The widespread use of antibiotics and non-antibiotic drugs contribute to colonic and endosymbiotic archaeal growth, climate change and still further more archaeal growth. This human colonic and endosymbiotic archaeal growth consequent to stress, internet fields, low level EMF, low dietary fibre intake, antibiotic use and non-antibiotic drug use producing methanogenesis is the principal cause of global warming. The widespread use of antibiotic and non-antibiotic drugs kills of the colonic microflora and replaces it with archaea resistant to antibiotic and non-antibiotic drugs. The low intake of dietary fibre promotes colonic archaeal and endosymbiotic archaeal growth. Stress produces catecholamine induced increased in colonic and endosymbiotic archaeal growth and breaching of the gut-blood barrier. The widespread use of internet and low level of EMF field of mobile phones promotes endosymbiotic archaeal growth. Thus endosymbiotic and colonic archaeal growth and methanogenesis producing global warming is an endogenous human biological event with human biological activity as a centre piece triggering the process. The widespread use of antibiotics and non-antibiotic drugs treat infections viral and bacterial as well as civilizational disease like cancer, schizophrenia, mood disorder, depression, autism, autoimmune disease and neurodegeneration promotes colonic and endosymbiotic archaeal growth which neanderthalises the species leading to still further transformation of the human system leading to induction of civilizational disease like cancer, metabolic syndrome X, neurodegeneration, autoimmunity and vascular disease. Thus, the antibiotic and non-antibiotic drugs used to treat infections and systemic diseases produces colonic and endosymbiotic archaeal growth leading to superbugs, superviruses, pandemics and epidemic of civilizational diseases mediated by colonic and endosymbiotic archaeal growth. Human medicine as we know it has reached a dead end. The antibiotic and non-antibiotic era of drugs had come to an end. The human earth, as we know it also undergoes catastrophic climate change leading to extinction of the human species and conditions for human habitations in earth. The modern medicine as we know of and the earth as we know of have ended. The principal cause of climate change and its catastrophic complications is contributed in part only by the destruction of the tropical Amazonian and rain forest. The deforestation of the gut microflora created over billions of years of human evolution is the principal cause of the climate change and a possible end of the earth as we know of and the human species as we know of. The revolutionary medicine and the golden earth with its habitable conditions have ended. The increase in dietary fibre intake upto 40 g/day can inhibit colonic archaeal growth and endosymbiosis putting a break on generation of superbugs, emerging viruses and climate change. Increase in dietary fibre intake of populations can prevent the state of climate change and gut microbial change putting human life and the life of the planet at risk. Dietary fibre can be considered as the elixir of life and endosymbiotic archaea the philosopher's stone.

Endosymbiotic actinidic archaea as having evolved from an early isoprenoid organisms by abiogenesis. An actinidc archaea/viroid mediated model of prokaryote, viral, eukaryotic, primate and human evolution is discussed.[1-4]. Actinidic archaea has been related to the pathogenesis of schizophrenia, malignancy, metabolic syndrome X, autoimmune disease and

neuronal degeneration[2]. Actinidic archaea have a mevalonate pathway and cholesterol catabolism[5-7]. Davies has put forward the concept of a shadow biosphere of organisms with alternate biochemistry present in earth itself[8]. An actinide dependent shadow biosphere of archaea and viroids in the above mentioned disease states is described[6]. Metal actinides in beach sands have been postulated to play a role in abiogenesis[6]. Actinide mineral like rutile, monazite and illmenite by surface metabolism would have contributed to abiogenesis[9]. A hypothesis of cholesterol as the primal prebiotic molecule synthesised on actinide surfaces with all other biomolecules arising from it and a self-replicating cholesterol lipid organism as the initial life form is presented. The actinidic archaea and viroids would have evolved from the primitive isoprenoid organism. The origin of viruses, prokaryotes, eukaryotes, primates and humans from the initial isoprenoid organism derived actinidic archaea is postulated.

Materials and Methods

Informed consent of the subjects and the approval of the Ethics Committee were obtained for the study. The following groups were included in the study:- endomyocardial fibrosis, Alzheimer's disease, multiple sclerosis, non-Hodgkin's lymphoma, metabolic syndrome X with cerebrovascular thrombosis and coronary artery disease, schizophrenia, autism, seizure disorder, Creutzfeldt-Jakob's disease and acquired immunodeficiency syndrome. There were 10 patients in each group and each patient had an age and sex matched healthy control selected randomly from the general population. The blood samples were drawn in the fasting state before treatment was initiated. Plasma from fasting heparinised blood was used and the experimental protocol was as follows:- (I) Plasma+phosphate buffered saline, (II) same as I+cholesterol substrate, (III) same as II+rutile 0.1 mg/ml, and (IV) same as II+ciprofloxacine and doxycycline each in a concentration of 1 mg/ml. Cholesterol substrate was prepared as described by Richmond[10]. Aliquots were withdrawn at zero time

immediately after mixing and after incubation at 37 °C for 1 hour. The following estimations were carried out:- Cytochrome F420, free RNA, free DNA, muramic acid, polycyclic aromatic hydrocarbon, hydrogen peroxide, serotonin, pyruvate, ammonia, glutamate, cytochrome C, hexokinase, ATP synthase, HMG CoA reductase, digoxin and bile acids[11-14]. Cytochrome F420 was estimated flourimetrically (excitation wavelength 420 nm and emission wavelength 520 nm). Polycyclic aromatic hydrocarbon was estimated by measuring hydrogen peroxide liberated by using glucose reagent. The statistical analysis was done by ANOVA.

Results

The parameters checked as indicated above were:- cytochrome F420, free RNA, free DNA, muramic acid, polycyclic aromatic hydrocarbon, hydrogen peroxide, serotonin, pyruvate, ammonia, glutamate, cytochrome C, hexokinase, ATP synthase, HMG CoA reductase, digoxin and bile acids. Plasma of control subjects showed increased levels of the above mentioned parameters with after incubation for 1 hour and addition of cholesterol substrate resulted in still further significant increase in these parameters. The plasma of patients showed similar results but the extent of increase was more. The addition of antibiotics to the control plasma caused a decrease in all the parameters while addition of rutile increased their levels. The addition of antibiotics to the patient's plasma caused a decrease in all the parameters while addition of rutile increased their levels but the extent of change was more in patient's sera as compared to controls.

Discussion

There was increase in cytochrome F420 indicating archaeal growth. The archaea can synthesize and use cholesterol as a carbon and energy source[15,16]. The archaeal origin of the enzyme activities was indicated by antibiotic induced

suppression. The study indicates the presence of actinide based archaea with an alternate actinide based enzymes or metalloenzymes in the system as indicated by rutile induced increase in enzyme activities[17]. There was also an increase in archaeal HMG CoA reductase activity indicating increased cholesterol synthesis by the archaeal mevalonate pathway. The archaeal beta hydroxyl steroid dehydrogenase activity indicating digoxin synthesis and archaeal cholesterol hydroxylase activity indicating bile acid synthesis were increased[7]. The archaeal cholesterol oxidase activity was increased resulting in generation of pyruvate and hydrogen peroxide[16]. The pyruvate gets converted to glutamate and ammonia by the GABA shunt pathway. The archaeal aromatization of cholesterol generating PAH, serotonin and dopamine was also detected[18]. The archaeal glycolytic hexokinase activity and archaeal extracellular ATP synthase activity were increased. The archaea can undergo magnetite and calcium carbonate mineralization and can exist as calcified nanoforms[19].

The metal actinides provide radiolytic energy, catalysis for oligomer formation and provide a coordinating ion for metalloenzymes all important in abiogenesis[6]. The metal actinide surfaces would by surface metabolism generate acetate which could get converted to acetyl CoA and then to cholesterol which functions as the primal prebiotic molecule self-organising into self-replicating supramolecular systems, the lipid organism[8,9,20]. Cholesterol by radiolysis by actinides would have formed PAH generating PAH aromatic organism[8]. Cholesterol radiolysis would generate pyruvate which would get converted to amino acids, sugars, nucleotides, porphyrins, fatty acids and TCA acids. Anastase and rutile surfaces can produce polymerization of amino acids, isoprenyl residues, PAH and nucleotides to generate the initial lipid organism, PAH organism, prions and RNA viroids which would have symbiosed to generate the archaeal protocell. The archaea evolved into gram negative and gram positive bacteria with a mevalonate pathway which had an evolutionary advantage. The symbiosis of archaea with gram negative organism generated

the eukaryotic cell[21]. The data supports the persistence of an actinide and cholesterol based shadow biosphere which throws light on the actinide based origin of life and cholesterol as the premier prebiotic molecule.

There was an increase in free RNA indicating self-replicating RNA viroids and free DNA indicating generation of viroid complementary DNA strands by archaeal reverse transcriptase activity. The actinides modulate RNA folding and catalyse its ribozymal action. Digoxin can cut and paste the viroidal strands by modulating RNA splicing generating RNA viroidal diversity. The viroids are evolutionarily escaped archaeal group I introns which have retrotransposition and self-splicing qualities. Archaeal pyruvate can produce histone deacetylase inhibition resulting in endogenous retroviral (HERV) reverse transcriptase and integrase expression. This can integrate the RNA viroidal complementary DNA into the non-coding region of eukaryotic non-coding DNA using HERV integrase as has been described for borna and ebola viruses[21]. The non-coding DNA is lengthened by integrating RNA viroidal complementary DNA with the integration going on as a continuing event. The archaea genome can also get integrated into human genome using integrase as has been described for trypanosomes. The integrated viroids and archaea can undergo vertical transmission and can exist as genomic parasites. This increases the length and alters the grammar of the non-coding region producing memes or memory of acquired characters. The viroidal complementary DNA can function as jumping genes producing a dynamic genome.[22-24]

The presence of muramic acid, HMG CoA reductase and cholesterol oxidase activity inhibited by antibiotics indicates the presence of bacteria with mevalonate pathway. The bacterial with mevalonate pathway include streptococcus, staphylococcus, actinomycetes, listeria, coxiella and borrelia. The bacteria and archaea with mevalonate pathway and cholesterol catabolism had an evolutionarily advantage and constitutes the isoprenoidal clade organism. The archaea evolved into mevalonate pathway gram positive and gram negative isoprenoid clade organism through horizontal gene transfer of viroidal and virus genes. The isoprenoidal clade prokaryotes develop into other groups of prokaryotes via viroidal/virus as well as eukaryotic horizontal gene transfer producing bacterial speciation[25-27].

The RNA viroids and its complementary DNA developed into cholesterol enveloped RNA and DNA viruses like herpes, retrovirus, influenza virus, borna virus, cytomegalo virus and ebstein barr virus by recombining with eukaryotic and human genes resulting in viral speciation. Bacterial and viral species are ill-defined and fuzzy with all of them forming one common genetic pool with frequent horizontal gene transfer and recombination.

Thus, the multi and unicellular eukaryote with its genes serves the purpose of prokaryotic and viral speciation. The multicellular eukaryote developed so that their endosymbiotic archaeal colonies could survive and forage better. The multicellular eukaryotes are like bacterial biofilms. The archaea and bacteria with a mevalonate pathway uses the extracellular RNA viroids and DNA viroids for quorum sensing and in the generation of symbiotic biofilm like structures which develop into multicellular eukaryotes. The endosymbiotic archaea and bacteria with mevalonate pathway still uses the RNA viroids and DNA viroids for the regulation of multicellular eukaryote.[28-31]

Pollution is a major inducer of evolutionary innovation. Pollution is induced by the primitive nanoarchaea and mevalonate pathway bacteria synthesized PAH and methane leading on to redox stress. Redox stress leads to sodium-potassium ATPase inhibition, inward movement of plasma membrane cholesterol, defective SREBP sensing, increased cholesterol synthesis and nanoarchaeal/mevalonate pathway bacterial growth. Redox stress leads on to viroidal and archaeal multiplication. Redox stress can also lead to HERV reverse transcriptase and integrase expression. The non-coding DNA is formed of integrating RNA viroidal complementary DNA and archaea with the integration going on as a continuing event. The archaeal pox like dsDNA virus forms evolutionarily the nucleus. The integrated viroidal, archaeal and mevalonate pathway bacterial sequences can undergo vertical transmission and can exist as genomic parasites. The genomic integrated archaea, mevalonate pathway bacteria and viroids form a genomic reserve of bacteria and viruses which can recombine with human and eukaryotic genes producing bacterial and viral speciation. The change in the length and grammar of the non-coding region produces eukaryotic speciation and individuality. The integration of nanoarchaea, mevalonate pathway prokaryotes and viroids in to the eukaryotic and human genome produces a chimera which can multiply producing biofilm like multicellular structures having a mixed archaeal, viroidal, prokaryotic and eukaryotic characters which is a regression from the multicellular eukaryotic tissue. This results in a new neuronal, metabolic, immune and tissue phenotype leading to human disease.[30-32]

Pollution would have been a major factor in eukaryotic speciation and primate/hominid evolution. The change in the length and grammar of the non-coding region produces eukaryotic speciation and individuality. It is the increase in non-coding region and HERV sequences of the genome that led to the evolution of the primate and the human brain and its attendant property of conscious and quantal perception. It is the non-coding region of the genome with its archaeal, RNA viroidal complementary DNA and HERV sequences that makes for the human qualities of the hominid brain. Changes in the length of non-coding region can lead on to disorders of consciousness like schizophrenia. A schizophrenia specific human endogenous retroviruses and change in the length and grammar of the non-coding region has been described in schizophrenia.[33,34]

An actinide dependent shadow biosphere of archaea and viroids in the above mentioned disease states is described. Metal actinides in beach sands

have been postulated to play a role in abiogenesis. Cholesterol is the primal prebiotic molecule synthesized on actinide surfaces with all other biomolecules arising from it. A self-replicating cholesterol lipid organism could be the initial life form. A cholesterol based abiogenesis is a more likely evolutionary option and the actinidic archaea and viroids would have evolved from it. The origin of viruses, prokaryotes, eukaryotes, primates and humans from the initial isoprenoid organism derived actinidic archaea is discussed.

References

1. Valiathan M.S., Somers, K., Kartha, C.C. (1993). *Endomyocardial Fibrosis.* Delhi: Oxford University Press.
2. Kurup R., Kurup, P.A. (2009). *Hypothalamic Digoxin, Cerebral Dominance and Brain Function in Health and Diseases*. New York: Nova Science Publishers.
3. Hanold D., Randies, J.W. (1991). Coconut cadang-cadang disease and its viroid agent, *Plant Disease,* 75, 330-335.
4. Edwin B.T., Mohankumaran, C. (2007). Kerala wilt disease phytoplasma: Phylogenetic analysis and identification of a vector, *Proutista moesta, Physiological and Molecular Plant Pathology,* 71(1-3), 41-47.
5. Eckburg P.B., Lepp, P.W., Relman, D.A. (2003). Archaea and their potential role in human disease, *Infect Immun,* 71, 591-596.
6. Adam Z. (2007). Actinides and Life's Origins, *Astrobiology,* 7, 6-10.
7. Schoner W. (2002). Endogenous cardiac glycosides, a new class of steroid hormones, *Eur J Biochem,* 269, 2440-2448.
8. Davies P.C.W., Benner, S.A., Cleland, C.E., Lineweaver, C.H., McKay, C.P., Wolfe-Simon, F. (2009). Signatures of a Shadow Biosphere, *Astrobiology,* 10, 241-249.
9. Wächtershäuser G. (1988). Before enzymes and templates: theory of surface metabolism, *Microbiol Rev,* 52(4), 452-84.
10. Richmond W. (1973). Preparation and properties of a cholesterol oxidase from nocardia species and its application to the enzymatic assay of total cholesterol in serum, *Clin Chem,* 19, 1350-1356.
11. Snell E.D., Snell, C.T. (1961). *Colorimetric Methods of Analysis.* Vol 3A. New York: Van NoStrand.
12. Glick D. (1971). *Methods of Biochemical Analysis.* Vol 5. New York: Interscience Publishers.
13. Colowick, Kaplan, N.O. (1955). *Methods in Enzymology.* Vol 2. New York: Academic Press.
14. Maarten A.H., Marie-Jose, M., Cornelia, G., van Helden–Meewsen, Fritz, E., Marten, P.H. (1995). Detection of muramic acid in human spleen, *Infection and Immunity,* 63(5), 1652 – 1657.

15. Smit A., Mushegian, A. (2000). Biosynthesis of isoprenoids via mevalonate in Archaea: the lost pathway, *Genome Res,* 10(10), 1468-84.

16. Van der Geize R., Yam, K., Heuser, T., Wilbrink, M.H., Hara, H., Anderton, M.C. (2007). A gene cluster encoding cholesterol catabolism in a soil actinomycete provides insight into Mycobacterium tuberculosis survival in macrophages, *Proc Natl Acad Sci USA,* 104(6), 1947-52.

17. Francis A.J. (1998). Biotransformation of uranium and other actinides in radioactive wastes, *Journal of Alloys and Compounds,* 271(273), 78-84.

18. Probian C., Wülfing, A., Harder, J. (2003). Anaerobic mineralization of quaternary carbon atoms: Isolation of denitrifying bacteria on pivalic acid (2,2-Dimethylpropionic acid), *Applied and Environmental Microbiology,* 69(3), 1866-1870.

19. Vainshtein M., Suzina, N., Kudryashova, E., Ariskina, E. (2002). New Magnet-Sensitive Structures in Bacterial and Archaeal Cells, *Biol Cell,* 94(1), 29-35.

20. Russell M.J., Martin, W. (2004). The rocky roots of the acetyl-CoA Pathway, *Trends in Biochemical Sciences,* 29, 7.

21. Margulis L. (1996). Archaeal-eubacterial mergers in the origin of Eukarya: phylogenetic classification of life, *Proc Natl Acad Sci USA,* 93, 1071-1076.

22. Tsagris E.M., de Alba, A.E., Gozmanova, M., Kalantidis, K. (2008). Viroids, *Cell Microbiol,* 10, 2168.

23. Horie M., Honda, T., Suzuki, Y., Kobayashi, Y., Daito, T., Oshida, T. (2010). Endogenous non-retroviral RNA virus elements in mammalian genomes, *Nature,* 463, 84-87.

24. Hecht M., Nitz, N., Araujo, P., Sousa, A., Rosa, A., Gomes, D. (2010). Genes from Chagas parasite can transfer to humans and be passed on to children. Inheritance of DNA Transferred from American Trypanosomes to Human Hosts, *PLoS ONE,* 5, 2-10.

25. Flam F. (1994). Hints of a language in junk DNA, *Science,* 266, 1320.

26. Horbach S., Sahm, H., Welle, R. (1993). Isoprenoid biosynthesis in bacteria: two different pathways? *FEMS Microbiol Lett,* 111, 135–140.

27. Gupta R.S. (1998). Protein phylogenetics and signature sequences: a reappraisal of evolutionary relationship among archaebacteria, eubacteria, and eukaryotes, *Microbiol Mol Biol Rev,* 62, 1435–1491.

28. Hanage W., Fraser, C., Spratt, B. (2005). Fuzzy species among recombinogenic bacteria, *BMC Biology,* 3, 6-10.

29. Whitchurch C.B., Tolker-Nielsen, T., Ragas, P.C., Mattick, J.S. (2002). Extracellular DNA Required for Bacterial Biofilm Formation. *Science,* 295(5559), 1487.

30. Webb J.S., Givskov, M., Kjelleberg, S. (2003). Bacterial biofilms: prokaryotic adventures in multicellularity, *Curr Opin Microbiol,* 6(6), 578–85.

31. Chen Y., Cai, T., Wang, H., Li, Z., Loreaux, E., Lingrel, J.B. (2009). Regulation of intracellular cholesterol distribution by Na/K-ATPase, *J Biol Chem,* 284(22), 14881-90.

32. Poole A.M. (2006). Did group II intron proliferation in an endosymbiont-bearing archaeon create eukaryotes? *Biol Direct,* 1, 36-40.

33. Villarreal L.P. (2006). How viruses shape the tree of life, *Future Virology,* 1(5), 587-595.

34. Lockwood M. (1989). *Mind, Brain and the Quantum*. Oxford: B. Blackwell.

CHAPTER 27

ENDOSYMBIOTIC ARCHAEAL RNA VIROIDS AND EMERGING VIRAL PANDEMICS – COVID 19 - INFECTIVE SYMBIOTIC TISSUE COLONIES OF PROTOZOA, ARCHAEA, VIRUSES AND BACTERIA – MITOCHONDRIAL DYSFUNCTION AND ANTI-GLYCOLYTIC ANTIBODY DISEASE

The plasma life based forms and the gravitational and anti-gravitational waves generated porphyrions would have been the most primitive of non-life forms. The porphyrin templates would have given rise to the isoprenoid organism, the RNA viroids, DNA viroids and prions which would have symbiosed to form the nanoarchaea. The archaea would have evolved into prokaryotes or bacteria. The archaeal symbiosis with prokaryotes would have generated the eukaryotes. The initial unicellular eukaryotes are the protozoa. The protozoa are capable of phagocytosis and would have had viruses, bacteria and archaea as their intracellular symbionts. The viruses of the protozoa like entamoeba and toxoplasma are giant viruses called megaviride. The protozoans like entamoeba with its viral, archaeal and bacterial endosymbionts would have evolved into the multicellular organisms or metazoa. The archaeal, bacterial and protozoal extracellular vesicles containing RNA and DNA viroids can by horizontal gene transfer lead to evolution of metazoan, plants, animals and human species. The metazoans would have evolved into plants, animals and humans by horizontal gene transfer generated by archaeal extracellular RNA and DNA viroid vesicles. The evidence of such a transformation is seen in the bacterial symbionts of protozoa like entamoeba which includes legionella, listeria, mycoplasma, mycobacteria and chlamydia. The symbionts of protozoan entamoeba include viruses like covid, HIV and cytomegalo virus. The protozoan entamoeba serves as a natural reservoir for covid infections, bacterial infections and HIV infections. The toxoplasma also serves as a covid reservoir. These

protozoan colonies of entamoeba and toxoplasma with its bacterial, archaeal and viral symbionts can be called as chimeric infective symbiotic tissue colonies which can exist in the air, water, earth, blood and tissues. The chimeric infective symbiotic tissue colonies containing bacteria, archaea and viruses can make the symbiotic organism multidrug resistant by horizontal gene transfer. The chimeric infective symbiotic tissue organism evolves as a consequence of global warming and the protozoal, bacterial, viral and archaeal symbiotic colonies are new forms of life. The chimeric infective symbiotic tissue organism composed of protozoa and its symbionts like virus, bacteria and archaea depend upon glycolysis for its energy needs. The glycolytic enzymes enolase, glyceraldehyde 3-phosphate dehydrogenase, hexokinase, phosphofructokinase, aldolase, pyruvic kinase and triose phosphate isomerase are moonlighting proteins having a cytoplasmic, membrane and nuclear functions. These glycolytic enzymes have a RNA binding motif which can modulate RNA transcription in the host and the symbiotic organism. These glycolytic enzymes with their RNA binding motif as well as DNA binding motif can bind to RNA and DNA viruses and viroids modulating their replication and life cycles. The chimeric infective symbiotic tissue organism composed of protozoa and its symbionts like virus, bacteria and archaea can enter the human host through the breaches of the gut-blood barrier and transform itself into the macrophages of the human host. The macrophages are involved in the phagocytosis and innate immunity against bacterial and viral infections. Paradoxically the macrophages like the protozoa can also serve as a reservoir where pathogens like mycobacteria tuberculosis, mycobacteria leprae, listeria, legionella, E. coli, shigella & salmonella, and viruses like HIV & covid can thrive and replicate. The modified chimeric infective symbiotic tissue organism or macrophage composed of protozoa and its symbionts like virus, bacteria and archaea can serve as a reservoir for horizontal gene transfer and generation of multidrug resistant bacteria and viruses. The infective chimeric symbiotic tissue organism composed of protozoa and its symbionts like virus, bacteria and archaea

depends upon moonlighting glycolytic enzymes for their energetics and replication. The development of antibodies against the infective chimeric symbiotic tissue organism composed of protozoa and its symbionts like virus, bacteria and archaea leads to the evolution of anti-brain antibody disease or anti-basal ganglia antibody disease. They present a threat to existence consequent to horiztontal gene transfer and generation of multidrug resistant bacteria, viruses, protozoa, mycobacteria and archaea.

The endosymbiotic and colonic human archaea secrete RNA and DNA viruses and serves as a reservoir for them. Viruses belonging to the pox family and African swine fever have been described in archaea. The archaeal viroids secrete enveloped viruses like SARS, MERS, corona and HIV. The archaeal virus secretion process is the same as that of enveloped viruses like SARS, MERS and Corona. The bats serve as a reservoir for the viruses like SARS, MERS and Corona. The bats contain archaea which can secrete RNA viroids and generate viruses. The colonic and endosymbiotic archaea also generates RNA viroids which are converted to DNA viroids by colonic epithelial HERV reverse transcriptase and gets integrated into the colonic genome. The RNA and DNA viroids hybridise with the viroidal and bacterial population of the gut microbiome and virobiome DNA and RNA of the gut leading to generation of new emerging viruses. The RNA and DNA viroids and gut bacteria can also hybridise with human genomic sequences for immunity, metabolic regulation, growth factors, cell growth, cell death, neurotransmitters and HERV sequences of the colonial epithelial cell. The hybridization of archaeal RNA viroids with human genomic HERV sequences can generate retroviruses. This leads to generation of new bacteria, RNA and DNA viruses producing pandemics and extermination of homo sapiens. The homo sapiens become an extinct species and the homo neanderthalis will dominate the earth.

The archaea can induce stem cell conversion and neanderthalisation of the human species. The archaea can produce permanent immune activation producing resistance to viral and bacterial infection. The archaeal cholesterol catabolism depletes tissue cholesterol producing vitamin D deficiency and immune activation. Thus archaeal overgrowth results in retroviral resistance and generation of the Neanderthal phenotype. The endosymbiotic archaea can secrete virus like RNA and DNA particles. The endosymbiotic archaea can induce uncoupling proteins inhibiting mitochondrial oxidative phosphorylation and generating ROS. The endosymbiotic archaeal magnetite can generate low level of EMF. The low level of EMF and ROS are genotoxic and produce breakages in hotspots of chromosome. It can also trigger rearrangements in hotspots of chromosome inhabited by retroviral and non-retroviral elements producing their expression. The archaeal secreted DNA and RNA viroids can recombine with the expressed retroviral, non-retroviral elements and other genomic segments of the human chromosome generating new RNA and DNA viruses. Thus the neanderthalised humans can serve as an origin for new RNA and DNA viruses as well as mutated retroviruses. The endosymbiotic archaeal digoxin can produce editing of RNA viruses and inactivating them in homo neanderthalis. The endosymbiotic archaea converts the Neanderthal cells to stem cells. The stem cells are resistant to immune attack. The stem cells can serve as a reservoir for this new RNA and DNA viruses. The stem cells and archaeal cells can also serve as a reservoir for viruses and bacteria belonging to other plants and animals. This helps to generate the species barrier jump in noted in recent emerging viral and bacterial infections. Thus the endosymbiotic archaeal growth produces neanderthalised version of homo sapiens which are retroviral resistant and resistant to other viral and bacterial infection consequent to immunotolerance of reservoir stem cell, archaeal induced immune activation and digoxin induced RNA editing. The endosymbiotic archaeal overgrowth mediated neanderthalised version of homo sapiens generates new mutated RNA and DNA viruses as well as retroviruses at the same time being resistant to them as in the case of the species bat. The homo sapiens do not have the Neanderthal mechanisms of immune activation as their archaeal load is meagre. They serve as fodder for infection from Neanderthal generated viruses and bacteria and suffer eventual extinction.[1-17] This paper studied the archaeal status in patients with recurrent viral infections and retroviral infections. The generation of RNA and DNA viroids from archaea was also studied.

Dietary fibre is the most important factor that modulates human biology and speciation. Dietary fibre and human drugs - antibiotic and non-antibiotic can modulate the gut flora producing colonic archaeal growth and endosymbiosis generating climate change, evolution of superbugs and emerging viruses and creating the disease phenotype. The high fibre diet is digested by the colonic flora leading to generation of short chain fatty acids and inhibits endosymbiotic and colonic archaeal growth. This leads to homo sapienisation of the species. A low fibre diet with increase in protein and fat leads to stimulation of endosymbiotic and colonic archaeal growth. This leads to neanderthalisation of the species. Thus a low fibre diet alters the gut microbiome in the colon with more of archaeal growth. The same mechanisms happen when populations are treated with antibiotics and other drugs like anti-hypertensives, anti-psychotics and analgesics. They kill the colonic bacterial flora and the extremophilic archaea dominates the colonic microbiome. Thus antibiotic and other non-antibiotic drugs promote colonic archaeal growth and endosymbiosis producing neanderthalisation of the species and generation of superviruses which kills of the homo sapien species. Thus, the generation of emerging viruses is promoted by a low fibre diet. A low fibre diet leads to colonic and endosymbiotic archaeal growth leading to neanderthalisation of the species and emerging viruses to which the neanderthalised species is resistant consequent to endogenous digoxin secretion. The neanderthalised species generated emerging viruses kills of the homo sapien species which are sensitive to emerging viruses and do not have resistance to it. Thus, the modern armamentarium of antibiotics, antipsychotics, analgesics, anti-hypertensives and anti-diabetic drugs promote colonic and endosymbiotic archaeal growth development of superbugs, neanderthalisation of species and induction of civilizational disease. Thus the drugs used to treat infections and human diseases tend to produce emerging virus infection. Thus, the drug armamentarium of antibiotics and non-antibiotic drugs used to treat human infections and diseases increases colonic and endosymbiotic archaeal growth, neanderthalises the species, generates emerging viruses. The gut microbiome and virobiome modulated by dietary fibre intake, high protein high fat diets, antibiotics and non-antibiotic drugs like anti-psychotics, anti-hypertensives, analgesics leads to alteration in gut microbial and viroidal bio-geography leading to generation of superviruses. The gut archaea highly resistant to conventional antibiotics and non-antibiotic drugs and proliferate when antibiotics and non-antibiotic drugs like anti-hypertensives, anti-psychotics and analgesics are used on a widespread basis. They generate superviruses exterminating the homo sapien population. The growth and dominance of the colonic and endosymbiotic archaea leads to generation of homo neanderthalic species which is resistant to superviruses and survive. Homo neanderthalis with its extremophilic archaeal endosymbiosis is hardy and resistant and can survive global warming and extremes of climate change as well as surviving extremophilic anaerobic conditions of other planets and galactic systems. The human methanogenic archaea are highly resistant to routine antibiotics and non-antibiotic drugs and the intake of such drugs kills the colonic microbiome leading to archaeal growth and endosymbiosis and neanderthalisation to produce a hardy homo galacticus species capable of surviving in extreme climates of earth and other planets. The human drugs and antibiotics induced modulation of the gut microbiome with increased colonic and endosymbiotic archaeal growth also contributes to global warming via the mechanism of archaeal methanogenesis. The widespread use of antibiotics and non-antibiotic drugs contribute to colonic and endosymbiotic archaeal growth, climate change and still further more archaeal growth. This human colonic and endosymbiotic archaeal growth consequent to stress, internet fields, low level EMF, low dietary fibre intake, antibiotic use and non-antibiotic drug use producing methanogenesis is the principal cause of global warming. The widespread use of antibiotic and non-antibiotic drugs kills of the colonic microflora and replaces it with archaea resistant to antibiotic and non-antibiotic drugs. The low intake of dietary fibre promotes colonic archaeal and endosymbiotic archaeal growth. Stress produces catecholamine induced increased in colonic and endosymbiotic archaeal growth and breaching of the gut-blood barrier. The widespread use of internet and low level of EMF field of mobile phones promotes endosymbiotic archaeal growth. Thus endosymbiotic and colonic archaeal growth and methanogenesis producing global warming is an endogenous human biological event with human biological activity as a centre piece triggering the process. The widespread use of antibiotics and non-antibiotic drugs treat infections viral and bacterial as well as civilizational disease like cancer, schizophrenia, mood disorder, depression, autism, autoimmune disease and neurodegeneration promotes colonic and endosymbiotic archaeal growth which neanderthalises the species leading to still further transformation of the human system leading to induction of civilizational disease like cancer, metabolic syndrome X, neurodegeneration, autoimmunity and vascular disease. Thus, the antibiotic and non-antibiotic drugs used to treat infections and systemic diseases produces colonic and endosymbiotic archaeal growth leading to superbugs, superviruses, pandemics. Human medicine as we know it has reached a dead end. The antibiotic and non-antibiotic era of drugs had come to an end. The human earth as we know it also undergoes catastrophic climate change leading to extinction of the human species and conditions for human habitations in earth. The modern medicine as we know of and the earth as we know of have ended. The principal cause of climate change and its catastrophic complications is contributed in part only by the destruction of the tropical Amazonian and rain forest. The deforestation of the gut microflora created over billions of years of human evolution is the principal cause of the climate change and a possible end of the earth as we know of and the human species as we know of. The revolutionary medicine and the golden earth with its habitable conditions have ended. The increase in dietary fibre intake upto 40 g/day can inhibit colonic archaeal growth and endosymbiosis putting a break on generation of superbugs, emerging viruses and climate change. Increase in dietary fibre intake of populations can prevent the state of climate change and gut microbial change putting human life and the life of the planet at risk. Dietary fibre can be considered as the elixir of life and endosymbiotic archaea the philosopher's stone.

Endosymbiotic actinidic archaea as having evolved from an early isoprenoid organisms by abiogenesis. An actinidc archaea/viroid mediated model of prokaryote, viral, eukaryotic, primate and human evolution is discussed.[1-4]. Actinidic archaea has been related to the pathogenesis of schizophrenia, malignancy, metabolic syndrome X, autoimmune disease and neuronal degeneration[2]. Actinidic archaea have a mevalonate pathway and cholesterol catabolism[5-7]. Davies has put forward the concept of a shadow biosphere of organisms with alternate biochemistry present in earth itself[8]. An actinide dependent shadow biosphere of archaea and viroids in the above mentioned disease states is described[6]. Metal actinides in beach sands have been postulated to play a role in abiogenesis[6]. Actinide mineral like rutile, monazite and illmenite by surface metabolism would have contributed to abiogenesis[9]. A hypothesis of cholesterol as the primal prebiotic molecule synthesized on actinide surfaces with all other biomolecules arising from it and a self-replicating cholesterol lipid organism as the initial life form is presented. The actinidic archaea and viroids would have evolved from the primitive isoprenoid organism. The origin of viruses, prokaryotes, eukaryotes, primates and humans from the initial isoprenoid organism derived actinidic archaea is postulated.

Materials and Methods

Informed consent of the subjects and the approval of the Ethics Committee were obtained for the study. The following groups were included in the study:- endomyocardial fibrosis, Alzheimer's disease, multiple sclerosis, non-Hodgkin's lymphoma, metabolic syndrome X with cerebrovascular thrombosis and coronary artery disease, schizophrenia, autism, seizure disorder, Creutzfeldt-Jakob's disease and acquired immunodeficiency syndrome. There were 10 patients in each group and each patient had an age and sex matched healthy control selected randomly from the general population. The blood samples were drawn in the fasting state before treatment was initiated. Plasma from fasting heparinised blood was used and the experimental protocol was as follows:- (I) Plasma+phosphate buffered saline, (II) same as I+cholesterol substrate, (III) same as II+rutile 0.1 mg/ml, and (IV) same as II+ciprofloxacine and doxycycline each in a concentration of 1 mg/ml. Cholesterol substrate was prepared as described by Richmond[10]. Aliquots were withdrawn at zero time immediately after mixing and after incubation at 37 ^{o}C for 1 hour. The following estimations were carried out:- Cytochrome F420, free RNA, free DNA, muramic acid, polycyclic aromatic hydrocarbon, hydrogen peroxide, serotonin, pyruvate, ammonia, glutamate, cytochrome C, hexokinase, ATP synthase, HMG CoA reductase, digoxin and bile acids[11-14]. Cytochrome F420 was estimated flourimetrically (excitation wavelength 420 nm and emission wavelength 520 nm). Polycyclic aromatic hydrocarbon was estimated by measuring hydrogen peroxide liberated by using glucose reagent. The statistical analysis was done by ANOVA.

Results

The parameters checked as indicated above were:- cytochrome F420, free RNA, free DNA, muramic acid, polycyclic aromatic hydrocarbon, hydrogen peroxide, serotonin, pyruvate, ammonia, glutamate, cytochrome C, hexokinase,

ATP synthase, HMG CoA reductase, digoxin and bile acids. Plasma of control subjects showed increased levels of the above mentioned parameters with after incubation for 1 hour and addition of cholesterol substrate resulted in still further significant increase in these parameters. The plasma of patients showed similar results but the extent of increase was more. The addition of antibiotics to the control plasma caused a decrease in all the parameters while addition of rutile increased their levels. The addition of antibiotics to the patient's plasma caused a decrease in all the parameters while addition of rutile increased their levels but the extent of change was more in patient's sera as compared to controls.

Discussion

There was increase in cytochrome F420 indicating archaeal growth. The archaea can synthesize and use cholesterol as a carbon and energy source[15,16]. The archaeal origin of the enzyme activities was indicated by antibiotic induced suppression. The study indicates the presence of actinide based archaea with an alternate actinide based enzymes or metalloenzymes in the system as indicated by rutile induced increase in enzyme activities[17]. There was also an increase in archaeal HMG CoA reductase activity indicating increased cholesterol synthesis by the archaeal mevalonate pathway. The archaeal beta hydroxyl steroid dehydrogenase activity indicating digoxin synthesis and archaeal cholesterol hydroxylase activity indicating bile acid synthesis were increased[7]. The archaeal cholesterol oxidase activity was increased resulting in generation of pyruvate and hydrogen peroxide[16]. The pyruvate gets converted to glutamate and ammonia by the GABA shunt pathway. The archaeal aromatization of cholesterol generating PAH, serotonin and dopamine was also detected[18]. The archaeal glycolytic hexokinase activity and archaeal extracellular ATP synthase activity were increased. The archaea can undergo magnetite and calcium carbonate mineralization and can exist as calcified nanoforms[19].

The metal actinides provide radiolytic energy, catalysis for oligomer formation and provide a coordinating ion for metalloenzymes all important in abiogenesis[6]. The metal actinide surfaces would by surface metabolism generate acetate which could get converted to acetyl CoA and then to cholesterol which functions as the primal prebiotic molecule self-organising into self-replicating supramolecular systems, the lipid organism[8,9,20]. Cholesterol by radiolysis by actinides would have formed PAH generating PAH aromatic organism[8]. Cholesterol radiolysis would generate pyruvate which would get converted to amino acids, sugars, nucleotides, porphyrins, fatty acids and TCA acids. Anastase and rutile surfaces can produce polymerization of amino acids, isoprenyl residues, PAH and nucleotides to generate the initial lipid organism, PAH organism, prions and RNA viroids which would have symbiosed to generate the archaeal protocell. The archaea evolved into gram negative and gram positive bacteria with a mevalonate pathway which had an evolutionary advantage. The symbiosis of archaea with gram negative organism generated the eukaryotic cell[21]. The data supports the persistence of an actinide and cholesterol based shadow biosphere which throws light on the actinide based origin of life and cholesterol as the premier prebiotic molecule.

There was an increase in free RNA indicating self-replicating RNA viroids and free DNA indicating generation of viroid complementary DNA strands by archaeal reverse transcriptase activity. The actinides modulate RNA folding and catalyse its ribozymal action. Digoxin can cut and paste the viroidal strands by modulating RNA splicing generating RNA viroidal diversity. The viroids are evolutionarily escaped archaeal group I introns which have retrotransposition and self-splicing qualities. Archaeal pyruvate can produce histone deacetylase inhibition resulting in endogenous retroviral (HERV) reverse transcriptase and integrase expression. This can integrate the RNA viroidal complementary DNA into the non-coding region of eukaryotic non-coding DNA using HERV integrase as has been described for borna and ebola viruses[21]. The non-coding DNA is lengthened by integrating RNA viroidal complementary DNA with the integration going on as a continuing event. The archaea genome can also get integrated into human genome using integrase as has been described for trypanosomes. The integrated viroids and archaea can undergo vertical transmission and can exist as genomic parasites. This increases the length and alters the grammar of the non-coding region producing memes or memory of acquired characters. The viroidal complementary DNA can function as jumping genes producing a dynamic genome.[22-24]

The presence of muramic acid, HMG CoA reductase and cholesterol oxidase activity inhibited by antibiotics indicates the presence of bacteria with mevalonate pathway. The bacterial with mevalonate pathway include streptococcus, staphylococcus, actinomycetes, listeria, coxiella and borrelia. The bacteria and archaea with mevalonate pathway and cholesterol catabolism had an evolutionarily advantage and constitutes the isoprenoidal clade organism. The archaea evolved into mevalonate pathway gram positive and gram negative isoprenoid clade organism through horizontal gene transfer of viroidal and virus genes. The isoprenoidal clade prokaryotes develop into other groups of prokaryotes via viroidal/virus as well as eukaryotic horizontal gene transfer producing bacterial speciation[25-27].

The RNA viroids and its complementary DNA developed into cholesterol enveloped RNA and DNA viruses like herpes, retrovirus, influenza virus, borna virus, cytomegalo virus and ebstein barr virus by recombining with eukaryotic and human genes resulting in viral speciation. Bacterial and viral species are ill-defined and fuzzy with all of them forming one common genetic pool with frequent horizontal gene transfer and recombination.

Thus, the multi and unicellular eukaryote with its genes serves the purpose of prokaryotic and viral speciation. The multicellular eukaryote developed so that their endosymbiotic archaeal colonies could survive and forage better. The multicellular eukaryotes are like bacterial biofilms. The archaea and bacteria with a mevalonate pathway uses the extracellular RNA viroids and DNA viroids for quorum sensing and in the generation of symbiotic biofilm like structures which develop into multicellular eukaryotes. The endosymbiotic archaea and bacteria with mevalonate pathway still uses the RNA viroids and DNA viroids for the regulation of multicellular eukaryote.[28-31]

Pollution is a major inducer of evolutionary innovation. Pollution is induced by the primitive nanoarchaea and mevalonate pathway bacteria synthesized PAH and methane leading on to redox stress. Redox stress leads to sodium-potassium ATPase inhibition, inward movement of plasma membrane cholesterol, defective SREBP sensing, increased cholesterol synthesis and nanoarchaeal/mevalonate pathway bacterial growth. Redox stress leads on to viroidal and archaeal multiplication. Redox stress can also lead to HERV reverse transcriptase and integrase expression. The non-coding DNA is formed of integrating RNA viroidal complementary DNA and archaea with the integration going on as a continuing event. The archaeal pox-like dsDNA virus forms evolutionarily the nucleus. The integrated viroidal, archaeal and mevalonate pathway bacterial sequences can undergo vertical transmission and can exist as genomic parasites. The genomic integrated archaea, mevalonate pathway bacteria and viroids form a genomic reserve of bacteria and viruses which can recombine with human and eukaryotic genes producing bacterial and viral speciation. The change in the length and grammar of the non-coding region produces eukaryotic speciation and individuality. The integration of nanoarchaea, mevalonate pathway prokaryotes and viroids in to the eukaryotic and human genome produces a chimera which can multiply producing biofilm like multicellular structures having a mixed archaeal, viroidal, prokaryotic and eukaryotic characters which is a regression from the multicellular eukaryotic tissue. This results in a new neuronal, metabolic, immune and tissue phenotype leading to human disease.[30-32]

Pollution would have been a major factor in eukaryotic speciation and primate/hominid evolution. The change in the length and grammar of the non-coding region produces eukaryotic speciation and individuality. It is the increase in non-coding region and HERV sequences of the genome that led to the evolution of the primate and the human brain and its attendant property of conscious and quantal perception. It is the non-coding region of the genome with its archaeal, RNA viroidal complementary DNA and HERV sequences that makes for the human qualities of the hominid brain. Changes in the length of non-coding region can lead on to disorders of consciousness like schizophrenia. A schizophrenia specific human endogenous retroviruses and change in the length and grammar of the non-coding region has been described in schizophrenia.[33,34]

An actinide dependent shadow biosphere of archaea and viroids in the above mentioned disease states is described. Metal actinides in beach sands have been postulated to play a role in abiogenesis. Cholesterol is the primal prebiotic molecule synthesized on actinide surfaces with all other biomolecules arising from it. A self-replicating cholesterol lipid organism could be the initial life form. A cholesterol based abiogenesis is a more likely evolutionary option and the actinidic archaea and viroids would have evolved from it. The origin of viruses, prokaryotes, eukaryotes, primates and humans from the initial isoprenoid organism derived actinidic archaea is discussed.

References

1. Valiathan M.S., Somers, K., Kartha, C.C. (1993). *Endomyocardial Fibrosis.* Delhi: Oxford University Press.
2. Kurup R., Kurup, P.A. (2009). *Hypothalamic Digoxin, Cerebral Dominance and Brain Function in Health and Diseases*. New York: Nova Science Publishers.
3. Hanold D., Randies, J.W. (1991). Coconut cadang-cadang disease and its viroid agent, *Plant Disease,* 75, 330-335.
4. Edwin B.T., Mohankumaran, C. (2007). Kerala wilt disease phytoplasma: Phylogenetic analysis and identification of a vector, *Proutista moesta, Physiological and Molecular Plant Pathology,* 71(1-3), 41-47.

5. Eckburg P.B., Lepp, P.W., Relman, D.A. (2003). Archaea and their potential role in human disease, *Infect Immun,* 71, 591-596.

6. Adam Z. (2007). Actinides and Life's Origins, *Astrobiology,* 7, 6-10.

7. Schoner W. (2002). Endogenous cardiac glycosides, a new class of steroid hormones, *Eur J Biochem,* 269, 2440-2448.

8. Davies P.C.W., Benner, S.A., Cleland, C.E., Lineweaver, C.H., McKay, C.P., Wolfe-Simon, F. (2009). Signatures of a Shadow Biosphere, *Astrobiology,* 10, 241-249.

9. Wächtershäuser G. (1988). Before enzymes and templates: theory of surface metabolism, *Microbiol Rev,* 52(4), 452-84.

10. Richmond W. (1973). Preparation and properties of a cholesterol oxidase from nocardia species and its application to the enzymatic assay of total cholesterol in serum, *Clin Chem,* 19, 1350-1356.

11. Snell E.D., Snell, C.T. (1961). *Colorimetric Methods of Analysis.* Vol 3A. New York: Van NoStrand.

12. Glick D. (1971). *Methods of Biochemical Analysis.* Vol 5. New York: Interscience Publishers.

13. Colowick, Kaplan, N.O. (1955). *Methods in Enzymology.* Vol 2. New York: Academic Press.

14. Maarten A.H., Marie-Jose, M., Cornelia, G., van Helden–Meewsen, Fritz, E., Marten, P.H. (1995). Detection of muramic acid in human spleen, *Infection and Immunity,* 63(5), 1652-1657.

15. Smit A., Mushegian, A. (2000). Biosynthesis of isoprenoids via mevalonate in Archaea: the lost pathway, *Genome Res,* 10(10), 1468-84.

16. Van der Geize R., Yam, K., Heuser, T., Wilbrink, M.H., Hara, H., Anderton, M.C. (2007). A gene cluster encoding cholesterol catabolism in a soil actinomycete provides insight into Mycobacterium tuberculosis survival in macrophages, *Proc Natl Acad Sci USA,* 104(6), 1947-52.

17. Francis A.J. (1998). Biotransformation of uranium and other actinides in radioactive wastes, *Journal of Alloys and Compounds,* 271(273), 78-84.

18. Probian C., Wülfing, A., Harder, J. (2003). Anaerobic mineralization of quaternary carbon atoms: Isolation of denitrifying bacteria on pivalic acid (2,2-Dimethylpropionic acid), *Applied and Environmental Microbiology,* 69(3), 1866-1870.

19. Vainshtein M., Suzina, N., Kudryashova, E., Ariskina, E. (2002). New Magnet-Sensitive Structures in Bacterial and Archaeal Cells, *Biol Cell,* 94(1), 29-35.

20. Russell M.J., Martin, W. (2004). The rocky roots of the acetyl-CoA Pathway, *Trends in Biochemical Sciences,* 29, 7.

21. Margulis L. (1996). Archaeal-eubacterial mergers in the origin of Eukarya: phylogenetic classification of life, *Proc Natl Acad Sci USA,* 93, 1071-1076.

22. Tsagris E.M., de Alba, A.E., Gozmanova, M., Kalantidis, K. (2008). Viroids, *Cell Microbiol,* 10, 2168.

23. Horie M., Honda, T., Suzuki, Y., Kobayashi, Y., Daito, T., Oshida, T. (2010). Endogenous non-retroviral RNA virus elements in mammalian genomes, *Nature,* 463, 84-87.

24. Hecht M., Nitz, N., Araujo, P., Sousa, A., Rosa, A., Gomes, D. (2010). Genes from Chagas parasite can transfer to humans and be passed on to children. Inheritance of DNA Transferred from American Trypanosomes to Human Hosts, *PLoS ONE,* 5, 2-10.

25. Flam F. (1994). Hints of a language in junk DNA, *Science,* 266, 1320.

26. Horbach S., Sahm, H., Welle, R. (1993). Isoprenoid biosynthesis in bacteria: two different pathways? *FEMS Microbiol Lett,* 111, 135–140.

27. Gupta R.S. (1998). Protein phylogenetics and signature sequences: a reappraisal of evolutionary relationship among archaebacteria, eubacteria, and eukaryotes, *Microbiol Mol Biol Rev,* 62, 1435–1491.

28. Hanage W., Fraser, C., Spratt, B. (2005). Fuzzy species among recombinogenic bacteria, *BMC Biology,* 3, 6-10.

29. Whitchurch C.B., Tolker-Nielsen, T., Ragas, P.C., Mattick, J.S. (2002). Extracellular DNA Required for Bacterial Biofilm Formation. *Science,* 295(5559), 1487.

30. Webb J.S., Givskov, M., Kjelleberg, S. (2003). Bacterial biofilms: prokaryotic adventures in multicellularity, *Curr Opin Microbiol,* 6(6), 578–85.

31. Chen Y., Cai, T., Wang, H., Li, Z., Loreaux, E., Lingrel, J.B. (2009). Regulation of intracellular cholesterol distribution by Na/K-ATPase, *J Biol Chem,* 284(22), 14881-90.

32. Poole A.M. (2006). Did group II intron proliferation in an endosymbiont-bearing archaeon create eukaryotes? *Biol Direct,* 1, 36-40.

33. Villarreal L.P. (2006). How viruses shape the tree of life, *Future Virology,* 1(5), 587-595.

34. Lockwood M. (1989). *Mind, Brain and the Quantum.* Oxford: B. Blackwell.

CHAPTER 28

INTERNET RELATED LOW LEVEL OF EMF POLLUTION - CLIMATE CHANGE - ENVIRONMENTAL EVOLUTIONARY HUMAN MICROBIOME METABOLIC DISORDER - INFECTIVE SYMBIOTIC TISSUE COLONIES OF PROTOZOA, ARCHAEA, VIRUSES AND BACTERIA

The plasma life based forms and the gravitational and anti-gravitational waves generated porphyrions would have been the most primitive of non-life forms. The porphyrin templates would have given rise to the isoprenoid organism, the RNA viroids, DNA viroids and prions which would have symbiosed to form the nanoarchaea. The archaea would have evolved into prokaryotes or bacteria. The archaeal symbiosis with prokaryotes would have generated the eukaryotes. The initial unicellular eukaryotes are the protozoa. The protozoa are capable of phagocytosis and would have had viruses, bacteria and archaea as their intracellular symbionts. The viruses of the protozoa like entamoeba and toxoplasma are giant viruses called megaviride. The protozoans like entamoeba with its viral, archaeal and bacterial endosymbionts would have evolved into the multicellular organisms or metazoa. The archaeal, bacterial and protozoal extracellular vesicles containing RNA and DNA viroids can by horizontal gene transfer lead to evolution of metazoan, plants, animals and human species. The metazoans would have evolved into plants, animals and humans by horizontal gene transfer generated by archaeal extracellular RNA and DNA viroid vesicles. The evidence of such a transformation is seen in the bacterial symbionts of protozoa like entamoeba which includes legionella, listeria, mycoplasma, mycobacteria and chlamydia. The symbionts of protozoan entamoeba include viruses like covid, HIV and cytomegalo virus. The protozoan entamoeba serves as a natural reservoir for covid infections, bacterial infections and HIV infections. The toxoplasma also serves as a covid reservoir. These

protozoan colonies of entamoeba and toxoplasma with its bacterial, archaeal and viral symbionts can be called as chimeric infective symbiotic tissue colonies which can exist in the air, water, earth, blood and tissues. The chimeric infective symbiotic tissue colonies containing bacteria, archaea and viruses can make the symbiotic organism multidrug resistant by horizontal gene transfer. The chimeric infective symbiotic tissue organism evolves as a consequence of global warming and the protozoal, bacterial, viral and archaeal symbiotic colonies are new forms of life. The chimeric infective symbiotic tissue organism composed of protozoa and its symbionts like virus, bacteria and archaea depend upon glycolysis for its energy needs. The glycolytic enzymes enolase, glyceraldehyde 3-phosphate dehydrogenase, hexokinase, phosphofructokinase, aldolase, pyruvic kinase and triose phosphate isomerase are moonlighting proteins having a cytoplasmic, membrane and nuclear functions. These glycolytic enzymes have a RNA binding motif which can modulate RNA transcription in the host and the symbiotic organism. These glycolytic enzymes with their RNA binding motif as well as DNA binding motif can bind to RNA and DNA viruses and viroids modulating their replication and life cycles. The chimeric infective symbiotic tissue organism composed of protozoa and its symbionts like virus, bacteria and archaea can enter the human host through the breaches of the gut-blood barrier and transform itself into the macrophages of the human host. The macrophages are involved in the phagocytosis and innate immunity against bacterial and viral infections. Paradoxically the macrophages like the protozoa can also serve as a reservoir where pathogens like mycobacteria tuberculosis, mycobacteria leprae, listeria, legionella, E. coli, shigella & salmonella, and viruses like HIV & covid can thrive and replicate. The modified chimeric infective symbiotic tissue organism or macrophage composed of protozoa and its symbionts like virus, bacteria and archaea can serve as a reservoir for horizontal gene transfer and generation of multidrug resistant bacteria and viruses. The infective chimeric symbiotic tissue organism composed of protozoa and its symbionts like virus, bacteria and archaea

depends upon moonlighting glycolytic enzymes for their energetics and replication. The development of antibodies against the infective chimeric symbiotic tissue organism composed of protozoa and its symbionts like virus, bacteria and archaea leads to the evolution of anti-brain antibody disease or anti-basal ganglia antibody disease. They present a threat to existence consequent to horiztontal gene transfer and generation of multidrug resistant bacteria, viruses, protozoa, mycobacteria and archaea.

The archaeal endosymbiosis in the brain produces increased magnetoperception and quantal perception of low level EMF producing frontal atrophy and cerebellar dominance. Changes in dietary habits leading to less consumption of fibre and more intake of protein and fat leads to increased colonic archaeal growth and archaeal endosymbiosis. The increase in endosymbiotic archaeal growth promotes Neanderthal metabolonomics and survival in extreme anaerobic conditions. These changes can be reversed by a higher intake of fibre.

Low level EMF pollution leads to endosymbiotic archaeal growth which has got magnetoperception and quantal perception and still further perception of low level of EMF. This leads to frontal cortical atrophy and cerebellar dominance. There is an endemic cerebellar cognitive affective disorder presenting as adult onset autism, impulsive behaviour, violence, introversion, solitude and aggression. The colonic archaeal growth and archaeal endosymbiosis can be modulated by higher intake of dietary fibre which inhibits colonic archaeal growth and endosymbiosis. This can modulate brain structure and function.

Modern civilization results in stress in societies leading to increased activation of the HPA axis leading to increased secretion of stress hormones the catecholamines producing increase in sympathetic tone. The increase in sympathetic tone leads to increase in stress hormones epinephrine and norepinephrine producing breeching of the gut-blood barrier. This leads to decrease in mucin secretion lining the gut wall and inflammatory changes in the colonic mucosa. Stress leads to increase in archaeal growth and colonic archaeal translocation into to the blood and tissues producing archaeal endosymbiosis. The archaeal endosymbiosis in the brain can lead to increased quantal and magnetoperception of low level EMF producing frontal atrophy and cerebellar dominance. Yogic pranayamic breathing techniques can modulate brain function and metabolonomics. Right nasal breathing stimulates the right hemisphere of the brain producing increased sympathetic activity and reduced parasympathetic activity. Left nasal breathing can stimulate the left hemisphere of the brain decreasing sympathetic activity and increasing parasympathetic activity. Right nasal breathing induced increased sympathetic tone occurs by stimulation of the right insula, mid-brain and periaqueductal regions. The left nasal breathing can lead to decrease in sympathetic tone, decreased amount of stress hormones, stability of the gut-blood barrier and reduction in colonic archaeal growth. This leads to reduction in archaeal endosymbiosis and preservation of the frontal cortex.

Global warming leads to archaeal endosymbiosis and dysautonomic syndrome. There is increased sympathetic activity and reduced parasympathetic tone. This results from activation of the insula, periaqueductal areas and upper mid-brain which leads to stimulation of opioid and cannabinoid receptors. The endosymbiotic archaeaons can form cell organelle called neurotransminoids which can synthesize morphine alkaloids and cannabinoids. The increase sympathetic tone and secretion of epinephrine and norepinephrine leads to uncoupling of oxidative phosphorylation. This produces increased generation of heat which can lead to survival in the cold anaerobic condition of other planets. This leads to a Warburg phenotype with energetics generated by glycolysis. The depletion of ATP leads to membrane sodium-potassium ATPase inhibition which is also accelerated by archaeaon induced digoxin synthesis. The membrane sodium-potassium ATPase inhibition leads to membrane ATP synthesis. The endosymbiotic archaeal mediated neanderthalic metabolonomics can lead to survival in extreme anaerobic conditions. This has been demonstrated in PET studies of the iceman's brain and metabolism.

Yogic pranayamic breathing techniques can modulate brain function and metabolonomics. Right nasal breathing stimulates the right hemisphere of the brain producing increased sympathetic activity and reduced parasympathetic activity. Left nasal breathing can stimulate the left hemisphere of the brain decreasing sympathetic activity and increasing parasympathetic activity. Right nasal breathing induced increased sympathetic tone occurs by stimulation of the right insula, mid-brain and periaqueductal regions. The right nasal breathing induced increased sympathetic tone can modulate metabolonomics driving it in the direction of anaerobic metabolism. The increase sympathetic tone and secretion of epinephrine and norepinephrine leads to uncoupling of oxidative phosphorylation. This produces increased generation of heat which can lead to survival in the cold anaerobic condition of other planets. This leads to a Warburg phenotype with energetics generated by glycolysis. The depletion of ATP leads to membrane sodium-potassium ATPase inhibition which is also accelerated by archaeaon induced digoxin synthesis. The membrane sodium-potassium ATPase inhibition leads to membrane ATP synthesis. Thus yogic pranayamic breathing exercises induced mediated neanderthalic metabolonomics can lead to survival in extreme anaerobic conditions. This has been demonstrated in PET studies of the iceman's brain and metabolism.

Meditation can modulate body metabolonomics and brain function. The mechanism is by induction of the heme oxygenase system. Meditation induced heme oxygenase which converts heme to carbon monoxide and bilirubin. Bilirubin and biliverdin are free radical scavengers and mops up free radicals. Free radicals are required for the function of the NMDA dependent thalamo-cortico-thalamic feedback reverberatory circuit crucial in consciousness. This circuit mediates working memory and focused attention. Free radicals also activate NMDA and by its capacity of diffusing freely through the brain systems can induce NMDA activity, synchronized burst firing of neurons in different parts of sensory areas producing perceptual synchronization. This mediates consciousness. Thus the scavenging of free radicals by heme oxygenase leads to suppression of consciousness and meditative trances. Heme oxygenase induction suppresses ALA synthase. Thus heme is depleted from the system. There is increased porphyrin synthesis leading on to porphyrinuria and porphyria. The stimulus for porphyrin synthesis comes from heme deficiency. Porphyrins can organise into self-replicating supramolecular structures called porphyrions which are induced by meditative practices. The porphyrins can self-organise to form macromolecular structures which can self-replicate to form a porphyrin organism. The photon induced transfer of electrons along the macromolecule can lead to light induced ATP synthesis. The porphyrins can form a template on which RNA and DNA can form generating viroids. The porphyrins can also form a template on which prions can form. They all can join together – RNA viroids, DNA viroids, prions – to form primitive archaea. Thus the archaea are capable of self-replication on porphyrin templates. The porphyrins are dipolar and the setting of membrane intercalated porphyrin mediated sodium-potassium ATPase inhibition produces a pumped phonon system and a Frohlich model of superconductivity mediating quantal perceptions and low temperature. Thus the porphyrions mediate quantal perception during meditative trances. The porphyrins have increased quantal perception leading to sensing of low level EMF fields causing cortical especially prefrontal cortex atrophy. The cerebellar dominance and a cerebellar cognitive affective disorder. This produces porphyrin related quantal perception. The porphyrins can have quantal perception of low level EMF fields leading to prefrontal cortex atrophy. This leads on to cortical dysfunction and lack of functioning of the conscious brain. The cerebellum dominates and the unconscious takes over. This leads on to neanderthalisation of the brain and schizophrenia and autism. Thus the increase in porphyrins leads to cortical dysfunction and prefrontal cortex atrophy. The porphyrins can destroy the human endogenous retroviruses and the jumping genes leading to lack of dynamicity of the genome. This leads on to maldevelopment of the prefrontal cortex. The cerebellar cortex becomes dominant as the cognitive organ with features of impulsivity, extrasensory perceptive mechanisms, robotic automatic programmes and autistic/schizophrenic features. This leads on to an induced cerebellar cognitive affective disorder with a sense of the quantal world and a feeling of oneness with the universe. This is the basis of the collective unconscious or spirituality.

Porphyrins are dipolar molecules and in the setting of porphyrin mediated membrane sodium-potassium ATPase inhibition induced pumped phonon system can produce a quantal perceptive state. Porphyrins are macromolecules which can have both a wave and particle existence and can bridge the particulate world and the quantal world. Membrane sodium-potassium ATPase inhibition induced dipolar porphyrin mediated pumped phonon system can lead on to a cellular plasma state and EMF signal transduction. Macromolecules like RNA, DNA, protein and the cell itself can have an EMF signature. This porphyrin generated macromolecular cellular EMF signature is important in regulation of cell function.

Meditation induced heme oxygenase can lead to metabolonomic and brain evolution. There is increased porphyrin synthesis from succinyl CoA and glycine. Defect in heme synthesis and heme depletion leads to deficiency of heme enzymes. Deficiency cytochrome C oxidase and aconitase leads to mitochondrial oxidative phosphorylation defects and TCA cycle defects. This leads to pyruvate dehydrogenase deficiency and defect in synthesis of acetyl

CoA. There is increase in glycolysis consequent to porphyrin photo-oxidation induced free radical generation and HIF alpha induction. This produces the Warburg phenotype. The increased level of pyruvate that is generated is converted to glutamate and ammonia. Thus there is hyperammonemia as a consequence of the metabolic defect. Since glycine is utilized for porphyrin synthesis serine is not synthesized leading on to deficiency of the substrate for synthesis of cystathionine. This leads to accumulation of homocysteine and homocystinuria. Deficiency of acetyl CoA leads to defects in the isoprenoid pathway and defective synthesis of cholesterol and ubiquinone. There is also deficiency of the heme containing cholesterol synthesizing enzyme lanosterol synthase. This leads to a cholesterol depleted state.

The heme deficiency leads to lack of synthesis of the heme enzyme cytochrome P420 dependent sex hormones and a widespread asexual state. The mitochondrial dysfunction leads on to insulin resistance and metabolic syndrome X. The Warburg phenotype and increased glycolysis leads to oncogenesis. The mitochondrial dysfunction can produce neurodegeneration. The increase in lymphocyte glycolysis can produce immune activation and autoimmune disease. Thus the meditation induced porphyria can lead to civilisational disease.

The porphyrins can self-organise to form macromolecular structures which can self-replicate to form a porphyrin organism. The photon induced transfer of electrons along the macromolecule can lead to light induced ATP synthesis. The porphyrins can form a template on which RNA and DNA can form generating viroids. The porphyrins can also form a template on which prions can form. They all can join together – RNA viroids, DNA viroids, prions – to form primitive archaea. Thus the archaea are capable of self-replication on porphyrin templates. The self-replicating archaea can sense gravity which gives rise to consciousness. They can also sense the antigravity fields which gives rise to the unconscious brain. Thus there can be both self-replicating archaea and anti-archaea regulating the conscious and unconscious brain. Thus the climate change stress mediated increased porphyrin synthesis leads to prefrontal cortex atrophy, cerebellar dominance, cerebellar cognitive affective disorder, quantal perception and neanderthalisation of the population. The porphyrions are self-replicating supramolecular organisms which forms the precursor template on which the viroids, prions and nanoarchaea originate. Meditative stress induced template directed abiogenesis of porphyrions, prions, viroids and archaea is a continuous process and can contribute to changes in brain structure and behaviour as well as disease process.

CHAPTER 29

CLIMATE CHANGE AND HUMAN SPECIES - HOMO NEANDERTHALIS, HOMO SAPIENS, HOMO SAPIEN EXTINCTUS AND HOMO NEONEANDERTHALIS - CLIMATE CHANGE AND EVOLUTION OF NEW SPECIES - INFECTIVE SYMBIOTIC TISSUE COLONIES OF PROTOZOA, ARCHAEA, VIRUSES AND BACTERIA

Endosymbiotic archaea and species evolution

The plasma life based forms and the gravitational and anti-gravitational waves generated porphyrions would have been the most primitive of non-life forms. The porphyrin templates would have given rise to the isoprenoid organism, the RNA viroids, DNA viroids and prions which would have symbiosed to form the nanoarchaea. The archaea would have evolved into prokaryotes or bacteria. The archaeal symbiosis with prokaryotes would have generated the eukaryotes. The initial unicellular eukaryotes are the protozoa. The protozoa are capable of phagocytosis and would have had viruses, bacteria and archaea as their intracellular symbionts. The viruses of the protozoa like entamoeba and toxoplasma are giant viruses called megaviride. The protozoans like entamoeba with its viral, archaeal and bacterial endosymbionts would have evolved into the multicellular organisms or metazoa. The archaeal, bacterial and protozoal extracellular vesicles containing RNA and DNA viroids can by horizontal gene transfer lead to evolution of metazoan, plants, animals and human species. The metazoans would have evolved into plants, animals and humans by horizontal gene transfer generated by archaeal extracellular RNA and DNA viroid vesicles. The evidence of such a transformation is seen in the bacterial symbionts of protozoa like entamoeba which includes legionella, listeria, mycoplasma, mycobacteria and chlamydia. The symbionts of protozoan entamoeba include viruses like covid, HIV and cytomegalo virus. The protozoan entamoeba serves as a natural reservoir for covid infections, bacterial infections

and HIV infections. The toxoplasma also serves as a covid reservoir. These protozoan colonies of entamoeba and toxoplasma with its bacterial, archaeal and viral symbionts can be called as chimeric infective symbiotic tissue colonies which can exist in the air, water, earth, blood and tissues. The chimeric infective symbiotic tissue colonies containing bacteria, archaea and viruses can make the symbiotic organism multidrug resistant by horizontal gene transfer. The chimeric infective symbiotic tissue organism evolves as a consequence of global warming and the protozoal, bacterial, viral and archaeal symbiotic colonies are new forms of life. The chimeric infective symbiotic tissue organism composed of protozoa and its symbionts like virus, bacteria and archaea depend upon glycolysis for its energy needs. The glycolytic enzymes enolase, glyceraldehyde 3-phosphate dehydrogenase, hexokinase, phosphofructokinase, aldolase, pyruvic kinase and triose phosphate isomerase are moonlighting proteins having a cytoplasmic, membrane and nuclear functions. These glycolytic enzymes have a RNA binding motif which can modulate RNA transcription in the host and the symbiotic organism. These glycolytic enzymes with their RNA binding motif as well as DNA binding motif can bind to RNA and DNA viruses and viroids modulating their replication and life cycles. The chimeric infective symbiotic tissue organism composed of protozoa and its symbionts like virus, bacteria and archaea can enter the human host through the breaches of the gut-blood barrier and transform itself into the macrophages of the human host. The macrophages are involved in the phagocytosis and innate immunity against bacterial and viral infections. Paradoxically the macrophages like the protozoa can also serve as a reservoir where pathogens like mycobacteria tuberculosis, mycobacteria leprae, listeria, legionella, E. coli, shigella & salmonella, and viruses like HIV & covid can thrive and replicate. The modified chimeric infective symbiotic tissue organism or macrophage composed of protozoa and its symbionts like virus, bacteria and archaea can serve as a reservoir for horizontal gene transfer and generation of multidrug resistant bacteria and viruses. The infective chimeric symbiotic tissue

organism composed of protozoa and its symbionts like virus, bacteria and archaea depends upon moonlighting glycolytic enzymes for their energetics and replication. The development of antibodies against the infective chimeric symbiotic tissue organism composed of protozoa and its symbionts like virus, bacteria and archaea leads to the evolution of anti-brain antibody disease or anti-basal ganglia antibody disease. They present a threat to existence consequent to horiztontal gene transfer and generation of multidrug resistant bacteria, viruses, protozoa, mycobacteria and archaea.

The global warming leads to endosymbiotic as well as colonic archaeal growth leading to alteration in the structure and function of the human body and system. The archaeal overgrowth within the cells leads to generation of new cellular organelle called archaeaons. The archaea have the shikimate pathway which can synthesize tyrosine and dopamine. Dopamine can be converted to dopachrome and epinephrine to adrenochrome. Dopachrome and adrenochrome can polymerize by oxidation generating melanin. The archaeaons secreting melanin can be called as archaeal melanosomes. The melanin in melanosomes has the wide range of absorption of the light spectra and gamma radiation and can transduct it to generate energy. This energy transduction can split water into H_2 and O_2 and generate protons modulating the proton gradient across the mitochondrial membrane synthesizing ATP. The melanin in the melanosome can absorb photons reducing ubiquinone to ubiquinol and generate ATP synthesis by oxidative phosphorylation. Thus the melanin in the archaeaons in the human cell can function as photosynthetic organelle. The archaeaons and their melanin can utilize gamma radiation to synthesize ATP and can exist in extreme conditions. Thus the archaeaons can produce a source of energy from light and electromagnetic waves and gamma radiation. The melanin is capable of transducting electromagnetic waves and low level electromagnetic fields and can be capable of quantal perception. Thus the melanin in the melanosomes is capable of information sensing and storage as well as energy production from electromagnetic waves and water. The human brain could have evolved by this mechanism. The humans are hairless as compared to other primates and are exposed to more of light inducing melanin induced photosynthesis and energy generation which could have contributed to the evolution of the human cortex and the complex human brain. The archaeaons melanosomes are capable of quenching free radicals and resist phagocytic destruction. The melanosomes can also resist radiation and UV light. The archaeaons are indestructible and eternal. The archaeaons have got magnetite and are capable of quantal perception and information storage. The melanin also serves the purpose of quantal perception and information storage. The archaeaon can also synthesize magnetite particles forming subcellular organelle called magnetosomes. Magnetite can interact with melanin forming supermolecular complex systems. The archaeaon can synthesize porphyrins which can self-organize to form self-replicating structures called porphyrions. Porphyrions can interact with melanin also forming supramolecular complex systems. Eumelanin pigments contain indole based tetramers that are arranged in porphyrin-like domains. The indole based structures can self-organize on porphyrin scaffolds to form tetrameric structures and melanin. The chemical structure of melanin on a macromolecular scale exhibits a tetrameric ring structure possibly because of self-organization on porphyrin scaffolds. Porphyrion can generate melanosome complexes and they can form self-organizing supramolecular complex systems. The archaeaon particles of melanosomes, magnetosomes and porphyrions forming complex colony network with specialized functions. It can function as a quantal computing system. The porphyrions and melanosomes can transducer energy and synthesize ATP functioning as primitive photosynthetic system. The magnetosome, porphyrions and melanosomes can function as information storage systems. Magnetosomes and porphyrions are dipolar and can have a quantal perceptive function based on sodium-potassium ATPase inhibition mediated pumped phonon system. The melanin can function as a superconductor for high frequency radiation and neurotransmission, as a semi-conductor for sound and heat, conduct body ionic charges and resonate for the frequencies of visible light. The archaeaon- magnetosome, porphyrions and melanosome network can function as a quantal computing brain reducing the human classical brain to a zombie brain. Thus the global warming induced archaeaon colony network and melanosomes are indestructible and eternal and takeover the human body. The human body metabolic programmes are suppressed including mitochondrial oxidative phosphorylation. The human body is reduced to a zombie or a framework for the archaeaon colony to thrive. The archaeaon induces stem cell transformation of the host human cells and change the metabolonomics of the human cells. The human cells oxidative phosphorylation is suppressed and it depends upon glycolysis for its energy needs. The human glycolytic pathway is taken over by the archaeaon for its needs. The glycolytic metabolites are channelled to the shikimic acid pathway and the D-xylulose phosphate pathway. The DXP pathway can synthesize cholesterol which is catabolized by the archaeaon for its energy. The cholesterol ring oxidases convert the cholesterol to pyruvate which then enters the GABA shunt pathway. The cholesterol side chain oxidases convert the side chain to short chain fatty acids and bile acids. The cholesterol aromatases converts the cholesterol ring to phenyl residues and synthesis of tyrosine and tryptophan. The shikimic acid pathway also utilizes substrates from the glycolytic pathway and generates tyrosine and tryptophan. The tyrosine that synthesize is converted to dopa, dopamine, dopachrome and oxidized to melanin. Melanin serves the purpose of capturing electromagnetic radiation, UV rays, Gamma radiation and light synthesizing ATP. Melanin can serve as a substrate for primitive archaeal photosynthesis. This leads to alteration in brain function and structure. The brain functions as an archaeaon melanosomal magnetite colony network capable of quantal perception, information storage and energy generation. This alters the brain function to an impulsive and anarchic mode of social function and functioning of the society as a group or collective organism. The quantal perception of the archaeaons also leads to evolution of a sort of communication with the quantal world creating a sort of universal personality or self. The human cell and system is converted to the stem cell colony which is immature and lacking functional differentiation becoming a zombie for the archaeal colony. The melanosome and melanin form a first line of defence against infection and is required for innate immunity. The melanosomes can kill the bacteria, viruses and other organisms as is evidenced by the albinism related Chediak Higashi syndrome and Griscelli syndrome. The archaeal melanin also protects it against high temperature, chemicals, oxygen radicals, oxidizing agents, UV radiation and heavy metals. The archaeal melanin makes the endosymbiotic archaea indestructible.

Intergalactic archaeal quantal computing cloud universalis

The intergalactic space contains micro-organism, especially extremophiles like archaea. The archaeal colony with its melanosomes, magnetosomes and porphyrions can form a giant quantal computing cloud in the intergalactic space functioning as an intergalactic superhuman intelligence. The porphyrions can form a template for the generation of RNA viroids, DNA viroids and prions which can self-organize to form archaeaons. The porphyrions themselves are capable of a wave-particle existence and self-replication. Thus, the quantal computing cloud of extraterrestrial intelligence can arise on its own from the quantal electromagnetic fields of the intergalactic space. This extraterrestrial intelligence of quantal computing cloud of archaeaons, magnetosomes, melanosomes and porphyrions in the intergalactic space can be called as intergalactic archaeal quantal computing cloud universalis. This forms the ubiquitous anthropomorphic observer creating the universe out of the quantal foam, itself arising out the quantal foam. The porphyrins can arise sui generis from a quantal foam and forms a template for the formation of RNA viroids. An interstellar cloud of RNA viroids forms. The RNA viroids later code for DNA viroids and prions. An isoprenoid organism can also arise in the porphyrin scaffold. The interstellar cloud of dominant RNA viroids gives rise to a form of universal consciousness or gravitational waves. The RNA viroids can generate electric currents by the piezoelectric effect where mechanical energy due to the shearing stress of RNA viroidal population is converted to electrical energy and this can give rise to gravitational waves and consciousness. The helical protein of the viruses has negative and positive charged ends and acts as a dipole. When they are squashed by shearing stress of viroidal population the rod shape of the viroids gets changed to oval and dipole becomes uneven. This generates electromagnetic forces and gravitational waves. The gravitational wave forms the basis of consciousness. The RNA viroidal population can have a silicon coating and can reach the earth by asteroidal hits and gives rise to endogenous retroviruses. The human endogenous retroviruses contribute to the plasticity the human genome and the development of synaptic connectivity important for the evolution of the prefrontal cortex. The RNA viroidal population best thrives in the presence of gravity and play an important role in the development of human cerebral cortex in homo sapiens. The homo sapien brain is cerebral cortical dominant with a fully developed human consciousness due to increase in HERV sequences which increases genomic plasticity and synaptic connectivity. The homo sapiens are creatures with dominant conscious function and are logical and rational. The interstellar RNA viroidal population contributes to consciousness and gravitational waves which are linked. The intergalactic dark matter and dark energy contributes to nearly 90 percent of the universe energy. The dark energy contributes to antigravity forces which are repulsive and contributes to expansion of the universe. The dark energy, dark matter and antigravity contribute to the collective unconscious and human unconscious. The dark matter is made up of melanotic archaeal networks which form huge clouds in the universe. The melanotic archaea arise abiogenetically from porphyrin scaffolds which get structured out of the quantal foam spontaneously. On this porphyrin scaffolds the RNA viroids, the DNA viroids, prions, melanin and isoprenoids organisms form which symbiose to form the melanotic archaea. Thus the porphyrion/RNA viroidal population which mediates gravity and consciousness gives rise to melanotic archaeal clouds and antigravity mediating the collective unconscious. Thus gravity gives rise to antigravity and consciousness gives rise to the unconsciousness. The melanotic archaea can use anti-gravitational waves, cosmic radiation and gamma radiation as energy source for ATP synthesis. The dark matter of melanotic archaea contributing to antigravity thrives and multiplies in zero gravity situations. The melanotic archaea contains magnetite which can repulse each other when properly aligned contributing to the repulsive antigravity. The antigravity is related to the collective unconscious in the world as well as the human unconscious which is structured in the cerebellum. The dark matter containing melanotic archaea gets transferred to Eurasian land mass and earth by asteroidal hits and forms giant colonies and networks evolving to homo neanderthalis. The homo neanderthalis brain has a cerebellar dominant structure and function and is impulsive with a predominant unconscious function. The conscious function and cerebral cortex is less developed in homo neanderthalis as they are retroviral resistant. The archaea induces stem cell conversion and secretes digoxin which makes the homo neanderthalis cell population retroviral resistant. The deficiency of HERV sequences leads to maldevelopment of the homo neanderthalis cerebral cortex. The homo neanderthalis are impulsive creatures of the unconscious modulated by anti-gravitational waves. This extraterrestrial intelligence of quantal computing cloud can see life in different parts of the galaxies via asteroids and meteors. The human species evolved out of the seeded archaeaons from the extraterrestrial intelligence of the quantal computing cloud formed of the archaeal colony of archaeaons - magnetosomes, melanosomes and porphyrions. This would have reached the earth by meteoric and asteroidal hits. The hits of the meteors and asteroids would have occurred first in the Eurasian land mass, especially in the northern Siberian tundra. The homo neanderthalis would have evolved in this Eurasian land mass. As the Siberian Eurasian land mass was cold and dark the homo neanderthalis were depigmented and fair-coloured, hairless with sparse red hair. They were deficient in melanin and melanin induced energy transduction and photosynthesis leading to synthesis of ATP. The homo neanderthalis was energy deprived and the neanderthalic cortex was primitively formed and the cerebellum dominated their cognitive function. The endosymbiotic archaeal network in the brain with its magnetosomes, melanosomes and porphyrions form a primitive quantal computing system. This functions as an information receptive and storage system in communication with the extraterrestrial intelligence of the quantal computing cloud in the intergalactic space. The homo neanderthalis owing to its lack of melanosomes and innate immunity became relatively extinct over a period of time with fossilized remnants in different parts of the world. The homo neanderthalis had quantal perception which created a feeling of oneness with gender and social equality in society. The society was gender-equal and matriarchal. The matriarchal societies of the Dravidians, Basque, Celts, Harappans, Sumerians and Jews were fossilized remnants of the homo neanderthalic species. The extremes of cold temperature of the ice age led to the growth of endosymbiotic archaea in the absence of melanosomes in the Neanderthal. The melanosomes function as the first line of defence against infection and is important in innate immunity. The absence of melanosomes would have led to defective innate immunity and eventual partial extinction of homo neanderthalis with preservation of fossilized matrilineal clusters. The fossilized matrilineal neanderthalic clusters are present in different parts of the world. The fossilised homo neanderthalis are susceptible to increased archaeal endosymbiosis consequent to global warming and related civilizational diseases of metabolic syndrome, schizophrenia, cancer, autoimmune disease and degeneration. The homo neanderthalis will become extinct owing to civilizational disease consequent to global warming induced endosymbiotic archaeal growth.

The Homo sapiens

The Homo sapiens evolved in the tropical hot African land mass. The first human species to evolve is the homo neanderthalis in the Eurasian steppes. The homo sapiens would have evolved out of the archaea secreted porphyrions and RNA viroids independently. The porphyrions could have been transmitted to the tropical African land mass and would have served as a substrate for the formation of RNA viroids, DNA viroids and prions which symbiosed to form the primitive eukaryotic cell. The high temperature of the African continent would have contributed to mutations in RNA viroids and DNA viroids leading on to rapid evolution. The sub-Saharan African soil is depleted of selenium. Selenium deficiency leads to RNA viroidal mutations. Thus extremes of temperature and selenium deficiency lead to RNA viroidal diversity. This RNA viroidal diversity would have led to rapid evolution of homo sapiens from the eukaryotic cell. This eukaryotic cell would have evolved into homo sapien species over a period of time. The RNA viroids are the basis of the HERV genes which contributes to the dynamicity of the homo sapien genome. The homo neanderthalis on the other hand are retroviral resistant while the homo sapiens is retroviral sensitive. The homo neanderthalis archaeaon secretes digoxin, a steroidal hormone which can destroy the retrovirus. The homo neanderthalis also has got endosymbiotic cholesterol catabolizing archaea which can alter the membrane sites for retroviral binding making the Neanderthal species resistant to retroviral infection. The homo neanderthalis have got a deficiency of HERV jumping genes in the genome and a rigid genome as compared to the HERV sequences mediated flexible genome of the homo sapiens. The homo sapiens as they evolved in the hot African savannah would have been exposed to heat and light. This would have related in increased melanogenesis and darker skin and plenty of hair in the evolved homo sapiens. The homo sapiens owing to their dark colour would have been energy surplus consequent to melanin induced energy transduction and ATP synthesis. This would have led to the evolution of the human cortex. The RNA viroids integrated into the genome would have function as jumping HERV genes contributing to the dynamicity of the genome. A dynamic and flexible genome is required for the development of synaptic connectivity and cerebral cortex. Thus the homo sapiens evolve the modern human cerebral cortex consequent to the surplus energy produced by melanin induced energy transduction and ATP synthesis. The increase in melanin and melanosomes increased the innate immunity of the homo sapiens making them resistant to endogenous archaeal endosymbiosis. The homo sapiens were resistant to endosymbiotic archaeal growth seen in extremes of climate of global warming and ice age. The homo sapiens which evolved out of hot tropical Africa had increased melanin content in the skin which inhibits archaeal endosymbiosis and neanderthalisation. The homo sapien species is thus protected against increased archaeal endosymbiosis consequent to global warming and related civilizational diseases of metabolic syndrome, schizophrenia, cancer, autoimmune disease and degeneration.

Homo sapien albino mutants and Homo neoneanderthalis

The homo sapiens developed albino mutants which lacked the tyrosinase enzyme. These albino homo sapien mutants could not survive in the hot African savannah due to lack of pigmentation and migrated to the southern European land mass. This evolved into the patrilineal homo sapien European civilization. The patrilineal homo sapien European civilization arose out of the homo sapien patrilineal African civilization. The albino mutants homo sapiens forming the European civilization are susceptible to endosymbiotic archaeal growth consequent to global warming. The albino mutants homo sapiens lack melanin and melanosomes important in innate immunity. This leads to fertile conditions for endosymbiotic archaeal growth in the albino mutants, Caucasoid population. The endosymbiotic archaeal growth in the Caucasoid population leads to the evolution of a new human species. The human zombie controlled by endosymbiotic melanotic magnetite archaeaon colony network can be called as a new species - Homo neoneanderthalis. Thus, the species change is occurring in the albino mutant homo sapien population of Europe and American consequent to global warming and endosymbiotic archaeal growth. The homo neoneanderthalic species and fossilized homo neanderthalis are susceptible to increased archaeal endosymbiosis consequent to global warming and related civilizational diseases of metabolic syndrome, schizophrenia, cancer, autoimmune disease and degeneration. The homo neanderthalis and homo neoneanderthalis will become extinct owing to civilizational disease consequent to global warming induced endosymbiotic archaeal growth.

Homo sapien extinctus

The homo neanderthalis and homo neoneanderthalis have endosymbiotic archaeal symbiosis. The endosymbiotic archaea secrete RNA viroids which can be acted upon by HERV reverse transcriptase generating corresponding DNA sequences which can be integrated into the genome by HERV integrase. The archaeal digoxin can edit the RNA viroids producing widespread diversity. The archaeal porphyrins can serve as a template for the generation of RNA viroids, DNA viroids and prions. The RNA viroids and DNA viroids can recombine with RNA and DNA viruses in the environment generating new RNA and DNA viruses. The RNA and DNA viroids can exchange their sequences with environmental bacteria generating new bacteria. Thus there can be endogenous generation of new RNA viruses, DNA viruses and bacteria in homo neanderthalis and homo neoneanderthalis consequent to endosymbiotic archaeal overgrowth as a result of global warming. The homo neanderthalis and homo neoneanderthalis are resistant to this newly generated RNA viruses, DNA viruses and bacteria and act as an environmental reservoir for them. The new evolved

RNA virus, DNA virus and bacteria generated from environmental reservoir of homo neanderthalis and homo neoneanderthalis infects the unprotected homo sapien species exterminating the homo sapien species. The homo sapien species is in decline as the homo sapien albino mutants are getting converted to homo neoneanderthalis and the African/ Asian homo sapiens are getting exterminated by epidemics of new RNA viral infection generated by Neanderthal reservoirs. This homo sapien species can be called as Homo sapien extinctus.

The archaea can induce stem cell conversion and neanderthalisation of the human species. The archaea catabolizes cholesterol generating digoxin which can modulate RNA editing and magnesium deficiency resulting in reverse transcriptase inhibition. The archaeal cholesterol catabolism can deplete the membrane rafts of the CD_4 cell of cholesterol impeding the entry of the retrovirus into the cell. The archaea can produce permanent immune activation producing resistance to viral and bacterial infection. The archaeal cholesterol catabolism depletes tissue cholesterol producing vitamin D deficiency and immune activation. Thus archaeal overgrowth results in retroviral resistance and generation of the Neanderthal phenotype. The endosymbiotic archaea can secrete virus like RNA and DNA particles. The endosymbiotic archaea can induce uncoupling proteins inhibiting mitochondrial oxidative phosphorylation and generating ROS. The endosymbiotic archaeal magnetite can generate low level of EMF. The low level of EMF and ROS are genotoxic and produce breakages in hotspots of chromosome. It can also trigger rearrangements in hotspots of chromosome inhabited by retroviral and non-retroviral elements producing their expression. The archaeal secreted DNA and RNA viroids can recombine with the expressed retroviral, non-retroviral elements and other genomic segments of the human chromosome generating new RNA and DNA viruses. Thus the neanderthalised humans can serve as an origin for new RNA and DNA viruses as well as mutated retroviruses. The endosymbiotic archaea converts the Neanderthal cells to stem cells. The stem cells are resistant to immune attack. The stem cells can serve as a reservoir for this new RNA and DNA viruses. The stem cells and archaeal cells can also serve as a reservoir for viruses and bacteria belonging to other plants and animals. This helps to generate the species barrier jump in noted in recent emerging viral and bacterial infections. Thus the endosymbiotic archaeal growth produces neanderthalised version of homo sapiens which are retroviral resistant and resistant to other viral and bacterial infection consequent to immune activation and digoxin induced RNA editing. The endosymbiotic archaeal overgrowth mediated neanderthalised version of homo sapiens generates new mutated RNA and DNA viruses as well as retroviruses at the same time being resistant to them as in the case of the species bat. The homo sapiens do not have the Neanderthal mechanisms of immune activation as their archaeal load is meagre. They serve as fodder for infection from Neanderthal generated viruses and bacteria and suffer eventual extinction.

Global warming and symbiotic evolution

Thus global warming leads to symbiotic evolution of the species. The extraterrestrial intergalactic quantal computing cloud of archaea forms an intelligent anthropomorphic observer. The quantal computing cloud of archaea seeds the archaea into the earth through meteoric and asteroidal impacts. The archaeal colonies eventually evolve into multicellular organism and further into homo neanderthalis. The homo neanderthalis can be conceived as a multicellular archaeal colony. The homo neanderthalis thus arises in earth in the Eurasian land mass out of the seeded archaeal colonies from the extraterrestrial intergalactic archaeal computing cloud. The homo neanderthalis is energy depleted. The homo neanderthalis secretes the archaeal steroidal trephone digoxin which modulates the neutral amino acid transporter increasing tryptophan transport over tyrosine. The homo neanderthalis is tyrosine depleted and deficient in melanin synthesis. There is no melanin induced ATP synthesis from electromagnetic waves and radiation transduction. The homo neanderthalis was energy depleted and therefore did not have the luxury for the development of a modern human cerebral cortex. The homo neanderthalis is also retroviral resistant. The homo neanderthalis were deficient in endogenous retroviral sequences contributing to a rigid and adynamic homo neanderthalic genome. This led to a reduction in synaptic connectivity and poor development of the homo neanderthalic cerebral cortex. The homo sapiens evolved out of terrestrial sources in Africa out of self-replicating porphyrin complexes. The self-replicating porphyrin complexes form a scaffold for supramolecular complexes of isoprenoid organism, RNA viroids, DNA viroids and prions to self-organise. The isoprenoid organism formed the cell container which symbiosed the RNA viroids, the DNA viroids and prions to form the primitive eukaryotic and prokaryotic cell. The eukaryotic organism developed into multicellular colonies and eventually evolved into homo sapiens in Africa. Thus, the homo sapiens is a multicellular eukaryotic colony which evolved over a period of time. In case of oncogenesis the homo sapiens reverts to the primitive eukaryotic or prokaryotic multicellular colony state. The homo sapiens in Africa thus evolved out of terrestrial abiogenetic sources. The homo sapiens owing to the harsh tropical environmental of Africa had increased melanin pigmentation in the skin for protection from UV rays as an evolutionary mechanism and were black. The homo sapien brain evolved out of the energy excess state produced by melanin. Melanin can transduce electromagnetic waves and radiation and produce ATP synthesis. The excess energy in homo sapiens led to the rapid evolution of the human cerebral cortex. The homo sapiens are also retroviral sensitive. The retroviral infection led to integration of retroviral genes into the homo sapien genome producing endogenous retroviral sequences functioning as jumping genes. The HERV gene contributes to dynamicity and flexibility of the homo sapien genome contributing to increased synaptic connectivity and formation of the human cerebral cortex. A tyrosinase mutation led to the evolution of homo sapien albino mutants. The homo sapien albino mutants being white were unable to withstand the hot climate of the African tropics and migrated to the cold European land mass. This created the homo sapien civilization in Europe. There was interbreeding between the homo sapien albino mutants and homo neanderthalis in southern Europe producing hybrids. The homo neanderthalis were matriarchal while homo sapiens albino mutants were patriarchal. The homo neanderthalis succumbed to civilizational diseases like metabolic syndrome X, tumours, autoimmune disease and neurodegeneration and became extinct leaving fossilized matrilineal societies like the Dravidians, Celts, Basques and Jews behind. The homo sapien albino mutants in the setting of global warming developed extremophilic endosymbiotic archaeal growth and gets converted to a homo neoneanderthalic species by the phenomenon of symbiotic evolution. The homo sapien species in Africa becomes liable to eventual extinction owing to infection by catastrophic epidemics of RNA viruses arising from homo neanderthalis and homo neoneanderthalis reservoirs. Endosymbiotic archaeal growth will lead to a species change and generation of two new species - Homo sapien extinctus and Homo neoneanderthalis. Death and aging indicates human endogenous archaeal overgrowth and takeover. This will lead to extinction of the human race as such and persistence as well as survival of the archaeaon colony of melanosomes, magnetosomes and porphyrions functioning as a quantal computing colony and intelligence. This will lead to the takeover of the world and the universe by the terrestrial and extraterrestrial archaeaon quantal computing clouds. The symbiotic evolution will eventually lead to extinction of all human species into eternal archaeal colonies which can have a wave-particle existence.

The human species - terrestrial and extraterrestrial origin

The homo sapiens evolved in earth from porphyrinoids generated abiogenetically. The porphyrinoid forms a template for the formation of RNA viroids, DNA viroids, isoprenoid organisms and prions which symbiosed to form the eukaryotic and prokaryotic cells. The eukaryotic multicellular colony evolved into homo sapiens. The prokaryotes can also form multicellular functional colonies called biofilms. The homo sapiens which evolved in the African savannah became pigmented owing to melanisation of the skin in response to the solar UV rays. The homo sapiens have skin melanin but owing to lack of endosymbiotic archaea are deficient in tissue melanin. The homo sapiens in view of the absence of endosymbiotic archaea and tissue melanin are susceptible to endogenous retroviral replication and a dynamic genome leading on to increased synaptic connectivity and evolution of the prefrontal cortex. The homo neanderthalis evolved in the Eurasian steppes out of extraterrestrial archaeal colonies hitting the earth by asteroidal impacts. The archaeal colonies evolved into multicellular structures and eventually homo neanderthalis. The endosymbiotic archaea have the shikimic acid pathway and melanin synthesis. The homo neanderthalis are rich in tissue melanin but having evolved in the cold Eurasian steppes are deficient in cutaneous melanin. The increase in tissue melanin inhibits endogenous retroviral replication. This decreases the density of endogenous retroviral jumping genes in the homo neanderthalis genome making it rigid and inflexible. This rigid inflexible genome leads to the reduction in synaptic connectivity and poor development of the cerebral cortex in the homo neanderthalis. The homo neanderthalis have a dominant cerebellar cortex and are impulsive in nature. The increased tissue melanin in homo neanderthalis is capable of energy transduction giving them a survival advantage in the extremes of the Eurasian north. The melanin is capable of sensing low EMF fields contributing to extrasensory perceptive capacity of the homo neanderthalis. The homo sapiens developed tyrosinase deficient albino mutants which could not survive in the tropical Africa and migrated to the European continent. The albino mutants lack melanin and are susceptible to endosymbiotic archaeal symbiosis leading to the genesis of homo neoneanderthalis from homo sapiens. Thus the human species can have a terrestrial origin as in the case of homo sapiens in Africa and also an extraterrestrial origin from intergalactic archaea as in the case of homo neanderthalis. There is also an intermediate species evolved in out of homo sapien albino mutants with endosymbiotic archaeal symbiosis called Homo neoneanderthalis.

References

1. Kurup, R.K. and Kurup, P.A. *Global Warming, Archaea and Viroid Induced Symbiotic Human Evolution - Retrovirus, Prions and Viroids – Porphyrinoids and Viroidelle*. New York: Open Science Publishers, 2016.

CHAPTER 30

CLIMATE CHANGE AND GLOBAL WARMING IS PRODUCED BY HUMAN ENDOSYMBIOTIC ARCHAEAL OVERGROWTH AND METHANOGENESIS - INFECTIVE SYMBIOTIC TISSUE COLONIES OF PROTOZOA, ARCHAEA, VIRUSES AND BACTERIA

Introduction

The plasma life based forms and the gravitational and anti-gravitational waves generated porphyrions would have been the most primitive of non-life forms. The porphyrin templates would have given rise to the isoprenoid organism, the RNA viroids, DNA viroids and prions which would have symbiosed to form the nanoarchaea. The archaea would have evolved into prokaryotes or bacteria. The archaeal symbiosis with prokaryotes would have generated the eukaryotes. The initial unicellular eukaryotes are the protozoa. The protozoa are capable of phagocytosis and would have had viruses, bacteria and archaea as their intracellular symbionts. The viruses of the protozoa like entamoeba and toxoplasma are giant viruses called megaviride. The protozoans like entamoeba with its viral, archaeal and bacterial endosymbionts would have evolved into the multicellular organisms or metazoa. The archaeal, bacterial and protozoal extracellular vesicles containing RNA and DNA viroids can by horizontal gene transfer lead to evolution of metazoan, plants, animals and human species. The metazoans would have evolved into plants, animals and humans by horizontal gene transfer generated by archaeal extracellular RNA and DNA viroid vesicles. The evidence of such a transformation is seen in the bacterial symbionts of protozoa like entamoeba which includes legionella, listeria, mycoplasma, mycobacteria and chlamydia. The symbionts of protozoan entamoeba include viruses like covid, HIV and cytomegalo virus. The protozoan entamoeba serves as a natural reservoir for covid infections, bacterial infections and HIV infections. The toxoplasma also serves as a covid reservoir. These

protozoan colonies of entamoeba and toxoplasma with its bacterial, archaeal and viral symbionts can be called as chimeric infective symbiotic tissue colonies which can exist in the air, water, earth, blood and tissues. The chimeric infective symbiotic tissue colonies containing bacteria, archaea and viruses can make the symbiotic organism multidrug resistant by horizontal gene transfer. The chimeric infective symbiotic tissue organism evolves as a consequence of global warming and the protozoal, bacterial, viral and archaeal symbiotic colonies are new forms of life. The chimeric infective symbiotic tissue organism composed of protozoa and its symbionts like virus, bacteria and archaea depend upon glycolysis for its energy needs. The glycolytic enzymes enolase, glyceraldehyde 3-phosphate dehydrogenase, hexokinase, phosphofructokinase, aldolase, pyruvic kinase and triose phosphate isomerase are moonlighting proteins having a cytoplasmic, membrane and nuclear functions. These glycolytic enzymes have a RNA binding motif which can modulate RNA transcription in the host and the symbiotic organism. These glycolytic enzymes with their RNA binding motif as well as DNA binding motif can bind to RNA and DNA viruses and viroids modulating their replication and life cycles. The chimeric infective symbiotic tissue organism composed of protozoa and its symbionts like virus, bacteria and archaea can enter the human host through the breaches of the gut-blood barrier and transform itself into the macrophages of the human host. The macrophages are involved in the phagocytosis and innate immunity against bacterial and viral infections. Paradoxically the macrophages like the protozoa can also serve as a reservoir where pathogens like mycobacteria tuberculosis, mycobacteria leprae, listeria, legionella, E. coli, shigella & salmonella, and viruses like HIV & covid can thrive and replicate. The modified chimeric infective symbiotic tissue organism or macrophage composed of protozoa and its symbionts like virus, bacteria and archaea can serve as a reservoir for horizontal gene transfer and generation of multidrug resistant bacteria and viruses. The infective chimeric symbiotic tissue organism composed of protozoa and its symbionts like virus, bacteria and archaea

depends upon moonlighting glycolytic enzymes for their energetics and replication. The development of antibodies against the infective chimeric symbiotic tissue organism composed of protozoa and its symbionts like virus, bacteria and archaea leads to the evolution of anti-brain antibody disease or anti-basal ganglia antibody disease. They present a threat to existence consequent to horiztontal gene transfer and generation of multidrug resistant bacteria, viruses, protozoa, mycobacteria and archaea.

The internet exposure leads to low level EMF induced heme oxygenase induction in the brain. The brain heme is depleted leading to increase in ALA synthase and porphyrin synthesis. The porphyrins self-aggregate to form supramolecular organisms called porphyrions. The porphyrin acts as a template for the formation of RNA viroids, DNA viroids, isoprenoid organisms and prions which symbiosed to form nanoarchaea. The nanoarchaea contain magnetite and are magnetotactic and can have quantal perception as well as low level EMF perception. This leads to more of brain endosymbiotic nanoarchaeal growth. The nanoarchaea are capable of methanogenesis which contributes to global warming. The global warming related to internet exposure can produce still further increase in endosymbiotic archaeal symbiosis.

The previous reports from this laboratory have demonstrated the existence and growth of endosymbiotic actinidic archaea in human population which has been related to disease states like schizophrenia, autism, metabolic syndrome, cancer, autoimmune disease and degenerations. The overgrowth of endosymbiotic actinidic archaea results in neanderthalisation of humans. The overgrowth of endosymbiotic archaea results from the growth of homo sapien civilization. The homo sapien civilization results in industrialization and production of carbon dioxide, a greenhouse gas. The homo sapien civilization also results in widespread use of electronic devices like mobile phones and internet producing interconnectivity and a globalised world. The resultant low level EMF pollution also results in endosymbiotic archaeal growth. Carbon dioxide is a major greenhouse gas whose effects are long term but moderate. The archaea are methanogenic organisms. Methanogenesis results from the production of methane from carbon dioxide and hydrogen. Methanogenesis can also occur from formate and acetate. Acetate is the end product of carbohydrate, protein and lipid catabolism in humans. The human nutritional sources get metabolically converted to acetate and acetyl CoA which can enter the citric acid cycle and mitochondrial oxidative phosphorylation. The presence of endosymbiotic actinidic archaea results in conversion of acetate to formate and methane. It also results in conversion of the ubiquitous carbon dioxide and hydrogen to methane. Thus the human body due to endosymbiosis by archaea becomes the principal source of methanogenesis. Methane is an important greenhouse gas. The effect of methane is short-term as compared to carbon dioxide. Methane being a larger molecule can produce absorption of long range radiation and its global warming potential is 29-times that of carbon dioxide. Thus the principal culprit for global warming and eventual catastrophic extinction of human society is methane produced by human endosymbiotic archaea. The archaeal overgrowth due to global warming can affect ocean beds and lakes. This results in warming of the ocean and instability of methane hydrates in the ocean bed releasing methane. The arctic permafrost decays releasing organic carbon which can be a source of methanogenesis by archaea.[1-17] The study was conducted to evaluate the growth of actinidic archaea in humans.

Materials and Methods

Cytochrome F420 levels were studied in the homo sapien population as well as human populations exhibiting the Neanderthal phenotype. Fifteen cases each of the above mentioned groups were chosen for the study. The blood samples were drawn in the fasting state. Cytochrome F420 was estimated flourimetrically (excitation wavelength 420 nm and emission wavelength 520 nm). The permission from the Ethics Committee of the Institute was obtained for this study.

Results

The study showed that there was increased cytochrome F420 level in the population with Neanderthal phenotype in the blood. This indicated the growth of archaeal endosymbionts in the Neanderthal phenotype. There was also increased cytochrome F420 level in the normal homo sapien population but the extent of increase was small. The blood samples were drawn in the fasting state.

Discussion

Internet exposure, endosymbiotic archaea, porphyrinogenesis and abiogenesis

The internet exposure leads to low level EMF induced heme oxygenase induction in the brain. The brain heme is depleted leading to increase in ALA synthase and porphyrin synthesis. The porphyrins self-aggregate to form supramolecular organisms called porphyrions. The porphyrin acts as a template for the formation of RNA viroids, DNA viroids, isoprenoid organisms and prions which symbiosed to form nanoarchaea. The nanoarchaea contain magnetite and are magnetotactic and can have quantal perception as well as low level EMF perception. This leads to more of brain endosymbiotic nanoarchaeal growth.

Internet exposure, endosymbiotic archaea, methanogenesis and global warming

The global warming results from increased carbon dioxide emissions which trap heat producing rise in global temperatures. There is also ubiquitous pollution with low levels of EMF from the internet and mobile phones. The global warming and EMF pollution can produce archaeal multiplication. The archaea are endosymbiotic and gut symbionts of humans. The archaea have methanogenic metabolism. Methane is a greenhouse gas. Methane is effective than carbon dioxide in trapping heat and long wave radiation. 28 percent of greenhouse gases is constituted by methane and methane is 20-times more effective than carbon dioxide in trapping heat. The methane gas can absorb and emit radiation in the thermal range. The sources of methane are wetlands, cow gut, termites, rice paddies and petroleum industries. The decrease in permafrost in the Arctic ocean consequent to global warming results in increased generation of organic carbon for bacterial action. This generates more of methane. Methane, as said before is more effective than carbon dioxide in producing global warming. Global warming results in more methanogenesis by increasing archaeal multiplication. The methane hydrate in the ocean beds is also a source of methane. Global warming results in heating up of the ocean water resulting in instability of methane hydrate stores in the ocean bed and release of methane. The concentration of methane in the atmosphere is 1.8 ppm and it is as said before 20 percent of the greenhouse gas. The global warming potential of methane as compared to carbon dioxide is 2.1. Methane is a bigger molecule as compared to carbon dioxide and can trap more of heat and long wave radiation. Methane remains more in the atmosphere over a longer period of time as compared to carbon dioxide. Methane is oxidized to formaldehyde and then to carbon dioxide and water as well as ozone. The staying power of methane is shorter compared to carbon dioxide. Carbon dioxide remains in the atmosphere more than methane over a longer period of time. The effect of global warming produced by carbon dioxide is really the effect of carbon dioxide produced decades back. The effect of methane is more immediate. Thus methane is more dangerous and more powerful greenhouse gas as compared to carbon dioxide. The endosymbiotic archaeal growth and its increase consequent to increase in global temperature is of paramount importance. The increase in methane emissions over the previous decades paralleled the increase in human population. Thus human endosymbiotic archaeal overgrowth due to warming consequent to industrialization and carbon dioxide emissions and EMF pollution consequent to widespread internet usage and the archaeal methanogenesis is the most important factor for global warming. Rising temperatures produce more human archaeal growth, more methanogenesis, still more rising temperatures and still more archaeal growth producing a catastrophic terminal cycle of destruction.[1-17]

Internet exposure, endosymbiotic archaea, neanderthalisation and retroviral resistance

The homo neanderthalis archaeal symbionts synthesized digoxin induced RNA editing, magnesium deficiency induced reverse transcriptase inhibition and cholesterol depletion modulation of the CD_4 receptor results in neanderthalic retroviral resistance. But the reduced amount of archaeal symbiosis in the homo sapiens produces retroviral sensitivity. The homo sapiens gets ravaged by retroviral infection and becomes extinct as is happening in the African continent. The Eurasian homo neanderthalis is retroviral resistant and thrives. At the same time the global warming and low level of EMF pollution due to constant internet and modern appliance usage results in archaeal overgrowth and neanderthalisation of the homo sapien species. The archaea will produce methane resulting in global warming producing still more archaeal growth. Archaea are extremophilic and global warming induces archaeal growth which produces more neanderthalisation. The retroviral resistant homo neanderthalis have a stem cell syndrome due to archaeal induction of uncoupling proteins and generation of the Warburg phenotype. The somatic cells become a stem cell variant and lose their function. The stem cells have uncontrolled proliferation producing cancer syndromes. The lymphocytes stem cells have uncontrolled proliferation producing autoimmune disease. The neuronal stem cells lose their function producing autism and schizophrenia as well as neuronal degeneration. The Warburg phenotype of the stem cells with inhibition of mitochondrial oxidative phosphorylation results in the metabolic syndrome. Thus the neanderthalic phenotype tends to have civilizational diseases. The civilizational diseases are less common in the African homo sapien phenotype which is more sensitive to retroviral infections and viral infections.[1-17]

Internet exposure, endosymbiotic archaea, HERV expression and neanderthalisation of the brain

The archaeal endosymbiosis and retroviral resistance can modulate brain development in Neanderthal species. The archaeal endosymbiosis and methanogenesis produce global warming and climate change. Akin to this, there is a phenomenon of mind change produced by archaeal endosymbiosis in the brain. The genomic HERV sequences function as jumping genes modulating the development of the human cerebral cortex. The retroviral resistance in homo neanderthalis produces cortical maldevelopment. The human cerebral cortex is the site of rationality and judgment. The homo neanderthalis becomes irrational and impulsive and can be described as a creation of senses. In the homo neanderthalis the primitive parts of the brain the cerebellar cortex and brain stem becomes dominant. Cerebellar cortex and brain stem are the sites of impulsivity and irrationality. It is the centre for the fear, fight and flight response. This results in a chaotic and irrational world where individual impulses dominate. This results in a functioning anarchy. The brain stem and cerebellum are the sites of dreams and hypnosis. The cerebral cortical conscious perception is inhibited. The cerebellar archaeal magnetite mediated extrasensory perception dominates. It produces the world of dreams and imagination resulting in abstract art and the magic realism of literature. The fear, flight and fight response results in an upsurge of violence and aggression which produces a sense of transcendence for the individual. This produces a modern cannibalistic culture of sadistic proportions. This produces the cult of violence and terrorism. The cerebral cortex right and left produces neuronal differentiation of sexual behaviour. The right hemisphere produces male behaviours and left hemisphere produces feminine behaviours. The loss of cerebral cortex function results in asexual behaviour and androgynous behaviour. The impulsivity of the primitive areas of the brain results in a cult of sexuality and obscenity. The primitive brain areas of cerebellum are concerned with ritualistic and obsessive behaviours. This results in a civilization of ritualism, obsessive neuroticity and selfishness. This can be described as the consumer culture where the world is homogenized. The dominance of the primitive brain areas results in impulsivity, individualism and selfish traits contributing to the origin of capitalism. The cerebellar dominant brain is an evil brain. The homo sapiens owing to retroviral sensitivity has HERV jumping genes contributing to the evolution of the homo sapien cortex. The homo sapiens develop a civil society with common goodness as the theme. There is no dreams, no magic realism, no creativity but only a society of altruistic common sensical behaviour. This homo sapien civilized society comes to an end replaced by the cruel impulsive, chaotic, anarchic neanderthalic world.[1-17]

Internet exposure, endosymbiotic archaea, neanderthalisation and symbiotic evolution

Neanderthalisation is a symbiotic event. As the homo sapien civilization reaches its prime there is increasing industrial production of carbon dioxide and greenhouse gases as well as EMF pollution. This results in increased archaeal growth in the ocean beds and as archaeal endosymbionts in homo sapiens. The homo sapien species gets converted to Neanderthals by archaeal symbiosis. This occurs in a more florid manner in remnant Neanderthal matrilineal societies like the Dravidians, Berbers, Basque and Celts. The archaeal growth and neanderthalisation becomes more extensive in remnant Neanderthal societies.

The endosymbiotic archaeal growth in humans results in generation of methane from carbon dioxide and hydrogen as well as synthesis of methane from formate and acetate. This produces increased methanogenesis and methane is the most potent greenhouse gas known which over a hundred year period is 29-times more potent than carbon dioxide. Methane has a large effect over a brief period while carbon dioxide has a small effect over long periods. It should be stated that the principal component of the global warming is the surge in methane production in human populations due to endosymbiotic archaeal growth. The surge in methane production and archaeal overgrowth in humans is triggered by the increased industrial carbon dioxide production and EMF pollution due to use of internet in civilized homo sapien societies. This results in archaeal overgrowth, methanogenesis and still further global warming. The growth of archaea in the ocean beds trigger further methanogenesis with the methane stored as methane hydrate. The global warming results in warming up of the oceans producing instability of methane hydrate stores and methane is released into the atmosphere. The global warming produces destruction of the permafrost of the Arctic and Antarctic. This produces increased availability of organic carbon for bacterial and archaeal action generating methane. There is also increased archaeal growth in the peats and wetlands producing increasing methanogenesis and global warming. This results in the population conversion of homo sapien species to Neanderthal species.[1-17]

Internet exposure, archaea, methanogenesis and catastrophic extinction

Thus the increased production of greenhouse gases predominantly methane is from human sources alone due to increased growth of endosymbiotic archaea consequent to global warming triggered by industrial overproduction of carbon dioxide and EMF pollution. The homo sapien industrialisation is a small trigger but is rapidly taken over and dominated by endosymbiotic archaeal growth in humans. The archaeal overgrowth and neanderthalisation of homo sapiens converts the somatic cells to stem cells leading to cancer, autoimmune disease, degeneration and autism/schizophrenia in the Neo-neanderthal species. The Warburg phenotype of stem cells also produces the metabolic syndrome. The Neo-neanderthals becomes prone to civilizational disease. The neanderthalised humans stem cells phenotypes are retroviral resistant due to digoxin induced RNA editing, reverse transcriptase inhibition due to magnesium deficiency and membrane raft changes due to cholesterol depletion. The neanderthalised human stem cells serve as a reservoir for other species virus and bacteria resulting in breakage of the species barrier for infection. The archaeal symbionts can secrete RNA and DNA virus like particles which can recombine with expressed viral remnants in the genome as well as parts of the human genome per se producing new viruses and bacteria. The neanderthalised humans stem cells are resistant to infection which ravages the sensitive homo sapien phenotype exterminating them. Thus archaeal symbiosis, global warming, generation of new emerging viruses, pandemics of viral infections, homo sapien extinction and homo neanderthalis dominance becomes the rule. The realm of the homo neanderthalis sets in. The archaeal overgrowth in the oceanic beds and oceanic warming results in instability of methane hydrates in the ocean bed releasing methane. This produces global catastrophe. It results in oceanic earthquakes, continental shifts, tsunamis and flooding leading to eventual extinction of the human race of both species.[1-17]

References

1. Weaver TD, Hublin JJ. Neandertal Birth Canal Shape and the Evolution of Human Childbirth. *Proc. Natl. Acad. Sci. USA* 2009; 106:8151–8156.
2. Kurup RA, Kurup PA. Endosymbiotic Actinidic Archaeal Mediated Warburg Phenotype Mediates Human Disease State. *Advances in Natural Science* 2012; 5(1):81-84.
3. Morgan E. The *Neanderthal* theory of *autism*, Asperger and ADHD; 2007, www.rdos.net/eng/asperger.htm.
4. Graves P. New Models and Metaphors for the Neanderthal Debate. *Current Anthropology* 1991; 32(5): 513-541.

5. Sawyer GJ, Maley B. Neanderthal Reconstructed. *The Anatomical Record Part B: The New Anatomist* 2005; 283B(1):23-31.
6. Bastir M, O'Higgins P, Rosas A. Facial Ontogeny in Neanderthals and Modern Humans. *Proc. Biol. Sci.* 2007; 274:1125–1132.
7. Neubauer S, Gunz P, Hublin JJ. Endocranial Shape Changes during Growth in Chimpanzees and Humans: A Morphometric Analysis of Unique and Shared Aspects. *J. Hum. Evol.* 2010; 59:555–566.
8. Courchesne E, Pierce K. Brain Overgrowth in Autism during a Critical Time in Development: Implications for Frontal Pyramidal Neuron and Interneuron Development and Connectivity. *Int. J. Dev. Neurosci.* 2005; 23:153–170.
9. Green RE, Krause J, Briggs AW, Maricic T, Stenzel U, Kircher M, Patterson N, Li H, Zhai W, *et al.* A Draft Sequence of the Neandertal Genome. *Science* 2010; 328:710–722.
10. Mithen SJ. *The Singing Neanderthals: The Origins of Music, Language, Mind and Body*; 2005, ISBN 0-297-64317-7.
11. Bruner E, Manzi G, Arsuaga JL. Encephalization and Allometric Trajectories in the Genus Homo: Evidence from the Neandertal and Modern Lineages. *Proc. Natl. Acad. Sci.* USA 2003; 100:15335–15340.
12. Gooch S. *The Dream Culture of the Neanderthals: Guardians of the Ancient Wisdom.* Inner Traditions, Wildwood House, London; 2006.
13. Gooch S. *The Neanderthal Legacy: Reawakening Our Genetic and Cultural Origins.* Inner Traditions, Wildwood House, London; 2008.
14. Kurtén B. *Den Svarta Tigern*, ALBA Publishing, Stockholm, Sweden; 1978.
15. Spikins P. Autism, the Integrations of 'Difference' and the Origins of Modern Human Behaviour. *Cambridge Archaeological Journal* 2009; 19(2):179-201.
16. Eswaran V, Harpending H, Rogers AR. Genomics Refutes an Exclusively African Origin of Humans. *Journal of Human Evolution* 2005; 49(1):1-18.
17. Ramachandran V.S. The Reith lectures, BBC London. 2012.

CHAPTER 31

NEANDERTHAL METABOLONOMICS AND ANDROGYNOUS BEHAVIOURAL PATTERNS - INFECTIVE SYMBIOTIC TISSUE COLONIES OF PROTOZOA, ARCHAEA, VIRUSES AND BACTERIA

Introduction

The plasma life based forms and the gravitational and anti-gravitational waves generated porphyrions would have been the most primitive of non-life forms. The porphyrin templates would have given rise to the isoprenoid organism, the RNA viroids, DNA viroids and prions which would have symbiosed to form the nanoarchaea. The archaea would have evolved into prokaryotes or bacteria. The archaeal symbiosis with prokaryotes would have generated the eukaryotes. The initial unicellular eukaryotes are the protozoa. The protozoa are capable of phagocytosis and would have had viruses, bacteria and archaea as their intracellular symbionts. The viruses of the protozoa like entamoeba and toxoplasma are giant viruses called megaviride. The protozoans like entamoeba with its viral, archaeal and bacterial endosymbionts would have evolved into the multicellular organisms or metazoa. The archaeal, bacterial and protozoal extracellular vesicles containing RNA and DNA viroids can by horizontal gene transfer lead to evolution of metazoan, plants, animals and human species. The metazoans would have evolved into plants, animals and humans by horizontal gene transfer generated by archaeal extracellular RNA and DNA viroid vesicles. The evidence of such a transformation is seen in the bacterial symbionts of protozoa like entamoeba which includes legionella, listeria, mycoplasma, mycobacteria and chlamydia. The symbionts of protozoan entamoeba include viruses like covid, HIV and cytomegalo virus. The protozoan entamoeba serves as a natural reservoir for covid infections, bacterial infections and HIV infections. The toxoplasma also serves as a covid reservoir. These protozoan colonies of entamoeba and toxoplasma with its bacterial, archaeal and

viral symbionts can be called as chimeric infective symbiotic tissue colonies which can exist in the air, water, earth, blood and tissues. The chimeric infective symbiotic tissue colonies containing bacteria, archaea and viruses can make the symbiotic organism multidrug resistant by horizontal gene transfer. The chimeric infective symbiotic tissue organism evolves as a consequence of global warming and the protozoal, bacterial, viral and archaeal symbiotic colonies are new forms of life. The chimeric infective symbiotic tissue organism composed of protozoa and its symbionts like virus, bacteria and archaea depend upon glycolysis for its energy needs. The glycolytic enzymes enolase, glyceraldehyde 3-phosphate dehydrogenase, hexokinase, phosphofructokinase, aldolase, pyruvic kinase and triose phosphate isomerase are moonlighting proteins having a cytoplasmic, membrane and nuclear functions. These glycolytic enzymes have a RNA binding motif which can modulate RNA transcription in the host and the symbiotic organism. These glycolytic enzymes with their RNA binding motif as well as DNA binding motif can bind to RNA and DNA viruses and viroids modulating their replication and life cycles. The chimeric infective symbiotic tissue organism composed of protozoa and its symbionts like virus, bacteria and archaea can enter the human host through the breaches of the gut-blood barrier and transform itself into the macrophages of the human host. The macrophages are involved in the phagocytosis and innate immunity against bacterial and viral infections. Paradoxically the macrophages like the protozoa can also serve as a reservoir where pathogens like mycobacteria tuberculosis, mycobacteria leprae, listeria, legionella, E. coli, shigella & salmonella, and viruses like HIV & covid can thrive and replicate. The modified chimeric infective symbiotic tissue organism or macrophage composed of protozoa and its symbionts like virus, bacteria and archaea can serve as a reservoir for horizontal gene transfer and generation of multidrug resistant bacteria and viruses. The infective chimeric symbiotic tissue organism composed of protozoa and its symbionts like virus, bacteria and archaea depends upon moonlighting glycolytic enzymes for their energetics and

replication. The development of antibodies against the infective chimeric symbiotic tissue organism composed of protozoa and its symbionts like virus, bacteria and archaea leads to the evolution of anti-brain antibody disease or anti-basal ganglia antibody disease. They present a threat to existence consequent to horiztontal gene transfer and generation of multidrug resistant bacteria, viruses, protozoa, mycobacteria and archaea.

Neanderthal genes have been described in the homo sapien population. The Neanderthal brain has a prominent cerebellar cortex and small prefrontal cortex. This results in defective vocalization, symbolic speech, impulsive behaviour, obsessive traits, intuition and extrasensory perception. The Neanderthal brain structure results in female dominance and matriarchal social patterns. It was considered plausible that Neanderthal genomics and metabolonomics could also contribute to androgynous behaviour. Autistic patients tend to have Neanderthal metabolonomics and phenotype. It has been demonstrated that Neanderthal phenotype is due to symbiosis by actinidic archaea using cholesterol as an energy substrate. The actinidic archaea catabolises cholesterol with ring A being oxidized to pyruvate which gets channeled to the GABA shunt pathway resulting in the formation of glycine and succinyl CoA. This results in porphyrin synthesis. The side chain oxidation results in generation of short chain fatty acids. Cholesterol is also converted to steroidogenic estrogens and testosterone. The increasing growth of actinidic archaea converts the body metabolites the cholesterol which is subsequently oxidized and depleted. Cholesterol is also converted by actinidic archaea to endogenous digoxin which helps to integrate the neuro-immuno-endocrine system. Digoxin produces sodium-potassium ATPase inhibition and increased in intracellular calcium inducing nitric oxide synthase and heme oxygenase generating gasotransmitters nitric oxide and carbon monoxide important in smooth muscle contraction and autonomic function. The study deals with assessment of Neanderthal metabolonomics in androgynous individuals.[1-16]

Materials and Methods

Fifty healthy individuals with androgynous behaviour and free of any disease were chosen for the study. Each individual had a normal age and sex matched control. The estimations done in the blood samples collected include cytochrome F420 activity, cholesterol oxidase activity - cholesterol ring oxidase activity, cholesterol side chain oxidase activity, digoxin, lactate, pyruvate, ALA levels and hexokinase activity. Neanderthal anthropometry was studied in the androgynous population. The statistical analysis was done by ANOVA. Informed consent and permission of the Ethics Committee was obtained.

Results

The results of the study were as follows. The androgynous individuals had increased cytochrome F420 activity, cholesterol oxidase activity, ring oxidase activity and digoxin synthesis. The androgynous had decreased PDH activity as indicated by increased pyruvate and lactate levels. The androgynous group had increased GABA shunt pathway as indicated by increased pyruvate. The androgynous group had increased porphyrin synthesis as indicated by increased ALA levels. They had increased hexokinase activity indicating a Warburg phenotype in this group. The androgynous group had features of Neanderthal metabolism as indicated by pyruvate dehydrogenase suppression. The androgynous group has the Neanderthal anthropometric phenotype with slanting forehead, large face, stubby nose, prominent mandibles, low 2D:4D ratio, large coarse trunk, macrocephaly and longer second toe as compared to big toe.

Discussion

The study indicates that androgynous individuals tend to have the Neanderthal phenotype with skeletal characteristics. The androgynous individuals may have more of Neanderthal genotype. The metabolonomics in androgyny is suggestive of Neanderthal phenotype. There is increased actinidic archaeal symbiosis as indicated by increase in cytochrome F420 activity. The actinidic archaea uses cholesterol as a metabolic substrate. There is ring oxidation of cholesterol generating pyruvate. The pyruvate enters the GABA shunt pathway producing glycine and succinyl CoA. This results in porphyrin synthesis. The cholesterol is also converted to steroidal glycoside digoxin. Digoxin and porphyrin intercalation in the cell membrane produces sodium-potassium ATPase inhibition and accumulation of intracellular calcium. The increase in intracellular calcium induces nitric oxide synthase, heme oxygenase and cystathione synthase generating nitric oxide, carbon monoxide and hydrogen sulphide. This results in vasodilation of the blood spaces in the corpora cavernosa and increasing autonomic function of the genitourinary system resulting in obsessive traits. The increasing cholesterol catabolism by actinidic archaea results in depletion of cholesterol from the body. This produces inhibition of estrogen and testosterone synthesis. This results in an asexual state and androgynous behaviour. The brain function depends on testosterone and estrogens. The sex hormones modulate hemispheric dominance. The estrogens produce left hemispheric dominance and testosterones produce right hemispheric dominance. The lack of estrogens and testosterones in androgyny results in equidominance. This leads to equal function of the right hemisphere and left hemisphere and a state of creativity mixed with practicality. The right hemisphere is concerned with creative behaviour and the left hemisphere is concerned with practical behaviour. Equidominance results in the generation of a new phenotype with dominance of both creativity and practicality. Equidominance and lack of estrogens and testosterones can contribute to the social state of matriarchy. There is female dominance in society. The behavioural patterns between the male and female section of the population becomes homogenized. This results in generation of matrilineal societies and the demise of patriarchy.

Porphyrinuria and porphyria are the hallmarks of androgyny. This contributes to neuro-immuno-endocrine regulation and disease states associated with androgyny. The cholesterol is catabolised to porphyrins. Porphyrins are dipolar molecules and can contribute to quantal perception which is more in androgyny contributing to creativity, spirituality and extrasensory perceptive modes of this phenotype. Low level electromagnetic fields and its porphyrin messengers can regulate the brain mediating conscious and quantal perception. Porphyrin microarrays serve the purpose of quantal and conscious perception. The archaea and viroids via porphyrin synthesis can regulate the nervous system including the NMDA/GABA thalamo-cortico-thalamic pathway mediating conscious perception. Porphyrin photo-oxidation can generate free radicals which can modulate NMDA transmission. Free radicals can increase NMDA transmission. Free radicals can induce GAD and increase GABA synthesis. ALA blocks GABA transmission and upregulates NMDA. Protoporphyrins bind to GABA receptor and promote GABA transmission. Thus porphyrins can modulate the thalamo-cortico-thalamic pathway of conscious perception. The dipolar porphyrins in the setting of digoxin induced sodium-potassium ATPase inhibition can produce a pumped phonon system mediated Frohlich model superconducting state inducing quantal perception with nanoarchaeal sensed gravity producing the orchestrated reduction of the quantal possibilities to the macroscopic world. ALA can produce sodium-potassium ATPase inhibition resulting in a pumped phonon system mediated quantal state involving dipolar porphyrins. Porphyrin molecules have a wave-particle existence and can bridge the dividing line between quantal state and particulate state. Thus, the porphyrins can mediate conscious and quantal perception. Porphyrins binding to proteins, nucleic acids and cell membranes can produce biophoton emission. Porphyrins by auto-oxidation can generate biophotons and are involved in quantal perception. Biophotons can mediate quantal perception. Cellular porphyrins photo-oxidation are involved in sensing of earth magnetic fields and low level biomagnetic fields. Thus porphyrin microarrays can function as a quantal computer mediating extrasensory perception. Porphyrin microarrays in human systems and brain owing to the wave-particle nature of porphyrins can bridge the quantal world and particulate world. The porphyrins can modulate hemispheric dominance. There is increased porphyrin synthesis and RHCD and decreased porphyrin synthesis in LHCD. The increase in archaeal porphyrins can contribute to the pathogenesis of schizophrenia and autism. Porphyria can lead to psychiatric disorders and seizures. Altered porphyrin metabolism has been described in autism. Porphyrin by modulating conscious and quantal perception is involved in the pathogenesis of schizophrenia and autism. Thus porphyrins microarrays can function as a quantal brain modulating extrasensory quantal perception. Porphyrin microarrays can function as a quantal brain in communication with digital world and geomagnetic fields. The dipolar porphyrins in the setting of digoxin induced sodium-potassium ATPase inhibition can produce a pumped phonon system mediated Frohlich model superconducting state inducing quantal perception with nanoarchaeal sensed gravity producing the orchestrated reduction of the quantal possibilities to the macroscopic world. ALA can produce sodium-potassium ATPase inhibition resulting in a pumped phonon system mediated quantal state involving dipolar porphyrins. Porphyrins by auto-oxidation can generate biophotons and are involved in quantal perception. Biophotons can mediate quantal perception. Porphyrin auto-oxidation is modulated by low level of electromagnetic fields and geomagnetic fields. Cellular porphyrins photo-oxidation are involved in sensing of earth magnetic fields and low level biomagnetic fields. Porphyrins can thus contribute to quantal perception. Low level electromagnetic fields and light can induce porphyrin synthesis. Low level EMF can produce ferrochelatase inhibition as well as heme oxygenase induction contributing to heme depletion, ALA synthase induction and increased porphyrin synthesis. Light also induces ALA synthase and porphyrin synthesis. The increased porphyrin synthesized can contribute to increased quantal perception and can modulate conscious perception. The human porphyrin microarrays induced biophotons and quantal fields can modulate the source from which low level EMF and photic fields were generated. Thus the porphyrin generated by extraneous low level EMF and photic fields can interact with the source of low level EMF and photic fields modulating it. Thus porphyrins can serve as a bridge between the human brain and the source of low level EMF and photic fields. This serves as a mode of communication between the human brain and digital EMF storage devices like internet. The porphyrins can also serve as the source of communication with the environment. Environmental EMF and chemicals produce heme oxygenase induction and heme depletion increasing porphyrin synthesis, quantal perception and two-way communication. Thus induction of porphyrin synthesis can serve as a mechanism of communication between human brain and the environment by extrasensory perception. Porphyrin microarrays can function as quantal computers storing information and can serve the purpose of extrasensory perception. Porphyrins can serve as a two-way communicating bridge between digital information storage systems generating low level electromagnetic fields and human systems. The low level of EMF produced by digital system enhances porphyrin synthesis and serves the purpose of two-way extrasensory perception and communication. The human porphyrin quantal computers can in turn by biophoton emission modulate digital information storage system.

Low level of electromagnetic fields and its porphyrin messengers can induce the Warburg phenotype. An actinide dependent shadow biosphere of archaea and viroids in the above mentioned disease states is described. The archaea can synthesize porphyrins and induce porphyrin synthesis. Porphyrins have been related to schizophrenia, metabolic syndrome X, malignancy, systemic lupus erythematosis, multiple sclerosis and Alzheimer's diseases. Porphyrins can mediate the effect of low level electromagnetic fields inducing the Warburg phenotype leading to the above mentioned disease states. The Warburg phenotype results in inhibition of pyruvate dehydrogenase and the TCA cycle. The pyruvate enters the GABA shunt pathway where it is converted to succinyl CoA. The glycolytic pathway is upregulated and the glycolytic metabolite phosphoglycerate is converted to serine and glycine. Glycine and succinyl CoA are the substrates for ALA synthesis. The archaea induces the enzyme heme oxygenase. Heme oxygenase converts heme to bilirubin and biliverdin. This depletes heme from the system and results in upregulation of ALA synthase activity resulting in porphyria. Heme inhibits HIF alpha. The heme depletion results in upregulation of HIF alpha activity and further strengthening of the Warburg phenotype. The porphyrin self-oxidation results in redox stress which activates HIF alpha and generates the Warburg phenotype. The Warburg phenotype results in channelling acetyl CoA for cholesterol synthesis as the TCA cycle and mitochondrial oxidative phosphorylation are blocked. The archaea uses cholesterol as an energy substrate. Porphyrin and ALA inhibits sodium-potassium ATPase. This increases cholesterol synthesis by acting upon intracellular SREBP. The cholesterol is metabolized to pyruvate and then the GABA shunt pathway for ultimate use in porphyrin synthesis. The porphyrins can self-organize and self-replicate into macromolecular arrays. The porphyrin arrays behave like an autonomous organism and can have intramolecular electron transport generating ATP. The porphyrin macroarrays can store information and can have quantal perception. The porphyrin macroarrays serves the purpose of archaeal energetics and sensory perception. The Warburg phenotype is associated with malignancy, autoimmune disease and metabolic syndrome X. Low level electromagnetic fields can induce the Warburg phenotype contributing to human disease.

The role of porphyrins and low level electromagnetic fields in regulation of cell functions and neuro-immuno-endocrine integration is discussed. Low levels of EMF fields can induce digoxin synthesis. Protoporphyrin binds to the peripheral benzodiazepine receptor regulating steroid and digoxin synthesis. Increased porphyrin metabolites can contribute to hyperdigoxinemia. Digoxin can modulate the neuro-immuno-endocrine system. Low level of EMF fields can modulate membrane, nucleic acid and protein structure and function via induction of porphyrin synthesis. Porphyrins can combine with membranes modulating membrane function. Porphyrins can combine with proteins oxidizing their tyrosine, tryptophan, cysteine and histidine residues producing crosslinking and altering protein conformation and function. Porphyrins can complex with DNA and RNA modulating their function. Porphyrin interpolating with DNA can alter transcription and generate HERV expression. Low level of EMF fields through modulation of porphyrin metabolism can produce heme deficiency by inhibiting heme oxygenase and ferrochelatase. Heme deficiency can also result in disease states. Heme deficiency results in deficiency of heme enzymes. There is deficiency of cytochrome C oxidase and mitochondrial dysfunction. The glutathione peroxidase is dysfunctional and the glutathione system of free radical scavenging does not function. The cytochrome P450 enzymes involved in steroid and bile acid synthesis have reduced activity leading to steroid- cortisol and sex hormones as well as bile acid deficiency states. The heme deficiency results in dysfunction of nitric oxide synthase, heme oxygenase and cystathione beta synthase resulting in lack of gasotransmitters regulating the vascular system and NMDA receptor- NO, CO and H_2S. Heme has got cytoprotective, neuroprotective, anti-inflammatory and anti-proliferative effects. Heme is also involved in the stress response. Heme deficiency leads to metabolic syndrome, immune disease, degenerations and cancer. Low level electromagnetic fields can modulate cell functions and neuro-immuno-endocrine-genetic integration via induction of porphyrin synthesis. Low level electromagnetic fields via modulating porphyrin metabolism can produce an autonomic neuropathy. Protoporphyrins block acetyl choline transmission producing a vagal neuropathy with sympathetic overactivity. Vagal neuropathy results in immune activation, vasospasm and vascular disease. A vagal neuropathy underlines neoplastic and autoimmune processes as well as metabolic syndrome X. Low level electromagnetic fields by modulating porphyrin metabolism can induce cell death. Porphyrin induced increased NMDA transmission and free radical injury can contribute to neuronal degeneration. Free radicals can produce mitochondrial PT pore dysfunction. This can lead to cyto C leak and activation of the caspase cascade leading to apoptosis and cell death. Altered porphyrin metabolism has been described in Alzheimer's disease. The increased porphyrin photo-oxidation generated free radicals mediated NMDA transmission can also contribute to epileptogenesis. The protoporphyrins binding to mitochondrial benzodiazepine receptors can regulate brain function and cell death. Low level electromagnetic fields by modulating porphyrin metabolism can generate redox stress to regulate cell functions. The porphyrins can undergo photo-oxidation and auto-oxidation generating free radicals. The archaeal porphyrins can produce free radical injury. Free radicals produce NFkB activation, open the mitochondrial PT pore resulting in cell death, produce oncogene activation, activate NMDA receptor and GAD enzyme regulating neurotransmission and generates the Warburg phenotypes activating glycolysis and inhibiting TCA cycle/oxphos. Porphyrins have been related to schizophrenia, metabolic syndrome X, malignancy, systemic lupus erythematosis, multiple sclerosis and Alzheimer's diseases. Low level electromagnetic fields by modulating porphyrin metabolism can regulate cell membrane sodium-potassium ATPase. The porphyrins can complex and intercalate with the cell membrane producing sodium-potassium ATPase inhibition adding on to digoxin mediated inhibition. Porphyrins can complex with proteins and nucleic acid producing biophoton emission. Low level electromagnetic fields by modulating porphyrin metabolism can regulate DNA, RNA and protein structure and function. Porphyrins complexing with proteins can modulate protein structure and function. Porphyrins complexing with DNA and RNA can modulate transcription and translation. Low level electromagnetic fields by modulating porphyrin metabolism can regulate mitochondrial function, peripheral benzodiazepine receptor and steroidogenesis. The porphyrin especially protoporphyrins can bind to peripheral benzodiazepine receptors in the mitochondria and modulate its function, mitochondrial cholesterol transport and steroidogenesis. Peripheral benzodiazepine receptor modulation by protoporphyrins can regulate cell death, cell proliferation, immunity and neural functions. Low level electromagnetic fields by modulating porphyrin metabolism and inducing redox stress can regulate enzyme systems. The porphyrin photo-oxidation generates free radicals which can modulate enzyme function. Redox stress modulated enzymes include pyruvate dehydrogenase, nitric oxide synthase, cystathione beta synthase and heme oxygenase. Free radicals can modulate mitochondrial PT pore function. Free radicals can modulate cell membrane function and inhibit sodium-potassium ATPase activity. Thus the porphyrins are key regulatory molecules modulating all aspects of cell function. Low level of electromagnetic fields by modulating porphyrin metabolism can induce viroidal and HERV expression. There was an increase in free RNA indicating self-replicating RNA viroids and free DNA indicating generation of viroid complementary DNA strands by archaeal reverse transcriptase activity. The actinides and porphyrins modulate RNA folding and catalyse its ribozymal action. Digoxin can cut and paste the viroidal strands by modulating RNA splicing generating RNA viroidal diversity. The viroids are evolutionarily escaped archaeal group I introns which have retrotransposition and self-splicing qualities. Porphyrin photo-oxidation induced redox stress can produce HDAC inhibition. Archaeal pyruvate producing histone deacetylase inhibition and porphyrins intercalating with DNA can produce endogenous retroviral (HERV) reverse transcriptase and integrase expression. This can integrate the RNA viroidal complementary DNA into the non-coding region of eukaryotic non-coding DNA using HERV integrase as has been described for borna and ebola viruses. The archaea and viroids can also induce cellular porphyrin synthesis. Bacterial and viral infections can precipitate porphyria. Thus porphyrins can regulate genomic function. The increased expression of HERV RNA can result in acquired immunodeficiency syndrome, autoimmune disease, neuronal degenerations, schizophrenia and malignancy. Low level electromagnetic fields by modulating porphyrin metabolism and generating redox stress can produce immune activation. The porphyrin photo-oxidation can generate free radicals which can activate NFkB. This can produce immune activation and cytokine mediated injury. The increase in archaeal porphyrins can lead to autoimmune disease like SLE and MS. A hereditary form of MS and SLE related to altered porphyrin metabolism has been described. The protoporphyrins binding to mitochondrial benzodiazepine receptors can modulate immune function. Porphyrins can combine with proteins oxidizing their tyrosine, tryptophan, cysteine and histidine residues producing crosslinking and altering protein conformation and function. Porphyrins can complex with DNA and RNA modulating their structure. Porphyrin complexed with proteins and nucleic acids are antigenic and can lead on to autoimmune disease. Low level electromagnetic fields by modulating porphyrin metabolism and inducing redox stress can produce insulin resistance. The porphyrin photo-oxidation mediated free radical injury can lead to insulin resistance and atherogenesis. Thus archaeal porphyrins can contribute to metabolic syndrome X. Glucose has got a negative effect upon ALA synthase activity. Therefore hyperglycemia may be reactive protective mechanism to increased archaeal porphyrin synthesis. The protoporphyrins binding to mitochondrial benzodiazepine receptors can modulate mitochondrial steroidogenesis and metabolism. Altered porphyrin metabolism has been described in the metabolic syndrome X. Porphyrias can lead on to vascular thrombosis. Low level electromagnetic fields by modulating porphyrin metabolism and inducing redox stress/heme deficiency can activate HIF alpha. The porphyrin photo-oxidation can generate free radicals inducing HIF alpha and producing oncogene activation. Heme deficiency can lead to activation of HIF alpha and oncogenesis. This can lead to oncogenesis. Hepatic porphyrias induced hepatocellular carcinoma. The protoporphyrins binding to mitochondrial benzodiazepine receptors can regulate cell proliferation. Low level electromagnetic fields by modulating porphyrin metabolism can regulate prion protein conformation. The porphyrin can combine with prion proteins modulating their conformation. This leads to abnormal prion protein conformation and degradation. Archaeal porphyrins can contribute to prion disease. Low level electromagnetic fields by modulating porphyrin metabolism can produce redox stress and regulate HERV expression. The porphyrins can also intercalate with DNA producing HERV expression. The HERV particles generated can contribute to the retroviral state associated with androgyny. The porphyrins in the blood can combine with bacteria and viruses and the photo-oxidation generated free radicals can kill them. Low level electromagnetic fields by modulating porphyrin metabolism can lead to increase predilection for viral and bacterial infections. The archaeal porphyrins can modulate bacterial and viral infections. The archaeal porphyrins are regulatory molecules keeping other prokaryotes and viruses on check.

Thus, the actinidic archaeal symbiosis results in neanderthalisation of the population and generation of androgyny. The actinidic archaeal overgrowth and symbiosis is a consequence of global warming. Archaea are extremophiles and increase in density during periods of climate change. The actinidic archaeal catabolism of cholesterol generates digoxin and increased intracellular calcium resulting in formation of excess of gasotransmitters important in autonomic function of structures like the corpora cavernosa. The cholesterol catabolism results in depletion of cholesterol and to a state of lack of sex hormone synthesis. This produces an asexual state resulting in a social system of matriarchy related to androgyny. The actinidic archaeal cholesterol catabolism generates porphyrins producing the extrasensory quantal perceptive state associated with androgyny. This contributes to the creativity of the androgynous state. The porphyrin synthesis associated with androgyny also contributes to the disease states associated with it. This includes autoimmune disease, cancer, degenerations, acquired immuno-deficiency syndrome, metabolic syndrome X and all civilisational disease.

References

1. Weaver TD, Hublin JJ. Neandertal Birth Canal Shape and the Evolution of Human Childbirth. *Proc. Natl. Acad. Sci. USA* 2009; 106:8151–8156.
2. Kurup RA, Kurup PA. Endosymbiotic Actinidic Archaeal Mediated Warburg Phenotype Mediates Human Disease State. *Advances in Natural Science* 2012; 5(1):81-84.
3. Morgan E. The *Neanderthal* theory of *autism*, Asperger and ADHD; 2007, www.rdos.net/eng/asperger.htm.
4. Graves P. New Models and Metaphors for the Neanderthal Debate. *Current Anthropology* 1991; 32(5): 513-541.
5. Sawyer GJ, Maley B. Neanderthal Reconstructed. *The Anatomical Record Part B: The New Anatomist* 2005; 283B(1):23-31.
6. Bastir M, O'Higgins P, Rosas A. Facial Ontogeny in Neanderthals and Modern Humans. *Proc. Biol. Sci.* 2007; 274:1125–1132.
7. Neubauer S, Gunz P, Hublin JJ. Endocranial Shape Changes during Growth in Chimpanzees and Humans: A Morphometric Analysis of Unique and Shared Aspects. *J. Hum. Evol.* 2010; 59:555–566.
8. Courchesne E, Pierce K. Brain Overgrowth in Autism during a Critical Time in Development: Implications for Frontal Pyramidal Neuron and Interneuron Development and Connectivity. *Int. J. Dev. Neurosci.* 2005; 23:153–170.
9. Green RE, Krause J, Briggs AW, Maricic T, Stenzel U, Kircher M, Patterson N, Li H, Zhai W, *et al.* A Draft Sequence of the Neandertal Genome. *Science* 2010; 328:710–722.
10. Mithen SJ. The *Singing Neanderthals: The Origins of Music, Language, Mind and Body*; 2005, ISBN 0-297-64317-7.
11. Bruner E, Manzi G, Arsuaga JL. Encephalization and Allometric Trajectories in the Genus Homo: Evidence from the Neandertal and Modern Lineages. *Proc. Natl. Acad. Sci. USA* 2003; 100:15335–15340.
12. Gooch S. *The Dream Culture of the Neanderthals: Guardians of the Ancient Wisdom.* Inner Traditions, Wildwood House, London; 2006.
13. Gooch S. *The Neanderthal Legacy: Reawakening Our Genetic and Cultural Origins.* Inner Traditions, Wildwood House, London; 2008.
14. Kurtén B. *Den Svarta Tigern*, ALBA Publishing, Stockholm, Sweden; 1978.
15. Spikins P. Autism, the Integrations of 'Difference' and the Origins of Modern Human Behaviour. *Cambridge Archaeological Journal* 2009; 19(2):179-201.
16. Eswaran V, Harpending H, Rogers AR. Genomics Refutes an Exclusively African Origin of Humans. *Journal of Human Evolution* 2005; 49(1):1-18.

CHAPTER 32

ARCHAEA INDUCED STEM CELL SYNDROME AND ANDROGYNOUS CREATIVE MATRIARCHAL CANNIBALISTIC CAPITALISTIC STATE - INFECTIVE SYMBIOTIC TISSUE COLONIES OF PROTOZOA, ARCHAEA, VIRUSES AND BACTERIA

Introduction

The plasma life based forms and the gravitational and anti-gravitational waves generated porphyrions would have been the most primitive of non-life forms. The porphyrin templates would have given rise to the isoprenoid organism, the RNA viroids, DNA viroids and prions which would have symbiosed to form the nanoarchaea. The archaea would have evolved into prokaryotes or bacteria. The archaeal symbiosis with prokaryotes would have generated the eukaryotes. The initial unicellular eukaryotes are the protozoa. The protozoa are capable of phagocytosis and would have had viruses, bacteria and archaea as their intracellular symbionts. The viruses of the protozoa like entamoeba and toxoplasma are giant viruses called megaviride. The protozoans like entamoeba with its viral, archaeal and bacterial endosymbionts would have evolved into the multicellular organisms or metazoa. The archaeal, bacterial and protozoal extracellular vesicles containing RNA and DNA viroids can by horizontal gene transfer lead to evolution of metazoan, plants, animals and human species. The metazoans would have evolved into plants, animals and humans by horizontal gene transfer generated by archaeal extracellular RNA and DNA viroid vesicles. The evidence of such a transformation is seen in the bacterial symbionts of protozoa like entamoeba which includes legionella, listeria, mycoplasma, mycobacteria and chlamydia. The symbionts of protozoan entamoeba include viruses like covid, HIV and cytomegalo virus. The protozoan entamoeba serves as a natural reservoir for covid infections, bacterial infections and HIV infections. The toxoplasma also serves as a covid reservoir. These protozoan colonies of entamoeba and toxoplasma with its bacterial, archaeal and

viral symbionts can be called as chimeric infective symbiotic tissue colonies which can exist in the air, water, earth, blood and tissues. The chimeric infective symbiotic tissue colonies containing bacteria, archaea and viruses can make the symbiotic organism multidrug resistant by horizontal gene transfer. The chimeric infective symbiotic tissue organism evolves as a consequence of global warming and the protozoal, bacterial, viral and archaeal symbiotic colonies are new forms of life. The chimeric infective symbiotic tissue organism composed of protozoa and its symbionts like virus, bacteria and archaea depend upon glycolysis for its energy needs. The glycolytic enzymes enolase, glyceraldehyde 3-phosphate dehydrogenase, hexokinase, phosphofructokinase, aldolase, pyruvic kinase and triose phosphate isomerase are moonlighting proteins having a cytoplasmic, membrane and nuclear functions. These glycolytic enzymes have a RNA binding motif which can modulate RNA transcription in the host and the symbiotic organism. These glycolytic enzymes with their RNA binding motif as well as DNA binding motif can bind to RNA and DNA viruses and viroids modulating their replication and life cycles. The chimeric infective symbiotic tissue organism composed of protozoa and its symbionts like virus, bacteria and archaea can enter the human host through the breaches of the gut-blood barrier and transform itself into the macrophages of the human host. The macrophages are involved in the phagocytosis and innate immunity against bacterial and viral infections. Paradoxically the macrophages like the protozoa can also serve as a reservoir where pathogens like mycobacteria tuberculosis, mycobacteria leprae, listeria, legionella, E. coli, shigella & salmonella, and viruses like HIV & covid can thrive and replicate. The modified chimeric infective symbiotic tissue organism or macrophage composed of protozoa and its symbionts like virus, bacteria and archaea can serve as a reservoir for horizontal gene transfer and generation of multidrug resistant bacteria and viruses. The infective chimeric symbiotic tissue organism composed of protozoa and its symbionts like virus, bacteria and archaea depends upon moonlighting glycolytic enzymes for their energetics and replication. The development of antibodies against the infective chimeric

symbiotic tissue organism composed of protozoa and its symbionts like virus, bacteria and archaea leads to the evolution of anti-brain antibody disease or anti-basal ganglia antibody disease. They present a threat to existence consequent to horiztontal gene transfer and generation of multidrug resistant bacteria, viruses, protozoa, mycobacteria and archaea.

The global warming produces extremes of temperature and accumulation of atmospheric carbon dioxide resulting in growth of symbiotic extremophiles like archaea. Archaea can induce dedifferentiation of somatic cells to stem cells. This involves the process of reverse aging. The differentiated somatic cells lose their function as they become stem cells. The archaeal magnetite induces quantal extrasensory perception of low level of EMF as the somatic neuronal cells lose their function. This results in low level of EMF effect on the brain producing cortical atrophy especially the prefrontal cortex. The primitive parts of the brain dominate with cerebellum and brain stem undergoing hypertrophy. The atrophy of the cortex results in behavioural changes. The cortex has different hemispheric dominance in males and females. The right hemisphere is a creative hemisphere and is male. The left hemisphere is the practical hemisphere and is female. When the cortex atrophies the hemispheric differentiation and the effect on behaviour is obliterated. The cortical effect on male and female behaviour is lost. Behaviour becomes uniform and single and is dominated by the primitive brain stem and cerebellar cortex. It results in impulsive behaviour dominated by the will to power and individuality.[1-17] This forms the basis of the androgynous state and alternate forms of sexuality. This hypothesis was studied in this paper by checking the archaeal growth in population with alternate sexual traits.

Materials and Methods

The blood samples were drawn from 15 normal individuals with alternate sexual traits and cytochrome F420 activity was studied. The estimations done in the blood samples collected blood lactate, pyruvate, hexokinase, cytochrome C, digoxin, bile acids, butyrate and propionate were estimated.

Results

The results showed that the individuals with alternate sexual traits had increased archaeal symbiosis and increased cytochrome F420 activity. They also had increased blood lactate and pyruvate, increased RBC hexokinase, increased serum cytochrome C and serum cytochrome F420, increased serum digoxin, bile acids, butyrate and propionate. The serum cytochrome C levels in the blood were increased. This suggested mitochondrial dysfunction. There was an increased in glycolysis as suggested by increased RBC hexokinase activity and lactic acidosis. Owing to the mitochondrial dysfunction and pyruvate dehydrogenase inhibition there was pyruvate accumulation. The pyruvate was converted to lactate by the Cori cycle and also to glutamate and ammonia. This metabolism is suggestive of the Warburg phenotype and stem cell conversion. The stem cells depend on Warburg anaerobic glycolysis for energetics and have a mitochondrial dysfunction. The lysosomal enzyme beta galactosidase activity was increased in the disease group and in creative artists and criminals suggesting stem cell conversion. This suggests that individuals with androgynous traits had stem cell metabolonomics and stem cell conversion.

Discussion

The cortical atrophy and cerebellar/brain stem dominance results in obliteration in hemispheric difference in sexual behaviour. The right hemisphere is creative and male in outlook while left hemisphere is practical and female in outlook. The primitive parts of the brain take over the function of regulating sexual behaviour. The cerebellum plays an important role and this results in impulsive sexual traits. The difference between male and female sexual behaviours induced by cerebral cortical function is lost. The archaeal cholesterol catabolism results in depletion of sex steroids and deficiency of testosterone and estrogens. The archaeal induced conversion of ovarian and testicular cells into stem cells results in loss of function and decreased secretion of male and female hormones. Behaviour becomes unisexual. This becomes non-inhibitory and impulsive in nature. It transcends all taboos and has got a reflection in culture and society affecting all manners of social interaction. The predominant form of brain perception is extrasensory or quantal. The primitive human impulses become unleashed and this results in a flood of primitive behavioural traits with violent, aggressive and obscene traits in society. The increased incidence of violent sexual behavioural traits is related to the dominance of the primitive areas of the brain- the cerebellum and brain stem. The dress code of the society also changes and results in metrosexual and unisexual garments. The mode of grooming of male and female changes and both becomes equal and the same. This creates the metrosexual world.[1-17]

The dominance of the primitive areas of the brain results in fear flight and fight response resulting in an epidemic of selfishness in society. Individualism takes over and there is no commitment to the society as such. Sexual behaviours were programmed for the benefit of the society so that the human population is replaced. The cortical atrophy and cerebellar dominance results in selfish sexual behavioural traits producing sexual behaviour for individual pleasure and gratification in animalistic sense. This results in loss of family values and declining population as is seen in European countries. The cerebral cortical atrophy and dominance of cerebellum result in selfishness and individuality contributing to an anarchic society. The cerebral cortical atrophy results from perception of low level of EMF resulting from increased archaeal magnetite as well as EMF pollution resulting from internet exposure. Society becomes globalized and anarchic fueled by the internet. This results in an acortical acephalic society with dominant primitive cerebellar function. There is no compassion, love, feeling of altruism or goodness. This is replaced by selfishness and individuality. The internet and social media becomes the common market place for interactions. The feeling of human touch and love is lost. Society becomes increasingly robotical and autistic. The realm of the senses takes over the kingdom of God. Everything becomes subsumed and sacrificed in the altar of selfishness, greed and pleasure. This produces an anarchic, unisexual and society of primitive impulses. The cortical atrophy and cerebellar dominance results in a play of primitive impulses resulting in violence and aggression. This results from a culture of selfishness. This produces terrorism and acts of war which are a form of transcendence. This also produces criminal behaviour where individuality and selfishness dominates. Society becomes dominated by ritualized and in some cases obscene behaviour.[1-17]

The cortical atrophy and dominance of cerebellum result in loss of cortical neuronal function and increased extrasensory perception mediated by archaeal magnetite. This results in dominant spiritual behaviours where one comes into contact with the eternal and archetypes. This results in a literature of transcendence. This produces what is called as magic realism of writers like Gabriel Marquez. The literature explores the evil depths of the human soul. This results in a dominance of sexual, violent, obscene and evil in literature as seen in post modern literature. This has also a reflection in art of painting, dance and music. Painting, dance and music become surreal and the rationality of the cortex regulating it is lost. This results in psychedelic and rock music as well as the surrealistic abstract art of Picasso. Dance forms also take violent, obscene, chaotic forms. This is art of the surrealistic acephalic irrational world in the realm of senses driven by obscenity. This type of art and literature correlates with the androgynous creativity.[1-17]

The prefrontal cortical atrophy and cerebellar dominance is due to archaeal growth which results in stem cell conversion. The stem cell syndrome can produce a proliferation of systemic diseases. The neuronal stem cell conversion results in loss of neuronal function and dominant extrasensory archaeal magnetite mediated perception. This produces an epidemic of schizophrenia and autism. The stem cells have the Warburg phenotype with mitochondrial dysfunction and glycolytic energetics. This results in metabolic syndrome X. The stem cells can proliferate resulting in cancer syndromes. The lymphocytic stem cells proliferate producing an autoimmune disease. The neuronal stem cells transformation and loss of function can lead to degenerations. Thus the systemic somatic and neuropsychiatric diseases correlate with alternate sexual traits and stem cell transformation.[1-17]

The archaeal symbiosis mediated brain changes producing cerebellar dominance and cortical atrophy results in an individualistic selfish society. This is the kernel of capitalistic growth and models which tend to fail because of the individualistic will to power and dominate at all cost. The society becomes more dictatorial and fascism and nazistic behaviour takes over. There is individualistic trait of selfishness and a primitive impulse to follow the leader. The civil society which is just, good, equal, socialistic, democratic and fair generated by cortical impulses becomes dead. The society which is governed by cerebellar function and unisexual tendencies becomes more matriarchal as men and women tend to have similar traits. Women also tend to be as aggressive if not more than men. The cortical hemispheric control over social and individual behaviour is lost. It becomes the primitive world of selfishness and individuality uninhibited by sexual mores.[1-17]

The archaeal overgrowth and digoxin synthesis can modulate retroviral growth. Digoxin can modulate RNA editing and retroviral replication. Digoxin can also produce intracellular magnesium deficiency resulting in reverse transcriptase inhibition. Thus the archaeal induced stem cell syndrome is retroviral resistant. This results in changes in the human genome as such. HERV sequences in the human genome functions as jumping genes producing dynamicity and flexibility of the human genome. This is required for the changes in cortical synaptic connectivity, HLA gene flexibility and developmental changes. The archaeal induced stem cell syndrome produces a rigid adynamic genome not able to cope with the complexities of the cortical connectivity, HLA gene rearrangements for immune response and gene changes for complex development. This neanderthalisation of the human body due to archaeal symbiosis can spell the death of the human species. The new human species which may be transient consequent to archaeal symbiosis produced by extremophilic climatic changes consequent to global warming can be called the human homo neoneanderthalis. It is androgynous, creative, psychedelic, artistic, spiritual, aggressive, violent, selfish, impulsive, anarchic, chaotic and individualistic.[1-17]

References

1. Weaver TD, Hublin JJ. Neandertal Birth Canal Shape and the Evolution of Human Childbirth. *Proc. Natl. Acad. Sci. USA* 2009; 106:8151–8156.
2. Kurup RA, Kurup PA. Endosymbiotic Actinidic Archaeal Mediated Warburg Phenotype Mediates Human Disease State. *Advances in Natural Science* 2012; 5(1):81-84.
3. Morgan E. The *Neanderthal* theory of *autism*, Asperger and ADHD 2007, www.rdos.net/eng/asperger.htm.
4. Graves P. New Models and Metaphors for the Neanderthal Debate. *Current Anthropology* 1991; 32(5): 513-541.
5. Sawyer GJ, Maley B. Neanderthal Reconstructed. The Anatomical Record Part B: *The New Anatomist* 2005; 283B(1):23-31.
6. Bastir M, O'Higgins P, Rosas A. Facial Ontogeny in Neanderthals and Modern Humans. *Proc. Biol. Sci.* 2007; 274:1125–1132.
7. Neubauer S, Gunz P, Hublin JJ. Endocranial Shape Changes during Growth in Chimpanzees and Humans: A Morphometric Analysis of Unique and Shared Aspects. *J. Hum. Evol.* 2010; 59:555–566.
8. Courchesne E, Pierce K. Brain Overgrowth in Autism during a Critical Time in Development: Implications for Frontal Pyramidal Neuron and Interneuron Development and Connectivity. *Int. J. Dev. Neurosci.* 2005; 23:153–170.
9. Green RE, Krause J, Briggs AW, Maricic T, Stenzel U, Kircher M, Patterson N, Li H, Zhai W, *et al.* A Draft Sequence of the Neandertal Genome. *Science* 2010; 328:710–722.

10. Mithen SJ. *The Singing Neanderthals: The Origins of Music, Language, Mind and Body*; 2005, ISBN 0-297-64317-7.

11. Bruner E, Manzi G, Arsuaga JL. Encephalization and Allometric Trajectories in the Genus Homo: Evidence from the Neandertal and Modern Lineages. *Proc. Natl. Acad. Sci. USA* 2003; 100:15335–15340.

12. Gooch S. *The Dream Culture of the Neanderthals: Guardians of the Ancient Wisdom.* Inner Traditions, Wildwood House, London; 2006.

13. Gooch S. *The Neanderthal Legacy: Reawakening Our Genetic and Cultural Origins.* Inner Traditions, Wildwood House, London; 2008.

14. Kurtén B. *Den Svarta Tigern*, ALBA Publishing, Stockholm, Sweden; 1978.

15. Spikins P. Autism, the Integrations of 'Difference' and the Origins of Modern Human Behaviour. *Cambridge Archaeological Journal* 2009; 19(2):179-201.

16. Eswaran V, Harpending H, Rogers AR. Genomics Refutes an Exclusively African Origin of Humans. *Journal of Human Evolution* 2005; 49(1):1-18.

17. Ramachandran V.S. The Reith lectures, BBC London. 2012.

CHAPTER 33

ARCHAEAL ENDOSYMBIOSIS, EMERGING VIRUSES AND BRAIN STRUCTURE - THE ARCHAEAL INDUCED SPIRITUAL AND EVIL BRAIN - INFECTIVE SYMBIOTIC TISSUE COLONIES OF PROTOZOA, ARCHAEA, VIRUSES AND BACTERIA

Introduction

The plasma life based forms and the gravitational and anti-gravitational waves generated porphyrions would have been the most primitive of non-life forms. The porphyrin templates would have given rise to the isoprenoid organism, the RNA viroids, DNA viroids and prions which would have symbiosed to form the nanoarchaea. The archaea would have evolved into prokaryotes or bacteria. The archaeal symbiosis with prokaryotes would have generated the eukaryotes. The initial unicellular eukaryotes are the protozoa. The protozoa are capable of phagocytosis and would have had viruses, bacteria and archaea as their intracellular symbionts. The viruses of the protozoa like entamoeba and toxoplasma are giant viruses called megaviride. The protozoans like entamoeba with its viral, archaeal and bacterial endosymbionts would have evolved into the multicellular organisms or metazoa. The archaeal, bacterial and protozoal extracellular vesicles containing RNA and DNA viroids can by horizontal gene transfer lead to evolution of metazoan, plants, animals and human species. The metazoans would have evolved into plants, animals and humans by horizontal gene transfer generated by archaeal extracellular RNA and DNA viroid vesicles. The evidence of such a transformation is seen in the bacterial symbionts of protozoa like entamoeba which includes legionella, listeria, mycoplasma, mycobacteria and chlamydia. The symbionts of protozoan entamoeba include viruses like covid, HIV and cytomegalo virus. The protozoan entamoeba serves as a natural reservoir for covid infections, bacterial infections and HIV infections. The toxoplasma also serves as a covid reservoir. These

protozoan colonies of entamoeba and toxoplasma with its bacterial, archaeal and viral symbionts can be called as chimeric infective symbiotic tissue colonies which can exist in the air, water, earth, blood and tissues. The chimeric infective symbiotic tissue colonies containing bacteria, archaea and viruses can make the symbiotic organism multidrug resistant by horizontal gene transfer. The chimeric infective symbiotic tissue organism evolves as a consequence of global warming and the protozoal, bacterial, viral and archaeal symbiotic colonies are new forms of life. The chimeric infective symbiotic tissue organism composed of protozoa and its symbionts like virus, bacteria and archaea depend upon glycolysis for its energy needs. The glycolytic enzymes enolase, glyceraldehyde 3-phosphate dehydrogenase, hexokinase, phosphofructokinase, aldolase, pyruvic kinase and triose phosphate isomerase are moonlighting proteins having a cytoplasmic, membrane and nuclear functions. These glycolytic enzymes have a RNA binding motif which can modulate RNA transcription in the host and the symbiotic organism. These glycolytic enzymes with their RNA binding motif as well as DNA binding motif can bind to RNA and DNA viruses and viroids modulating their replication and life cycles. The chimeric infective symbiotic tissue organism composed of protozoa and its symbionts like virus, bacteria and archaea can enter the human host through the breaches of the gut-blood barrier and transform itself into the macrophages of the human host. The macrophages are involved in the phagocytosis and innate immunity against bacterial and viral infections. Paradoxically the macrophages like the protozoa can also serve as a reservoir where pathogens like mycobacteria tuberculosis, mycobacteria leprae, listeria, legionella, E.coli, shigella & salmonella, and viruses like HIV & covid can thrive and replicate. The modified chimeric infective symbiotic tissue organism or macrophage composed of protozoa and its symbionts like virus, bacteria and archaea can serve as a reservoir for horizontal gene transfer and generation of multidrug resistant bacteria and viruses. The infective chimeric symbiotic tissue organism composed of protozoa and its symbionts like virus, bacteria and archaea

depends upon moonlighting glycolytic enzymes for their energetics and replication. The development of antibodies against the infective chimeric symbiotic tissue organism composed of protozoa and its symbionts like virus, bacteria and archaea leads to the evolution of anti-brain antibody disease or anti-basal ganglia antibody disease. They present a threat to existence consequent to horiztontal gene transfer and generation of multidrug resistant bacteria, viruses, protozoa, mycobacteria and archaea.

The global warming produces increased acidity and atmospheric carbon dioxide resulting in extremophilic archaeal symbiosis in humans. The archaeal symbiosis results in neanderthalisation of humans. The archaea induced uncoupling proteins producing the primitive Warburg phenotype and stem cell metabolonomics. The archaeal metabolites of cholesterol digoxin, bile acids and short chain fatty acids induce uncoupling proteins. The lysosomal enzymes a marker of stem cell conversion are markedly increased along with genesis of the archaeal phenotype in metabolic syndrome X, degenerations, autoimmune diseases, cancer, schizophrenia and autism. In all these systemic diseases there is somatic cell transformation to stem cell and lose of function. The neurons become immature and lose their dendritic spines and connectivity. This results in loss of neuronal function and reversion to archaeal magnetite mediated extrasensory perception of low level of EMF. Exposure to low level of EMF results in brain changes. This results in prefrontal cortex atrophy. The primitive brain areas of cerebellum and brain stem become hypertrophic. The somatic and neuronal cell proliferates and there is neanderthalisation of the brain and body.[1-17]

The idea of goodness is based on reason and logic. Reason judgment and logic is a function of the cerebral cortex especially the prefrontal lobe. Prefrontal lobe function needs dynamic synaptic connectivity which is produced by jumping genes mediated by human endogenous retroviral sequences. Goodness is correlated with heaven. The idea of evil is based on the unconscious and the impulsive behaviour related to subcortical areas especially the cerebellum. The cerebellum is the site of impulsive behaviour and the unconscious behaviour. The cerebellar and subcortical brain connections are predominantly archaeal colony networks. The idea of evil is related to hell. The idea of conscious judgmental acts and unconscious impulsive acts, heaven and hell, goodness and evil are juxtapositions. The global warming and exposure to low level of EMF leads to actinidic archaeal growth in the brain and increased archaeal magnetite mediated perception of low level of EMF. This leads to prefrontal cortex atrophy and cerebellar dominance. The conscious becomes minimal and unconscious brain takes over. The study assessed archaeal growth as assessed by cytochrome F420 activity and stem cell type metabolonomics in systemic diseases, neuropsychiatric disorders and normal individuals with differing psychological profile - prisoners, creative individuals and common sense modulated business men.[1-17] The results are presented in this paper.

Materials and Methods

The blood samples were drawn from four groups of psychological different population spiritually inclined, criminal prisoners, creative artists and business men. There were 15 members in each group. The blood samples were also drawn from 15 cases each of metabolic syndrome, degenerations - Alzheimer's disease, autoimmune disease- SLE, cancer- brain glioma, schizophrenia and autism. The estimations done in the blood samples collected include cytochrome F420 activity. Blood lactate, pyruvate, hexokinase, cytochrome C, cytochrome F420, digoxin, bile acids, butyrate and propionate were estimated.

Results

The results showed that the spiritual, artistic creative individuals and criminal prisoners had increased cytochrome F420 activity and RBC digoxin levels. The results showed that the businessmen had decreased cytochrome F420 activity and RBC digoxin levels. The blood samples of Alzheimer's disease, autoimmune disease- SLE, cancer- brain glioma, schizophrenia and autism had increased blood lactate and pyruvate, increased RBC hexokinase, increased serum cytochrome C and serum cytochrome F420, increased serum digoxin, bile acids, butyrate and propionate. The disease state had increased cytochrome F420 activity. The serum cytochrome C levels in the blood were increased. This suggested mitochondrial dysfunction. There was an increased in glycolysis as suggested by increased RBC hexokinase activity and lactic acidosis. Owing to the mitochondrial dysfunction and pyruvate dehydrogenase inhibition there was pyruvate accumulation. The pyruvate was converted to lactate by the Cori cycle and also to glutamate and ammonia. This metabolism is suggestive of the Warburg phenotype and stem cell conversion. The stem cells depend on Warburg anaerobic glycolysis for energetics and have a mitochondrial dysfunction. The lysosomal enzyme beta galactosidase activity was increased in the disease group and in creative artists and criminals suggesting stem cell conversion. This suggests that artistic creative, criminal prisoners as well as spiritual individuals tend to have stem cell metabolonomics and stem cell conversion.

Discussion

Archaeal symbiosis, Neanderthal metabolonomics and stem cell transformation

The systemic diseases and neuropsychiatric disorders tend to have a predominant anaerobic glycolytic metabolism and mitochondrial oxidative phosphorylation is suppressed. The metabolism is similar to the metabolism of the stem cell. The pyruvate and lactate levels are increased with a decrease in acetyl coenzyme A and ATP. The glycolytic pathway and hexokinase is increased. This indicates a Warburg phenotype depending upon anaerobic glycolysis for energetics. The lysosomal enzymes beta galactosidase a stem cell marker is increased. The cytochrome F420 is also increased as well as the archaeal catabolite digoxin which suppresses sodium-potassium ATPase. Bacteria and archaea are supposed to induce stem cell transformation. The induction of uncoupling proteins leads to stem cell transformation. The uncoupling proteins inhibit oxidative phosphorylation and the substrates are directed to anaerobic glycolysis. Digoxin by inhibiting sodium-potassium ATPase can increase intracellular calcium, induce mitochondrial permeability transient pore function and uncouple oxidative phosphorylation. The side chain of cholesterol is catabolized by archaea to butyric acid and propionic acid which uncouple oxidative phosphorylation. The archaeal side chain hydroxylase convert cholesterol to bile acids which uncouple oxidative phosphorylation. Thus archaeal symbiosis in the cell results in cholesterol catabolism and the catabolites digoxin, bile acids and short chain fatty acids uncouple oxidative phosphorylation, inhibit mitochondrial function and promote anaerobic glycolysis. The conversion of somatic cells to stem cell helps in archaeal persistence within the cell and symbiosis. Mycobacterium leprae infection can convert Schwann cells to stem cells. Archaeal infection produces somatic cell conversion to stem cells for archaeal persistence. The conversion to stem cell results in proliferation and loss of function resulting in systemic disease and neuropsychiatric disorders. Stem cell conversion of neurons and loss of function results in development of a new psychological phenotype.[1-17]

Endosymbiotic brain archaea and altered brain function

The systemic and neuronal cell in metabolic syndrome X, cancer, autoimmune disease, degenerations, schizophrenia and autism behaves like the stem cell. It is plausible to hypothesize a somatic cell conversion to stem cell in these disorders. The differentiated cells by archaeal induction get converted to stem cell. The stem cell is an immature cell with loss of function. The neurons lose their dendritic spines and loss of connectivity. The brain function becomes primitive. The neurons are adendritic and disconnected. This results in complex brain structures like the modern cerebral cortex and prefrontal cortex atrophy. The primitive parts of the brain the brain stem and cerebellum hypertrophies. This results in neanderthalisation of the brain with a prominent occipital bun and atrophied prefrontal cortex. The prefrontal cortex atrophy results in loss of logic, judgment, reasoning and executive functions. The hypertrophy of the cerebellum and brain stem results in dominance of impulsive behaviour. The difference between reality and dreams is lost. The brain is ruled by the senses and impulses. The brain becomes dysfunctional with more of violent, aggressive and cannibalistic behaviour. The art becomes more abstract and related to the unconscious. The world of the unconscious brain with its archetypes takes over. There is loss of the world of reasoning, logic and judgment. It is a world of impulsiveness in which primitive tendencies with relation to the unconscious becomes dominant. This produces more of ritualized behaviour, violent and aggressive tendencies, terrorism, war, sexual obscenities and alternate sexuality. It is a world of the senses. It is also intensely evil as well as spiritual. The inhibition of the conscious due to loss of cortical functions and the dominance of the unconscious leads to mystical experience. There is a overflowing of spirituality. The paradoxical side of this behaviour also dominates. The violence, aggression, obsessive sexuality, magic realism in literature, abstract painting, rock music and dance and modern poetry as well as literature produces transcendence of a different kind. This results in surrealism and syntheism. The loss of function of the neurons results in schizophrenia, autism and degenerations. The increased archaeal induced proliferation of stem cells results in a big sized brain and trunk as in Neanderthals. This archaeal symbiosis produces neanderthalisation and a stem cell syndrome. This produces reverse aging which can be called as an epidemic Benjamin Button syndrome. The lymphocytic stem cells have uncontrolled proliferation and results in autoimmune diseases. The stem cell proliferation results in oncogenesis. The stem cell metabolonomics with inhibited mitochondrial function and anaerobic glycolysis results in metabolic syndrome X. Stem cell markers are increased in schizophrenia and autism and the neurons lack dendritic spines. Stem cell markers are also increased in autoimmune disease. The diabetic metabolism is akin to stem cell metabolism. The cancer cell behaves like the stem cell.[1-17]

Endosymbiotic archaea, brain function and the origin of spirituality and evil

In the metaphysics of evil the unconscious dominates and the behaviour is impulsive dictated by primitive thoughts. The unconscious modulated by the cerebellum is responsible for automatic acts producing what is called as psychic automatism. The unconscious parallels what Jung described as the archetypes of the collective unconscious. The metaphysics of evil leads to a syntheistic brain with the dominance of the willpower. The primitive archetypes produce concepts of abstract painting, psychedelic music and dance and postmodern literature or magical realism. All these are modes of connecting with the unconscious. The unconscious produces primitive selfish tendencies leading to individualism and capitalism. The unconscious helps to transcend taboos and creates the surrealistic world. The collective unconscious also produces a sense of spirituality and oneness. It is an impulsive brain with fixations and primitive obsessions. There is cerebellar psychic automatism. This leads to ritualized behaviours. The dominance of the collective unconscious results in ritualized behaviours characteristic of religious worship. The collective unconscious also leads to the creation of obscene art and literature as well as violence which is a form of transcendence. Coprolalic religious ritual ceremonies had been described in some parts of the world. Terrorism and acts of violence are also a type of transcendence. The same phenomena occur in ritual sacrifices in religion, the violence of war and the acquisitiveness of capitalism. The primitive unconscious leads to the will to power. This produces greedy capitalism, dictatorship and fascism. The will to power results in worship of the powerful. It is an individualistic, anarchic, selfish world. The cerebellar world is the primitive world of archetypes in the collective unconscious. The abstract paintings have links with the collective unconscious. The rock music or modern music contains rhythmic primitive chaotic sounds coming out the collective unconscious. The primitive collective unconscious links up post modern literature or magic realism with violence, love, hate, evil, obscenities and death. Thus literature, music, dance and painting helps to overcome reality and rationality producing transcendence. The unconscious brain is formed of an archaeal colony network and is adynamic and inflexible. There is an epidemic of autism and schizophrenia. The loss of function of neurons leads to increased extrasensory perception via archaeal magnetite. This can lead to the lack of development of speech and ritualized behaviours of autism. This also produces the thought disorder, hallucinations and delusions of schizophrenia. It looks like an epidemic cerebellar cognitive, affective disorder.[1-17]

The goodness is related to conscious brain localized in the cortical areas. The cortical areas mediate moralistic, functionally atheistic, civil society behaviour. The civil society depends upon common good. The cortical world is a world of morality, rationality, altruism, civility and decencies. This needs inhibitory power of the cerebral cortex. Such a society is non-capitalistic and works for the common good. It tends to be non-creative. The primitive collective spirituality and oneness is lost. It is replaced by goodness based on judgment, reasoning and morality. It is a moralistic world where taboos are banned. This requires synaptic plasticity and is modulated by HERV mediated jumping genes. This needs a dynamic brain and the human cerebral cortex evolved due to the jumping genes generated from human endogenous retroviral sequences. The cerebellar world comparatively is impulsive, criminal, violent, terroristic with love of war, selfish, acquisitive, spiritual, autistic, obsessive, schizophrenic, obscene, evil, ritualized, artistic, illogical and cruel. It is mediated by the archaeal colony network. The stem cell transformation of somatic cells results in HERV resistance and retroviral resistance. Archaeal digoxin inhibits reverse transcriptase by producing magnesium deficiency as well as modulates RNA viral editing inhibiting retroviral replication. This produces lack of HERV jumping genes in this stem cell brain and lack of synaptic plasticity and dynamicity. The stem cell syndrome is characterized by retroviral resistance. Archaeal symbiosis inhibits retroviral infection. The homo sapiens with less of archaeal symbiosis becomes susceptible to retroviral and other RNA viral infection and gets wiped out. The homo neoneanderthalis are resistance to retroviral and other RNA viral infection and persists. The homo neoneanderthalis dominates all over the world. But the homo neoneanderthalis are prone to civilizational disease like malignancy, autoimmune disease, neurodegeneration, metabolic syndrome and neuropsychiatric disorders. The homo neoneanderthalis becomes extinct after a period of time.[1-17]

The archaeal induced stem cell syndrome or neanderthalisation is due to global warming and acid rains resulting in increased extremophilic archaeal symbiosis. The archaea catabolises cholesterol and generates digoxin, bile acids and short chain fatty acids which produce induction of uncoupling proteins. This produces mitochondrial dysfunction and the cell obtains its energetics from glycolysis. Archaeal digoxin produces membrane sodium-potassium ATPase inhibition which also contributes to stem cell conversion. The whole body somatic and brain undergoes stem cell conversion and becomes a stem cell phenotype with Warburg metabolic phenotype. The generalized acidity due to global warming and increased atmospheric carbon dioxide also facilitates archaeal growth and stem cell transformation. The acidic pH due to the Warburg phenotype and increased atmospheric carbon dioxide also results in stem cell conversion. The somatic differentiated cell getting converted to stem cells lose their function and become dysfunctional metabolically, neurologically, immunologically and endocrine-wise. This produces the epidemic Benjamin button syndrome and the human species becomes neanderthalic and a collection of immature stem cells. This results in epidemic metabolic syndrome X, degenerations, cancer, autoimmune disease, autism and schizophrenia. The brain becomes converted to a collection of stem cells which are dedifferentiated with loss of function and is like an archaeal colony network. The perception becomes extrasensory and quantal depending on archaeal magnetite. The increased amount of low level EMF perception results in prefrontal cortical atrophy. It also produces cerebellar hypertrophy and the cerebellar cognitive function takes over. This also results in societal changes where evil and spirituality dominates. The world of the logical civil society of the Christian world comes to end and paganistic behaviour takes over. The society becomes selfish and dominated by impulsive consumerism and acquisitive capitalism. The world becomes cruel, violent, aggressive and terroristic. Art becomes chaotic and abstract in line with the senses and unconscious. There is a predominance of obsessive and alternate sexuality. Criminal behaviour and cruelty dominates. The world is impulsive psychopathic, creative autistic with features of idiotic savants, ritualistic, chaotic, sexual, ugly, anarchic, violent, evil, paganistic, obscene, atheistically spiritual as well as selfish. It mimics the Nietzschean world, the deconstructed world of Derrida, the surrealistic world of Bataille and the nihilistic, anarchic world. There is the death of the individual and life becomes a social value. It is an acephalistic world of Freud and Jung. The art is abstract, the literature is magically real, the music is rock and the dance chaotic. All these result from the extinction of rationality and the dominance of primitive impulsive behaviour. A civilization of the senses dominated by the unconscious takes over. The will to goodness given by the cerebral cortex is lost. This results in development of a new homo neoneanderthal human species with its dominant evilly spiritual cerebellar brain. It produces a surrealistic evil brain with realm of the senses, archetypes, evil spirituality and impulsiveness taking over. It is a kingdom of the collective unconscious and selfish capitalism with the will to power and the realm of the senses.[1-17]

References

1. Weaver TD, Hublin JJ. Neandertal Birth Canal Shape and the Evolution of Human Childbirth. *Proc. Natl. Acad. Sci. USA* 2009; 106:8151–8156.
2. Kurup RA, Kurup PA. Endosymbiotic Actinidic Archaeal Mediated Warburg Phenotype Mediates Human Disease State. *Advances in Natural Science* 2012; 5(1):81-84.
3. Morgan E. The *Neanderthal* theory of *autism*, Asperger and ADHD; 2007, www.rdos.net/eng/asperger.htm.
4. Graves P. New Models and Metaphors for the Neanderthal Debate. *Current Anthropology* 1991; 32(5): 513-541.
5. Sawyer GJ, Maley B. Neanderthal Reconstructed. *The Anatomical Record Part B: The New Anatomist* 2005; 283B(1):23-31.
6. Bastir M, O'Higgins P, Rosas A. Facial Ontogeny in Neanderthals and Modern Humans. *Proc. Biol. Sci.* 2007; 274:1125–1132.
7. Neubauer S, Gunz P, Hublin JJ. Endocranial Shape Changes during Growth in Chimpanzees and Humans: A Morphometric Analysis of Unique and Shared Aspects. *J. Hum. Evol.* 2010; 59:555–566.
8. Courchesne E, Pierce K. Brain Overgrowth in Autism during a Critical Time in Development: Implications for Frontal Pyramidal Neuron and Interneuron Development and Connectivity. *Int. J. Dev. Neurosci.* 2005; 23:153–170.
9. Green RE, Krause J, Briggs AW, Maricic T, Stenzel U, Kircher M, Patterson N, Li H, Zhai W, *et al.* A Draft Sequence of the Neandertal Genome. *Science* 2010; 328:710–722.
10. Mithen SJ. *The Singing Neanderthals: The Origins of Music, Language, Mind and Body;* 2005, ISBN 0-297-64317-7.
11. Bruner E, Manzi G, Arsuaga JL. Encephalization and Allometric Trajectories in the Genus Homo: Evidence from the Neandertal and Modern Lineages. *Proc. Natl. Acad. Sci. USA* 2003; 100:15335–15340.
12. Gooch S. *The Dream Culture of the Neanderthals: Guardians of the Ancient Wisdom.* Inner Traditions, Wildwood House, London; 2006.
13. Gooch S. *The Neanderthal Legacy: Reawakening Our Genetic and Cultural Origins.* Inner Traditions, Wildwood House, London; 2008.
14. Kurtén B. *Den Svarta Tigern*, ALBA Publishing, Stockholm, Sweden; 1978.
15. Spikins P. Autism, the Integrations of 'Difference' and the Origins of Modern Human Behaviour. *Cambridge Archaeological Journal* 2009; 19(2):179-201.
16. Eswaran V, Harpending H, Rogers AR. Genomics Refutes an Exclusively African Origin of Humans. *Journal of Human Evolution* 2005; 49(1):1-18.
17. Ramachandran V.S. The Reith lectures, BBC London. 2012.

CHAPTER 34
CLIMATE CHANGE, EMERGING VIRUSES AND NEUROBIOLOGY OF SOCIO-POLITICAL, SPIRITUAL, SEXUAL AND CULTURAL IDENTITY - INFECTIVE SYMBIOTIC TISSUE COLONIES OF PROTOZOA, ARCHAEA, VIRUSES AND BACTERIA

Introduction

The plasma life based forms and the gravitational and anti-gravitational waves generated porphyrions would have been the most primitive of non-life forms. The porphyrin templates would have given rise to the isoprenoid organism, the RNA viroids, DNA viroids and prions which would have symbiosed to form the nanoarchaea. The archaea would have evolved into prokaryotes or bacteria. The archaeal symbiosis with prokaryotes would have generated the eukaryotes. The initial unicellular eukaryotes are the protozoa. The protozoa are capable of phagocytosis and would have had viruses, bacteria and archaea as their intracellular symbionts. The viruses of the protozoa like entamoeba and toxoplasma are giant viruses called megaviride. The protozoans like entamoeba with its viral, archaeal and bacterial endosymbionts would have evolved into the multicellular organisms or metazoa. The archaeal, bacterial and protozoal extracellular vesicles containing RNA and DNA viroids can by horizontal gene transfer lead to evolution of metazoan, plants, animals and human species. The metazoans would have evolved into plants, animals and humans by horizontal gene transfer generated by archaeal extracellular RNA and DNA viroid vesicles. The evidence of such a transformation is seen in the bacterial symbionts of protozoa like entamoeba which includes legionella, listeria, mycoplasma, mycobacteria and chlamydia. The symbionts of protozoan entamoeba include viruses like covid, HIV and cytomegalo virus. The protozoan entamoeba serves as a natural reservoir for covid infections, bacterial infections and HIV infections. The toxoplasma also serves as a covid reservoir. These

protozoan colonies of entamoeba and toxoplasma with its bacterial, archaeal and viral symbionts can be called as chimeric infective symbiotic tissue colonies which can exist in the air, water, earth, blood and tissues. The chimeric infective symbiotic tissue colonies containing bacteria, archaea and viruses can make the symbiotic organism multidrug resistant by horizontal gene transfer. The chimeric infective symbiotic tissue organism evolves as a consequence of global warming and the protozoal, bacterial, viral and archaeal symbiotic colonies are new forms of life. The chimeric infective symbiotic tissue organism composed of protozoa and its symbionts like virus, bacteria and archaea depend upon glycolysis for its energy needs. The glycolytic enzymes enolase, glyceraldehyde 3-phosphate dehydrogenase, hexokinase, phosphofructokinase, aldolase, pyruvic kinase and triose phosphate isomerase are moonlighting proteins having a cytoplasmic, membrane and nuclear functions. These glycolytic enzymes have a RNA binding motif which can modulate RNA transcription in the host and the symbiotic organism. These glycolytic enzymes with their RNA binding motif as well as DNA binding motif can bind to RNA and DNA viruses and viroids modulating their replication and life cycles. The chimeric infective symbiotic tissue organism composed of protozoa and its symbionts like virus, bacteria and archaea can enter the human host through the breaches of the gut-blood barrier and transform itself into the macrophages of the human host. The macrophages are involved in the phagocytosis and innate immunity against bacterial and viral infections. Paradoxically the macrophages like the protozoa can also serve as a reservoir where pathogens like mycobacteria tuberculosis, mycobacteria leprae, listeria, legionella, E. coli, shigella & salmonella, and viruses like HIV & covid can thrive and replicate. The modified chimeric infective symbiotic tissue organism or macrophage composed of protozoa and its symbionts like virus, bacteria and archaea can serve as a reservoir for horizontal gene transfer and generation of multidrug resistant bacteria and viruses. The infective chimeric symbiotic tissue organism composed of protozoa and its symbionts like virus, bacteria and archaea

depends upon moonlighting glycolytic enzymes for their energetics and replication. The development of antibodies against the infective chimeric symbiotic tissue organism composed of protozoa and its symbionts like virus, bacteria and archaea leads to the evolution of anti-brain antibody disease or anti-basal ganglia antibody disease. They present a threat to existence consequent to horiztontal gene transfer and generation of multidrug resistant bacteria, viruses, protozoa, mycobacteria and archaea.

Actinidic archaea has been related to global warming and human diseases especially neuropsychiatric disorder. The growth of endosymbiotic actinidic archaea in relation to climate change and global warming leads to neanderthalisation of the human mind-body system. Neanderthal anthropometry and metabolonomics has been described in neuropsychiatric disorder especially the Warburg phenotype and hyperdigoxinemia. Digoxin produced by archaeal cholesterol catabolism produces neanderthalisation. Prefrontal cortical atrophy and cerebellar hyperplasia has been related to neuropsychiatric disorder. Digoxin produces intracellular magnesium deficiency and reverse transcriptase inhibition. This blocks retroviral replication and its integration into the genome. This reduces the flexibility and dynamicity of the genome. Endogenous retroviral sequence induced jumping genes contributes to the flexibility and dynamicity of the genome required for generation of synaptic connectivity in the prefrontal cortex. The inhibition of endogenous retroviral expression and integration by digoxin leads to atrophy of the prefrontal cortex. The cerebellum becomes dominant and can be considered as an endosymbiotic archaeal network. The cerebellar dominance leads to more of extrasensory perception, quantal perception, impulsive behaviour and sense of oneness with the world around us. Cerebellar dominance also leads to dysautonomia with sympathetic hyperactivity and parasympathetic neuropathy in these disorders. Actinidic archaeal related cerebellar dominance leads to changes in brain function. This produces a new socio-political, spiritual, sexual and cultural phenotype.[1-16] The data is described in this paper.

Materials and Methods

Fifteen cases, each of neuropsychiatric disorder, alternate sexuality and internet addicts were selected for the study. Each case had an age and sex matched control. Neanderthal anthropometric and phenotypic measurements which included protruding supra-orbital ridges, dolichocephalic skull, small mandible, prominent mid face and nose, short upper and lower limbs, prominent trunk, low index finger-ring finger ratio and fair complexion were evaluated in the cases study. Autonomic function tests were done to assess the sympathetic and parasympathetic system in each case. CT scan of the head was done to have a volumetric assessment of the prefrontal cortex and cerebellum. Blood cytochrome F420 activity was assessed by spectrophotometric measurement.

Results

All the case groups studied had higher percentage of Neanderthal anthropometric and phenotypic measurements. There was low index finger-ring finger ratio suggestive of high testosterone levels in all the patient population studied. In all the case groups studied, there also was prefrontal cortex atrophy and cerebellar hyperplasia. Similarly in the all the case groups studied, there was dysautonomia with sympathetic overactivity and parasympathetic neuropathy. Cytochrome F420 was detected in the entire case group studied showing endosymbiotic archaeal overgrowth.

Discussion

Neanderthal metabolonomics contribute to the generation of a new sociopolitical, spiritual, sexual and cultural phenotype and pathogenesis of schizophrenia/autism. There were Neanderthal phenotypic features in all the case groups studied as well as low index finger-ring finger ratios suggestive of increased testosterone levels. Neanderthalisation of the mind-body system occurs due to increased growth of actinidic archaea as a consequence of global warming. Digoxin produces intracellular magnesium deficiency and reverse transcriptase inhibition. This blocks retroviral replication and its integration into the genome. This reduces the flexibility and dynamicity of the genome. Endogenous retroviral sequence induced jumping genes contributes to the flexibility and dynamicity of the genome required for generation of synaptic connectivity in the prefrontal cortex. The inhibition of endogenous retroviral expression and integration by digoxin leads to atrophy of the prefrontal cortex. The cerebellum becomes dominant and can be considered as an endosymbiotic archaeal network. The cerebellar dominance leads to more of extrasensory perception, quantal perception, impulsive behaviour and sense of oneness with the world around us. Neanderthalisation of the mind thus leads to cerebellar dominance and prefrontal cortex atrophy. This leads to dysautonomia with parasympathetic neuropathy and sympathetic hyperactivity. This produces a new sociopolitical, spiritual, sexual and cultural phenotype.

Global warming and the ice age produces increased growth of extremophiles. This leads to increased growth of actinidic archaeal endosymbiosis in humans. There is archaeal proliferation in the gut which enters the cerebellum and brain stem by reverse axonal transport via the vagus. The cerebellum and brain stem can be considered as an archaeal colony. The archaea are cholesterol catabolizing and use cholesterol as a carbon and energy source. The actinidic archaea activates the toll receptor HIF alpha inducing the Warburg phenotype resulting in increased glycolysis with generation of glycine as well as pyruvate dehydrogenase suppression. The accumulated pyruvate enters the GABA shunt generating of succinyl CoA and glycine. The archaeal catabolism of cholesterol produces ring oxidation and generation of pyruvate which also enters the GABA shunt scheme producing glycine and succinyl

CoA. This leads to increased synthesis of porphyrins. In the setting of digoxin induced sodium-potassium ATPase inhibition the dipolar porphyrins produce a pumped phonon system resulting in the Frohlich model Bose-Einstein's condensate and quantal perception of low level EMF. Low level EMF pollution is common with internet usage. Perception of low level of EMF leads to neanderthalisation of the brain with prefrontal cortex atrophy and cerebellar hyperplasia. The archaea which reaches the cerebellum from the gut via the vagus nerve proliferates and makes the cerebellum dominant with resultant suppression and atrophy of the prefrontal cortex. This leads to widespread autistic and schizophrenic traits in population. The actinidic archaea induces the Warburg phenotype with increased glycolysis, PDH inhibition and mitochondrial suppression. This produces neanderthalisation of the mind-body system. The actinidic archaea secretes RNA viroids which block HERV expression by RNA interference. The HERV suppression contributes to the inhibition of prefrontal cortex development in Neanderthals and cerebellar dominance. Archaeal digoxin produces sodium-potassium ATPase inhibition and magnesium depletion causing reverse transcriptase inhibition and decreased generation of HERV. The HERV contributes to the dynamicity of the genome and are required for the development of the prefrontal cortex. The HERV suppression contributes to retroviral resistance in Neanderthals. The actinidic archaea catabolizes cholesterol leading to cholesterol depleted state. Cholesterol depletion also leads to poor synaptic connectivity and decreased development of prefrontal cortex. This is not genetic change but a form of symbiotic change with endosymbiotic actinidic archaeal growth in the body and brain.

The neanderthalic brain had predominant quantal perception and extrasensory communication leading to oneness of society. This contributed to a just and equal society with compassion and altruism. This can be thought of as a beginning of a primitive socialistic or communistic society with equal rights for all and sharing of wealth in the community. The Neo-neanderthal society had a collective unconscious contributing to an equal, liberal and socialistic society. The archaeal cholesterol metabolism leads to depletion of cholesterol and deficiency of sex hormones. This leads to an asexual state with equality of both males and females or a female dominance. This leads on to androgynous behaviour and new alternate sexual identities. The Neo-neanderthal society is matriarchal with female leadership and dominance. Matriarchy and feminine dominance is the hall mark of Neoneanderthal society. The dominant extrasensory perception and quantal perception leads to appreciation of the vacancy of the universe and a sense of god. This fulfils the concept of Maya and oneness of everything in the universe according to Hindu Philosophy. The vacancy consequent to quantal perception gives a feeling of perception of god. The god concept probably evolved out of this quantal perception. The information stored in the brain functioning as a quantal computer exists as multiple possibilities in multiverse universes. The extinction of one possibility in earth does not foreclose the existence of the other possibilities in the quantal state in other universes. This leads to the concept of universal eternal existence and reincarnation described in Hindu philosophy and the illusory nature of death. There is eternal existence in the multiverse quantal universe. This leads on to an extreme spiritual society with a feeling of oneness generated by the collective unconscious consequent to quantal perception. The oneness is also shared with the environment contributing to eco-spirituality and environmental consciousness.

Internet use and low level EMF pollution is common in this century. This results in increased low level EMF perception by the brain by the digoxin-porphyrin mediated pumped phonon system created Bose-Einstein's condensate contributing to prefrontal cortex atrophy and cerebellar dominance. Cerebellar dominance leads to schizophrenia and autism. There is an epidemic of autism and schizophrenia in the present day community. The porphyrin mediated extrasensory perception can contribute to communication among Neanderthals.

Neanderthals did not have a language and used extrasensory perception as a form of group communication. Because of dominant extrasensory quantal perception, the Neanderthals did not have individual identity but only group identity. Cerebellar dominance results in creativity consequent to quantal perception and group perception. The neanderthalic traits contribute to innovation and creativity. Cerebellar dominance results in development of a symbolic language. The Neanderthals used dance and music as a form of communication. Painting as a form of communication was also common in Neanderthals. Neanderthal behaviour was robotic. Robotic behaviour is characteristic of cerebellar dominance. Robotic, symbolic and ritualistic behaviour is common with cerebellar dominance and is seen in autistic traits. The cerebellar dominance in Neanderthals leads to intuitive intelligence and a hypnotic quality to communication. The increased extrasensory quantal perception leads to more communion with nature and a form of eco-spirituality. The increasing use of dance and music as a form of communication and eco-spirituality is common in the modern century along with increased incidence of autism. The cholesterol depletion leads to bile acid deficiency and generation of small social groups in Neanderthals. Bile acid binds to olfactory receptors and contributes to group identity. This can also contributes to the generation of autistic features in Neanderthals.

The Neanderthal population was predominantly autistic and schizophrenic. The modern population is a hybrid of homo sapiens and homo neanderthalis. This contributes to 10 to 20 percent dominant hybrids who tend to have schizophrenic and autistic qualities and contributes to creativity of civilization. This can be called as a schizophrenic or autistic human tribe. The Neanderthals tend to be innovative and chaotic. They tend to be creative in art, literature, dance, spirituality and science. Eighty percent of less dominant hybrids are stable and contribute to a stabilizing influence leading to growth of civilization. The homo sapiens were stable and non-creative over a long period of their existence. There was a burst of creativity with generation of music, dance, painting, ornaments, the creation of concept of God and compassionate group behaviour around 10,000 years ago in the homo sapien community. This correlated with the generation of Neanderthal hybrids when the Eurasian Neanderthal male mated with homo sapiens African females. The extrasensory/quantal perception due to dipolar porphyrins and digoxin induced sodium-potassium ATPase inhibition and the generated pumped phonon system mediated quantal perception leads to the globalisation phenomena and feeling of the world being a global village. The archaeal cholesterol catabolism leads to increased synthesis of digoxin. Digoxin promotes tryptophan transport over tyrosine. Tyrosine deficiency leads to dopamine deficiency and morphine deficiency. This leads to a morphine deficiency syndrome in Neanderthals. This contributes to addiction traits and creativity. The increased tryptophan levels produce increased alkaloids like LSD contributing to ecstasy and spirituality of Neanderthal population. Addictive, ADHD and autistic features are related to the morphine deficiency state. The ketogenic diet consumed by the meat eating Neanderthals leads on to increased generation of hydroxy butyric acid which produces ecstasy and a dissociative type of anaesthesia contributing to the Neanderthal psychology. The dopamine deficiency leads to decreased melanin synthesis and fairness of the population. This was responsible for the fair colour of the Neanderthals.

The Neanderthals were essentially meat eaters taking a ketogenic diet. The acetoacetic acid is converted to acetyl CoA which enters the TCA cycle. When the Neanderthal hybrids consume a glucogenic diet owing to the spread of settled civilization it produces pyruvate accumulation owing to PDH suppression in Neanderthals. The increased archaeal growth activates the toll receptor and induces HIF alpha resulting in increased glycolysis, PDH suppression and mitochondrial dysfunction - the Warburg phenotype. The pyruvate enters the GABA shunt pathway producing glutamate, ammonia and porphyrins resulting in neuropathology of autism and schizophrenia. Neanderthals consuming a ketogenic diet produces more of GABA an inhibitory neurotransmitter resulting in the docile quiet nature of the Neanderthals. There is less production of glutamate the predominant excitatory neurotransmitter of the prefrontal cortex and consciousness pathways. This leads on to dominance of cerebellar function. The Neanderthal hybrids have cerebellar dominance and less of conscious behaviour. Cerebellum is responsible for intuitive, unconscious behaviour as well as creativity and spirituality. The cerebellum is the site of extrasensory perception, magical acts and hypnosis. The predominant homo sapiens had prefrontal cortex dominance over the cerebellum resulting in more of conscious behaviour.

The Neanderthals consuming a glucogenic diet produces increased glycolysis in the setting of PDH inhibition. This produces the Warburg phenotype. The glycolytic intermediate 3-phosphoglycerate is converted to glycine resulting in NMDA excitotoxicity contributing to schizophrenia and autism. Cerebellar dominance is reported in schizophrenia and autism. The cerebellar hyperplasia results in sympathetic hyperactivity and parasympathetic neuropathy. Cerebellar dominance and cerebellar cognitive affective dysfunction can contribute to schizophrenia and autism. The increased porphyrin synthesis resulting from succinyl CoA generated by GABA shunt and glycine generated by glycolysis contributes to increased extrasensory perception important in schizophrenia and autism. The archaeal cholesterol catabolism generates digoxin which produces sodium-potassium ATPase inhibition and increase in intracellular calcium and decrease in intracellular magnesium. The increase in intracellular calcium can modulate the neurotransmitter release from presynaptic vesicles. This can modulate neurotransmission. Digoxin induced magnesium depletion can remove the magnesium block on the NMDA receptor resulting in NMDA excitotoxicity. Digoxin can modulate the glutamatergic thalamo-cortico-thalamic pathway and consciousness resulting in schizophrenia and autism. Digoxin induced magnesium depletion can inhibit reverse transcriptase activity and HERV generation modulating the dynamicity of the genome. This can produce digoxin induced neuro-immuno-endocrine integration. Digoxin functions as a Neanderthal master hormone.

The actinidic archaea are cholesterol catabolizing and leads to low levels of testosterone and estrogen. This leads on to asexual features and low reproductive rates of the Neanderthal population. The Neanderthals consume a low fibre diet with low lignan content. The actinidic archaea has cholesterol catabolizing enzymes generating more of testosterone than estrogens. This contributes to estrogen deficiency and testosterone overactivity. The Neanderthal population is hypermales with concomitant right hemispheric dominance and cerebellar dominance. Testosterone suppresses left hemispheric function. The high testosterone levels in Neanderthals contribute to a bigger brain. The Neanderthals males as well as females had a higher level of testosterone contributing to gender equality and gender-neutral states. There was group identity and group motherhood with no differences between roles of both males and females. This also resulted in matrilinearity. The higher testosterone levels in males as well as females led to alternate type of sexuality and aberrant behaviour. The homo sapiens eat a high fibre diet with low cholesterol and high lignan content contributing to estrogen dominance, left hemispheric dominance and cerebellar hypoplasia. Homo sapiens had higher reproductive rates and overtook the Neanderthal population resulting in its extinction. The homo sapien population was conservative with normal sexual mores, family values and patriarchal type of behaviour. The role of females the homo sapien community was inferior to males. The increasing generation of Neanderthal hybrids due to climate change mediated archaeal overgrowth leads to gender equality and equidominance of male and female in this century. This is the basis of Neo-neanderthal matriarchy.

Thus global warming and increased endosymbiotic actinidic archaeal growth leads to cholesterol catabolism and generation of the Warburg phenotype resulting in increased porphyrin synthesis, extrasensory low EMF perception, prefrontal cortex atrophy, insulin resistance and cerebellar dominance. This leads on to neanderthalisation of the body and brain. This produces a new sociopolitical, spiritual, sexual and cultural phenotype.

References

1. Weaver TD, Hublin JJ. Neandertal Birth Canal Shape and the Evolution of Human Childbirth. *Proc. Natl. Acad. Sci. USA* 2009; 106:8151–8156.
2. Kurup RA, Kurup PA. Endosymbiotic Actinidic Archaeal Mediated Warburg Phenotype Mediates Human Disease State. *Advances in Natural Science* 2012; 5(1):81-84.
3. Morgan E. The *Neanderthal* theory of *autism*, Asperger and ADHD; 2007, www.rdos.net/eng/asperger.htm.
4. Graves P. New Models and Metaphors for the Neanderthal Debate. *Current Anthropology 1991;* 32(5): 513-541.
5. Sawyer GJ, Maley B. Neanderthal Reconstructed. *The Anatomical Record Part B: The New Anatomist* 2005; 283B(1):23-31.
6. Bastir M, O'Higgins P, Rosas A. Facial Ontogeny in Neanderthals and Modern Humans. *Proc. Biol. Sci.* 2007; 274:1125–1132.
7. Neubauer S, Gunz P, Hublin JJ. Endocranial Shape Changes during Growth in Chimpanzees and Humans: A Morphometric Analysis of Unique and Shared Aspects. *J. Hum. Evol.* 2010; 59:555–566.
8. Courchesne E, Pierce K. Brain Overgrowth in Autism during a Critical Time in Development: Implications for Frontal Pyramidal Neuron and Interneuron Development and Connectivity. *Int. J. Dev. Neurosci.* 2005; 23:153–170.
9. Green RE, Krause J, Briggs AW, Maricic T, Stenzel U, Kircher M, Patterson N, Li H, Zhai W, *et al.* A Draft Sequence of the Neandertal Genome. *Science* 2010; 328:710–722.
10. Mithen SJ. *The Singing Neanderthals: The Origins of Music, Language, Mind and Body*; 2005, ISBN 0-297-64317-7.
11. Bruner E, Manzi G, Arsuaga JL. Encephalization and Allometric Trajectories in the Genus Homo: Evidence from the Neandertal and Modern Lineages. *Proc. Natl. Acad. Sci. USA* 2003; 100:15335–15340.
12. Gooch S. *The Dream Culture of the Neanderthals: Guardians of the Ancient Wisdom. Inner Traditions*, Wildwood House, London; 2006.
13. Gooch S. *The Neanderthal Legacy: Reawakening Our Genetic and Cultural Origins. Inner Traditions*, Wildwood House, London; 2008.
14. Kurtén B. *Den Svarta Tigern*, ALBA Publishing, Stockholm, Sweden; 1978.
15. Spikins P. Autism, the Integrations of 'Difference' and the Origins of Modern Human Behaviour. *Cambridge Archaeological Journal* 2009; 19(2):179-201.
16. Eswaran V, Harpending H, Rogers AR. Genomics Refutes an Exclusively African Origin of Humans. *Journal of Human Evolution* 2005; 49(1):1-18.

CHAPTER 35

THE END OF THE WORLD - ENDOSYMBIOTIC ACTINIDIC ARCHAEA, GALACTIC EVOLUTION, GLOBAL WARMING AND CATASTROPHIC EVOLUTIONARY CYCLES - INFECTIVE SYMBIOTIC TISSUE COLONIES OF PROTOZOA, ARCHAEA, VIRUSES AND BACTERIA

Introduction

The plasma life based forms and the gravitational and anti-gravitational waves generated porphyrions would have been the most primitive of non-life forms. The porphyrin templates would have given rise to the isoprenoid organism, the RNA viroids, DNA viroids and prions which would have symbiosed to form the nanoarchaea. The archaea would have evolved into prokaryotes or bacteria. The archaeal symbiosis with prokaryotes would have generated the eukaryotes. The initial unicellular eukaryotes are the protozoa. The protozoa are capable of phagocytosis and would have had viruses, bacteria and archaea as their intracellular symbionts. The viruses of the protozoa like entamoeba and toxoplasma are giant viruses called megaviride. The protozoans like entamoeba with its viral, archaeal and bacterial endosymbionts would have evolved into the multicellular organisms or metazoa. The archaeal, bacterial and protozoal extracellular vesicles containing RNA and DNA viroids can by horizontal gene transfer lead to evolution of metazoan, plants, animals and human species. The metazoans would have evolved into plants, animals and humans by horizontal gene transfer generated by archaeal extracellular RNA and DNA viroid vesicles. The evidence of such a transformation is seen in the bacterial symbionts of protozoa like entamoeba which includes legionella, listeria, mycoplasma, mycobacteria and chlamydia. The symbionts of protozoan entamoeba include viruses like covid, HIV and cytomegalo virus. The protozoan entamoeba serves as a natural reservoir for covid infections, bacterial infections

and HIV infections. The toxoplasma also serves as a covid reservoir. These protozoan colonies of entamoeba and toxoplasma with its bacterial, archaeal and viral symbionts can be called as chimeric infective symbiotic tissue colonies which can exist in the air, water, earth, blood and tissues. The chimeric infective symbiotic tissue colonies containing bacteria, archaea and viruses can make the symbiotic organism multidrug resistant by horizontal gene transfer. The chimeric infective symbiotic tissue organism evolves as a consequence of global warming and the protozoal, bacterial, viral and archaeal symbiotic colonies are new forms of life. The chimeric infective symbiotic tissue organism composed of protozoa and its symbionts like virus, bacteria and archaea depend upon glycolysis for its energy needs. The glycolytic enzymes enolase, glyceraldehyde 3-phosphate dehydrogenase, hexokinase, phosphofructokinase, aldolase, pyruvic kinase and triose phosphate isomerase are moonlighting proteins having a cytoplasmic, membrane and nuclear functions. These glycolytic enzymes have a RNA binding motif which can modulate RNA transcription in the host and the symbiotic organism. These glycolytic enzymes with their RNA binding motif as well as DNA binding motif can bind to RNA and DNA viruses and viroids modulating their replication and life cycles. The chimeric infective symbiotic tissue organism composed of protozoa and its symbionts like virus, bacteria and archaea can enter the human host through the breaches of the gut-blood barrier and transform itself into the macrophages of the human host. The macrophages are involved in the phagocytosis and innate immunity against bacterial and viral infections. Paradoxically the macrophages like the protozoa can also serve as a reservoir where pathogens like mycobacteria tuberculosis, mycobacteria leprae, listeria, legionella, E. coli, shigella & salmonella, and viruses like HIV & covid can thrive and replicate. The modified chimeric infective symbiotic tissue organism or macrophage composed of protozoa and its symbionts like virus, bacteria and archaea can serve as a reservoir for horizontal gene transfer and generation of multidrug resistant bacteria and viruses. The infective chimeric symbiotic tissue

organism composed of protozoa and its symbionts like virus, bacteria and archaea depends upon moonlighting glycolytic enzymes for their energetics and replication. The development of antibodies against the infective chimeric symbiotic tissue organism composed of protozoa and its symbionts like virus, bacteria and archaea leads to the evolution of anti-brain antibody disease or anti-basal ganglia antibody disease. They present a threat to existence consequent to horiztontal gene transfer and generation of multidrug resistant bacteria, viruses, protozoa, mycobacteria and archaea.

A hypothesis regarding cholesterol based abiogenesis and its role in the evolution of universe is elucidated. Endomyocardial fibrosis along with the root wilt disease of coconut is endemic to Kerala with its radioactive actinide beach sands. Actinides like rutile producing intracellular magnesium deficiency due to rutile-magnesium exchange sites in the cell membrane has been implicated in the etiology of EMF[1,2]. Organisms like phytoplasmas and viroids have also been demonstrated to play a role in the etiology of these diseases[3,4]. Actinidic archaea has been related to the pathogenesis of schizophrenia, malignancy, metabolic syndrome X, autoimmune disease and neuronal degeneration. An actinide dependent shadow biosphere of archaea and viroids in the above mentioned disease states is described. Actinide based primitive organism like archaea have a mevalonate pathway and cholesterol catabolism[5-7]. Davies has put forward the concept of a shadow biosphere of organisms with alternate biochemistry present in earth itself[8]. This points to cholesterol as the primal prebiotic molecule and evolution of actinidic archaea and viroids from a primitive isoprenoid organism.

Metal actinides in beach sands have been postulated to play a role in abiogenesis[6]. Actinide mineral like rutile, monazite and illmenite by surface metabolism would have contributed to abiogenesis[9]. A hypothesis of cholesterol as the primal prebiotic molecule synthesized on actinide surfaces with all other biomolecules arising from it and a self-replicating cholesterol lipid organism as the initial life form is presented. The role of actinidic archaea in the genesis of the interstellar polycyclic aromatic hydrocarbons as well as the interstellar magnetic fields important in the evolution of the universe is hypothesized. The role of actinidic archaea in global warming and evolutionary cycles is discussed.

Materials and Methods

Informed consent of the subjects and the approval of the Ethics Committee were obtained for the study. The following groups were included in the study:- endomyocardial fibrosis, Alzheimer's disease, multiple sclerosis, non-Hodgkin's lymphoma, metabolic syndrome X with cerebrovascular thrombosis and coronary artery disease, schizophrenia, autism, seizure disorder, Creutzfeldt-Jakob's disease and acquired immunodeficiency syndrome. There were 10 patients in each group and each patient had an age and sex matched healthy control selected randomly from the general population. The blood samples were drawn in the fasting state before treatment was initiated. Plasma from fasting heparinised blood was used and the experimental protocol was as follows:- (I) Plasma+phosphate buffered saline, (II) same as I+cholesterol substrate, (III) same as II+rutile 0.1 mg/ml, and (IV) same as II+ciprofloxacine and doxycycline each in a concentration of 1 mg/ml. Cholesterol substrate was prepared as described by Richmond[10]. Aliquots were withdrawn at zero time immediately after mixing and after incubation at 37 °C for 1 hour. The following estimations were carried out:- Cytochrome F420, free RNA, free DNA, muramic acid, polycyclic aromatic hydrocarbon, hydrogen peroxide, serotonin, pyruvate, ammonia, glutamate, cytochrome C, hexokinase, ATP synthase, HMG CoA reductase, digoxin and bile acids[11-14]. Cytochrome F420 was estimated flourimetrically (excitation wavelength 420 nm and emission wavelength 520 nm). Polycyclic aromatic hydrocarbon was estimated by measuring hydrogen peroxide liberated by using glucose reagent. The statistical analysis was done by ANOVA.

Results

The parameters checked as indicated above were:- cytochrome F420, free RNA, free DNA, muramic acid, polycyclic aromatic hydrocarbon, hydrogen peroxide, serotonin, pyruvate, ammonia, glutamate, cytochrome C, hexokinase, ATP synthase, HMG CoA reductase, digoxin and bile acids. Plasma of control subjects showed increased levels of the above mentioned parameters with after incubation for 1 hour and addition of cholesterol substrate resulted in still further significant increase in these parameters. The plasma of patients showed similar results but the extent of increase was more. The addition of antibiotics to the control plasma caused a decrease in all the parameters while addition of rutile increased their levels. The addition of antibiotics to the patient's plasma caused a decrease in all the parameters while addition of rutile increased their levels but the extent of change was more in patient's sera as compared to controls.

Discussion

There was increase in cytochrome F420 indicating archaeal growth. The archaea can synthesize and use cholesterol as a carbon and energy source[15,16]. The archaeal origin of the enzyme activities was indicated by antibiotic induced suppression. The study indicates the presence of actinide based archaea with an alternate actinide based enzymes or metalloenzymes in the system as indicated by rutile induced increase in enzyme activities[17]. There was also an increase in archaeal HMG CoA reductase activity indicating increased cholesterol synthesis by the archaeal mevalonate pathway. The archaeal beta hydroxyl steroid dehydrogenase activity indicating digoxin synthesis and archaeal cholesterol hydroxylase activity indicating bile acid synthesis were increased[7]. The archaeal cholesterol oxidase activity was increased resulting in generation of pyruvate and hydrogen peroxide[16]. The pyruvate gets converted to glutamate and ammonia by the GABA shunt pathway. The archaeal aromatization of cholesterol generating PAH, serotonin and dopamine was also detected[18]. The archaeal glycolytic hexokinase activity and archaeal extracellular ATP synthase activity were increased. The archaea can undergo magnetite and calcium carbonate mineralization and can exist as calcified nanoforms[19].

There was an increase in free RNA indicating self-replicating RNA viroids and free DNA indicating generation of viroid complementary DNA strands by archaeal reverse transcriptase activity. The actinides modulate RNA folding and catalyse its ribozymal action. Digoxin can cut and paste the viroidal strands by modulating RNA splicing generating RNA viroidal diversity. The viroids are evolutionarily escaped archaeal group I introns which have retrotransposition and self-splicing qualities[20]. Archaeal pyruvate can produce histone deacetylase inhibition resulting in endogenous retroviral (HERV) reverse transcriptase and integrase expression. This can integrate the RNA viroidal complementary DNA into the non-coding region of eukaryotic non-coding DNA using HERV integrase as has been described for borna and ebola viruses[21]. The non-coding DNA is lengthened by integrating RNA viroidal complementary DNA with the integration going on as a continuing event. The archaea genome can also get integrated into human genome using integrase as has been described for trypanosomes[22]. The integrated viroids and archaea can undergo vertical transmission and can exist as genomic parasites[21,22]. This increases the length and alters the grammar of the non-coding region producing memes or memory of acquired characters[23]. The viroidal complementary DNA can function as jumping genes producing a dynamic genome.

The presence of muramic acid, HMG CoA reductase and cholesterol oxidase activity inhibited by antibiotics indicates the presence of bacteria with mevalonate pathway. The bacterial with mevalonate pathway include streptococcus, staphylococcus, actinomycetes, listeria, coxiella and borrelia[24]. The bacteria and archaea with mevalonate pathway and cholesterol catabolism had a evolutionarily advantage and constitutes the isoprenoidal clade organism with the archaea evolving into mevalonate pathway gram positive and gram negative organism through horizontal gene transfer of viroidal and virus genes[25]. The isoprenoidal clade prokaryotes develop into other groups of prokaryotes via viroidal/virus as well as eukaryotic horizontal gene transfer producing bacterial speciation[26]. The RNA viroids and its complementary DNA developed into cholesterol enveloped RNA and DNA viruses like herpes, retrovirus, influenza virus, borna virus, cytomegalo virus and ebstein barr virus by recombining with eukaryotic and human genes resulting in viral speciation. Bacterial and viral species are ill-defined and fuzzy with all of them forming one common genetic pool with frequent horizontal gene transfer and recombination. Thus the multi and unicellular eukaryote with its genes serves the purpose of prokaryotic and viral speciation. The multicellular eukaryote developed so that their endosymbiotic archaeal colonies could survive and forage better. The multicellular eukaryotes are like bacterial biofilms. The archaea and bacteria with a mevalonate pathway uses the extracellular RNA viroids and DNA viroids for quorum sensing and in the generation of symbiotic biofilm like structures which develop into multicellular eukaryotes[27,28]. The endosymbiotic archaea and bacteria with mevalonate pathway still uses the RNA viroids and DNA viroids for the regulation of multicellular eukaryote.

Pollution is induced by the primitive nanoarchaea and mevalonate pathway bacteria synthesized PAH and methane leading on to redox stress. Redox stress leads to sodium-potassium ATPase inhibition, inward movement of plasma membrane cholesterol, defective SREBP sensing, increased cholesterol synthesis and nanoarchaeal/mevalonate pathway bacterial growth[29]. Redox stress leads on to viroidal and archaeal multiplication. Redox stress can also lead to HERV reverse transcriptase and integrase expression. The non-coding DNA is formed of integrating RNA viroidal complementary DNA and archaea with the integration going on as a continuing event. The archaeal pox like dsDNA virus forms evolutionarily the nucleus. The integrated viroidal, archaeal and mevalonate pathway bacterial sequences can undergo vertical transmission and can exist as genomic parasites. The genomic integrated archaea, mevalonate pathway bacteria and viroids form a genomic reserve of bacteria and viruses which can recombine with human and eukaryotic genes producing bacterial and viral speciation. The change in the length and grammar of the non-coding region produces eukaryotic speciation and individuality[30]. The integration of nanoarchaea, mevalonate pathway prokaryotes and viroids in to the eukaryotic and human genome produces a chimera which can multiply producing biofilm like multicellular structures having a mixed archaeal, viroidal, prokaryotic and eukaryotic characters which is a regression from the multicellular eukaryotic tissue. This results in a new neuronal, metabolic, immune and tissue phenotype leading to human disease.[31-34]

The metal actinides provide radiolytic energy, catalysis for oligomer formation and provide a coordinating ion for metalloenzymes all important in abiogenesis[6]. The metal actinide surfaces would by surface metabolism generate acetate which could get converted to acetyl CoA and then to cholesterol which functions as the primal prebiotic molecule self-organizing into self-replicating supramolecular systems, the lipid organism[8,9,35]. Cholesterol by radiolysis by actinides would have formed PAH generating PAH aromatic organism[8]. Cholesterol radiolysis would generate pyruvate which would get converted to amino acids, sugars, nucleotides, porphyrins, fatty acids and TCA acids. Anastase and rutile surfaces can produce polymerization of amino acids, isoprenyl residues, PAH and nucleotides to generate the initial lipid organism, PAH organism, prions and RNA viroids which would have symbiosed to generate the archaeal protocell. The archaea evolved into gram negative and gram positive bacteria with a mevalonate pathway which had an evolutionary advantage and the symbiosis of archaea with gram negative organism generated the eukaryotic cell[36]. The data supports the persistence of an actinide and cholesterol based shadow biosphere which throws light on the actinide based origin of life and cholesterol as the premier prebiotic molecule.

The archaea can synthesize magnetite by biomineralisation. The archaeal cholesterol catabolism can generate PAH. The archaea can exist as nanoarchaea and can have calcified nanoforms. The actinidic magnetotactic nanoarchaea and its secreted PAH organisms are extremophiles and survive in the interstellar space and can contribute to the interstellar grains and magnetic fields which play a role in the formation of the galaxies and star systems[37]. The cosmic dust grains occupy the intergalactic space and are thought to be formed of magnetotactic bacteria identified according to their spectral signatures. According to the Hoyle's hypothesis, the cosmic dust magnetotactic bacteria play a role in the formation of the intergalactic magnetic field. A magnetic field equal in strength to about one millionth part of the magnetic field of earth exists throughout much of our galaxy. The magnetic files can be used to trace the spiral arms of the galaxy following a pattern of field lines that connect young stars and dust in which new stars are formed at a rapid rate. Studies have shown that a fraction of the dust particles have elongated shape similar to bacilli and they are systematically lined up in our galaxy. Moreover, the direction of alignment is such that the long axes of the dust tend to be at right angles to the direction of the galactic magnetic field at every point. Magnetotactic bacteria have the property to affect the degree of alignment that is observed. The fact that the magnetotactic bacteria appear to be connected to the magnetic field lines that thread through the spiral arms of the galaxy connecting one region of star formation to another support a role for them in star formation and in the mass distribution and rotation of stars. The nutrient supply for a population of interstellar bacteria comes from mass flows out of supernovas populating the galaxy. Giants arising in the evolution of such stars experience a phenomenon in which material containing nitrogen, carbon monoxide, hydrogen, helium, water and trace elements essential for life flows continuously outward into space. The interstellar bacteria need liquid water. Water exists only as vapour or solid in the interstellar space and only through star formation leading to associated planets and cometary bodies can there be access to liquid water. To control conditions leading to star formation is of paramount importance in cosmic biology. The rate of star formation is controlled by two factors. Too high a rate of star formation produces a destructive effect of UV radiation and destroys cosmic biology. Star formation as stated before produces water crucial for bacterial growth. Cosmic biology of magnetotactic bacteria and star formation are thus closely interlinked. Systems like solar systems do not arise in random condensation of blobs of interstellar gas. Only by a rigorous control of rotation of various parts of the system would galaxies and solar system evolved. The key to maintaining control over rotation seems to lie in the intergalactic magnetic field as indeed the whole phenomena of star formation. The intergalactic magnetic fields owes its origin to the lining up of magnetotactic bacteria and the cosmic biology of interstellar bacteria can prosper only by maintaining a firm grip on the interstellar magnetic field and hence on the rate of star formation and type of star system produced. This points to a cosmic intelligence or brain capable of computation, analysis and exploration of the universe at large - of magnetotactic bacterial networks. The interstellar PAH aromatic organism is formed from nanoarchaeal cholesterol catabolism. The PAH and cholesterol are the interconvertible primal prebiotic molecules. PAH aromatic organism and nanoarchaeal magnetite can have a wave-particle existence and bridge the world of bosons and fermions. The nanoarchaea can form biofilms and the PAH aromatic organism can form a molecular quantum computing cloud in the biofilm which forms an interstellar intelligence regulating the formation of star systems and galaxies. The magnetite loaded nanoarchaeal biofilms and PAH aromatic organism quantal computing cloud can bridge the wave-particle world functioning as the anthropic observer sensing gravity which orchestrates the reduction of the quantal world of possibilities in to the macroscopic world.

The origin of life on earth according to the Hoyle's hypothesis would be by seeding of bacteria from the outer intergalactic space. Comets carrying micro-organisms would have interacted with the earth. A thin skin of graphitized material around a single bacteria or clumps of bacteria can shield the interior from destruction by UV light. The sudden surge and diversification of species of plants and animals and their equally sudden extinction has seen from fossil records point to sporadic evolution produced by induction of fresh cometary genes with the arrival of each major new crop of comets[38,39].

The actinide based nanoarchaea can regulate the earth's carbon cycle by methanogenesis, nitrogen cycle by ammonia oxidation and rain formation by contributing the seeding nucleus. The earth's temperature and global warming and cooling are regulated by nanoarchaeal synthesized PAH from cholesterol and methanogenesis. The archaeal synthesis of PAH and methane may be the principal contribution to global warming. Global warming and pollution are pivotal inducers of evolutionary innovation.

Catastrophic evolutionary cycles may be related to extensive nanoarchaeal growth in the ocean beds. The increased nanoarchaeal growth in ocean beds and soil leads to increased methane production and movement of the earth's crust producing tsunamis and massive earthquake leading to catastrophic mass extinction[40]. The eternal nanoarchaea survive and start the cycle of evolution once more. The actinide based nanoarchaea regulates the human system and biological universe.

References

1. Valiathan M.S., Somers, K., Kartha, C.C. (1993). *Endomyocardial Fibrosis.* Delhi: Oxford University Press.

2. Kurup R., Kurup, P.A. (2009). *Hypothalamic Digoxin, Cerebral Dominance and Brain Function in Health and Diseases*. New York: Nova Science Publishers.

3. Hanold D., Randies, J.W. (1991). Coconut cadang-cadang disease and its viroid agent, *Plant Disease,* 75, 330-335.

4. Edwin B.T., Mohankumaran, C. (2007). Kerala wilt disease phytoplasma: Phylogenetic analysis and identification of a vector, *Proutista moesta, Physiological and Molecular Plant Pathology,* 71(1-3), 41-47.

5. Eckburg P.B., Lepp, P.W., Relman, D.A. (2003). Archaea and their potential role in human disease, *Infect Immun,* 71, 591-596.

6. Adam Z. (2007). Actinides and Life's Origins, *Astrobiology,* 7, 6-10.

7. Schoner W. (2002). Endogenous cardiac glycosides, a new class of steroid hormones, *Eur J Biochem,* 269, 2440-2448.

8. Davies P.C.W., Benner, S.A., Cleland, C.E., Lineweaver, C.H., McKay, C.P., Wolfe-Simon, F. (2009). Signatures of a Shadow Biosphere, *Astrobiology,* 10, 241-249.

9. Wächtershäuser G. (1988). Before enzymes and templates: theory of surface metabolism, *Microbiol Rev,* 52(4), 452-84.

10. Richmond W. (1973). Preparation and properties of a cholesterol oxidase from nocardia species and its application to the enzymatic assay of total cholesterol in serum, *Clin Chem,* 19, 1350-1356.

11. Snell E.D., Snell, C.T. (1961). *Colorimetric Methods of Analysis.* Vol 3A. New York: Van NoStrand.

12. Glick D. (1971). *Methods of Biochemical Analysis.* Vol 5. New York: Interscience Publishers.

13. Colowick, Kaplan, N.O. (1955). *Methods in Enzymology.* Vol 2. New York: Academic Press.

14. Maarten A.H., Marie-Jose, M., Cornelia, G., van Helden–Meewsen, Fritz, E., Marten, P.H. (1995). Detection of muramic acid in human spleen, *Infection and Immunity,* 63(5), 1652 – 1657.

15. Smit A., Mushegian, A. (2000). Biosynthesis of isoprenoids via mevalonate in Archaea: the lost pathway, *Genome Res,* 10(10), 1468-84.

16. Van der Geize R., Yam, K., Heuser, T., Wilbrink, M.H., Hara, H., Anderton, M.C. (2007). A gene cluster encoding cholesterol catabolism in a soil actinomycete provides insight into Mycobacterium tuberculosis survival in macrophages, *Proc Natl Acad Sci USA,* 104(6), 1947-52.

17. Francis A.J. (1998). Biotransformation of uranium and other actinides in radioactive wastes, *Journal of Alloys and Compounds,* 271(273), 78-84.

18. Probian C., Wülfing, A., Harder, J. (2003). Anaerobic mineralization of quaternary carbon atoms: Isolation of denitrifying bacteria on pivalic acid (2,2-Dimethylpropionic acid), *Applied and Environmental Microbiology,* 69(3), 1866-1870.

19. Vainshtein M., Suzina, N., Kudryashova, E., Ariskina, E. (2002). New Magnet-Sensitive Structures in Bacterial and Archaeal Cells, *Biol Cell,* 94(1), 29-35.

20. Tsagris E.M., de Alba, A.E., Gozmanova, M., Kalantidis, K. (2008). Viroids, *Cell Microbiol,* 10, 2168.

21. Horie M., Honda, T., Suzuki, Y., Kobayashi, Y., Daito, T., Oshida, T. (2010). Endogenous non-retroviral RNA virus elements in mammalian genomes, *Nature,* 463, 84-87.

22. Hecht M., Nitz, N., Araujo, P., Sousa, A., Rosa, A., Gomes, D. (2010). Genes from Chagas parasite can transfer to humans and be passed on to children. Inheritance of DNA Transferred from American Trypanosomes to Human Hosts, *PLoS ONE,* 5, 2-10.

23. Flam F. (1994). Hints of a language in junk DNA, *Science,* 266, 1320.

24. Horbach S., Sahm, H., Welle, R. (1993). Isoprenoid biosynthesis in bacteria: two different pathways? *FEMS Microbiol Lett,* 111, 135–140.

25. Gupta R.S. (1998). Protein phylogenetics and signature sequences: a reappraisal of evolutionary relationship among archaebacteria, eubacteria, and eukaryotes, *Microbiol Mol Biol Rev,* 62, 1435–1491.

26. Hanage W., Fraser, C., Spratt, B. (2005). Fuzzy species among recombinogenic bacteria, *BMC Biology,* 3, 6-10.

27. Whitchurch C.B., Tolker-Nielsen, T., Ragas, P.C., Mattick, J.S. (2002). Extracellular DNA Required for Bacterial Biofilm Formation. *Science,* 295(5559), 1487.

28. Webb J.S., Givskov, M., Kjelleberg, S. (2003). Bacterial biofilms: prokaryotic adventures in multicellularity, *Curr Opin Microbiol,* 6(6), 578–85.

29. Chen Y., Cai, T., Wang, H., Li, Z., Loreaux, E., Lingrel, J.B. (2009). Regulation of intracellular cholesterol distribution by Na/K-ATPase, *J Biol Chem,* 284(22), 14881-90.

30. Poole A.M. (2006). Did group II intron proliferation in an endosymbiont-bearing archaeon create eukaryotes? *Biol Direct,* 1, 36-40.

31. Lockwood M. (1989). *Mind, Brain and the Quantum*. Oxford: B. Blackwell.

32. Eberl M., Hintz, M., Reichenberg, A., Kollas, A., Wiesner, J., Jomaa, H. (2010). Microbial isoprenoid biosynthesis and human γδ T cell activation, *FEBS Letters,* 544(1), 4-10.

33. Wallace D.C. (2005). Mitochondria and Cancer: Warburg Addressed, *Cold Spring Harbor Symposia on Quantitative Biology,* 70, 363-374.

34. Lefebvre P., Cariou, B., Lien, F., Kuipers, F., Staels, B. (2009). Role of Bile Acids and Bile Acid Receptors in Metabolic Regulation, *Physiol Rev,* 89(1), 147 - 191.

35. Russell M.J., Martin, W. (2004). The rocky roots of the acetyl-CoA Pathway, *Trends in Biochemical Sciences,* 29, 7.

36. Margulis L. (1996). Archaeal-eubacterial mergers in the origin of Eukarya: phylogenetic classification of life, *Proc Natl Acad Sci USA,* 93, 1071-1076.

37. Tielens A.G.G.M. (2008). Interstellar Polycyclic Aromatic Hydrocarbon Molecules, *Annual Review of Astronomy and Astrophysics,* 46, 289-337.

38. Wickramasinghe C. (2004). The universe: a cryogenic habitat for microbial life, *Cryobiology,* 48(2), 113-125.

39. Hoyle F., Wickramasinghe, C. (1988). *Cosmic Life-Force.* London: J.M. Dent and Sons Ltd.

40. Dun D. (2005). *The Black Silent*. New York: Pinnacle Books.

CHAPTER 36

A NEW HUMAN SPECIES - CLIMATE CHANGE, GLOBAL WARMING AND NEONEANDERTHALS - THE AGE OF HOMO NEONEANDERTHALIS - INFECTIVE SYMBIOTIC TISSUE COLONIES OF PROTOZOA, ARCHAEA, VIRUSES AND BACTERIA

Introduction

The plasma life based forms and the gravitational and anti-gravitational waves generated porphyrions would have been the most primitive of non-life forms. The porphyrin templates would have given rise to the isoprenoid organism, the RNA viroids, DNA viroids and prions which would have symbiosed to form the nanoarchaea. The archaea would have evolved into prokaryotes or bacteria. The archaeal symbiosis with prokaryotes would have generated the eukaryotes. The initial unicellular eukaryotes are the protozoa. The protozoa are capable of phagocytosis and would have had viruses, bacteria and archaea as their intracellular symbionts. The viruses of the protozoa like entamoeba and toxoplasma are giant viruses called megaviride. The protozoans like entamoeba with its viral, archaeal and bacterial endosymbionts would have evolved into the multicellular organisms or metazoa. The archaeal, bacterial and protozoal extracellular vesicles containing RNA and DNA viroids can by horizontal gene transfer lead to evolution of metazoan, plants, animals and human species. The metazoans would have evolved into plants, animals and humans by horizontal gene transfer generated by archaeal extracellular RNA and DNA viroid vesicles. The evidence of such a transformation is seen in the bacterial symbionts of protozoa like entamoeba which includes legionella, listeria, mycoplasma, mycobacteria and chlamydia. The symbionts of protozoan entamoeba include viruses like covid, HIV and cytomegalo virus. The protozoan entamoeba serves as a natural reservoir for covid infections, bacterial infections and HIV infections. The toxoplasma also serves as a covid reservoir. These

protozoan colonies of entamoeba and toxoplasma with its bacterial, archaeal and viral symbionts can be called as chimeric infective symbiotic tissue colonies which can exist in the air, water, earth, blood and tissues. The chimeric infective symbiotic tissue colonies containing bacteria, archaea and viruses can make the symbiotic organism multidrug resistant by horizontal gene transfer. The chimeric infective symbiotic tissue organism evolves as a consequence of global warming and the protozoal, bacterial, viral and archaeal symbiotic colonies are new forms of life. The chimeric infective symbiotic tissue organism composed of protozoa and its symbionts like virus, bacteria and archaea depend upon glycolysis for its energy needs. The glycolytic enzymes enolase, glyceraldehyde 3-phosphate dehydrogenase, hexokinase, phosphofructokinase, aldolase, pyruvic kinase and triose phosphate isomerase are moonlighting proteins having a cytoplasmic, membrane and nuclear functions. These glycolytic enzymes have a RNA binding motif which can modulate RNA transcription in the host and the symbiotic organism. These glycolytic enzymes with their RNA binding motif as well as DNA binding motif can bind to RNA and DNA viruses and viroids modulating their replication and life cycles. The chimeric infective symbiotic tissue organism composed of protozoa and its symbionts like virus, bacteria and archaea can enter the human host through the breaches of the gut-blood barrier and transform itself into the macrophages of the human host. The macrophages are involved in the phagocytosis and innate immunity against bacterial and viral infections. Paradoxically the macrophages like the protozoa can also serve as a reservoir where pathogens like mycobacteria tuberculosis, mycobacteria leprae, listeria, legionella, E.coli, shigella & salmonella, and viruses like HIV & covid can thrive and replicate. The modified chimeric infective symbiotic tissue organism or macrophage composed of protozoa and its symbionts like virus, bacteria and archaea can serve as a reservoir for horizontal gene transfer and generation of multidrug resistant bacteria and viruses. The infective chimeric symbiotic tissue organism composed of protozoa and its symbionts like virus, bacteria and archaea

depends upon moonlighting glycolytic enzymes for their energetics and replication. The development of antibodies against the infective chimeric symbiotic tissue organism composed of protozoa and its symbionts like virus, bacteria and archaea leads to the evolution of anti-brain antibody disease or anti-basal ganglia antibody disease. They present a threat to existence consequent to horiztontal gene transfer and generation of multidrug resistant bacteria, viruses, protozoa, mycobacteria and archaea.

Actinidic archaea has been related to global warming and human diseases especially autoimmune disease, neurodegeneration, neuropsychiatric disorder, neoplasm and metabolic syndrome X. The growth of endosymbiotic actinidic archaea in relation to climate change and global warming leads to neanderthalisation of the human mind-body system. Neanderthal anthropometry and metabolonomics has been described in autoimmune disease, neurodegeneration, neuropsychiatric disorder, neoplasm and metabolic syndrome X, especially the Warburg phenotype and hyperdigoxinemia. Digoxin produced by archaeal cholesterol catabolism produces neanderthalisation. Prefrontal cortical atrophy and cerebellar hyperplasia has been related to autoimmune disease, neurodegeneration, neuropsychiatric disorder, neoplasm and metabolic syndrome X in this communication. This leads on to dysautonomia with sympathetic hyperactivity and parasympathetic neuropathy in these disorders. Actinidic archaeal related cerebellar dominance leads to changes in brain function.[1-16]

Materials and Methods

Fifteen cases, each of autoimmune disease, neurodegeneration, neuropsychiatric disorder, neoplasm, metabolic syndrome X and internet addicts were selected for the study. Each case had an age and sex matched control. Neanderthal anthropometric and phenotypic measurements which included protruding supra-orbital ridges, dolichocephalic skull, small mandible, prominent mid face and nose, short upper and lower limbs, prominent trunk, low index finger-ring finger ratio and fair complexion were evaluated in the cases study. Autonomic function tests were done to assess the sympathetic and parasympathetic system in each case. CT scan of the head was done to have a volumetric assessment of the prefrontal cortex and cerebellum. Blood cytochrome F420 activity was assessed by spectrophotometric measurement.

Results

All the case groups studied had higher percentage of Neanderthal anthropometric and phenotypic measurements. There was low index finger-ring finger ratio suggestive of high testosterone levels in all the patient population studied. In all the case groups studied, there also was prefrontal cortex atrophy and cerebellar hyperplasia. Similarly in the all the case groups studied, there was dysautonomia with sympathetic overactivity and parasympathetic neuropathy. Cytochrome F420 was detected in the entire case group studied showing endosymbiotic archaeal overgrowth.

Discussion

Global warming, neanderthalisation and mind-body system

Neanderthal metabolonomics contribute to the pathogenesis of these disorders. There were Neanderthal phenotypic features in all the case groups studied as well as low index finger-ring finger ratios suggestive of increased testosterone levels. Neanderthalisation of the mind-body system occurs due to increased growth of actinidic archaea as a consequence of global warming. Neanderthalisation of the mind leads to cerebellar dominance and prefrontal cortex atrophy. This leads to dysautonomia with parasympathetic neuropathy and sympathetic hyperactivity.

Global warming and the ice age produces increased growth of extremophiles. This leads to increased growth of actinidic archaeal endosymbiosis in humans. There is archaeal proliferation in the gut which enters the cerebellum and brain stem by reverse axonal transport via the vagus. The cerebellum and brain stem can be considered as an archaeal colony. The archaea are cholesterol catabolising and use cholesterol as a carbon and energy source. The actinidic archaea activates the toll receptor HIF alpha inducing the Warburg phenotype resulting in increased glycolysis with generation of glycine as well as pyruvate dehydrogenase suppression. The accumulated pyruvate enters the GABA shunt generating of succinyl CoA and glycine. The archaeal catabolism of cholesterol produces ring oxidation and generation of pyruvate which also enters the GABA shunt scheme producing glycine and succinyl CoA. This leads to increased synthesis of porphyrins. In the setting of digoxin induced sodium-potassium ATPase inhibition the dipolar porphyrins produce a pumped phonon system resulting in the Frohlich model Bose-Einstein's condensate and quantal perception of low level EMF. Low level EMF pollution is common with internet usage. Perception of low level of EMF leads to neanderthalisation of the brain with prefrontal cortex atrophy and cerebellar hyperplasia. The archaea which reaches the cerebellum from the gut via the vagus nerve proliferates and makes the cerebellum dominant with resultant suppression and atrophy of the prefrontal cortex. This leads to widespread autistic and schizophrenic traits in population. The actinidic archaea induces the Warburg phenotype with increased glycolysis, PDH inhibition and mitochondrial suppression. This produces neanderthalisation of the mind-body system. The actinidic archaea secretes RNA viroids which block HERV expression by RNA interference. The HERV suppression contributes to the inhibition of prefrontal cortex development in Neanderthals and cerebellar dominance. Archaeal digoxin produces sodium-potassium ATPase inhibition and magnesium depletion causing reverse transcriptase inhibition and decreased generation of HERV. The HERV contributes to the dynamicity of the genome and are required for the development of the prefrontal cortex. The HERV suppression contributes to retroviral resistance in Neanderthals. The actinidic archaea catabolises cholesterol leading to cholesterol depleted state. Cholesterol depletion also leads to poor synaptic connectivity and decreased development of prefrontal cortex. This is not genetic change but a form of symbiotic change with endosymbiotic actinidic archaeal growth in the body and brain.

Endosymbiotic archaea, internet use and brain function- neanderthalisation

Internet use and low level EMF pollution is common in this century. This results in increased low level EMF perception by the brain by the digoxin-porphyrin mediated pumped phonon system created Bose-Einstein's condensate contributing to prefrontal cortex atrophy and cerebellar dominance. Cerebellar dominance leads to schizophrenia and autism. There is an epidemic of autism and schizophrenia in the present day community. The porphyrin mediated extrasensory perception can contribute to communication among Neanderthals. Neanderthals did not have a language and used extrasensory perception as a form of group communication. Because of dominant extrasensory quantal perception, the Neanderthals did not have individual identity but only group identity. Cerebellar dominance results in creativity consequent to quantal perception and group perception. The neanderthalic traits contribute to innovation and creativity. Cerebellar dominance results in development of a symbolic language. The Neanderthals used dance and music as a form of communication. Painting as a form of communication was also common in Neanderthals. Neanderthal behaviour was robotic. Robotic behaviour is characteristic of cerebellar dominance. Robotic, symbolic and ritualistic behaviour is common with cerebellar dominance and is seen in autistic traits. The cerebellar dominance in Neanderthals leads to intuitive intelligence and a hypnotic quality to communication. The increased extrasensory quantal perception leads to more communion with nature and a form of eco-spirituality. The increasing use of dance and music as a form of communication and eco-spirituality is common in the modern century along with increased incidence of autism. The cholesterol depletion leads to bile acid deficiency and generation of small social groups in Neanderthals. Bile acid binds to olfactory receptors and contributes to group identity. This can also contributes to the generation of autistic features in Neanderthals.

Global warming, neanderthalisation and social behaviour

The Neanderthal population was predominantly autistic and schizophrenic. The modern population is a hybrid of Homo sapiens and Homo neanderthalis. This contributes to 10 to 20 percent dominant hybrids who tend to have schizophrenic and autistic qualities and contributes to creativity of civilization. The Neanderthals tend to be innovative and chaotic. They tend to be creative in art, literature, dance, spirituality and science. Eighty percent of less dominant hybrids are stable and contribute to a stabilizing influence leading to growth of civilization. The homo sapiens were stable and non-creative over a long period of their existence. There was a burst of creativity with generation of music, dance, painting, ornaments, the creation of concept of God and compassionate group behaviour around 10,000 years ago in the homo sapien community. This correlated with the generation of Neanderthal hybrids when the Eurasian Neanderthal male mated with homo sapien African females. The extrasensory/quantal perception due to dipolar porphyrins and digoxin induced sodium-potassium ATPase inhibition and the generated pumped phonon system mediated quantal perception leads to the globalisation phenomena and feeling of the world being a global village. The archaeal cholesterol catabolism leads to increased synthesis of digoxin. Digoxin promotes tryptophan transport over tyrosine. Tyrosine deficiency leads to dopamine deficiency and morphine deficiency. This leads to a morphine deficiency syndrome in Neanderthals. This contributes to addiction traits and creativity. The increased tryptophan levels produce increased alkaloids like LSD contributing to ecstasy and spirituality of Neanderthal population. Addictive, ADHD and autistic features are related to the morphine deficiency state. The ketogenic diet consumed by the meat eating Neanderthals leads on to increased generation of hydroxy butyric acid which produces ecstasy and a dissociative type of anaesthesia contributing to the Neanderthal psychology. The dopamine deficiency leads to decreased melanin synthesis and fairness of the population. This was responsible for the fair colour of the Neanderthals.

Global warming, neanderthalisation and Warburg phenotype

The Neanderthals were essentially meat eaters taking a ketogenic diet. The acetoacetic acid is converted to acetyl CoA which enters the TCA cycle. When the Neanderthal hybrids consume a glucogenic diet owing to the spread of settled civilization it produces pyruvate accumulation owing to PDH suppression in Neanderthals. The increased archaeal growth activates the toll receptor and induces HIF alpha resulting in increased glycolysis, PDH suppression and mitochondrial dysfunction - the Warburg phenotype. The pyruvate enters the GABA shunt pathway producing glutamate, ammonia and porphyrins resulting in neuropathology of autism and schizophrenia. Neanderthals consuming a ketogenic diet produces more of GABA an inhibitory neurotransmitter resulting in the docile quiet nature of the Neanderthals. There is less production of glutamate the predominant excitatory neurotransmitter of the prefrontal cortex and consciousness pathways. This leads on to dominance of cerebellar function. The Neanderthal hybrids have cerebellar dominance and less of conscious behaviour. Cerebellum is responsible for intuitive, unconscious behaviour as well as creativity and spirituality. The cerebellum is the site of extrasensory perception, magical acts and hypnosis. The predominant homo sapiens had prefrontal cortex dominance over the cerebellum resulting in more of conscious behaviour.

The Neanderthals consuming a glucogenic diet produces increased glycolysis in the setting of PDH inhibition. This produces the Warburg phenotype. There is increased lymphocytic glycolysis producing autoimmune diseases and immune activation. The increased levels of GAPD result in nuclear cell death and neurodegeneration. The predominance of glycolysis and suppression of mitochondrial function results in glycemia and metabolic syndrome X. The increased mitochondrial PT pore hexokinase leads to cell proliferation and oncogenesis. The glycolytic intermediate 3-phosphoglycerate is converted to glycine resulting in NMDA excitotoxicity contributing to schizophrenia and autism. Cerebellar dominance is reported in schizophrenia and autism.

Neanderthalisation and autonomic dysfunction

The cerebellar hyperplasia results in sympathetic hyperactivity and parasympathetic neuropathy. This contributes to cell proliferation and oncogenesis. Vagal neuropathy results in immune activation and autoimmune disease. Vagal neuropathy and sympathetic overactivity can contribute to glycogenolysis and lipolysis resulting in metabolic syndrome X. Cerebellar dominance and cerebellar cognitive affective dysfunction can contribute to schizophrenia and autism. The increased porphyrin synthesis resulting from succinyl CoA generated by GABA shunt and glycine generated by glycolysis contributes to increased extrasensory perception important in schizophrenia and autism. Sympathetic overactivity and parasympathetic neuropathy can contribute to neurodegeneration.

Digoxin - the Neanderthal hormone

The archaeal cholesterol catabolism generates digoxin which produces sodium-potassium ATPase inhibition and increase in intracellular calcium and decrease in intracellular magnesium. The increase in intracellular calcium produces oncogene activation and NFkB activation resulting in malignancies and autoimmune diseases. The increase in intracellular calcium opens the mitochondrial PT pore resulting in cell death and neurodegeneration. The increase in intracellular calcium can modulate the neurotransmitter release from presynaptic vesicles. This can modulate neurotransmission. Digoxin induced magnesium depletion can remove the magnesium block on the NMDA receptor resulting in NMDA excitotoxicity. Digoxin can modulate the glutamatergic thalamo-cortico-thalamic pathway and consciousness resulting in schizophrenia and autism. Digoxin induced magnesium depletion can inhibit reverse transcriptase activity and HERV generation modulating the dynamicity of the genome. Digoxin induced intracellular calcium accumulation and magnesium depletion can modulate G-protein and protein tyrosine kinase dependent neurotransmitter and endocrine receptors. This can produce digoxin induced neuro-immuno-endocrine integration. Digoxin functions as a Neanderthal master hormone.

Neanderthalisation and androgyny

The actinidic archaea are cholesterol catabolising and leads to low levels of testosterone and estrogen. This leads on to asexual features and low reproductive rates of the Neanderthal population. The Neanderthals consume a low fibre diet with low lignan content. The actinidic archaea has cholesterol catabolising enzymes generating more of testosterone than estrogens. This contributes to estrogen deficiency and testosterone overactivity. The Neanderthal population are hypermales with concomitant right hemispheric dominance and cerebellar dominance. Testosterone suppresses left hemispheric function. The high testosterone levels in Neanderthals contribute to a bigger brain. The Neanderthals males as well as females had a higher level of testosterone contributing to gender equality and gender-neutral states. There was group identity and group motherhood with no differences between roles of both males and females. This also resulted in matrilinearity. The higher testosterone levels in males as well as females led to alternate type of sexuality and aberrant behaviour. The homo sapiens eat a high fibre diet with low cholesterol and high lignan content contributing to estrogen dominance, left hemispheric dominance and cerebellar hypoplasia. Homo sapiens had higher reproductive rates and overtook the Neanderthal population resulting in its extinction. The homo sapien population was conservative with normal sexual mores, family values and patriarchial type of behaviour. The role of females the homo sapien community was inferior to males. The increasing generation of Neanderthal hybrids due to climate change mediated archaeal overgrowth leads to gender equality and equidominance of male and female in this century.

Endosymbiotic archaea and cholesterol catabolism

The cholesterol catabolism results in cholesterol depletion and bile acid deficiency. Bile acids bind to VDR and are immunomodulatory. Bile acid deficiency leads to immune activation and autoimmune disease. Bile acids bind to FXR, LXR and PXR modulating lipid and carbohydrate metabolism. This leads to metabolic syndrome X in the presence of bile acid deficiency. Bile acid uncouples oxidative phosphorylation and its deficiency leads to obesity of metabolic syndrome X. Bile acids bind to olfactory receptors and are important in group identity. Bile acid deficiency leads to formation of small social groups in Neanderthals and genesis of autism. Cholesterol depletion also leads to vitamin D deficiency. Vitamin D binds to VDR and produces immunomodulation. Vitamin D deficiency leads to immune activation and autoimmune diseases. Vitamin D deficiency can also produce rickets and contribute to the phenotypic features of Neanderthals. Vitamin D deficiency can contribute to brain development resulting in macrocephaly. Vitamin D deficiency contributes to insulin resistance and truncal obesity of Neanderthals. Vitamin D deficiency contributes to the fairness of the Neanderthal skin as a phenotypic adaptation. The Neanderthal phenotypic features are due to vitamin D deficiency and insulin resistance.

Thus global warming and increased endosymbiotic actinidic archaeal growth leads to cholesterol catabolism and generation of the Warburg phenotype resulting in increased porphyrin synthesis, extrasensory low EMF perception, prefrontal cortex atrophy, insulin resistance and cerebellar dominance. This leads on to neanderthalisation of the body and brain.

References

1. Weaver TD, Hublin JJ. Neandertal Birth Canal Shape and the Evolution of Human Childbirth. *Proc. Natl. Acad. Sci. USA* 2009; 106:8151–8156.
2. Kurup RA, Kurup PA. Endosymbiotic Actinidic Archaeal Mediated Warburg Phenotype Mediates Human Disease State. *Advances in Natural Science* 2012; 5(1):81-84.
3. Morgan E. The Neanderthal theory of autism, Asperger and ADHD; 2007, www.rdos.net/eng/asperger.htm.
4. Graves P. New Models and Metaphors for the Neanderthal Debate. *Current Anthropology* 1991; 32(5): 513-541.
5. Sawyer GJ, Maley B. Neanderthal Reconstructed. *The Anatomical Record Part B: The New Anatomist* 2005; 283B(1):23-31.
6. Bastir M, O'Higgins P, Rosas A. Facial Ontogeny in Neanderthals and Modern Humans. *Proc. Biol. Sci.* 2007; 274:1125–1132.
7. Neubauer S, Gunz P, Hublin JJ. Endocranial Shape Changes during Growth in Chimpanzees and Humans: A Morphometric Analysis of Unique and Shared Aspects. *J. Hum. Evol.* 2010; 59:555–566.
8. Courchesne E, Pierce K. Brain Overgrowth in Autism during a Critical Time in Development: Implications for Frontal Pyramidal Neuron and Interneuron Development and Connectivity. *Int. J. Dev. Neurosci.* 2005; 23:153–170.
9. Green RE, Krause J, Briggs AW, Maricic T, Stenzel U, Kircher M, Patterson N, Li H, Zhai W, *et al.* A Draft Sequence of the Neandertal Genome. *Science* 2010; 328:710–722.
10. Mithen SJ. *The Singing Neanderthals: The Origins of Music, Language, Mind and Body*; 2005, ISBN 0-297-64317-7.
11. Bruner E, Manzi G, Arsuaga JL. Encephalization and Allometric Trajectories in the Genus Homo: Evidence from the Neandertal and Modern Lineages. *Proc. Natl. Acad. Sci. USA* 2003; 100:15335–15340.
12. Gooch S. *The Dream Culture of the Neanderthals: Guardians of the Ancient Wisdom.* Inner Traditions, Wildwood House, London; 2006.
13. Gooch S. *The Neanderthal Legacy: Reawakening Our Genetic and Cultural Origins.* Inner Traditions, Wildwood House, London; 2008.
14. Kurtén B. *Den Svarta Tigern*, ALBA Publishing, Stockholm, Sweden; 1978.
15. Spikins P. Autism, the Integrations of 'Difference' and the Origins of Modern Human Behaviour. *Cambridge Archaeological Journal* 2009; 19(2):179-201.
16. Eswaran V, Harpending H, Rogers AR. Genomics Refutes an Exclusively African Origin of Humans. *Journal of Human Evolution* 2005; 49(1):1-18.

CHAPTER 37

CATASTROPHIC VIRAL PANDEMICS AS A CONTINUING EVOLUTIONARY CRO-MAGNON HOMO SAPIEN–NEANDERTHAL CONFLICT - INFECTIVE SYMBIOTIC TISSUE COLONIES OF PROTOZOA, ARCHAEA, VIRUSES AND BACTERIA

Introduction

The plasma life based forms and the gravitational and anti-gravitational waves generated porphyrions would have been the most primitive of non-life forms. The porphyrin templates would have given rise to the isoprenoid organism, the RNA viroids, DNA viroids and prions which would have symbiosed to form the nanoarchaea. The archaea would have evolved into prokaryotes or bacteria. The archaeal symbiosis with prokaryotes would have generated the eukaryotes. The initial unicellular eukaryotes are the protozoa. The protozoa are capable of phagocytosis and would have had viruses, bacteria and archaea as their intracellular symbionts. The viruses of the protozoa like entamoeba and toxoplasma are giant viruses called megaviride. The protozoans like entamoeba with its viral, archaeal and bacterial endosymbionts would have evolved into the multicellular organisms or metazoa. The archaeal, bacterial and protozoal extracellular vesicles containing RNA and DNA viroids can by horizontal gene transfer lead to evolution of metazoan, plants, animals and human species. The metazoans would have evolved into plants, animals and humans by horizontal gene transfer generated by archaeal extracellular RNA and DNA viroid vesicles. The evidence of such a transformation is seen in the bacterial symbionts of protozoa like entamoeba which includes legionella, listeria, mycoplasma, mycobacteria and chlamydia. The symbionts of protozoan entamoeba include viruses like covid, HIV and cytomegalo virus. The protozoan entamoeba serves as a natural reservoir for covid infections, bacterial infections and HIV infections. The toxoplasma also serves as a covid reservoir. These

protozoan colonies of entamoeba and toxoplasma with its bacterial, archaeal and viral symbionts can be called as chimeric infective symbiotic tissue colonies which can exist in the air, water, earth, blood and tissues. The chimeric infective symbiotic tissue colonies containing bacteria, archaea and viruses can make the symbiotic organism multidrug resistant by horizontal gene transfer. The chimeric infective symbiotic tissue organism evolves as a consequence of global warming and the protozoal, bacterial, viral and archaeal symbiotic colonies are new forms of life. The chimeric infective symbiotic tissue organism composed of protozoa and its symbionts like virus, bacteria and archaea depend upon glycolysis for its energy needs. The glycolytic enzymes enolase, glyceraldehyde 3-phosphate dehydrogenase, hexokinase, phosphofructokinase, aldolase, pyruvic kinase and triose phosphate isomerase are moonlighting proteins having a cytoplasmic, membrane and nuclear functions. These glycolytic enzymes have a RNA binding motif which can modulate RNA transcription in the host and the symbiotic organism. These glycolytic enzymes with their RNA binding motif as well as DNA binding motif can bind to RNA and DNA viruses and viroids modulating their replication and life cycles. The chimeric infective symbiotic tissue organism composed of protozoa and its symbionts like virus, bacteria and archaea can enter the human host through the breaches of the gut-blood barrier and transform itself into the macrophages of the human host. The macrophages are involved in the phagocytosis and innate immunity against bacterial and viral infections. Paradoxically the macrophages like the protozoa can also serve as a reservoir where pathogens like Mycobacteria tuberculosis, Mycobacteria leprae, listeria, legionella, E. coli, shigella & salmonella and viruses like HIV & covid can thrive and replicate. The modified chimeric infective symbiotic tissue organism or macrophage composed of protozoa and its symbionts like virus, bacteria and archaea can serve as a reservoir for horizontal gene transfer and generation of multidrug resistant bacteria and viruses. The infective chimeric symbiotic tissue organism composed of protozoa and its symbionts like virus, bacteria and archaea

depends upon moonlighting glycolytic enzymes for their energetics and replication. The development of antibodies against the infective chimeric symbiotic tissue organism composed of protozoa and its symbionts like virus, bacteria and archaea leads to the evolution of anti-brain antibody disease or anti-basal ganglia antibody disease. They present a threat to existence consequent to horiztontal gene transfer and generation of multidrug resistant bacteria, viruses, protozoa, mycobacteria and archaea.

The extremes of climate change produce endosymbiotic archaeal growth. The archaea are cholesterol catabolising organism. This results in neanderthalisation of the human species. This occurred during the ice age and is possibly a continuing phenomenon during the periods of global warming. The homo neanderthalis are matrilineal and the residual matrilineal societies of the Dravidians, Semites, Basques, Celts and Berbers are neanderthalic. The global warming produces endosymbiotic archaeal growth and neanderthalisation. This produces brain changes with the cerebral cortex becoming dysfunctional and cerebellum becoming dominant. This is due to increased perception of low level EMF by archaeal magnetite. This produces changes in human society, behaviour and disease patterns.[1-17] The homo neanderthalic population is susceptible to civilizational diseases like schizophrenia, autism, metabolic syndrome X, diabetes mellitus, coronary artery disease, stroke, neurodegeneration and autoimmune disease. The homo sapien population suffers from endemic depression and lack of creativity with a propensity to develop recurrent infections. The homo sapien population is relatively protected from civilizational diseases. The Neanderthals consequent to archaeal symbiosis can generate new viruses serving as viral reservoirs. The archaeal endosymbiosis in Neanderthals produces stem cell transformation protecting the stem cell reservoirs of viruses from immune attack. These newly generated viruses in homo neanderthalis to which they are immune can attack and exterminate the homo sapien population.

There is an eternal conflict between Neanderthals and Cro-magnon Homo sapiens. The Cro-magnon tried to exterminate the Neanderthals but they survived as the Jews, Arabs, the lower caste Indians, Aboriginals and Native Americans. These are the people which the Cro-magnon excluded from society. The underclass of Indian and European civilization was neanderthalic. With the advent of global warming and increasing archaeal symbiosis the neanderthalic population becomes activated and they try to exterminate the Cro-magnon section of population. The symbiotic archaea generates new viruses which infects the non immune Cro-magnon and tries to exterminate them. The hot spots of global conflict and terrorism can be localized to neanderthalic areas. The Neanderthals dominate three world religions- Jews, Muslims and Hindus. The Cro-magnons are predominantly the Africans and the Europeans. They follow the Christian religion. World conflicts are basically between the neanderthalic races and the Cro-magnon races.

There is a high incidence of autism and Neanderthal anthropometric phenotypes in the Nair community of Kerala. The nair community is matrilineal and is one of the few functional matriarches in the world and speaks the Dravidian language with similarities to Celtic, Scythian, Berber and Basque societies. The autistic brain is comparable to the large sized Neanderthal brain. Autistic and matrilineal societies like Nair can be considered as fossilized remnants of the Neanderthal population. Endosymbiotic actinidic archaea using cholesterol as an energy substrate has been described in systemic disease from our laboratory. The autistic and Nair population were studied for actinide dependent cytochrome F420 activity suggestive of endosymbiotic archaeal growth.[1-17] This hypothesis was studied by evaluating the endosymbiotic archaeal growth in populations derived from matrilineal societies.

Materials and Methods

Three groups, 25 numbers in each group were chosen for the study - the autistic population diagnosed according to DSM criteria, the normal Nair population and the normal non-Nair population. The matrilineal characteristics and Neanderthal anthropometric characteristics of normal Nair and non-Nair population as well as autistic population were studied. The blood samples were drawn in the fasting state before treatment was initiated. The estimations done in the blood samples collected include cytochrome F420 activity, Cytochrome F420 was estimated flourimetrically (excitation wavelength 420 nm and emission wavelength 520 nm). The statistical analysis was done by ANOVA.

Results

The results of the study were as follows. The Nair and autistic and civilisational disease group had increased cytochrome F420 activity.

Discussion

Neanderthalisation is a symbiotic event due to archaeal symbiosis. The Neanderthals had increased symbiotic actinidic archaeal growth. This occurs in extremes of climate like Ice Age and global warming. The homo neanderthalis evolved from the bonobo primates consequent to this symbiosis. There is increased neanderthalisation of homo sapiens during global warming consequent to increased actinidic archaeal growth. The homo neanderthalis never became extinct but survives as matrilineal societies in the lower Eurasian region. The initial matrilineal neanderthalic civilizations were the Harappan, Sumerian - Akkadian, Assyrian, Etruscan, Minoan, Celtic, Basque, Semitic, Jewish, Arabic, Australian Aboriginal civilization. The civilizations are all matrilineal. The initial neanderthalic civilization survives as the lower caste Shudras of India, Dravidians, Australian Aboriginals, the Persians, the Semitic Arabs, the Semitic Jews, the Berbers, the Basque, Greeks, Celts and Native Americans. The people inhabiting these civilizations are religious, intuitive, feminine, child-like, dreamy, somnolent, communal conscious, primitive socialistic, more sexual groups. The body habitus of these populations are shorter, sloping forehead, recessive chin and more fairer in colour. This is opposed to the Cro-magnon population in the northern part of Eurasia and Africa. These populations are scientific, logical minded, patriarchal, more adult-like, more wakeful, fascist and less sexual. The neanderthalic populations inhabit the Indian ocean rim in Southern Asia, West Asia as well as in the Perimediterranean region. The Neanderthals originated initially from the mythical Lemurian supercontinent in the Indian ocean. The earthquakes and tsunamis in the Indian ocean led to the breakage of the supercontinent and migration of Neanderthals to Harappa, Sumeria, Egypt and Basque. The Harappan civilization was predominantly neanderthalic. They are the Asuras described in the Rig Veda. Most of the descriptions in the Rig Veda pertain to the Asuras with the Rig vedic gods being predominantly asuric. Sanskrit was possibly the Harappan language. The Devas described in the Rig Veda were the Cro-magnon Aryan invaders. The Rig Veda describes continuing conflict between the Asuras and the Devas. Finally, the neanderthalic Harappan asuras were subdued and conquered. The cro-magnonic Aryans who conquered Harappa became the upper caste Hindu elite and the Harappan Asuras became the lower caste Shudras. The Cro-magnon Aryans took over the asuric gods, Vedas and language and made it their own. The Harappan civilization of the asuras was extremely advanced and the Cro-magnon Aryans were a primitive nomadic tribe. The Cro-magnon originated in Africa and migrated to Eurasia. The Cro-magnon population subdued the neanderthalic population and tried to exterminate them. There was also interbreeding and intermixing between the Cro-magnon and neanderthalic population. The modern neanderthalic societies are in the peri-Indian ocean area of India, Iran and Semitic Arabs. They also inhabit the peri-mediterranean area as Semitic Jews, Berbers, Basque and Celts. The predominant African and north European population is Cro-magnon.

There is an eternal conflict between Neanderthals and Cro-magnon. The Cro-magnon tried to exterminate the Neanderthals but they survived as the Jews, Arabs, the lower caste Indians, Aboriginals and Native Americans. These are the people which the Cro-magnon excluded from society. The underclass of Indian and European civilization was neanderthalic. With the advent of global warming an increasing archaeal symbiosis the neanderthalic population becomes activated and they try to exterminate the Cro-magnon. The symbiotic archaea generates new viruses which infects the non immune Cro-magnon and tries to exterminate them. The hot spots of global conflict and terrorism can be localized to neanderthalic areas. The Neanderthals dominate three world religions - Jews, Muslims and Hindus. The Cro-magnons are predominantly the Africans and the Europeans. They follow the Christian religion. World conflicts are basically between the neanderthalic races and the Cro-magnon races. This is exemplified by the Jewish leadership of the Russian and French Revolutions with its idea of liberty, equality and fraternity. The neanderthalic ideas basically tried to create an equal society. The Buddhist movement and religion among the religious lower caste of India can be thought of as a neanderthalic uprising against the Aryan Cro-magnon domination. The present rumblings in the Muslim Semitic world manifesting as global terrorism is a reflection of the neanderthalic Cro-magnon conflict. The conflict is basically between the Cro-magnon ideas of colonization, capitalism, free market globalization, rightist, fascist, nazi ideas and the neanderthalic ideas of equality, democracy, freedom and socialism. The cro-magnic civilization produces increased greenhouse gases leading to increased endosymbiotic archaeal growth. Endosymbiotic archaeal growth is the basis of neanderthalisation. Neanderthalisation is a symbiotic event and not a genetic change. This results in expansion of the existing neanderthalic societies - the Semites, the Dravidians and southern Europeans and extinction of the Cro-magnon Aryan phenotype. The present neanderthalic areas include South Europe, India, Iran, the Arab Peninsula, the Jewish homeland and the Australian Aboriginals. The Cro-magnon areas include Europe and Africa.

The Neanderthals were cerebellar dominant. The cerebellum is concerned with intuition and extrasensory perceptive phenomena. The Neanderthals were retroviral resistant. The archaea metabolises cholesterol and generates digoxin which produces membrane sodium-potassium ATPase inhibition and intracellular magnesium deficiency. Magnesium deficiency produces reverse transcriptase inhibition. Digoxin itself modulates RNA editing. The retroviral resistance leads to a deficiency of endogenous retroviral sequences. The endogenous retroviral sequences function as jumping genes required for the dynamicity of synaptic connectivity. Dynamic synaptic connectivity is required for cortical function. The cerebral cortex is dysfunctional in Neanderthals leading to cerebellar dominance. The Neanderthals inhabit a cerebellar world. The neanderthalic population is psychedelic, spiritual, dreamy, more feminine, intuitive, equal and female dominant. They had a communal life. They were hypersexual and promiscuous. They can be compared to bonobo monkeys. They were matriarchal and female dominant. They are child-like have dreamy sleep, somnolent, altruistic and docile. The neanderthalic population believed in communal living and was of hypersexual behaviour. The unconscious mind was dominant in Neanderthals. They had pre-cognition and post-cognition. They had telepathy and clairvoyance. They could have mediumistic possession and could go into hypnotic regression. They had poltergeist phenomena, group personality, multiple personality, split personality alien abduction phenomena, memory of past life, incubus and succubus. They had a magical civilization of dreams. They were subjective, personal, emotional, irrational and dreamy. They preferred the dark and nights. They had more of autism and schizophrenia. They had more of attention deficit hyperactivity and addiction. They were magical, had dominant art and religion were sexual and believed in things without proof. The belief was intuitive. They had shamanistic and magical consciousness. The Neanderthals were left-handed and right hemisphere/cerebellar dominant. They were creatures of the senses and created a spiritual dreamy civilization. They were children of the dark. The self-old brain of vampires, troglodytes, demons and the occult belongs to the Neanderthals. The cerebellar dominance and hypertrophy leads to cerebellar dysfunction and ataxia of speech as well as motor movements. Ataxic speech leads to the evolution of music. Ataxia of motor movements leads to abstract art. Thus the Neanderthal brain with its extrasensory perception is extremely artistic. Digoxin and dipolar magnetite in the setting of membrane sodium-potassium ATPase inhibition produces a pumped phonon system modulating quantal perception. Quantal perceptive phenomena are dominant in Neanderthals. This leads to increased extrasensory perception. This also produces a feeling of oneness and equality called the collective unconscious. This produces the socialistic equal Neanderthal society. The Neanderthals were also more spiritual and unconscious dominant. The cortical dysfunction leads to loss of hemispheric differentiation and sexual differentiation. Right hemisphere is predominantly masculine and the left hemisphere feminine. This results in asexual behaviours and cerebellar dominance leads to hypersexuality. The Cro-magnon population believed in pair bonding and family patterns. They were more violent and aggressive. They were patriarchal and male dominant. They were adult-like and logical. They had rightist and fascist tendencies. They were conservative in their sexual practices. They were conscious, egoistic, wakeful, male dominant, favoured the light, objective, impersonal and cruel. The conscious logical brain dominated. They depended upon proofs, logic were detached, asexual and male dominant. The Cro-magnons were predominantly left hemisphere dominant and right-handed practical people. They created a material civilization. They had a rational consciousness. They were children of the light.

The global warming produces endosymbiotic archaeal growth and neanderthalisation of homo sapiens. All these produce a dualistic consciousness. The left wing versus right wing and the conservative versus liberal. It produces a double self and divided self. It results in a Cain and Abel as well as Jekyll and Hyde personality. The Neanderthals had sloping forehead, small jaw, occipital bun and large cranium. They were shorter in height and the body weight was bigger. The brain size of Neanderthals was larger. The second toe of the feet was bigger than the big toe. They had the simian crease. The homo sapiens had a smaller brain and smaller cranium. They were taller.[1-17]

References

1. Weaver TD, Hublin JJ. Neandertal Birth Canal Shape and the Evolution of Human Childbirth. *Proc. Natl. Acad. Sci. USA* 2009; 106:8151–8156.
2. Kurup RA, Kurup PA. Endosymbiotic Actinidic Archaeal Mediated Warburg Phenotype Mediates Human Disease State. *Advances in Natural Science* 2012; 5(1):81-84.
3. Morgan E. The *Neanderthal* theory of *autism*, Asperger and ADHD; 2007, www.rdos.net/eng/asperger.htm.
4. Graves P. New Models and Metaphors for the Neanderthal Debate. *Current Anthropology* 1991; 32(5): 513-541.
5. Sawyer GJ, Maley B. Neanderthal Reconstructed. The Anatomical Record Part B: *The New Anatomist* 2005; 283B(1):23-31.
6. Bastir M, O'Higgins P, Rosas A. Facial Ontogeny in Neanderthals and Modern Humans. *Proc. Biol. Sci.* 2007; 274:1125–1132.
7. Neubauer S, Gunz P, Hublin JJ. Endocranial Shape Changes during Growth in Chimpanzees and Humans: A Morphometric Analysis of Unique and Shared Aspects. *J. Hum. Evol.* 2010; 59:555–566.
8. Courchesne E, Pierce K. Brain Overgrowth in Autism during a Critical Time in Development: Implications for Frontal Pyramidal Neuron and Interneuron Development and Connectivity. *Int. J. Dev. Neurosci.* 2005; 23:153–170.
9. Green RE, Krause J, Briggs AW, Maricic T, Stenzel U, Kircher M, Patterson N, Li H, Zhai W, *et al.* A Draft Sequence of the Neandertal Genome. *Science* 2010; 328:710–722.
10. Mithen SJ. *The Singing Neanderthals: The Origins of Music, Language, Mind and Body*; 2005, ISBN 0-297-64317-7.
11. Bruner E, Manzi G, Arsuaga JL. Encephalization and Allometric Trajectories in the Genus Homo: Evidence from the Neandertal and Modern Lineages. *Proc. Natl. Acad. Sci. USA* 2003; 100:15335–15340.
12. Gooch S. *The Dream Culture of the Neanderthals: Guardians of the Ancient Wisdom.* Inner Traditions, Wildwood House, London; 2006.
13. Gooch S. *The Neanderthal Legacy: Reawakening Our Genetic and Cultural Origins.* Inner Traditions, Wildwood House, London; 2008.
14. Kurtén B. *Den Svarta Tigern*, ALBA Publishing, Stockholm, Sweden; 1978.
15. Spikins P. Autism, the Integrations of 'Difference' and the Origins of Modern Human Behaviour. *Cambridge Archaeological Journal* 2009; 19(2):179-201.
16. Eswaran V, Harpending H, Rogers AR. Genomics Refutes an Exclusively African Origin of Humans. *Journal of Human Evolution* 2005; 49(1):1-18.
17. Ramachandran V.S. The Reith lectures, BBC London. 2012.

CHAPTER 38

CLIMATE CHANGE AND HUMAN SPECIES - HOMO NEANDERTHALIS, HOMO SAPIENS, HOMO SAPIEN EXTINCTUS AND HOMO NEONEANDERTHALIS - CLIMATE CHANGE AND EVOLUTION OF NEW SPECIES/ORGANELLE - INFECTIVE SYMBIOTIC TISSUE COLONIES OF PROTOZOA, ARCHAEA, VIRUSES AND BACTERIA

Endosymbiotic archaea and species evolution

The plasma life based forms and the gravitational and anti-gravitational waves generated porphyrions would have been the most primitive of non-life forms. The porphyrin templates would have given rise to the isoprenoid organism, the RNA viroids, DNA viroids and prions which would have symbiosed to form the nanoarchaea. The archaea would have evolved into prokaryotes or bacteria. The archaeal symbiosis with prokaryotes would have generated the eukaryotes. The initial unicellular eukaryotes are the protozoa. The protozoa are capable of phagocytosis and would have had viruses, bacteria and archaea as their intracellular symbionts. The viruses of the protozoa like entamoeba and toxoplasma are giant viruses called megaviride. The protozoans like entamoeba with its viral, archaeal and bacterial endosymbionts would have evolved into the multicellular organisms or metazoa. The archaeal, bacterial and protozoal extracellular vesicles containing RNA and DNA viroids can by horizontal gene transfer lead to evolution of metazoan, plants, animals and human species. The metazoans would have evolved into plants, animals and humans by horizontal gene transfer generated by archaeal extracellular RNA and DNA viroid vesicles. The evidence of such a transformation is seen in the bacterial symbionts of protozoa like entamoeba which includes legionella, listeria, mycoplasma, mycobacteria and chlamydia. The symbionts of protozoan entamoeba include viruses like covid, HIV and cytomegalo virus. The protozoan

entamoeba serves as a natural reservoir for covid infections, bacterial infections and HIV infections. The toxoplasma also serves as a covid reservoir. These protozoan colonies of entamoeba and toxoplasma with its bacterial, archaeal and viral symbionts can be called as chimeric infective symbiotic tissue colonies which can exist in the air, water, earth, blood and tissues. The chimeric infective symbiotic tissue colonies containing bacteria, archaea and viruses can make the symbiotic organism multidrug resistant by horizontal gene transfer. The chimeric infective symbiotic tissue organism evolves as a consequence of global warming and the protozoal, bacterial, viral and archaeal symbiotic colonies are new forms of life. The chimeric infective symbiotic tissue organism composed of protozoa and its symbionts like virus, bacteria and archaea depend upon glycolysis for its energy needs. The glycolytic enzymes enolase, glyceraldehyde 3-phosphate dehydrogenase, hexokinase, phosphofructokinase, aldolase, pyruvic kinase and triose phosphate isomerase are moonlighting proteins having a cytoplasmic, membrane and nuclear functions. These glycolytic enzymes have a RNA binding motif which can modulate RNA transcription in the host and the symbiotic organism. These glycolytic enzymes with their RNA binding motif as well as DNA binding motif can bind to RNA and DNA viruses and viroids modulating their replication and life cycles. The chimeric infective symbiotic tissue organism composed of protozoa and its symbionts like virus, bacteria and archaea can enter the human host through the breaches of the gut-blood barrier and transform itself into the macrophages of the human host. The macrophages are involved in the phagocytosis and innate immunity against bacterial and viral infections. Paradoxically the macrophages like the protozoa can also serve as a reservoir where pathogens like mycobacteria tuberculosis, mycobacteria leprae, listeria, legionella, E. coli, shigella & salmonella, and viruses like HIV & covid can thrive and replicate. The modified chimeric infective symbiotic tissue organism or macrophage composed of protozoa and its symbionts like virus, bacteria and archaea can serve as a reservoir for horizontal gene transfer and generation of

multidrug resistant bacteria and viruses. The infective chimeric symbiotic tissue organism composed of protozoa and its symbionts like virus, bacteria and archaea depends upon moonlighting glycolytic enzymes for their energetics and replication. The development of antibodies against the infective chimeric symbiotic tissue organism composed of protozoa and its symbionts like virus, bacteria and archaea leads to the evolution of anti-brain antibody disease or anti-basal ganglia antibody disease. They present a threat to existence consequent to horiztontal gene transfer and generation of multidrug resistant bacteria, viruses, protozoa, mycobacteria and archaea.

The global warming leads to endosymbiotic as well as colonic archaeal growth leading to alteration in the structure and function of the human body and system. The archaeal overgrowth within the cells leads to generation of new cellular organelle called archaeaons. The archaea have the shikimate pathway which can synthesize tyrosine and dopamine. Dopamine can be converted to dopachrome and epinephrine to adrenochrome. Dopachrome and adrenochrome can polymerize by oxidation generating melanin. The archaeaons secreting melanin can be called as archaeal melanosomes. The melanin in melanosomes has the wide range of absorption of the light spectra and gamma radiation and can transduct it to generate energy. This energy transduction can split water into H_2 and O_2 and generate protons modulating the proton gradient across the mitochondrial membrane synthesizing ATP. The melanin in the melanosome can absorb photons reducing ubiquinone to ubiquinol and generate ATP synthesis by oxidative phosphorylation. Thus the melanin in the archaeaons in the human cell can function as photosynthetic organelle. The archaeaons and their melanin can utilize gamma radiation to synthesize ATP and can exist in extreme conditions. Thus the archaeaons can produce a source of energy from light and electromagnetic waves and gamma radiation. The melanin is capable of transducting electromagnetic waves and low level electromagnetic fields and can be capable of quantal perception. Thus the melanin in the melanosomes is capable of information sensing and storage as well as energy production from electromagnetic waves and water. The human brain could have evolved by this mechanism. The humans are hairless as compared to other primates and are exposed to more of light inducing melanin induced photosynthesis and energy generation which could have contributed to the evolution of the human cortex and the complex human brain. The archaeaons melanosomes are capable of quenching free radicals and resist phagocytic destruction. The melanosomes can also resist radiation and UV light. The archaeaons are indestructible and eternal. The archaeaons have got magnetite and are capable of quantal perception and information storage. The melanin also serves the purpose of quantal perception and information storage. The archaeaon can also synthesize magnetite particles forming subcellular organelle called magnetosomes. Magnetite can interact with melanin forming supermolecular complex systems. The archaeaon can synthesize porphyrins which can self-organize to form self-replicating structures called porphyrions. Porphyrions can interact with melanin also forming supramolecular complex systems. Eumelanin pigments contain indole based tetramers that are arranged in porphyrin-like domains. The indole based structures can self-organize on porphyrin scaffolds to form tetrameric structures and melanin. The chemical structure of melanin on a macromolecular scale exhibits a tetrameric ring structure possibly because of self-organization on porphyrin scaffolds. Porphyrion can generate melanosome complexes and they can form self-organizing supramolecular complex systems. The archaeaon particles of melanosomes, magnetosomes and porphyrions forming complex colony network with specialized functions. It can function as a quantal computing system. The porphyrions and melanosomes can transducer energy and synthesize ATP functioning as primitive photosynthetic system. The magnetosome, porphyrions and melanosomes can function as information storage systems. Magnetosomes and porphyrions are dipolar and can have a quantal perceptive function based on sodium-potassium ATPase inhibition mediated pumped phonon system. The melanin can function as a superconductor for high frequency radiation and neurotransmission, as a semi-conductor for sound and heat, conduct body ionic charges and resonate for the frequencies of visible light. The archaeaon - magnetosome, porphyrions and melanosome network can function as a quantal computing brain reducing the human classical brain to a zombie brain. Thus the global warming induced archaeaon colony network and melanosomes are indestructible and eternal and takeover the human body. The human body metabolic programmes are suppressed including mitochondrial oxidative phosphorylation. The human body is reduced to a zombie or a framework for the archaeaon colony to thrive. The archaeaon induces stem cell transformation of the host human cells and change the metabolonomics of the human cells. The human cells oxidative phosphorylation is suppressed and it depends upon glycolysis for its energy needs. The human glycolytic pathway is taken over by the archaeaon for its needs. The glycolytic metabolites are channelled to the shikimic acid pathway and the D-xylulose phosphate pathway. The DXP pathway can synthesize cholesterol which is catabolized by the archaeaon for its energy. The cholesterol ring oxidases convert the cholesterol to pyruvate which then enters the GABA shunt pathway. The cholesterol side chain oxidases convert the side chain to short chain fatty acids and bile acids. The cholesterol aromatases converts the cholesterol ring to phenyl residues and synthesis of tyrosine and tryptophan. The shikimic acid pathway also utilizes substrates from the glycolytic pathway and generates tyrosine and tryptophan. The tyrosine that synthesize is converted to dopa, dopamine, dopachrome and oxidized to melanin. Melanin serves the purpose of capturing electromagnetic radiation, UV rays, Gamma radiation and light synthesizing ATP. Melanin can serve as a substrate for primitive archaeal photosynthesis. This leads to alteration in brain function and structure. The brain functions as an archaeaon melanosomal magnetite colony network capable of quantal perception, information storage and energy generation. This alters the brain function to an impulsive and anarchic mode of social function and functioning of the society as a group or collective organism. The quantal perception of the archaeaons also leads to evolution of a sort of communication with the quantal world creating a sort of universal personality or self. The human cell and system is converted to the stem cell colony which is immature and lacking functional differentiation becoming a zombie for the archaeal colony. The melanosome and melanin form a first line of defence against infection and is required for innate immunity. The melanosomes can kill the bacteria, viruses and other organisms as is evidenced by the albinism related Chediak Higashi syndrome and Griscelli syndrome. The archaeal melanin also protects it against high temperature, chemicals, oxygen radicals, oxidizing agents, UV radiation and heavy metals. The archaeal melanin makes the endosymbiotic archaea indestructible.

Intergalactic archaeal quantal computing cloud universalis

The intergalactic space contains micro-organism, especially extremophiles like archaea. The archaeal colony with its melanosomes, magnetosomes and porphyrions can form a giant quantal computing cloud in the intergalactic space functioning as an intergalactic superhuman intelligence. The porphyrions can form a template for the generation of RNA viroids, DNA viroids and prions which can self-organize to form archaeaons. The porphyrions themselves are capable of a wave-particle existence and self-replication. Thus the quantal computing cloud of extraterrestrial intelligence can arise on its own from the quantal electromagnetic fields of the intergalactic space. This extraterrestrial intelligence of quantal computing cloud of archaeaons, magnetosomes, melanosomes and porphyrions in the intergalactic space can be called as intergalactic archaeal quantal computing cloud universalis. This forms the ubiquitous anthropomorphic observer creating the universe out of the quantal foam, itself arising out the quantal foam. The porphyrins can arise sui generis from a quantal foam and forms a template for the formation of RNA viroids. An interstellar cloud of RNA viroids forms. The RNA viroids later code for DNA viroids and prions. An isoprenoid organism can also arise in the porphyrin scaffold. The interstellar cloud of dominant RNA viroids gives rise to a form of universal consciousness or gravitational waves. The RNA viroids can generate electric currents by the piezoelectric effect where mechanical energy due to the shearing stress of RNA viroidal population is converted to electrical energy and this can give rise to gravitational waves and consciousness. The helical protein of the viruses has negative and positive charged ends and acts as a dipole. When they are squashed by shearing stress of viroidal population the rod shape of the viroids gets changed to oval and dipole becomes uneven. This generates electromagnetic forces and gravitational waves. The gravitational wave forms the basis of consciousness. The RNA viroidal population can have a silicon coating and can reach the earth by asteroidal hits and gives rise to endogenous retroviruses. The human endogenous retroviruses contribute to the plasticity the human genome and the development of synaptic connectivity important for the evolution of the prefrontal cortex. The RNA viroidal population best thrives in the presence of gravity and play an important role in the development of human cerebral cortex in homo sapiens. The homo sapien brain is cerebral cortical dominant with a fully developed human consciousness due to increase in HERV sequences which increases genomic plasticity and synaptic connectivity. The homo sapiens are creatures with dominant conscious function and are logical and rational. The interstellar RNA viroidal population contributes to consciousness and gravitational waves which are linked. The intergalactic dark matter and dark energy contributes to nearly 90% of the universe energy. The dark energy contributes to antigravity forces which are repulsive and contributes to expansion of the universe. The dark energy, dark matter and antigravity contribute to the collective unconscious and human unconscious. The dark matter is made up of melanotic archaeal networks which form huge clouds in the universe. The melanotic archaea arise abiogenetically from porphyrin scaffolds which get structured out of the quantal foam spontaneously. On this porphyrin scaffolds the RNA viroids, the DNA viroids, prions, melanin and isoprenoids organisms form which symbiose to form the melanotic archaea. Thus the porphyrion/RNA viroidal population which mediates gravity and consciousness gives rise to melanotic archaeal clouds and antigravity mediating the collective unconscious. Thus gravity gives rise to antigravity and consciousness gives rise to the unconsciousness. The melanotic archaea can use anti-gravitational waves, cosmic radiation and gamma radiation as energy source for ATP synthesis. The dark matter of melanotic archaea contributing to antigravity thrives and multiplies in zero gravity situations. The melanotic archaea contains magnetite which can repulse each other when properly aligned contributing to the repulsive antigravity. The antigravity is related to the collective unconscious in the world as well as the human unconscious which is structured in the cerebellum. The dark matter containing melanotic archaea gets transferred to Eurasian land mass and earth by asteroidal hits and forms giant colonies and networks evolving to homo neanderthalis. The homo neanderthalis brain has a cerebellar dominant structure and function and is impulsive with a predominant unconscious function. The conscious function and cerebral cortex is less developed in homo neanderthalis as they are retroviral resistant. The archaea induces stem cell conversion and secretes digoxin which makes the homo neanderthalis cell population retroviral resistant. The deficiency of HERV sequences leads to maldevelopment of the homo neanderthalis cerebral cortex. The homo neanderthalis are impulsive creatures of the unconscious modulated by anti-gravitational waves. This extraterrestrial intelligence of quantal computing cloud can see life in different parts of the galaxies via asteroids and meteors. The human species evolved out of the seeded archaeaons from the extraterrestrial intelligence of the quantal computing cloud formed of the archaeal colony of archaeaons- magnetosomes, melanosomes and porphyrions. This would have reached the earth by meteoric and asteroidal hits. The hits of the meteors and asteroids would have occurred first in the Eurasian land mass, especially in the northern Siberian tundra. The homo neanderthalis would have evolved in this Eurasian land mass. As the Siberian Eurasian land mass was cold and dark the homo neanderthalis were depigmented and fair-coloured, hairless with sparse red hair. They were deficient in melanin and melanin induced energy transduction and photosynthesis leading to synthesis of ATP. The homo neanderthalis was energy deprived and the neanderthalic cortex was primitively formed and the cerebellum dominated their cognitive function. The endosymbiotic archaeal network in the brain with its magnetosomes, melanosomes and porphyrions form a primitive quantal computing system. This functions as an information receptive and storage system in communication with the extraterrestrial intelligence of the quantal computing cloud in the intergalactic space. The homo neanderthalis owing to its lack of melanosomes and innate immunity became relatively extinct over a period of time with fossilized remnants in different parts of the world. The homo neanderthalis had quantal perception which created a feeling of oneness with gender and social equality in society. The society was gender-equal and matriarchal. The matriarchal societies of the Dravidians, Basque, Celts, Harappans, Sumerians and Jews were fossilized remnants of the homo neanderthalic species. The extremes of cold temperature of the ice age led to the growth of endosymbiotic archaea in the absence of melanosomes in the Neanderthal. The melanosomes function as the first line of defence against infection and is important in innate immunity. The absence of melanosomes would have led to defective innate immunity and eventual partial extinction of homo neanderthalis with preservation of fossilized matrilineal clusters. The fossilized matrilineal neanderthalic clusters are present in different parts of the world. The fossilised homo neanderthalis are susceptible to increased archaeal endosymbiosis consequent to global warming and related civilizational diseases of metabolic syndrome, schizophrenia, cancer, autoimmune disease and degeneration. The homo neanderthalis will become extinct owing to civilizational disease consequent to global warming induced endosymbiotic archaeal growth.

The Homo sapiens

The Homo sapiens evolved in the tropical hot African land mass. The first human species to evolve is the homo neanderthalis in the Eurasian steppes. The homo sapiens would have evolved out of the archaea secreted porphyrions and RNA viroids independently. The porphyrions could have been transmitted to the tropical African land mass and would have served as a substrate for the formation of RNA viroids, DNA viroids and prions which symbiosed to form the primitive eukaryotic cell. The high temperature of the African continent would have contributed to mutations in RNA viroids and DNA viroids leading on to rapid evolution. The sub-Saharan African soil is depleted of selenium. Selenium deficiency leads to RNA viroidal mutations. Thus extremes of temperature and selenium deficiency lead to RNA viroidal diversity. This RNA viroidal diversity would have led to rapid evolution of homo sapiens from the eukaryotic cell. This eukaryotic cell would have evolved into homo sapien species over a period of time. The RNA viroids are the basis of the HERV genes which contributes to the dynamicity of the homo sapien genome. The homo neanderthalis on the other hand are retroviral resistant while the homo sapiens is retroviral sensitive. The homo neanderthalis archaeaon secretes digoxin, a steroidal hormone which can destroy the retrovirus. The homo neanderthalis also has got endosymbiotic cholesterol catabolizing archaea which can alter the membrane sites for retroviral binding making the Neanderthal species resistant to retroviral infection. The homo neanderthalis have got a deficiency of HERV jumping genes in the genome and a rigid genome as compared to the HERV sequences mediated flexible genome of the homo sapiens. The homo sapiens as they evolved in the hot African savannah would have been exposed to heat and light. This would have related in increased melanogenesis and darker skin and plenty of hair in the evolved homo sapiens. The homo sapiens owing to their dark colour would have been energy surplus consequent to melanin induced energy transduction and ATP synthesis. This would have led to the evolution of the human cortex. The RNA viroids integrated into the genome would have function as jumping HERV genes contributing to the dynamicity of the genome. A dynamic and flexible genome is required for the development of synaptic connectivity and cerebral cortex. Thus, the homo sapiens evolve the modern human cerebral cortex consequent to the surplus energy produced by melanin induced energy transduction and ATP synthesis. The increase in melanin and melanosomes increased the innate immunity of the homo sapiens making them resistant to endogenous archaeal endosymbiosis. The homo sapiens were resistant to endosymbiotic archaeal growth seen in extremes of climate of global warming and ice age. The homo sapiens which evolved out of hot tropical Africa had increased melanin content in the skin which inhibits archaeal endosymbiosis and neanderthalisation. The homo sapien species is thus protected against increased archaeal endosymbiosis consequent to global warming and related civilizational diseases of metabolic syndrome, schizophrenia, cancer, autoimmune disease and degeneration.

Homo sapien albino mutants and Homo neoneanderthalis

The Homo sapiens developed albino mutants which lacked the tyrosinase enzyme. These albino homo sapien mutants could not survive in the hot African savannah due to lack of pigmentation and migrated to the southern European land mass. This evolved into the patrilineal homo sapien European civilization. The patrilineal homo sapien European civilization arose out of the homo sapien patrilineal African civilization. The albino mutants homo sapiens forming the European civilization are susceptible to endosymbiotic archaeal growth consequent to global warming. The albino mutants homo sapiens lack melanin and melanosomes important in innate immunity. This leads to fertile conditions for endosymbiotic archaeal growth in the albino mutants, Caucasoid population. The endosymbiotic archaeal growth in the Caucasoid population leads to the evolution of a new human species. The human zombie controlled by endosymbiotic melanotic magnetite archaeaon colony network can be called as a new species - homo neoneanderthalis. Thus, the species change is occurring in the albino mutant homo sapien population of Europe and American consequent to global warming and endosymbiotic archaeal growth. The homo neoneanderthalic species and fossilized homo neanderthalis are susceptible to increased archaeal endosymbiosis consequent to global warming and related civilizational diseases of metabolic syndrome, schizophrenia, cancer, autoimmune disease and degeneration. The homo neanderthalis and homo neoneanderthalis will become extinct owing to civilizational disease consequent to global warming induced endosymbiotic archaeal growth.

Homo sapien extinctus

The Homo neanderthalis and Homo neoneanderthalis have endosymbiotic archaeal symbiosis. The endosymbiotic archaea secrete RNA viroids which can be acted upon by HERV reverse transcriptase generating corresponding DNA sequences which can be integrated into the genome by HERV integrase. The archaeal digoxin can edit the RNA viroids producing widespread diversity. The archaeal porphyrins can serve as a template for the generation of RNA viroids, DNA viroids and prions. The RNA viroids and DNA viroids can recombine with RNA and DNA viruses in the environment generating new RNA and DNA viruses. The RNA and DNA viroids can exchange their sequences with environmental bacteria generating new bacteria. Thus, there can be endogenous generation of new RNA viruses, DNA viruses and bacteria in homo neanderthalis and homo neoneanderthalis consequent to endosymbiotic archaeal overgrowth as a result of global warming. The homo neanderthalis and homo neoneanderthalis are resistant to this newly generated RNA viruses, DNA viruses and bacteria and act as an environmental reservoir for them. The new evolved RNA virus, DNA virus and bacteria generated from environmental reservoir of homo neanderthalis and homo neoneanderthalis infects the unprotected homo sapien species exterminating the homo sapien species. The homo sapien species is in decline as the homo sapien albino mutants are getting converted to homo neoneanderthalis and the African/Asian homo sapiens are getting exterminated by epidemics of new RNA viral infection generated by Neanderthal reservoirs. This homo sapien species can be called as homo sapien extinctus.

The archaea can induce stem cell conversion and neanderthalisation of the human species. The archaea catabolizes cholesterol generating digoxin which can modulate RNA editing and magnesium deficiency resulting in reverse transcriptase inhibition. The archaeal cholesterol catabolism can deplete the membrane rafts of the CD_4 cell of cholesterol impeding the entry of the retrovirus into the cell. The archaea can produce permanent immune activation producing resistance to viral and bacterial infection. The archaeal cholesterol catabolism depletes tissue cholesterol producing vitamin D deficiency and immune activation. Thus archaeal overgrowth results in retroviral resistance and generation of the Neanderthal phenotype. The endosymbiotic archaea can secrete virus-like RNA and DNA particles. The endosymbiotic archaea can induce uncoupling proteins inhibiting mitochondrial oxidative phosphorylation and generating ROS. The endosymbiotic archaeal magnetite can generate low level of EMF. The low level of EMF and ROS are genotoxic and produce breakages in hotspots of chromosome. It can also trigger rearrangements in hotspots of chromosome inhabited by retroviral and non-retroviral elements producing their expression. The archaeal secreted DNA and RNA viroids can recombine with the expressed retroviral, non-retroviral elements and other genomic segments of the human chromosome generating new RNA and DNA viruses. Thus the neanderthalised humans can serve as an origin for new RNA and DNA viruses as well as mutated retroviruses. The endosymbiotic archaea converts the Neanderthal cells to stem cells. The stem cells are resistant to immune attack. The stem cells can serve as a reservoir for this new RNA and DNA viruses. The stem cells and archaeal cells can also serve as a reservoir for viruses and bacteria belonging to other plants and animals. This helps to generate the species barrier jump in noted in recent emerging viral and bacterial infections. Thus the endosymbiotic archaeal growth produces neanderthalised version of homo sapiens which are retroviral resistant and resistant to other viral and bacterial infection consequent to immune activation and digoxin induced RNA editing. The endosymbiotic archaeal overgrowth mediated neanderthalised version of homo sapiens generates new mutated RNA and DNA viruses as well as retroviruses at the same time being resistant to them as in the case of the species bat. The homo sapiens do not have the Neanderthal mechanisms of immune activation as their archaeal load is meagre. They serve as fodder for infection from Neanderthal generated viruses and bacteria and suffer eventual extinction.

Global warming and symbiotic evolution

Thus global warming leads to symbiotic evolution of the species. The extraterrestrial intergalactic quantal computing cloud of archaea forms an intelligent anthropomorphic observer. The quantal computing cloud of archaea seeds the archaea into the earth through meteoric and asteroidal impacts. The archaeal colonies eventually evolve into multicellular organism and further into homo neanderthalis. The homo neanderthalis can be conceived as a multicellular archaeal colony. The homo neanderthalis thus, arises in earth in the Eurasian land mass out of the seeded archaeal colonies from the extraterrestrial intergalactic archaeal computing cloud. The homo neanderthalis is energy depleted. The homo neanderthalis secretes the archaeal steroidal trephone digoxin which modulates the neutral amino acid transporter increasing tryptophan transport over tyrosine. The homo neanderthalis is tyrosine depleted and deficient in melanin synthesis. There is no melanin induced ATP synthesis from electromagnetic waves and radiation transduction. The homo neanderthalis was energy depleted and therefore did not have the luxury for the development of a modern human cerebral cortex. The homo neanderthalis is also retroviral resistant. The homo neanderthalis were deficient in endogenous retroviral sequences contributing to a rigid and adynamic homo neanderthalic genome. This led to a reduction in synaptic connectivity and poor development of the homo neanderthalic cerebral cortex. The homo sapiens evolved out of terrestrial sources in Africa out of self-replicating porphyrin complexes. The self-replicating porphyrin complexes form a scaffold for supramolecular complexes of isoprenoid organism, RNA viroids, DNA viroids and prions to self-organise. The isoprenoid organism formed the cell container which symbiosed the RNA viroids, the DNA viroids and prions to form the primitive eukaryotic and prokaryotic cell. The eukaryotic organism developed into multicellular colonies and eventually evolved into homo sapiens in Africa. Thus the homo sapiens is a multicellular eukaryotic colony which evolved over a period of time. In case of oncogenesis the homo sapiens reverts to the primitive eukaryotic or prokaryotic multicellular colony state. The homo sapiens in Africa thus evolved out of terrestrial abiogenetic sources. The homo sapiens owing to the harsh tropical environmental of Africa had increased melanin pigmentation in the skin for protection from UV rays as an evolutionary mechanism and were black. The homo sapien brain evolved out of the energy excess state produced by melanin. Melanin can transduce electromagnetic waves and radiation and produce ATP synthesis. The excess energy in homo sapiens led to the rapid evolution of the human cerebral cortex. The homo sapiens are also retroviral sensitive. The retroviral infection led to integration of retroviral genes into the homo sapien genome producing endogenous retroviral sequences functioning as jumping genes. The HERV gene contributes to dynamicity and flexibility of the homo sapien genome contributing to increased synaptic connectivity and formation of the human cerebral cortex. A tyrosinase mutation led to the evolution of homo sapien albino mutants. The homo sapien albino mutants being white were unable to withstand the hot climate of the African tropics and migrated to the cold European land mass. This created the homo sapien civilization in Europe. There was interbreeding between the homo sapien albino mutants and homo neanderthalis in southern Europe producing hybrids. The homo neanderthalis were matriarchal while homo sapiens albino mutants were patriarchal. The homo neanderthalis succumbed to civilizational diseases like metabolic syndrome X, tumours, autoimmune disease and neurodegeneration and became extinct leaving fossilized matrilineal societies like the Dravidians, Celts, Basques and Jews behind. The homo sapien albino mutants in the setting of global warming developed extremophilic endosymbiotic archaeal growth and gets converted to a homo neoneanderthalic species by the phenomenon of symbiotic evolution. The homo sapien species in Africa becomes liable to eventual extinction owing to infection by catastrophic epidemics of RNA viruses arising from homo neanderthalis and homo neoneanderthalis reservoirs. Endosymbiotic archaeal growth will lead to a species change and generation of two new species - Homo sapien extinctus and Homo neoneanderthalis. Death and aging indicates human endogenous archaeal overgrowth and takeover. This will lead to extinction of the human race as such and persistence as well as survival of the archaeaon colony of melanosomes, magnetosomes and porphyrions functioning as a quantal computing colony and intelligence. This will lead to the takeover of the world and the universe by the terrestrial and extraterrestrial archaeaon quantal computing clouds. The symbiotic evolution will eventually lead to extinction of all human species into eternal archaeal colonies which can have a wave-particle existence.

The human species - terrestrial and extraterrestrial origin

The homo sapiens evolved in earth from porphyrinoids generated abiogenetically. The porphyrinoid forms a template for the formation of RNA viroids, DNA viroids, isoprenoid organisms and prions which symbiosed to form the eukaryotic and prokaryotic cells. The eukaryotic multicellular colony evolved into homo sapiens. The prokaryotes can also form multicellular functional colonies called biofilms. The homo sapiens which evolved in the African savannah became pigmented owing to melanisation of the skin in response to the solar UV rays. The homo sapiens have skin melanin but owing to lack of endosymbiotic archaea are deficient in tissue melanin. The homo sapiens in view of the absence of endosymbiotic archaea and tissue melanin are susceptible to endogenous retroviral replication and a dynamic genome leading on to increased synaptic connectivity and evolution of the prefrontal cortex. The homo neanderthalis evolved in the Eurasian steppes out of extraterrestrial archaeal colonies hitting the earth by asteroidal impacts. The archaeal colonies evolved into multicellular structures and eventually homo neanderthalis. The endosymbiotic archaea have the shikimic acid pathway and melanin synthesis. The homo neanderthalis are rich in tissue melanin but having evolved in the cold Eurasian steppes are deficient in cutaneous melanin. The increase in tissue melanin inhibits endogenous retroviral replication. This decreases the density of endogenous retroviral jumping genes in the homo neanderthalis genome making it rigid and inflexible. This rigid inflexible genome leads to the reduction in synaptic connectivity and poor development of the cerebral cortex in the homo neanderthalis. The homo neanderthalis have a dominant cerebellar cortex and are impulsive in nature. The increased tissue melanin in homo neanderthalis is capable of energy transduction giving them a survival advantage in the extremes of the Eurasian north. The melanin is capable of sensing low EMF fields contributing to extrasensory perceptive capacity of the homo neanderthalis. The homo sapiens developed tyrosinase deficient albino mutants which could not survive in the tropical Africa and migrated to the European continent. The albino mutants lack melanin and are susceptible to endosymbiotic archaeal symbiosis leading to the genesis of homo neoneanderthalis from homo sapiens. Thus the human species can have a terrestrial origin as in the case of homo sapiens in Africa and also an extraterrestrial origin from intergalactic archaea as in the case of homo neanderthalis. There is also an intermediate species evolved in out of homo sapien albino mutants with endosymbiotic archaeal symbiosis called Homo neoneanderthalis.

References

1. Kurup, R.K. and Kurup, P.A. *Global Warming, Archaea and Viroid Induced Symbiotic Human Evolution - Retrovirus, Prions and Viroids – Porphyrinoids and Viroidelle.* New York: Open Science Publishers, 2016.

CHAPTER 39

CLIMATE CHANGE AND SPECIES EXTINCTION - THE EXTINCTION OF HOMO SAPIENS AND SYMBIOTIC NEANDERTHALISATION - HOMO SAPIEN EXTINCTUS - INFECTIVE SYMBIOTIC TISSUE COLONIES OF PROTOZOA, ARCHAEA, VIRUSES AND BACTERIA

Introduction

The plasma life based forms and the gravitational and anti-gravitational waves generated porphyrions would have been the most primitive of non-life forms. The porphyrin templates would have given rise to the isoprenoid organism, the RNA viroids, DNA viroids and prions which would have symbiosed to form the nanoarchaea. The archaea would have evolved into prokaryotes or bacteria. The archaeal symbiosis with prokaryotes would have generated the eukaryotes. The initial unicellular eukaryotes are the protozoa. The protozoa are capable of phagocytosis and would have had viruses, bacteria and archaea as their intracellular symbionts. The viruses of the protozoa like entamoeba and toxoplasma are giant viruses called megaviride. The protozoans like entamoeba with its viral, archaeal and bacterial endosymbionts would have evolved into the multicellular organisms or metazoa. The archaeal, bacterial and protozoal extracellular vesicles containing RNA and DNA viroids can by horizontal gene transfer lead to evolution of metazoan, plants, animals and human species. The metazoans would have evolved into plants, animals and humans by horizontal gene transfer generated by archaeal extracellular RNA and DNA viroid vesicles. The evidence of such a transformation is seen in the bacterial symbionts of protozoa like entamoeba which includes legionella, listeria, mycoplasma, mycobacteria and chlamydia. The symbionts of protozoan entamoeba include viruses like covid, HIV and cytomegalo virus. The protozoan entamoeba serves as a natural reservoir for covid infections, bacterial infections

and HIV infections. The toxoplasma also serves as a covid reservoir. These protozoan colonies of entamoeba and toxoplasma with its bacterial, archaeal and viral symbionts can be called as chimeric infective symbiotic tissue colonies which can exist in the air, water, earth, blood and tissues. The chimeric infective symbiotic tissue colonies containing bacteria, archaea and viruses can make the symbiotic organism multidrug resistant by horizontal gene transfer. The chimeric infective symbiotic tissue organism evolves as a consequence of global warming and the protozoal, bacterial, viral and archaeal symbiotic colonies are new forms of life. The chimeric infective symbiotic tissue organism composed of protozoa and its symbionts like virus, bacteria and archaea depend upon glycolysis for its energy needs. The glycolytic enzymes enolase, glyceraldehyde 3-phosphate dehydrogenase, hexokinase, phosphofructokinase, aldolase, pyruvic kinase and triose phosphate isomerase are moonlighting proteins having a cytoplasmic, membrane and nuclear functions. These glycolytic enzymes have a RNA binding motif which can modulate RNA transcription in the host and the symbiotic organism. These glycolytic enzymes with their RNA binding motif as well as DNA binding motif can bind to RNA and DNA viruses and viroids modulating their replication and life cycles. The chimeric infective symbiotic tissue organism composed of protozoa and its symbionts like virus, bacteria and archaea can enter the human host through the breaches of the gut-blood barrier and transform itself into the macrophages of the human host. The macrophages are involved in the phagocytosis and innate immunity against bacterial and viral infections. Paradoxically the macrophages like the protozoa can also serve as a reservoir where pathogens like mycobacteria tuberculosis, mycobacteria leprae, listeria, legionella, E. coli, shigella & salmonella, and viruses like HIV & covid can thrive and replicate. The modified chimeric infective symbiotic tissue organism or macrophage composed of protozoa and its symbionts like virus, bacteria and archaea can serve as a reservoir for horizontal gene transfer and generation of multidrug resistant bacteria and viruses. The infective chimeric symbiotic tissue

organism composed of protozoa and its symbionts like virus, bacteria and archaea depends upon moonlighting glycolytic enzymes for their energetics and replication. The development of antibodies against the infective chimeric symbiotic tissue organism composed of protozoa and its symbionts like virus, bacteria and archaea leads to the evolution of anti-brain antibody disease or anti-basal ganglia antibody disease. They present a threat to existence consequent to horiztontal gene transfer and generation of multidrug resistant bacteria, viruses, protozoa, mycobacteria and archaea.

Prion proteins have been implicated in systemic disorders like neurodegenerations, cancer and metabolic syndrome. The beta amyloid in Alzheimer's disease, alpha synuclein in Parkinson's disease, the TAR protein in frontotemporal dementia and copper-zinc dismutase in motor neuron disease behaves like prion proteins. Prion proteins like behaviour is also seen in the tumour suppressor P_{53} protein in cancer and the islet cell associated amyloid in diabetes mellitus. Prion diseases are conformational diseases. The abnormal prion protein seeded into the system converts the normal proteins with prion like domains to abnormal configuration. This abnormal protein resists digestion by lysosomal enzymes after its half life is over and results in deposition of amyloid plaques. This produces organ dysfunction. Prion phenomena were initially described for Creutzfeldt-Jakob's disease, but now it is found to be widespread in chronic disease pathogenesis. Ribonucleoproteins are well-known to behave like prion proteins and form amyloid. We have demonstrated actinidic archaea which secretes RNA viroids in metabolic syndrome, neurodegenerations, cancer, autoimmune disease, schizophrenia, autism and CJD. The RNA viroids can bind with normal proteins with prion like domains, eg. superoxide dismutase and produce a ribonucleoprotein resulting in prion phenomena and amyloidogenesis. The actinidic archaeal growth results in increased digoxin synthesis and phenotypic conversion of homo sapiens to homo Neanderthals as reported earlier. The increased actinidic archaeal growth is due to global warming and this results in neanderthalisation. Homo neanderthalis tend to have more of civilizational diseases like metabolic syndrome, neurodegenerations, cancer, autoimmune disease, schizophrenia, autism and CJD. Actinidic archaeal secreted RNA viroids may play a crucial role in amyloid formation and pathogenesis of these disorders.[1-16]

Materials and Methods

The following groups were included in the study:- Alzheimer's disease, multiple sclerosis, non-Hodgkin's lymphoma, metabolic syndrome X with cerebrovascular thrombosis and coronary artery disease, schizophrenia, autism, seizure disorder, Creutzfeldt-Jakob's disease and acquired immunodeficiency syndrome. There were 10 patients in each group and each patient had an age and sex matched healthy control selected randomly from the general population. The blood samples were drawn in the fasting state before treatment was initiated. Plasma from fasting heparinised blood was used and the experimental protocol was as follows:- (I) Plasma+phosphate buffered saline, (II) same as I+cholesterol substrate, (III) same as II+cerium 0.1 mg/ml, and (IV) same as II+ciprofloxacine and doxycycline each in a concentration of 1 mg/ml. Cholesterol substrate was prepared as described by Richmond. Aliquots were withdrawn at zero time immediately after mixing and after incubation at 37 °C for 1 hour. The following estimations were carried out:- Cytochrome F420, free RNA, Cytochrome F420 was estimated flourimetrically (excitation wavelength 420 nm and emission wavelength 520 nm). Plasma of control subjects showed increased levels of the above mentioned parameters with after incubation for 1 hour and addition of cholesterol substrate resulted in still further significant increase in these parameters. The plasma of patients showed similar results but the extent of increase was more. The addition of antibiotics to the control plasma caused a decrease in all the parameters while addition of cerium increased their levels. The addition of antibiotics to the patient's plasma caused a decrease in all the parameters while addition of cerium increased their levels but the extent of change was more in patient's sera as compared to controls.

Results

The results show that there was increase in cytochrome F420 in CJD and other disease groups indicating increased archaeal growth. There was also an increase in free RNA indicating self-replicating RNA viroids in CJD and other disease groups. The RNA viroid generation was catalysed by actinides. The RNA viroids can bind with proteins having prion like domains forming ribonucleoproteins. These ribonucleoproteins can give an abnormal conformation to the protein resulting in generation of abnormal prions. The abnormal prions can act as a template to convert normal proteins with normal configuration to abnormal conformation. This can result in amyloidogenesis. The abnormal configured proteins will resist lysosomal digestion and accumulate as amyloid.

Discussion

Archaeal RNA viroids, prions and amyloid

There was increase in cytochrome F420 indicating archaeal growth. The archaea can synthesize and use cholesterol as a carbon and energy source. The archaeal origin of the self-replicating RNA was indicated by antibiotic induced suppression. The study indicates the presence of actinide based archaea with an alternate actinide based enzymes or metalloenzymes in the system as indicated by cerium induced increase in enzyme activities. There was an increase in free RNA indicating self-replicating RNA viroids. The actinides modulate RNA folding and catalyse its ribozymal action. Digoxin can cut and paste the viroidal strands by modulating RNA splicing generating RNA viroidal diversity. The viroids are evolutionarily escaped archaeal group I introns which have retrotransposition and self-splicing qualities. The RNA viroids can bind with proteins having prion like domains forming ribonucleoproteins. These ribonucleoproteins can give an abnormal conformation to the protein resulting in generation of abnormal prions. The abnormal prions can act as a template to convert normal proteins with normal configuration to abnormal conformation. This can result in amyloidogenesis. The abnormal configured proteins will resist lysosomal digestion and accumulate as amyloid.

Amyloidogenesis has been implicated in systemic disorders. The beta amyloid in Alzheimer's disease, alpha synuclein in Parkinson's disease, the TAR protein in frontotemporal dementia and copper-zinc dismutase in motor neuron disease behaves like prion proteins. Prion proteins like behaviour is also seen in the tumour suppressor P_{53} protein in cancer and the islet cell associated amyloid in diabetes mellitus. Prion diseases are conformational diseases.

The RNA viroids generated from actinidic archaea can bind to proteins with prion like domains resulting in generation of ribonucleoproteins. Ribonucleoproteins with abnormal conformation can act as a template for normal proteins with prion like domains to change to abnormal conformation. This results in generation of prion proteins with abnormal conformation resisting lysosomal digestion and generating amyloid. These systemic diseases are due to actinidic archaeal generated RNA viroid induced prion protein generation and amyloidogenesis. Prion proteins have been implicated in systemic disorders like neurodegenerations, cancer and metabolic syndrome. The beta amyloid in Alzheimer's disease, alpha synuclein in Parkinson's disease, the TAR protein in frontotemporal dementia and copper-zinc dismutase in motor neuron disease behaves like prion proteins. Prion proteins like behaviour is also seen in the tumour suppressor P_{53} protein in cancer and the islet cell associated amyloid in diabetes mellitus. The present study shows that the same prion protein mechanism can operate in schizophrenia, autism and autoimmune diseases. Sporadic CJD is also induced by actinidic archaea induced RNA viroids. Actinidic archaeal induced RNA viroids generated prions can be transferred between individuals indicating the infective nature of neurodegenerations, cancer, metabolic syndrome, autoimmune disease and neuropsychiatric diseases.

Archaeal porphyrins, amyloid and prions

The archaeal porphyrins can modulate amyloid formation. The archaeal cholesterol oxidase activity was increased resulting in generation of pyruvate and hydrogen peroxide. The pyruvate gets converted to glutamate and ammonia by the GABA shunt pathway. The pyruvate is converted to glutamate by serum glutamate pyruvate transaminase. The glutamate gets acted upon by glutamate dehydrogenase to generate alpha ketoglutarate and ammonia. Alanine is most commonly produced by the reductive amination of pyruvate via alanine transaminase. This reversible reaction involves the interconversion of alanine and pyruvate, coupled to the interconversion of alpha ketoglutarate (2-oxoglutarate) and glutamate. Alanine can contribute to glycine. Glutamate is acted upon by glutamic acid decarboxylase to generate GABA. GABA is converted to succinic semialdehyde by GABA transaminase. Succinic semialdehyde is converted to succinic acid by succinic semialdehyde dehydrogenase. Glycine combines with succinyl CoA to generate delta aminolevulinic acid catalysed by the enzyme ALA synthase. There was upregulated archaeal porphyrin synthesis in the patient population which was archaeal in origin as indicated by actinide catalysis of the reactions. The cholesterol oxidase pathway generated pyruvate which entered the GABA shunt pathway. This resulted in synthesis of succinate and glycine which are substrates for ALA synthase. Glycine and succinyl CoA are the substrates for ALA synthesis. Porphyrin and ALA inhibits sodium-potassium ATPase. This increases cholesterol synthesis by acting upon intracellular SREBP. The cholesterol is metabolized to pyruvate and then the GABA shunt pathway for ultimate use in porphyrin synthesis. The porphyrins can self-organize and self-replicate into macromolecular arrays. The porphyrin arrays behave like an autonomous organism and can have intramolecular electron transport generating ATP. The porphyrin macroarrays can store information and can have quantal perception. The porphyrin macroarrays serves the purpose of archaeal energetics and sensory perception. Protoporphyrine binds to the peripheral benzodiazepine receptor regulating steroid and digoxin synthesis. Increased porphyrin metabolites can contribute to hyperdigoxinemia. Digoxin can modulate the neuro-immuno-endocrine system.

The global warming results in increased growth of actinidic archaea and neanderthalisation of the homo sapien species. The actinidic archaea secreted viroids can generate ribonucleoproteins by binding to proteins with prion like domains. This generates amyloidogenesis and systemic diseases like neurodegenerations, cancer, metabolic syndrome, autoimmune disease and neuropsychiatric diseases. The widespread incidence of these systemic diseases leads to extinction of the neanderthalised species.

References

1. Weaver TD, Hublin JJ. Neandertal Birth Canal Shape and the Evolution of Human Childbirth. *Proc. Natl. Acad. Sci. USA* 2009; 106:8151–8156.
2. Kurup RA, Kurup PA. Endosymbiotic Actinidic Archaeal Mediated Warburg Phenotype Mediates Human Disease State. *Advances in Natural Science* 2012; 5(1):81-84.
3. Morgan E. The *Neanderthal* theory of *autism*, Asperger and ADHD; 2007, www.rdos.net/eng/asperger.htm.
4. Graves P. New Models and Metaphors for the Neanderthal Debate. *Current Anthropology* 1991; 32(5): 513-541.
5. Sawyer GJ, Maley B. Neanderthal Reconstructed. *The Anatomical Record Part B: The New Anatomist* 2005; 283B(1):23-31.
6. Bastir M, O'Higgins P, Rosas A. Facial Ontogeny in Neanderthals and Modern Humans. *Proc. Biol. Sci.* 2007; 274:1125–1132.
7. Neubauer S, Gunz P, Hublin JJ. Endocranial Shape Changes during Growth in Chimpanzees and Humans: A Morphometric Analysis of Unique and Shared Aspects. *J. Hum. Evol.* 2010; 59:555–566.
8. Courchesne E, Pierce K. Brain Overgrowth in Autism during a Critical Time in Development: Implications for Frontal Pyramidal Neuron and Interneuron Development and Connectivity. *Int. J. Dev. Neurosci.* 2005; 23:153–170.
9. Green RE, Krause J, Briggs AW, Maricic T, Stenzel U, Kircher M, Patterson N, Li H, Zhai W, *et al.* A Draft Sequence of the Neandertal Genome. *Science* 2010; 328:710–722.
10. Mithen SJ. *The Singing Neanderthals: The Origins of Music, Language, Mind and Body*; 2005, ISBN 0-297-64317-7.

11. Bruner E, Manzi G, Arsuaga JL. Encephalization and Allometric Trajectories in the Genus Homo: Evidence from the Neandertal and Modern Lineages. *Proc. Natl. Acad. Sci. USA* 2003; 100:15335–15340.

12. Gooch S. *The Dream Culture of the Neanderthals: Guardians of the Ancient Wisdom.* Inner Traditions, Wildwood House, London; 2006.

13. Gooch S. *The Neanderthal Legacy: Reawakening Our Genetic and Cultural Origins.* Inner Traditions, Wildwood House, London; 2008.

14. Kurtén B. *Den Svarta Tigern*, ALBA Publishing, Stockholm, Sweden; 1978.

15. Spikins P. Autism, the Integrations of 'Difference' and the Origins of Modern Human Behaviour. *Cambridge Archaeological Journal* 2009; 19(2):179-201.

16. Eswaran V, Harpending H, Rogers AR. Genomics Refutes an Exclusively African Origin of Humans. *Journal of Human Evolution* 2005; 49(1):1-18

CHAPTER 40

CLIMATE CHANGE, MATRIARCHY AND THE TRIBE OF MALE EUNUCHS - INFECTIVE SYMBIOTIC TISSUE COLONIES OF PROTOZOA, ARCHAEA, VIRUSES AND BACTERIA

The plasma life based forms and the gravitational and anti-gravitational waves generated porphyrions would have been the most primitive of non-life forms. The porphyrin templates would have given rise to the isoprenoid organism, the RNA viroids, DNA viroids and prions which would have symbiosed to form the nanoarchaea. The archaea would have evolved into prokaryotes or bacteria. The archaeal symbiosis with prokaryotes would have generated the eukaryotes. The initial unicellular eukaryotes are the protozoa. The protozoa are capable of phagocytosis and would have had viruses, bacteria and archaea as their intracellular symbionts. The viruses of the protozoa like entamoeba and toxoplasma are giant viruses called megaviride. The protozoans like entamoeba with its viral, archaeal and bacterial endosymbionts would have evolved into the multicellular organisms or metazoa. The archaeal, bacterial and protozoal extracellular vesicles containing RNA and DNA viroids can by horizontal gene transfer lead to evolution of metazoan, plants, animals and human species. The metazoans would have evolved into plants, animals and humans by horizontal gene transfer generated by archaeal extracellular RNA and DNA viroid vesicles. The evidence of such a transformation is seen in the bacterial symbionts of protozoa like entamoeba which includes legionella, listeria, mycoplasma, mycobacteria and chlamydia. The symbionts of protozoan entamoeba include viruses like covid, HIV and cytomegalo virus. The protozoan entamoeba serves as a natural reservoir for covid infections, bacterial infections and HIV infections. The toxoplasma also serves as a covid reservoir. These protozoan colonies of entamoeba and toxoplasma with its bacterial, archaeal and viral symbionts can be called as chimeric infective symbiotic tissue colonies

which can exist in the air, water, earth, blood and tissues. The chimeric infective symbiotic tissue colonies containing bacteria, archaea and viruses can make the symbiotic organism multidrug resistant by horizontal gene transfer. The chimeric infective symbiotic tissue organism evolves as a consequence of global warming and the protozoal, bacterial, viral and archaeal symbiotic colonies are new forms of life. The chimeric infective symbiotic tissue organism composed of protozoa and its symbionts like virus, bacteria and archaea depend upon glycolysis for its energy needs. The glycolytic enzymes enolase, glyceraldehyde 3-phosphate dehydrogenase, hexokinase, phosphofructokinase, aldolase, pyruvic kinase and triose phosphate isomerase are moonlighting proteins having a cytoplasmic, membrane and nuclear functions. These glycolytic enzymes have a RNA binding motif which can modulate RNA transcription in the host and the symbiotic organism. These glycolytic enzymes with their RNA binding motif as well as DNA binding motif can bind to RNA and DNA viruses and viroids modulating their replication and life cycles. The chimeric infective symbiotic tissue organism composed of protozoa and its symbionts like virus, bacteria and archaea can enter the human host through the breaches of the gut-blood barrier and transform itself into the macrophages of the human host. The macrophages are involved in the phagocytosis and innate immunity against bacterial and viral infections. Paradoxically the macrophages like the protozoa can also serve as a reservoir where pathogens like mycobacteria tuberculosis, mycobacteria leprae, listeria, legionella, E. coli, shigella & salmonella, and viruses like HIV & covid can thrive and replicate. The modified chimeric infective symbiotic tissue organism or macrophage composed of protozoa and its symbionts like virus, bacteria and archaea can serve as a reservoir for horizontal gene transfer and generation of multidrug resistant bacteria and viruses. The infective chimeric symbiotic tissue organism composed of protozoa and its symbionts like virus, bacteria and archaea depends upon moonlighting glycolytic enzymes for their energetics and replication. The development of antibodies against the infective chimeric

symbiotic tissue organism composed of protozoa and its symbionts like virus, bacteria and archaea leads to the evolution of anti-brain antibody disease or anti-basal ganglia antibody disease. They present a threat to existence consequent to horiztontal gene transfer and generation of multidrug resistant bacteria, viruses, protozoa, mycobacteria and archaea.

Global warming leads to endosymbiotic archaeal growth. The endosymbiotic archaea produces methane by methanogenesis contributing to global warming. The endosymbiotic archaea develops into cell organelle called archaeaons which can function as vitaminocytes, neurotransminoids, steroidelle and viroidelle creating newer cell organelle and endogenous synthetic biology. This changes the cell and tissue function and structure leading on to a new human phenotype called Homo neoneanderthalis. The homo neoneanderthalis is a hardy species which can resist climate change and thrive in environments of high temperature, high atmospheric carbon dioxide and water shortage. The new organelle that are developed can withstand climate change and fix carbon dioxide and ammonia leading to endogenous synthesis of carbohydrates and protein. The homo neoneanderthalis is a new metabolic phenotype that can survive in extremes of climate produced by global warming. The archaea forming new cellular organelle called archaeaons are extremophilic and survive in extremes of climate.

The endosymbiotic archaea secrete RNA viroids which can be converted to corresponding DNA sequences by endogenous retroviral reverse transcriptase and can get incorporated into the genome by endogenous retroviral integrase. Thus there is a constant addition of genomic sequences derived from archaeal RNA viroids producing a dynamic human neoneanderthalic genome adaptable to extremes of climate produced by global warming. Thus the human neoneanderthalis is a new genotype. The neoneanderthalic archaeaons are capable of fixing carbon dioxide and oxygen, using ammonia for energetics and capable of biological fission and fusion reactions. They are also capable of photoinduction of membrane ATP synthesis by sodium-potassium ATPase. The fixing of carbon dioxide leads to water as a by-product. This generates endogenous water in the neoneanderthalic system helping to survive extremophilic climatic conditions. This creates a new system of energetics helping the Neoneanderthal survives in extremes of climate.

The neoneanderthalic archaeaons have got magnetite capable of magnetoperception and quantal perception. The quantal perception of low level of EMF leads to frontal atrophy and cerebellar dominance producing a new neuronal phenotype with features of endemic cerebellar cognitive affective disorder manifesting as epidemic autism. This forms a new autistic behavioural phenotype which is introverted, non-compassionate, non-empathic, psychopathic and with extrasensory perceptual abilities of telepathy, telekinesis and time travel leading to communication with other galactic civilizations. The new behavioural autistic robotic neoneanderthalic phenotype will have the necessary intelligence and mental abilities cope with climate change. The neoneanderthalic archaeaons are capable of quantal perception and can communicate quantally with other star systems and multiverses. This leads on to the evolution of a galactic civilization. The neoneanderthalic archaeaons produce a metabolic phenotype capable of surviving in extremophilic environments of other star systems and can survive as civilization in other stars and exoplanets. The neoneanderthalic autistic phenotype with its extremophilic archaeaon organelle is suited for space colonisation.

The neoneanderthalic archaeaons use cholesterol oxidation as a form of energetics. This produces depletion of sex hormones producing asexual phenotypes. This leads to sexual phenotypes with alternate sexuality and parthenogenesis as a form of reproduction. This produces change in family and society as well as organisation of the community.

The neoneanderthalic archaeaons secrete RNA viroids which get converted to DNA viroids by human endogenous retroviral reverse transcriptase. The neoneanderthalic archaeaon generated RNA and DNA viroids can hybridize with other RNA and DNA viruses producing new RNA and DNA viruses to which the neoneanderthalic species is resistant. Therefore, there exist two sets of population with differing variety of archaeal symbiosis. The homo neanderthalic species has persisted as fossilised matrilineal societies in Europe and Asia which includes the Celts, the Dravidians, the Berbers and the Sumerians. These fossilised remnants of homo neanderthalic species are ideal substrates for archaeal endosymbiosis in the setting of global warming and extremophilic climatic conditions. These fossilised neanderthalic matrilineal societies will expand in the period of global warming and extremophilic conditions due to endosymbiotic archaeal symbiosis to which they are already acclimatised and exposed to. The human species homo neanderthalis is a consequence of archaeal symbiosis in the extremophilic conditions of ice age and are RNA viral/retroviral resistant consequent to endogenous digoxin synthesis and by archaeal RNA viroidal induced RNA interference and viral inactivation. The homo sapiens which evolved in Africa and migrated later to the Eurasian land mass are retroviral sensitive and other RNA viral sensitive and have a low load of endosymbiotic archaea growth. The homo sapiens owing to conditions of evolution is not an ideal substrate for global warming related endosymbiotic archaeal symbiosis. The homo neanderthalis archaeaons secrete RNA viroids which are converted to DNA viroids with both of them hybridizing with other RNA and DNA viruses in nature producing new viruses to which the homo neanderthalic and homo neoneanderthalic host is resistant. The new RNA and DNA viruses produced by homo neoneanderthalis and neanderthalis in global warming will infect the susceptible homo sapien species and wipe them out from earth. The homo sapien species is not a good substrate for global warming related endosymbiotic archaeal growth and would not evolve new cellular organelle like steroidelle, viroidelle, neurotransminoids and vitaminocyte which would help them to survive in extremophilic conditions of global warming. The homo sapiens also lack archaeaon organelle which help them to fix carbon dioxide and generate water, synthesis ATP by photo-induction mediated by sodium-potassium ATPase and archaeaon mediated fission and fusion reaction. The homo sapien species therefore without evolving archaeaon organelle cannot survive in the harsh environment of global warming induced climate change and is exterminated by homo neanderthalis and homo neoneanderthalis. The earth gets dominated by homo neanderthalis and homo neoneanderthalis in the setting of global warming and extremophilic climate change and they go on to contact other galactic civilizations by quantal perceptions as well as colonising other exoplanets, planets and star systems. This produces a civilization of superhuman homo neanderthalis and homo neoneanderthalis with new metabolic, endocrine, neuronal, psychological and genotypic features. The homo neoneanderthalis and homo neanderthalis with its cellular archaeaons and porphyrions can interlink with the internet and computer systems as well as robots producing man-machine hybrids by quantal perception. The artificial intelligence and robotic systems gets merged with homo neanderthalis and homo neoneanderthalis producing superhuman man-machine hybrids. The new homo neanderthalis and homo neoneanderthalis superhuman autistic man-machine hybrids will survive, prosper and flourish in the setting of global warming related extremophilic climate change and evolving to contact by quantal perception other galactic forms of life and civilization as well as colonising other planets and star systems.

CHAPTER 41

COVID 19 PANDEMIC - A WINDOW TO GENDER EVOLUTION – THE NEANDERTHALIC WOMEN AS AN ETERNAL ENTITY AND THE HOMO SAPIEN MALE A TRANSFORMED ENTITY FROM THE NEANDERTHALIC MALE - THE POST-COVID WORLD OF AMAZONIAN WOMEN AND MALE EUNUCHS - INFECTIVE SYMBIOTIC TISSUE COLONIES OF PROTOZOA, ARCHAEA, VIRUSES AND BACTERIA

The plasma life based forms and the gravitational and anti-gravitational waves generated porphyrions would have been the most primitive of non-life forms. The porphyrin templates would have given rise to the isoprenoid organism, the RNA viroids, DNA viroids and prions which would have symbiosed to form the nanoarchaea. The archaea would have evolved into prokaryotes or bacteria. The archaeal symbiosis with prokaryotes would have generated the eukaryotes. The initial unicellular eukaryotes are the protozoa. The protozoa are capable of phagocytosis and would have had viruses, bacteria and archaea as their intracellular symbionts. The viruses of the protozoa like entamoeba and toxoplasma are giant viruses called megaviride. The protozoans like entamoeba with its viral, archaeal and bacterial endosymbionts would have evolved into the multicellular organisms or metazoa. The archaeal, bacterial and protozoal extracellular vesicles containing RNA and DNA viroids can by horizontal gene transfer lead to evolution of metazoan, plants, animals and human species. The metazoans would have evolved into plants, animals and humans by horizontal gene transfer generated by archaeal extracellular RNA and DNA viroid vesicles. The evidence of such a transformation is seen in the bacterial symbionts of protozoa like entamoeba which includes legionella, listeria, mycoplasma, mycobacteria and chlamydia. The symbionts of protozoan entamoeba include viruses like covid, HIV and cytomegalo virus. The protozoan

entamoeba serves as a natural reservoir for covid infections, bacterial infections and HIV infections. The toxoplasma also serves as a covid reservoir. These protozoan colonies of entamoeba and toxoplasma with its bacterial, archaeal and viral symbionts can be called as chimeric infective symbiotic tissue colonies which can exist in the air, water, earth, blood and tissues. The chimeric infective symbiotic tissue colonies containing bacteria, archaea and viruses can make the symbiotic organism multidrug resistant by horizontal gene transfer. The chimeric infective symbiotic tissue organism evolves as a consequence of global warming and the protozoal, bacterial, viral and archaeal symbiotic colonies are new forms of life. The chimeric infective symbiotic tissue organism composed of protozoa and its symbionts like virus, bacteria and archaea depend upon glycolysis for its energy needs. The glycolytic enzymes enolase, glyceraldehyde 3-phosphate dehydrogenase, hexokinase, phosphofructokinase, aldolase, pyruvic kinase and triose phosphate isomerase are moonlighting proteins having a cytoplasmic, membrane and nuclear functions. These glycolytic enzymes have a RNA binding motif which can modulate RNA transcription in the host and the symbiotic organism. These glycolytic enzymes with their RNA binding motif as well as DNA binding motif can bind to RNA and DNA viruses and viroids modulating their replication and life cycles. The chimeric infective symbiotic tissue organism composed of protozoa and its symbionts like virus, bacteria and archaea can enter the human host through the breaches of the gut-blood barrier and transform itself into the macrophages of the human host. The macrophages are involved in the phagocytosis and innate immunity against bacterial and viral infections. Paradoxically the macrophages like the protozoa can also serve as a reservoir where pathogens like Mycobacteria tuberculosis, Mycobacteria leprae, listeria, legionella, E. coli, shigella & salmonella and viruses like HIV & covid can thrive and replicate. The modified chimeric infective symbiotic tissue organism or macrophage composed of protozoa and its symbionts like virus, bacteria and archaea can serve as a reservoir for horizontal gene transfer and generation of

multidrug resistant bacteria and viruses. The infective chimeric symbiotic tissue organism composed of protozoa and its symbionts like virus, bacteria and archaea depends upon moonlighting glycolytic enzymes for their energetics and replication. The development of antibodies against the infective chimeric symbiotic tissue organism composed of protozoa and its symbionts like virus, bacteria and archaea leads to the evolution of anti-brain antibody disease or anti-basal ganglia antibody disease. They present a threat to existence consequent to horiztontal gene transfer and generation of multidrug resistant bacteria, viruses, protozoa, mycobacteria and archaea.

Women are resistant to COVID 19 and the incidence of COVID 19 is less in women. Women also can fight back COVID 19 and mortality rates are low in women. Men have a high incidence of COVID 19 and high COVID 19 mortality. The question arises whether women are neanderthalic and have archaeal endosymbiosis creating RNA viroids which hybridize with human genomic elements and gut viruses to generate COVID 19. The women have endosymbiotic archaea which secrete digoxin and keep the virus in balance and prevent them from destroying the woman host. The women who harbor the virus serve as superspreaders with themselves having a mild disease. The men have more of homo sapiens substrates and do not generate the virus, but get infected by the virus generated by the woman. The men do not have significant endosymbiotic archaea to generate the virus and archaeal digoxin to kill the virus. Therefore, the incidence of COVID 19 infection is high in men and the mortality to COVID 19 is also high. This raises the evolutionary question as to whether the women of the species had a predominantly neanderthalic origin and evolved first. The epidemic of neanderthalic archaeal RNA viroidal generated retroviral infection in Africa created changes in the brain producing massive brain expansion especially in the frontal cortex to generate homo sapiens. The retrovirus got integrated into the human male genome and formed human endogenous retroviral sequences in the genome creating the ultimate genomic parasite which functions as jumping genes. This created dynamicity in the homo sapien genome creating the structure for the formation of wide synaptic networks required for the modern brain function. Homo neanderthalis is resistant to retroviral infection and consequently has an ill-developed human frontal cortex. The homo neanderthalis is predominantly cerebellar dominant and has got a cerebellar cognitive affective disorder. The homo neanderthalis changed into homo sapiens in a linear pattern consequent to retroviral infection. The neanderthalic women retained archaeal symbiosis and archaeal digoxin with neanderthalic features and remained the same eternally. The neanderthalic men who got infected with retrovirus itself created by archaeal RNA viroids became dominant with HERV sequences integrated in the Y chromosome. This modified the neanderthalic male Y chromosome and created the homo sapien male. Human endogenous retroviruses which mutated killed the endosymbiotic archaea in men and they lost the capacity to synthesize digoxin. Thus homo neanderthalis homo sapien transformation was linear precipitated by retroviral infection and the women who were resistant to archaeal created retroviruses retained their archaeal endosymbiosis and neanderthality. The men who were infected by the archaeal RNA viroids generated retroviruses had it integrated in the genome in an extensive way and the HERV particles generated became toxic to endosymbiotic archaeal survival and digoxin synthesis. Thus there was no homo neanderthalis extinction or homo sapien evolution occurring in an abrupt cut-off way, but a slow linear transformation with both surviving - the Neanderthals as women and homo sapiens as men. Human illnesses affect men and women differently. In general, both the proportion of individuals infected and the severity of infection are higher in males than females for viral, bacterial, fungal and parasitic diseases because of lack of archaeal digoxin which kills viruses, bacteria and fungi.

The female chromosomal pattern is XX and the male chromosomal pattern is XY. The X chromosome of the Neanderthals possibly has archaeal genomic sequences integrated to it using HERV integrase. There can be expression of the integrated archaeal sequences and their reassembly to form endosymbiotic archaea. The neanderthalic female being XX had the high density of genomic archaeal sequences and increased expression of endosymbiotic archaea. The endosymbiotic archaea secreted digoxin which modulates viral RNA editing. The endosymbiotic archaeal RNA viroids integrate with human genomic sequences and gut viral genomic sequences to generate RNA viruses like retroviruses. The retroviruses generated by endosymbiotic archaea were preferentially integhrated into the Y chromosome of neanderthalic male. This resulted in modification of the Y chromosome of the neanderthalic male and eventual neanderthalic male extinction with homo sapien evolution. The Y chromosome has a high density of retroviral sequences and is the graveyard of the retroviruses. The density of retroviruses in the X chromosome is less compared to Y chromosome. The integration of retroviruses into the Y chromosome and less so in the X chromosome led to the evolution of the homo sapien male genome and homo sapien male. The neanderthalic genome and more of endosymbiotic archaeal genomic integration and its expression to form endosymbiotic archaea which secreted archaeal digoxin killing the RNA viruses like retroviruses and COVID 19 generated by the archaea. There was less of archaeal RNA viroid generated retroviral integration into the neanderthalic female genome. The homo sapiens male had only one X chromosome and limited number of endogenous archaeal genomic sequences and endosymbiotic archaeal expression. This resulted in decreased archaeal digoxin synthesis in homo sapiens to kill the RNA viruses like the retroviruses and COVID 19 virus. This resulted in increased expression of endogenous RNA virus and retroviral sequences to form viruses like COVID 19 and HIV. The endogenous RNA viral expression was more as in COVID 19 producing a higher incidence of COVID 19 in the homo sapien male and higher mortality due to lack of archaeal digoxin. The female with XX had increased archaeal genomic integration into both their X chromosome and their expression forming endosymbiotic archaea which secrete archaeal digoxin. The archaeal RNA viroid generated RNA viruses like COVID 19 and retroviruses are killed by archaeal digoxin itself. Therefore, the females with XX chromosome with archaeal integration and expression had increased generation of emerging RNA viruses like COVID 19 but less of severe disease as the RNA viruses are killed by the archaeal digoxin itself. The female with XX chromosomal pattern and archaeal integration remained functionally neanderthalic and generated endosymbiotic archaea and archaeal RNA viroid generated RNA viruses like COVID 19 and SARS by hybridization with human genomic elements and gut viruses. The female of the species with XX chromosomal pattern and archaeal integration into the genome has essentially remain neanderthalic throughout centuries without any change. The male of the species with XY chromosomal pattern and retroviral integration into the Y chromosome resulting in modification of the preexisting neanderthalic Y chromosome formed the new homo sapien species. The original Y chromosome of the Neanderthal was modified out of existence. Thus, it can be postulated that the female of the human species is neanderthalic from its origin itself while the male of the species following retroviral integration into Y chromosome became homo sapien centuries later.

The Homo neanderthalis endosymbiotic archaeal viroids hybridize with human genomic and gut virus sequences generating the COVID 19 virus. Homo neanderthalis archaeal RNA viroids are the progenitors of the viruses. The archaea can also secrete porphyrions which forms templates for RNA and DNA viroids to form. The porphyrions are the most simplest of the organism. The porphyrions are capable of self-organization and self-replication on pre-existing templates to form prions, primitive RNA viroids and DNA viroids. The archaeal endosymbiont of homo neanderthalis synthesizes digoxin from the isoprenoid pathway. Digoxin is the Neanderthal evolutionary hormone. Digoxin is a small molecule that can act at the RNA level splicing, cutting and editing RNA and can destroy the RNA viroids per se. The archaea can induce stem cell transformation of the human cells via kynurenine pathway of tryptophan catabolism. Archaeal induced Warburg phenotype can generate stem cells. The stem cells are protected from immune attack and the archaea generated viruses can survive in the stem cells. The archaeal digoxin will kill a percentage of the viruses and preventing them from expanding in their clones interbodies. Studies have shown that the anti-viral effect of sodium-potassium ATPase inhibition. Cardiac glycosides not only serve the pump function of sodium-potassium ATPase but also activate intracellular signal transduction which are important in many biological processes. Recently anti-viral effects have been described as a novel feature of sodium-potassium ATPase inhibition with the use of cardiac glycosides. Cardiac glycosides are reported to be effective against both DNA virsues such as cytomegalovirus and herpes simplex virus and RNA viruses such as influenza, chikungunya, corona virus and respiratory syncytial virus. Consequently cardiac glycosides have emerged as potential broad spectrum anti-viral drugs with great advantage of targeting host cell proteins which help to minimize resistance to anti-viral treatment making them a very promising strategy against human viral infection.

Sodium-potassium ATPase classical role is to maintain the electrolyte homeostasis of the cells by pumping cations in and out of the cell using the energy obtained from hydrolysis of ATP. However, sodium-potassium ATPase is also a key scaffolding protein that is capable of interacting with signaling proteins such a protein kinase C (PKC) and phosphoionositide 3-kinase (PI3K). It also works as cardiac receptor in which the binding of the cardiac glycoside induces activation of tyrosine kinase (SRC) and downstream signaling cascade independent of intracellular ions. Viruses are intracellular parasites whose life cycle is dependent on hijacking celluar functions like protein synthesis and intracellular transport of molecules to promote their replication of spreading.

The infection begins with the attachement of viral particle to the surface exposed cellular molecules and the entry into the cell by endocytosis. Once the viral components are in the intracellular environment, the viral protein that synthesized using the host cell translational machinery and new viral particles are transported to the surface to be released and infect other cells. Targeting host cell components such as sodium-potassium ATPase is an effective anti-viral strategy. The human expression of parvovirus B19 capsid protein VP1 decreases sodium-potassium ATPase activity. Sodium-potassium ATPase is down regulated in alveolar epithelial cells infected with influenza A, H1N1 and H5N1 strains. Sodium-potassium ATPase activity can also be decreased by the sindbis virus and enterovirus coxsackie B infection. Caridac glycosides inhibit CMV infection as well as herpes virus replication. They do so by inhibiting the expression the viral genes with the anti-viral action correlated with the potency against sodium-potassium ATPase. Digitoxin inhibits HSV replication 50 percentage. Digitoxin impairs the HSV viral life cycle at two different steps - viral DNA synthesis and viral release from host cell. Digoxin also impairs the adenovirus genome replication by altering the host pre RNA splicing machinery. Chickungunya virus is susceptible to treatment with digoxin as also other RNA viruses like sinbis virus, vesicular stomatitis virus. Cornoa virus, MERS CoV, SARS CoV are repressed sodium-potassium ATPase alpha subunit is silenced or inhibited by low dose of the cardiac glycoside. The cardiac glycoside antiviral effect is relieved by the inhibition of the SRC pathway. Inhibition of sodium-potassium ATPase by cardiac glycosides decreases influenza virus replication by host translational cell machinery via a decrease in intracellular potassium. The HIV virus is also susceptible to cardiac glycoside treatment. Wang et al. showed that digoxin inhibits the replication of clinical strains of HIV1 by impairing the splicing mechanism of viral RNA and consequently the structural protein synthesis of the virsus. Low concentration of cardiac glycosides inhibit HIV1 gene expression through modulation of the mitogen activated protein kinases- extracellular signal regulated kinase (MEK½ -ERK½) signaling via interaction with sodium-potassium ATPase independent of alteration in intracellular calcium. Either by activating signaling or by altering the intracellular concentration of ions the binding of cardiac glycosides sodium-potassium ATPase determines the environment that is unfavourable for viral replications and impaired synthesis of viral proteins by interfering with the host translational machinery. The cardiac glycosides can inhibit the CME virus and there is reduced levels of nuclear factor KB 12 hour post infection and decreased expression of intermediate early viral proteins IE1 and IE2. Influenza virus is impaired by the ouabin at nanomolar concentration by inhibiting the translation of viral proteins through a decrease in the concentration of intracellular potassium. SRC signaling is also reported to be involved in the inhibition of corona virus and vesicular stomatis virus infection as the inhibition of the virus can be rescued by inhibitors of SRC. Cardiac glycosides were supposed to affect the entry stage of virus by interfering with the endocytosis. The ouabain can impair viral genomic replication or protein synthesis suggesting that this drug targets host projects that is essential for virus to achieve a successful replication cycle. Thus, the homo neanderthalis endosymbiotic archaea secrete RNA viroids which get converted to DNA viroids by human genomic HERV transcriptase. The archaeal secreted RNA and DNA viroids can hybridize with the other parts of human virobiome as well as human genomic sequences producing new emerging catastrophic viruses. The homo neanderthalis with its endosymbiotic archaea synthesizes digoxin by isoprenoid pathway which keeps the viral population in the cell in check. The homo neanderthalis by the action of endosymbiotic archaea can have an increasing percentage of somatic to stem cell transformation. The homo neanderthalic stem cells are a reservoir for the archaeal generated viruses as the stem cells are not attacked by the homo neanderthalis immune cells. The homo neanderthalis are like the ubiquitous bat which carries the virus and serves as a reservoir, but is not affected by it. The homo neanderthalis goes step further with its endogenous archaea generating the virus and serving as a reservoir with the Neanderthal hormone digoxin keeping the viral population under check. The homo neanderthalis with its endosymbiotic archaea and digoxin are susceptible to civilizational diseases like diabetes, CAD, stroke, autoimmune diseases, cancer and neurodegeneration. The homo neanderthalis develops increased colonic and endosymbiotic archaeal growth consequent to global warming and the newly grown endosymbiotic archaea in its turn produces methane and PAH increasing global warming a win-win situation leading to more archaeogenesis and emerging viruses.

The Cro-magnon homo sapiens, on the other hand are not a fertile ground for endosymbiotic archaeal growth and proliferation. They get infected by strange emerging viruses to which they are not immune to and gets slaughtered and become extinct. The Cro-magnon homo sapiens inhabits parts of Africa, Asia and Europe and they are susceptible to recurrent infection occurring as pandemics. These pandemics are created by the homo neanderthalic elite to keep the homo sapien population in check. The homo neanderthalis have a cholesterol depleted state with low sex hormones and an androgynous culture. The homo neanderthalis exist as small groups. The homo sapiens on other hand, has high cholesterol levels and increased amount of sex hormones and a higher level of population which fills the earth which needs culling to create a balance.

The human response to COVID 19 on the basis of gender shows that women have predominantly a homo neanderthalic origin and are resistant to COVID 19 infection due to archaeal symbiosis and archaeal secreted digoxin. It also shows that the men have predominantly a homo sapien origin and lack archaeal endosymbiosis and archaeal digoxin producing a high rate of infection and mortality. If the trend continues there is a chance of the homo sapien male population becoming extinct and a world dominated by Amazonian neanderthalic women taking its place. This has an evolutionary basis also. The female dominant model of human evolution raises the question of who evolved first - the male or the female. The original fossils of human species are predominantly female and the male fossils evolved after billions of years. The original human species would have been a cluster of female bipedals in swampy waters feeding on tubers of water lilies and lotus as well as fish, mussels and shell fish. The women can reproduce by parthenogenesis like lower animals. Therefore, it is natural for the female of the species to evolve first. The sexual relationship in such female only societies in primodial times was lesbian. The evolution of males occurred at a later date. The macho model of human evolution with male hunter and male toolmaker and a female accomplice evolving together is highly unlikely. The next stage of human evolution has been postulated to be interspecies hybrids. This was put forward by Eugene McCarthy. McCarthy pointed to several features of men of humans similar to pigs. Pig organs can be transplanted to humans without rejection. The pigs-like humans are hairless, have thick layer of subcutaneous fat, protruding nose and heavy eye lashes. The pig genetic sequence contains similar SINE element ALU as humans. The phenomenon of crossing the species barrier is represented by the generation of the swine flu epidemic in humans. The early bipedal female only human species would have generated human pig interspecies hybrids. This would have generated interspecies hybrids of males and females. The male sexual organ corresponds to the tail of mammals. Similarly, sex organs of species like snakes correspond to limb buds. The human embryo in its various stages of development can be compared to fishes, amphibians and lower animals. Interspecies hybrid would have led to the development of male and female of the species. The water mammals like cows, bulls, pigs, elephants, rhinoceros, turtles, crocodiles, water snakes and giant tortoises would have contributed to the generation of interspecies hybrids. This would have generated a population of both male and female bipedals in the swamps with different type of sexual interactions - heterosexual, bisexual, homo sexual and lesbian. Genetic diversity is required for a species to survive. Thus heterosexuality as a mode of sexual behaviour would have become acceptable to society as such. Bacterial and archaeal conjugation with human cells have been described. Chimeras of humans and animals have been produced in labs. Interspecies hybrids have been generated in human labs and populations have been produced. Interspecies hybrids include the dzo between yak and cattle, zubron between cow and bison, cama between camel and illama, yakulo between yak and buffalo, sheep-goat and mules. This exemplified by the Plant of the Apes. The interspecies hybrids would have been protected from pre-zygotic and post-zygotic isolation and destruction by the phenomena of immunosuppression and immune escape mediated by tryptophan catabolite kynurenine. A fish diet in swampy waters is rich in tryptophan. The Hindu myth of creation of Matsya the fish, Koorma the tortoise, Varaha the boar and Narasimha the lion point to the generation of interspecies hybrids as the main lynch point of evolution. The generation of human brain structure also depends upon interspecies hybridisation. The reptilian complex of the brain is dominant in Neanderthals. The reptilian complex is seen in amniotes which include mammals, reptiles and birds. The reptilian complex of the reptilian brain includes the basal ganglia, the brain stem and cerebellum. This forms the basis of the cerebellar cognitive affective disorder in Neanderthals. The reptilian brain is the site of imagination, intuition, instinct, compulsivity and dreams. It communicates by symbols and archetypes. It is the site of obsessive compulsive disorder, superstition, ritualism, slavishness and conformation to all way of doing things. It is the site of territoriality, aggression, racism, violence and hypersexuality. The reptilian complex is dominant in amphibians, fishes and reptiles. The Neanderthal brain has got cerebral cortical atrophy and cerebellar dominance. The reptilian brain and reptilian genes are dominant in Neanderthals pointing to interspecies hybridisation of the first evolving Neanderthal females and matrilineal female only Neanderthal society. The interspecies hybrids are signified by the importance of even toed ungulates like pig and cattle in human culture and religion. The even toed ungulates include cattle, pig, deer, camel, sheep, goat and hippopotamus. The odd toed ungulates include rhinoceros and horses. The aquatic cetaceans like whales, dolphins and purpoises evolved from even toed ungulates. Dolphins can communicate with humans. The aquatic cetaceans and even toed ungulates together form a family called cetardiodactyla. The even toed ungulates form large social groups with hierarchy, harem groups and bachelor groups. They mark their territory through glandular secretions. The ungulates can swim and whales and dolphins can gallop in water. The pig antigens have great similarity with human antigens and pig organs can be xenotransplanted into humans. The rejection is due to the presence of retrovirus in the pigs which infects humans. The targeted deletion of retrovirus related DNA from pigs makes pig a valuable source for organ transplantation. The retroviral deleted pig has been cloned and developed into embryos and implanted into sows. The pigs serve as a reservoir for human organ transplant. Hog organs can be transplanted to humans if immunized against human serum. Horse serum is used to develop tetanus antibodies and snake anti-venom. The consumption of beef leads to the development of prion disease in humans. Porcine insulin can be injected into humans and porcine heart valves can be transplanted to humans. Cow products like milk, dung and urine are used as human medicines. Human stem cells injected into pig embryos have resulted in the development of human pig chimeras. This human pig chimera can be used for organ transplantation. The development of human-ungulate chimera embryos points to the importance of interspecies hybridization in human evolution. This is signified by the worship of cow as Kamadhenu or Mother Goddess in Hindu culture, the worship of Varaha, the boar as one of the avatars of Vishnu and the religious relationship between Satan and pigs in Semitic religions. The Hindu star signs in religion and astrology have a figurative animal representation. This points to interspecies hybridization playing an important role in evolution. The woman of the species would have evolved first creating the terminology- the descent of woman and the neanderthalic woman would have remain eternal throughout centuries until today. The homo sapien man evolved later consequent to neanderthalic endosymbiotic archaeal generated RNA viroid transformed retroviruses getting integrated into Y chromosome and modifying it. The present day population is formed of homo sapien male and homo neanderthalis female. The homo neanderthalis female generates the RNA viruses like COVID 19 and acts as a superspreader spreading it to the homo sapien male. The homo neanderthalic female endosymbiotic archaea generates archaeal digoxin killing RNA viruses like COVID 19 and protects the homo neanderthalic female. The homo sapien male lacking endosymbiotic archaea and archaeal digoxin has a high incidence of high mortality COVID 19 infection.

The homo sapien testis is involved in COVID 19 infection and has the ACE2 receptor. The homo sapien male testes from COVID-19 patients exhibited significant seminiferous tubular injury, reduced Leydig cells, and mild lymphocytic inflammation. Studies show that the serum sex-related hormones ratio of testosterone to luteinizing hormone and the ratio of follicle stimulating hormone to luteinizing hormone, to some extent represent reproductive ability, has a significant discrepancy between infectors and healthy control. The testosterone levels falls and the FSH and LH levels rises. This produces male infertility and sexual impotence on a massive scale resulting in the population of homo sapien male eunuchs. The low fertility rate may result in decline in and possible extinction of world population. The killing of the homo sapien male population and dominance of the eternally neanderthalic female population creating an Amazonian world dominated by women.

CHAPTER 42

HUMAN ENDOSYMBIOTIC ARCHAEAL AND PORPHYRIN MEDIATED COVID 19 ORIGIN & RESISTANCE TO COVID 19 DISEASE STATE – ARCHAEAL SYMBIOSIS AND METABOLIC ENGINEERING TO MODULATE COVID 19 DISEASE STATE – A NEANDERTHALIC SUBSTRATE MEDIATES THE HIGH COVID 19 INCIDENCE WITH LOW MORTALITY IN PENINSULAR INDIA - INFECTIVE SYMBIOTIC TISSUE COLONIES OF PROTOZOA, ARCHAEA, VIRUSES AND BACTERIA

The plasma life based forms and the gravitational and anti-gravitational waves generated porphyrions would have been the most primitive of non-life forms. The porphyrin templates would have given rise to the isoprenoid organism, the RNA viroids, DNA viroids and prions which would have symbiosed to form the nanoarchaea. The archaea would have evolved into prokaryotes or bacteria. The archaeal symbiosis with prokaryotes would have generated the eukaryotes. The initial unicellular eukaryotes are the protozoa. The protozoa are capable of phagocytosis and would have had viruses, bacteria and archaea as their intracellular symbionts. The viruses of the protozoa like entamoeba and toxoplasma are giant viruses called megaviride. The protozoans like entamoeba with its viral, archaeal and bacterial endosymbionts would have evolved into the multicellular organisms or metazoa. The archaeal, bacterial and protozoal extracellular vesicles containing RNA and DNA viroids can by horizontal gene transfer lead to evolution of metazoan, plants, animals and human species. The metazoans would have evolved into plants, animals and humans by horizontal gene transfer generated by archaeal extracellular RNA and DNA viroid vesicles. The evidence of such a transformation is seen in the bacterial symbionts of protozoa like entamoeba which includes legionella, listeria, mycoplasma, mycobacteria and chlamydia. The symbionts of protozoan

entamoeba include viruses like covid, HIV and cytomegalo virus. The protozoan entamoeba serves as a natural reservoir for covid infections, bacterial infections and HIV infections. The toxoplasma also serves as a covid reservoir. These protozoan colonies of entamoeba and toxoplasma with its bacterial, archaeal and viral symbionts can be called as chimeric infective symbiotic tissue colonies which can exist in the air, water, earth, blood and tissues. The chimeric infective symbiotic tissue colonies containing bacteria, archaea and viruses can make the symbiotic organism multidrug resistant by horizontal gene transfer. The chimeric infective symbiotic tissue organism evolves as a consequence of global warming and the protozoal, bacterial, viral and archaeal symbiotic colonies are new forms of life. The chimeric infective symbiotic tissue organism composed of protozoa and its symbionts like virus, bacteria and archaea depend upon glycolysis for its energy needs. The glycolytic enzymes enolase, glyceraldehyde 3-phosphate dehydrogenase, hexokinase, phosphofructokinase, aldolase, pyruvic kinase and triose phosphate isomerase are moonlighting proteins having a cytoplasmic, membrane and nuclear functions. These glycolytic enzymes have a RNA binding motif which can modulate RNA transcription in the host and the symbiotic organism. These glycolytic enzymes with their RNA binding motif as well as DNA binding motif can bind to RNA and DNA viruses and viroids modulating their replication and life cycles. The chimeric infective symbiotic tissue organism composed of protozoa and its symbionts like virus, bacteria and archaea can enter the human host through the breaches of the gut-blood barrier and transform itself into the macrophages of the human host. The macrophages are involved in the phagocytosis and innate immunity against bacterial and viral infections. Paradoxically the macrophages like the protozoa can also serve as a reservoir where pathogens like Mycobacteria tuberculosis, Mycobacteria leprae, listeria, legionella, E.coli, shigella & salmonella and viruses like HIV & covid can thrive and replicate. The modified chimeric infective symbiotic tissue organism or macrophage composed of protozoa and its symbionts like virus, bacteria and

archaea can serve as a reservoir for horizontal gene transfer and generation of multidrug resistant bacteria and viruses. The infective chimeric symbiotic tissue organism composed of protozoa and its symbionts like virus, bacteria and archaea depends upon moonlighting glycolytic enzymes for their energetics and replication. The development of antibodies against the infective chimeric symbiotic tissue organism composed of protozoa and its symbionts like virus, bacteria and archaea leads to the evolution of anti-brain antibody disease or anti-basal ganglia antibody disease. They present a threat to existence consequent to horiztontal gene transfer and generation of multidrug resistant bacteria, viruses, protozoa, mycobacteria and archaea.

The corona virus produces defects in oxygenation of hemoglobin. The lung lesions produced by the corona virus is not an adult respiratory distress syndrome but a syndrome akin to high altitude sickness due to oxygen deprivation. The corona virus is a lipid envelope RNA virus and gets transcribed into DNA segments using the HERV reverse transcriptase. The corona virus RNA that gets transcribed to DNA can get integrated into the non-coding sequence of the human DNA and can function as jumping genes. The corona viral proteins can be synthesized by the human cell. The corona virus can displace iron from the beta chain of hemoglobin rendering hemoglobin and oxygen transport dysfunctional. The human body is subjected to a hypoxic ischemic injury consequent to defect in oxygen transport as occurs in high altitude mountain sickness, cyanide poisoning and carbon monoxide poisoning. The corona virus can release iron from hemoglobin and increase the ferritin levels and free iron levels to catastrophic levels. This produces cytokine activation due to increase in free radicals. The corona virus can deplete heme from hemoglobin and other heme containing enzymes. This includes cytochrome C oxidase which produces a mitochondrial dysfunction and hypoxic injury. It can bind with heme enzymes like nitric oxide synthase which produces nitric oxide, cystathionine beta synthase which produces defective synthesis of hydrogen sulphide and heme oxygenase which produces defective carbon monoxide production. Thus the endogenous vasodilator neurotransmitters are depleted producing generalized tissue vasoconstriction. The heme containing enzymes involved in glutathione metabolism - glutathione synthetase and glutathione reductase are also made dysfunctioning inhibiting free radical scavenging. The TCA cycle enzyme aconitase is also a heme enzyme and the corona virus inactivates it. The cytochrome P450 enzymes involved in steroid synthesis are also heme enzymes and they are inactivated producing acute cortisol, aldosterone and sex steroid depletion. The corona virus also produces a protein that inactivates furin involved in activating various protein enzymes.

The corona virus infection produces a metabolic disorder with defects in oxygenation and the syndrome of acquired acute porphyria with secondary iron overload state. The treatment of corona virus infection does not depend upon ventilation at high PEEP low tidal volume but at high FIO2. Hyperbaric oxygen therapy can also correct. This is ineffective as there is no hemoglobin transport oxygen. Packed cell transfusions will be required to correct the hemoglobin deficiency and oxygenation defects. Increased amounts of free iron and high ferritin can produce free radical injury and a secondary ARDS like picture which may need ventilation. Ventilation as such would not correct the primary defect in heme oxygenation. The high levels of ferritin may need iron detoxification using drugs like Desferal which chelates out the iron overload state. Excess iron can lead to free radical generation, tissue injury, immune activation and cytokine injury. The heme deficiency related cytochrome C oxidase deficiency and aconitase deficiency will have to be treated by mitochondrial supplements like CoQ and carnitine. The corona virus produces an acute porphyria and porphyrin synthesis will have to be inhibited by giving IV heme infusions. Heme inhibits porphyrin synthesis. The symptoms of porphyria are due to defects in inhibitory GABA transmission. In porphyria there is increase in delta amino levulinic acid synthesis and decrease in GABA transmission. ALA will displace GABA from its binding sites. Porphyric crisis are treated by giving gamma amino butyric acid. IV dextrose infusions will also help to inhibit ALA synthase and porphyrin synthesis. Porphyrins can also be removed by plasmapheresis. The glutathione deficiency has to be corrected by giving IV glutathione infusions. The deficiency in nitric oxide can be corrected by inhale nitric oxide preparations. The cytochrome P450 deficiency mediated reductions in steroids will also have to be corrected.

Archaeal symbiosis is seen in Aryo-Dravidian Neanderthalic population. The Aryo-Dravidian Neanderthalic substratum of Indian civilisation includes peninsular India and the Dravidian tribals like Gonds of central India. The Dravidian languages are first attested in the second century BCE as Tamil-Brahmi script inscribed on the cave walls in the Madurai and Tirunelveli districts of Tamil Nadu. The Dravidian languages with the most speakers are Telugu, Tamil, Kannada and Malayalam, all of which have long literary traditions. Smaller literary languages are Tulu and Kodava. There are also small groups of Dravidian-speaking scheduled tribes, who live outside Dravidian-speaking areas, such as the Kurukh in Eastern India and Gondi in Central India. Only two Dravidian languages are spoken exclusively outside the post-1947 state of India: Brahui in the Balochistan region of Pakistan and Afghanistan and Dhangar, a dialect of Kurukh, in parts of Nepal and Bhutan. Dravidian place names along the Arabian Sea coasts and Dravidian grammatical influence such as clusivity in the Indo-Aryan languages, namely Marathi, Konkani, Gujarati, Marwari and Sindhi, suggest that Dravidian languages were once spoken more widely across the Indian subcontinent. The colonic and endosymbiotic archaea also generates RNA viroids which are converted to DNA viroids by colonic epithelial HERV reverse transcriptase and gets integrated into the human genome. The RNA and DNA viroids hybridise with the viroidal and bacterial population of the gut microbiome and virobiome DNA and RNA of the gut leading to generation of new emerging viruses like COVID 19. A low fibre diet with increase in protein and fat leads to stimulation of endosymbiotic and colonic archaeal growth. This leads to neanderthalisation of the species. Thus a low fibre diet alters the gut microbiome in the colon with more of archaeal growth. The Aryo-Dravidian neanderthalic population in South India and parts of Eurasia has archaeal induced acquired porphyria which can lead to a human origin of emerging viruses like COVID 19 as well as resistance to emerging viruses like COVID 19 infection. The porphyrion can form a template for further corona virus abiogenetic replication. The porphyrin COVID 19 complex can produce protection of the human tissues from COVID 19 and prevent the emergence of infection and death. The emerging viruses and COVID 19 can be detected in the Aryo-Dravidian neanderthalic and similar populations but they tend to be free of COVID 19 disease state leading to death. Heme like porphyrins can ameliorate the COVID 19 disease state. Thus emerging viral infection can be modulated by symbiotic metabolonomics and human metabolic engineering induced by archaeal symbiosis. This produces a high incidence of COVID 19 with low COVID 19 mortality.

The corona virus binds to porphyrins and takes out porphyrins from the human system producing an acquired porphyria. When the porphyrins are taken away from heme, ALA synthase is activated producing an acquired porphyria. Such an acquired porphyria has been described in porphyria variegate and viral infections like SARS, HIV, HCV. The abnormal increase in porphyrin synthesis produces an acquired porphyria in corona virus infection with neuropsychiatric manifestations, autonomic dysfunction, visceral crisis, decreased inhibitory GABA transmission and dysfunction of all heme containing enzyme systems - mitochondrial cytochrome C oxidase, cytochrome P450 enzyme systems involved in steroid and fatty acid metabolism, glutathione metabolism enzymes, TCA cycle aconitase and enzymes involving gasotransmitter synthesis. This produces a high incidence of COVID 19 with low COVID 19 mortality.

The corona virus induced increase in porphyrin synthesis and porphyria has got other major catastrophic consequences. The porphyrins can self-organise and self-replicate to form porphyrions which are primitive organisms. The porphyrions are the initial primitive abiogenetic organisms. The porphyrions can self-replicate and can have a primitive electron transport chain for ATP synthesis. The porphyrion can form a template for further corona virus abiogenetic replication. Thus the corona virus can induce an acquired porphyria, form porphyrions which can self-replicate and self-organize and can have porphyrion template induced abiogenetic corona virus replication. Thus the potential for corona virus replication by abiogenetic porphyrin induced templates is non-stoppable and continues probably eternal. The corona virus does not need the human host to replicate and survive. The corona virus RNA can get transcribed to DNA using HERV reverse transcriptase and colonize to human genome as the ultimate parasite taking over the function of the human body. The corona virus will thus never leave the human system. The Aryo-Dravidian Neanderthalic population in South India and parts of Eurasia has archaeal induced acquired porphyria which can lead to a human origin of emerging viruses like COVID 19 as well as resistance to emerging viruses like COVID 19 infection. The porphyrin COVID 19 complex can produce protection of the human tissues from COVID 19 and prevent the emergence of infection and death. The emerging viruses and COVID 19 can be detected in the Aryo-Dravidian Neanderthalic and similar.

Climate change leads to increased porphyrin synthesis and porphyrin template mediated replication of viruses, prions and archaea. There is increased porphyrin synthesis leading on to porphyrinuria and porphyria. The stimulus for porphyrin synthesis comes from heme deficiency. Heme suppresses ALA synthase. Stress induces heme oxygenase which converts heme to carbon monoxide and bilirubin. Thus heme is depleted from the system. Heme oxygenase is induced by environmental stress. Climatic changes of global warming and ice age can induce heme oxygenase. Heme oxygenase is also induced by EMF pollution of the environment. Thus there is increased porphyrin synthesis from succinyl CoA and glycine. The porphyrins can self-organize and self-replicate to form primitive organisms called porphyrions. The porphyrions have a primitive electron transport chain and energetics. The porphyrions are capable of information transfer and storage by quantal perception. The porphyrions form a template for the formation of RNA viruses and viroids, DNA viruses and viroids, isoprenoid organisms and prions which all can self-organize and associate to form primitive archaea. Thus, the climate change induced global increase in porphyrin synthesis in humans lead to increased abiogenetic formation of archaea which produces more of methane and methanogenesis leading to more of global warming. Global warming leads to more heme oxygenase induction and porphyrinogenesis. The porphyrins can form a template for abiogenetic replication of RNA viruses like corona viruses. The porphyrins can intercalate between cellular and organelle membranes and the porphyrin corona virus complex can enter all cell types and organelle with these. The porphyrins will serve for abiogenetic template mediated self-replication of the corona virus. The corona virus RNA can get transcribed to DNA sequences by using human HERV reverse transcriptase. The corona virus DNA sequences can get integrated into the human genome non-coding area producing dynamic changes in the human genome. The corona virus can thus persists and replicate biologically and abiogenetically in the human system forever eternally.

Defect in heme synthesis and heme depletion leads to deficiency of heme enzymes. Deficiency cytochrome C oxidase and aconitase leads to mitochondrial oxidative phosphorylation defects and TCA cycle defects. This leads to pyruvate dehydrogenase deficiency and defect in synthesis of acetyl CoA. There is increase in glycolysis consequent to porphyrin photo-oxidation induced free radical generation and HIF alpha induction. This produces the

Warburg phenotype. The increased level of pyruvate that is generated is converted to glutamate and ammonia. Thus there is hyperammonemia as a consequence of the metabolic defect. Since glycine is utilized for porphyrin synthesis serine is not synthesized leading on to deficiency of the substrate for synthesis of cystathionine. This leads to accumulation of homocysteine and homocystinuria. Deficiency of acetyl CoA leads to defects in the isoprenoid pathway and defective synthesis of cholesterol and ubiquinone. There is also deficiency of the heme containing cholesterol synthesizing enzyme lanosterol synthase. This leads to a cholesterol depleted state.

The increase in porphyrins leads to cortical dysfunction and prefrontal cortex atrophy. The porphyrins can destroy the human endogenous retroviruses and the jumping genes leading to lack of dynamicity of the genome. This leads on to maldevelopment of the prefrontal cortex. This leads on to cerebellar dominance and a cerebellar cognitive affective disorder. This produces porphyrin related quantal perception. Porphyrins are dipolar molecules and in the setting of porphyrin mediated membrane sodium-potassium ATPase inhibition induced pumped phonon system can produce a quantal perceptive state. Porphyrins are macromolecules which can have both a wave and particle existence and can bridge the particulate world and the quantal world. Membrane sodium-potassium ATPase inhibition induced dipolar porphyrin mediated pumped phonon system can lead on to a cellular plasma state and EMF signal transduction. Macromolecules like RNA, DNA, protein and the cell itself can have an EMF signature. This porphyrin generated macromolecular cellular EMF signature is important in regulation of cell function. The porphyrins can have quantal perception of low level EMF fields leading to prefrontal cortex atrophy. This leads on to cortical dysfunction and lack of functioning of the conscious brain. The cerebellum dominates and the unconscious takes over. This leads on to neanderthalisation of the brain and schizophrenia and autism. The heme deficiency leads to lack of synthesis of the heme enzyme cytochrome P420 dependent sex hormones and a widespread asexual state. The mitochondrial dysfunction leads on to insulin resistance and metabolic syndrome X. The Warburg phenotype and increased glycolysis leads to oncogenesis. The mitochondrial dysfunction can produce neurodegeneration. The increase in lymphocyte glycolysis can produce immune activation and autoimmune disease. Thus, the stress induced porphyria due to climatic change and environmental pollution can lead to civilisational disease.

The porphyrins can self-organize to form macromolecular structures which can self-replicate to form a porphyrin organism. The photon induced transfer of electrons along the macromolecule can lead to light induced ATP synthesis. The porphyrins can form a template on which RNA and DNA can form generating viroids. The porphyrins can also form a template on which prions can form. They all can join together – RNA viroids, DNA viroids, prions – to form primitive archaea. Thus the archaea are capable of self-replication on porphyrin templates. The self-replicating archaea can sense gravity which gives rise to consciousness. They can also sense the antigravity fields which gives rise to the unconscious brain. Thus, there can be both self-replicating archaea and anti-archaea regulating the conscious and unconscious brain. Thus the climate change stress mediated increased porphyrin synthesis leads to prefrontal cortex atrophy, cerebellar dominance, cerebellar cognitive affective disorder, quantal perception and neanderthalisation of the population. The porphyrions are self-replicating supramolecular organisms which forms the precursor template on which the viroids, prions and nanoarchaea originate. Stress induced template directed abiogenesis of porphyrions, prions, viroids and archaea is a continuous process and can contribute to changes in brain structure and behaviour as well as disease process.

The corona virus is possibly an extraterrestrial virus which came into earth following a meteoric impact in Wuhan region of China and circulates in the stratosphere of earth directed by winds in the latitudinal region N 40-70. The universe is biological in origin. The metal actinides provide radiolytic energy, catalysis for oligomer formation and provide a coordinating ion for metalloenzymes all important in abiogenesis. The metal actinide surfaces would by surface metabolism generate porphyrins from simple compounds like succinic acid and glycine. Porphyrins can exist as wave forms and particulate forms and can bridge the dividing line between the quantal world and particulate world. Porphyrin molecules can self-organize into organisms with energy transduction, ATP synthesis and information storage with replicating capacity. A self-replicating porphyrin micro-organism may have played a role in the origin of life. Porphyrins can form templates on which macromolecules like polysaccharides, protein and nucleic acids can form. The macromolecules generated on actinidic porphyrins templates would have contributed to the actinidic nanoarchaea and the original organisms on earth. The data supports the persistence of an actinidic archaeal shadow biosphere which throws light on the actinide based origin of life and porphyrins as the premier prebiotic molecule.

Porphyrins play an important role in the genesis of the biological universe. The porphyrin macroarrays can form in the interstellar space on its own as porphyrins can exist both as particles and waves. Porphyrins form the bridging connection between the quantal world and the particulate world. The self-generated porphyrins from the quantal foam can self-organize to form macroarrays, can store information and self-replicate. This can be called as an abiotic porphyrin organism. The porphyrin template would have generated nucleic acids, proteins, polysaccharides and isoprenoids. This would have generated actinidic nanoarchaea in the interstellar space. The porphyrins have magnetic properties and the interstellar porphyrin organism can contribute to the interstellar grains and interstellar magnetic fields.[37] The cosmic dust grains of porphyrin macroarrays/nanoarchaeal organism occupy the intergalactic space and are thought to be formed of magnetotactic bacteria identified according to their spectral signatures. According to the Hoyle's hypothesis, the cosmic dust magnetotactic porphyrin macroarrays/nanoarchaeal organism plays a role in the formation of the intergalactic magnetic field. A magnetic field equal in strength to about one millionth part of the magnetic field of earth exists throughout much of our galaxy. The magnetic files can be used to trace the spiral arms of the galaxy following a pattern of field lines that connect young stars and dust in which new stars are formed at a rapid rate. Studies have shown that a fraction of the dust particles have elongated shape similar to bacilli and they are systematically lined up in our galaxy. Moreover, the direction of alignment is such that the long axes of the dust tend to be at right angles to the direction of the galactic magnetic field at every point. Magnetotactic porphyrin macroarrays/nanoarchaeal organisms have the property to affect the degree of alignment that is observed. The fact that the magnetotactic porphyrin macroarrays/nanoarchaeal organisms appear to be connected to the magnetic field lines that thread through the spiral arms of the galaxy connecting one region of star formation to another support a role for them in star formation and in the mass distribution and rotation of stars. The nutrient supply for a population of interstellar porphyrin macroarrays/nanoarchaeal organisms comes from mass flows out of supernovas populating the galaxy. Giants arising in the evolution of such stars experience a phenomenon in which material containing nitrogen, carbon monoxide, hydrogen, helium, water and trace elements essential for life flows continuously outward into space. The interstellar organisms need liquid water. Water exists only as vapour or solid in the interstellar space and only through star formation leading to associated planets and cometary bodies can there be access to liquid water. To control conditions leading to star formation is of paramount importance in cosmic biology. The rate of star formation is controlled by two factors. Too high a rate of star formation produces a destructive effect of UV radiation and destroys cosmic biology. Star formation as stated before produces water crucial for organism growth. Cosmic biology of magnetotactic organisms and star formation are thus closely interlinked. Systems like solar systems do not arise in random condensation of blobs of interstellar gas. Only by a rigorous control of rotation of various parts of the system would galaxies and solar system evolved. The key to maintaining control over rotation seems to lie in the intergalactic magnetic field as indeed the whole phenomena of star formation. The intergalactic magnetic fields owes its origin to the lining up of magnetotactic porphyrin macroarrays/ nanoarchaeal organisms and the cosmic biology of interstellar organisms can prosper only by maintaining a firm grip on the interstellar magnetic field and hence on the rate of star formation and type of star system produced. This points to a cosmic intelligence or brain capable of computation, analysis and exploration of the universe at large- of magnetotactic porphyrin macroarrays/nanoarchaeal organism networks. The origin of life on earth according to the Hoyle's hypothesis would be by seeding of porphyrin macroarrays/nanoarchaeal organism from the outer intergalactic space. The porphyrin organism can also be generated on actinidic surfaces in earth. Comets carrying porphyrin organisms would have interacted with the earth. A thin skin of graphitized material around a single porphyrin macroarrays/nanoarchaeal organism or clumps of organism can shield the interior from destruction by UV light. The sudden surge and diversification of species of plants and animals and their equally sudden extinction has seen from fossil records point to sporadic evolution produced by induction of fresh cometary genes with the arrival of each major new crop of comets[38,39]. The porphyrin macroarrays organism can have a wave-particle existence and bridge the world of bosons and fermions. The porphyrin macroarrays/nanoarchaeal organism can form biofilms and the porphyrin organism can form a molecular quantum computing cloud in the biofilm which forms an interstellar intelligence regulating the formation of star systems and galaxies. The porphyrin macroarrays/nanoarchaeal organism quantal computing cloud can bridge the wave-particle world functioning as the anthropic observer sensing gravity which orchestrates the reduction of the quantal world of possibilities in to the macroscopic world. The actinide based porphyrin macroarrays/nanoarchaeal organism regulates the human system and biological universe.[19-21]

Porphyrins also have evolutionary significance since porphyria is related to Scythian races and contributes to the behavioural and intellectual characteristics of this group of population. Porphyrins can intercalate into DNA and produce HERV expression. HERV RNA can get converted to DNA by reverse transcriptase which can get integrated into DNA by integrase. This tends to increase the length of the non-coding region of the DNA. The increase in non-coding region of the DNA is involved in primate and human evolution. Thus, increased rates of porphyrin synthesis would correlate with increase in non-coding DNA length. The alteration in the length of the non-coding region of the DNA contributes to the dynamic nature of the genome. Thus genetic and acquired porphyrias can lead to alteration in the non-coding region of the genome. The alteration of the length of the non-coding region of the DNA contributes to the racial and individual differences in populations. An increased length of non-coding region as well as increased porphyrin synthesis leads to increased cognitive and creative neuronal function. Porphyrins are involved in quantal perception and regulation of the thalamo-cortico-thalamic pathway of conscious perception. Thus genetic and acquired porphyrias contribute to higher cognitive and creative capacity of certain races. Porphyrias are common among Eurasian Scythian races who have assumed leadership roles in communities and groups. Porphyrins have contributed to human and primate evolution.[3,4]

The dipolar porphyrins, PAH and archaeal magnetite in the setting of digoxin induced sodium-potassium ATPase inhibition can produce a pumped phonon system mediated Frohlich model superconducting state inducing quantal perception with nanoarchaeal sensed gravity producing the orchestrated reduction of the quantal possibilities to the macroscopic world. ALA can produce sodium-potassium ATPase inhibition resulting in a pumped phonon system mediated quantal state involving dipolar porphyrins. Porphyrins by auto-oxidation can generate biophotons and are involved in quantal perception. Biophotons can mediate quantal perception. Cellular porphyrins photo-oxidation are involved in sensing of earth magnetic fields and low level biomagnetic fields. Porphyrins can thus contribute to quantal perception. Low level electromagnetic fields and light can induce porphyrin synthesis. Low level EMF can produce ferrochelatase inhibition as well as heme oxygenase induction contributing to heme depletion, ALA synthase induction and increased porphyrin synthesis. Light also induces ALA synthase and porphyrin synthesis. The increased porphyrin synthesized can contribute to increased quantal perception and can modulate conscious perception. The porphyrin induced biophotons and quantal fields can modulate the source from which low level EMF and photic fields were generated. Thus the porphyrin generated by extraneous low level EMF and photic fields can interact with the source of low level EMF and photic fields modulating it. Thus porphyrins can serve as a bridge between the human brain and the source of low level EMF and photic fields. This serves as a mode of communication between the human brain and EMF storage devices like internet. The porphyrins can also serve as the source of communication with the environment. Environmental EMF and chemicals produce heme oxygenase induction and heme depletion increasing porphyrin synthesis, quantal perception and two-way communication. Thus induction of porphyrin synthesis can serve as a mechanism of communication between human brain and the environment by extrasensory perception. The porphyrions are self-replicating supramolecular organisms which forms the precursor template on which the viroids, prions and nanoarchaea originate. Stress induced template directed abiogenesis of porphyrions, prions, viroids and archaea is a continuous process and can contribute to changes in brain structure and behaviour as well as disease process.

Life in earth would have evolved by seeding of viroids and archaea from the outer space consequent to asteroidal hits. The universe would have a biological origin and it is plausible that is there extraterrestrial intelligence in the form of a quantal computing cloud formed of archaea, viroids and porphyrions controlling and creating the universe. This could be considered as a biological god. The quantal computing cloud in the intergalactic space made up of archaea, viroids and porphyrions can seed organisms into the earth off and all through asteroidal impacting infusing a fresh set of genes into the planet and regulating the planet organization and function. The evolution of life and species on earth would have occurred by the phenomenon of panspermia. The various pandemics that occurred in the various phases of the history of the earth would have an extraterrestrial source. The theory of panspermia has been widely propagated by the astronomer Fred Holye in recent times. Holye also postulated a microbiological origin of the universe formed of intergalactic magnetic fields generated by bacteria. The pandemics occurring in earth at recurrent periods of time would be generated by the extraterrestrial quantal computing biological cloud due to asteroidal impacts. The asteroidal impacts have been described before the Spanish flu pandemic of 1915. Wickramasinghe has claimed that the corona virus came from space by meteor and the prevailing winds are spreading the disease to the hardest hit countries. The virus has spread by the prevailing winds along a global band about 40-60° N latitude. A meteor was seen as a blazing fireball flying across sky north-east of China in October 2019. The cornoa virus outbreak occurred in the same region as China as the fireball was fallen. He belives that the meteor contained embedded within it the nCoV virus particles that survived the interior of the incandescent meteor. Infectious agents are present in space carried on comets and can fall towards the earth to the troposphere. They bring about human epidemics and pandemics. The seeding of the earth by the extraterrestrial quantal computing biological cloud can be a deliberate act to regulate the organization and functioning of the earth. It could be a form of contact with an extraterrestrial alien which could be hostile or friendly. The extraterrestrial quantal computing biological cloud can produce seeding of archaea and viroids generating changes in society. It can produce changes in the human species and neanderthalisation, generates asexual phenotypes and autistic phenotypes. The extraterrestrial quantal computing biological cloud can create a surrealistic, syntheistic quantal perceptive and spiritual brain. This can also be lead on to globalization and anarchy. Extraterrestrial quantal computing biological intelligence can thus modulate the structure of society on earth. The archaeal endosymbiosis induced by the extraterrestrial quantal computing biological cloud can generate a homo galacticus species capable of surviving in extraterrestrial environment and communicating by extrasensory quantal perception. This can lead to the evolution of a galactic civilization and intergalactic communication modulated by the extraterrestrial quantal computing biological cloud of porphyrions, archaea and viroids.

The Fermi paradox says that there are inhabitory planetary systems in different galaxies and it is paradoxical that human beings are not come into contact with alien intelligence. The Drake equation helps us to calculate the number of intelligent civilizations in different galaxies of the universe. The intergalactic archaeal quantal computing clouds serves as the master of the universe and as the ubiquitous universal observer. The quantal computing archaeal clouds seed life in different planetary systems of galaxies. The universe can exist in the particulate and quantal state with swings between the two states determined by the master archaeal quantal computing cloud which functions as ubiquitous observer and intelligence controlling, creating and regulating the universe. Each civilization has a particulate and quantal counterpart. The particulate part of the civilization is not sensed by another civilization. The quantal computing cloud serving as the master intelligence is a universal observer and prevents particulate civilizations from coming to contact. There is a permanence of the quantal universe. The quantal civilizations inhabiting multiverses are in contact with each other through worm holes.

Every civilization is a dark forest. Every civilizations aim is to survive. Contact with another civilization results in annihilation. Civilizations hide their signals. Just like animals and hunters in the forest. Paranoid and suspicious about intentions. Finite resources in universe. Each civilization seeks to expand. Galactic civilizations are like the dark forest and hide from each other for the fear of mutual annihilation. Therefore, contact with other galactic civilizations is an impossibility. The civilization in earth would have by its technological advances, galactic travels and radiation sources penetrated the dark forest and came into contact with the extraterrestrial quantal computing cloud serving as the master intelligence and universal observer creating the observed world.

Gravity is related to consciousness which is structured in the cerebral cortex and antigravity to unconsciousness which is structured in the cerebellum. Quantal computing involving topological superconductivity can make use of cubits of graviton and antigraviton which can lead on to the creation of a universal quantal computing cloud - the supercomputer god. Panpsychism considers the universe as conscious and consciousness as an intrinsic feature of all sub-atomic particles, living things and non-living things. This gives continuity between the quantal world of existence and macroscopic world. The quantal superpositions collapse or undergo decoherance due to the observation of universal protoconsciousness field. The universal protoconsciousness field can be considered as gravity and antigravity. The universal protoconsciousness field permeates the whole galaxy and is intrinsic to all matter. Matter is a creation of consciousness. Extraterrestrial intelligence can exist as a superorganism controlling and creating the universe. A network of extraterrestrial porphyrin molecules or porphyrions fits in well with this concept. Porphyrins have got a wave-particle existence and can arise from the quantal foam of gravitational waves and anti-gravitational waves which creates human consciousness and unconsciousness, respectively. This can form a quantal computing god cloud of porphyrions with quantal computing involving topological superconductivity can make use of cubits of graviton and anti-graviton.

Extraterrestrial intelligence can exist as a superorganism controlling and creating the universe. In this regard, the concept of black cloud of gaseous extraterrestrial superorganism put forward by Fred Hoyle is significant. The black clouds of interstellar gas are made up of hydrogen. There are small pieces of solid matter at the centre and are linked to form a complex neurological system. Clouds have the capacity to increase the brain size and complexity and become intelligent. A network of extraterrestrial porphyrin molecules fits in well with this concept. Porphyrins have got a wave-particle existence and can arise from the quantal foam of gravitational waves and anti-gravitational waves which creates human consciousness and unconsciousness, respectively. A large organized self-replicating information system of porphyrins can function as an extraterrestrial intelligent organism. The intelligent black cloud when it finds a lifeless interstellar cloud creates simple brain system by transfer to the lifeless cloud. This gives a lifeless cloud life and intelligence. Black clouds never disappear, never age and never reproduce. Black clouds can communicate with other black clouds by radio wave brain waves. Black clouds need star light to metabolize chemical substances. The porphyrin arrays can utilize light to self-replicate and propagate. Being clouds of gas the superorganism can assume any shape. When the food supplies of the black cloud in the interstellar space falls the black cloud approaches a star and assumes a disc-like shape to absorb energy from the star. The cloud has to leave the star within a short span of time or it will condense to become the planet of the star. This can be compared to the molecular quantum computing clouds of Tamulis. These are massive immobile extraterrestrial intelligence which sprawls and stretches for hundreds and thousands of miles in the interstellar space. They can be compared with group of neurons which are interconnected. Each node in the network is a brain. This gaseous extraterrestrial intelligence can arise from an emission nebula. An emission nebula is one of the huge clouds of gas and dust that floats around in the interstellar space. In this cloud, chemical molecules of life carbon monoxide, formaldehyde, methanol, hydrogen sulphide, sulphur oxide, sulphur dioxide and hydrogen cyanide are seen. Life arises in this interstellar chemical cloud of gas. Porphyrins arrives which can self-replicate can arise in this chemical interstellar clouds. This creates a superorganism which can form superintelligence. The interstellar gas is mostly made up of hydrogen. The higher dense regions of the interstellar gas can condense to form stars and planets. The coldest clouds form low mass stars and giant molecular clouds forms large mass stars. This form of gaseous extraterrestrial intelligence is seen in atmospheres of other planets in the cloudy organic molecular atmosphere and interstellar space. This fits in with the concepts of human evolution in earth where the protoplasmic body can get converted to the robotic body and then the energy body. The cool or hot interstellar gas can generate networks of extremophilic archaea which can survive on carbon dioxide and hydrogen. They are called the halophilic archaea or the methanogenic archaea. They can live in extremes of cold, heat, acid and pressure. The porphyrin network can form a template for the generation of viroids and prions which can symbioze together to form an archaeal and huge extraterrestrial interstellar gas archaeal network with intelligence can arise. This can condense to generate stars and planets as well as evolving life in planetary systems. The archaeal networks are capable of nuclear fusion and fission reaction and can power star systems like the sun. It is extremely plausible that extraterrestrial intelligence is an interstellar archaeal colony network arising abiogenically on porphyrin templates which can be generated sui generis from the quantal foam of gravitational waves and anti-gravitational waves. The interstellar dust forms stars and planets and the magnetotactic archaea generated intergalactic magnetic fields plays a role in the generation of star and planetary systems. The giant molecular porphyrin clouds and archaeal networks are extraterrestrial intelligence that plays a crucial role in star formation and formation of planetary systems. The extraterrestrial intelligence porphyrin network and archaeal network can communicate with any part of the universe by the phenomenon of quantal entanglement. Interstellar gas is formed of grains of calcified nanoarchaea. Tamulis has postulated molecular quantal computing clouds that absorb magnetic and light emissions from planet and stars, compute information and move in space by using light pressure. The clouds use photoactive molecules for molecular quantal computing. They can convert light emissions to magnetic flops. The magnetic flops interact and control the central generating elements of 10 quantum bits of the molecular quantum computing clouds. Such clouds can exist in a coherent state at low temperature of the interstellar gas. Porphyrin networks are photoactive molecules and can generate molecular quantal computing clouds which can absorb magnetic and light emissions from planet and stars, compute information and move in space by using light pressure. The porphyrin networks can abiogenically generate archaeal bacterial massive extraterrestrial network intelligence. This formless mass living in space obtains its energy from the light of the stars and communicates with itself and other forms of lights by radio waves. The interstellar gas is analogous to blood, the heard and kidneys are electromagnetic and there is a complex neurological system of the brain. This cloud of interstellar gas porphyrin networks and archaeal colonies can convert into stars and planetary system by condensation. The energetics for star systems comes from biological fusion and fission reaction occurring in archaeal networks. The archaeal networks of extraterrestrial intelligence can engineer changes in human mind and civilization.

The extraterrestrial intelligence is an archaeal colony network. It arises on porphyrins which are generated sui generis on the quantal form. The porphyrin networks can self-replicate and can form a template for the formation of viroids and prions. The archaeal colony networks are extremophilic networks and can survive in the environment of extreme climates of Mars and Venus as well as in interstellar gas. The extremophilic archaeal extraterrestrial network can secrete RNA viroids which form RNA viroidal quasi-species consortia. The RNA viroidal quasi-species consortia can be edited by archaeal digoxin giving identity and specialization to the network. The archaeal extraterrestrial intelligence impacts on earth by meteoritic showers which give new cosmic genes to the earth. This is the form of cosmic gene transfer. The earth is populated by homo sapiens and the transfer of archaeal and RNA viroidal genes generates the homo neanderthalis. They can be compared to the hybrids described in Biblical literature and other religious texts as the fallen angels, nephilm or the Sumerian Annunaki. These are generated by transfer of cosmic genes by archaeal networks mediated by meteoritic impacts. The Neanderthal hybrids that are generated are susceptible to diseases like degeneration, autoimmune disease, metabolic syndrome, cancer and new viral infections. The hybrids are matriarchal, gender-equal, have alternate sexual practices, autistic, altruistic, spiritual and with extrasensory powers. These can be called as god astronauts of ancient literature. The first episode of transfer of archaeal and RNA viroidal genes occurred in the ice age and second transfer is occurring in the age of global warming. The Neo-neanderthal species of global warming age is produced by cometary transfer of archaeal and RNA viroidal genes. The extraterrestrial intelligence of extensive self-replicating porphyrin networks and archaeal networks generated abiogenically in the interstellar space is comparable to an external God-like force. The porphyrin networks can exist both as wave and particle and can arise from the quantal foam of gravitational waves of consciousness and anti-gravitational waves of unconsciousness which can later generate extensive extraterrestrial intelligence of nanoarchaeal networks. The extraterrestrial intelligence of porphyrin networks generated sui generis from the quantal form can be called the lotus of Brahma. The gravitational waves and anti-gravitational waves of consciousness and unconsciousness create matter by the phenomenon of soni-luminescence. This forms what is called as the collective unconscious of Jung or the monads of Leibnitz or the panpsychism where there is no distinction between the consciousness of living and non-living put forward by the Greeks, Hindus and recently expounded by Chalmers.

The extraterrestrial intelligence is an archaeal colony network created on a porphyrin network that arises spontaneously by wave-particle transformation from the quantal foam of gravitational waves of consciousness and anti-gravitational waves of unconsciousness. The whole process is abiogenetic and is a creation without a creator. The porphyrin arrays can self-replicate, store information and harvest magnetic and light energy from the stars. It can convert light energy to magnetic flops. The magnetic flops interact and control the central generating element of the 10 quantum bit of the quantum computer. These are analogous to the black cloud of Fred Hoyle and the molecular quantal computing clouds of Tamulis. Long lasting coherent quantal states are possible in the low temperature of the interplanetary dust. The porphyrin network and archaeal network extraterrestrial intelligence can communicate with itself and others by radio waves. The interstellar gas with this porphyrin network and archaeal network can condense at dense sites into stars and planets and give rise to life forms in those systems. The porphyrin network and archaeal network can survive in the extremes of temperature of planetary systems like Mars and Venus and can influence the behaviour and future of human civilization with which they can come into contact via transfer by meteoric impacts. Such meteoric impacts occurred in the ice age and at other frequent epochs in the planet's history. The impact in the ice age gave rise to Neanderthal hybrids with archaeal and RNA viroidal symbiosis by parasitizing with homo sapiens in the Eurasian steppes. The Neanderthal hybrids had defective cortical function and have a dominant cerebellar function with a different behavioural code. They were matriarchal in the sense the newer genes were of cometary origin. Males and paternal roles were insignificant. This gave rise to the female dominant society. The Neanderthal hybrids because of intermingling with archaeal and RNA viroidal genes were susceptible to autoimmune infection, degenerations, tumours, metabolic syndrome, autism, schizophrenia and deviant sexual behaviour. This resulted in rapid elimination of the Neanderthal hybrids from the society of homo sapiens. The Neanderthal hybrids owing to the archaeal and RNA viroidal genes were creative, autistic, savantic, spiritual, inventive, intuitive, scientific, artistic and were the dominant drivers of society. They became the leaders of society and God-like characters directing the flow of civilization. This was possibly mediated by the intervention of the interstellar porphyrin network and archaeal colony extraterrestrial intelligence in the intergalactic space which controls the universe and has a God-like function. Such porphyrin networks and archaeal colony networks of gigantic proportions are feasible in the atmosphere of other planets including earth. There is increased growth of endosymbiotic archaea in homo sapien population owing to global warming. The human body functions as a zombie or caste for the growth of the archaeal symbionts who take over the function of the human brain and its actions. This has resulted in the recent changes in society with globalization, terrorism, the rise of spirituality, anarchism, domination by certain states, lack of democracy, the rise of creativity in technology, art and culture. A new generation of Neo-neanderthal hybrids are being generated which are an army of zombies for an extended archaeal terrestrial intelligent network. The archaeal terrestrial intelligent network as an extension of the extraterrestrial interstellar space porphyrin and archaeal network is taking over the world and its culture. This will have far reaching consequences to human existence and for the destruction of the human zombies as we see ourselves today. The archaeal colony terrestrial network will survive and the global warming will destroy the human zombies. The human zombie created or engineered behavioural and life style change leading to global warming leads to extension of gigantic growth of the terrestrial intelligent archaeal network. The growth of the terrestrial intelligent archaeal network in the oceans will lead to release of methane gas, rise of ocean levels, tsunamis, earthquakes, destruction of the earth's ecology by global warming and a civilization of zombies as such. The homo sapiens can be considered as zombies controlled by a matrix of terrestrial and extraterrestrial porphyrin and archaeal colony network. This is the world of maya and the human zombies can cut through the illusions and connect with larger world of terrestrial and extraterrestrial intelligence called gods. We can become psychonauts and connect with the world of extraterrestrial and terrestrial intelligence. Extraterrestrial and terrestrial intelligence of porphyrin networks and archaeal colony network created on porphyrin templates are itself illusory. The porphyrin and template mediated nanoarchaeal colony network have a wave-particle existence and can merge into the quantal foam of nothingness with its gravitational waves generating consciousness and anti-gravitational waves generating unconsciousness. The terrestrial and extraterrestrial porphyrin and archaeal colony network with its RNA viroids quasi-species consortia are variable from time to time. The archaea secretes digoxin which edits the RNA and change the character of the network in response to environment. The network of terrestrial and extraterrestrial intelligence owing to the extreme variability of the RNA quasi-species consortia mediated editing forms intelligence and consciousness which changes from time to time in consonance with the hundred thousand or more superimposed possibilities of the quantal foam. They are the momentary consciousness and not a single wholesome structure. The consciousness of the universe of all objects living and non-living is interconnected forming what is called as a collective unconscious or the world of Maya or panpsychism. There is no death and everything is universal and eternal. Change in consciousness occurs every second and everything is momentary and at the same time long lasting. All the momentary stages of universal consciousness or Samsaras are fixed and are not susceptible to change. The changes in the quantal foam of the universal consciousness get reflected in the particulate terrestrial and extraterrestrial intelligence of porphyrin and archaeal colony network which is subjected to continuous uninterrupted editing by the changes in the codon feature of the RNA viroids. Thus, the quantal possibilities of the universal consciousness parallel the ever changing structural features of the terrestrial and extraterrestrial intelligence of porphyrions, archaea and RNA viroidal network.

The constant transfer of cometary genes from the interstellar extraterrestrial intelligence of archaea and RNA viroids created on a porphyrin templates generated from the quantal foam of consciousness which can be described as god due to impact of meteors changes the culture, civilization, behaviour, sexuality, politics, economics and environment of the homo sapiens and hybrid population in planet earth. This cometary gene interacts with the terrestrial bacteria and viruses creating new bacteria and viruses and new epidemics which define populations and give them identity. The new bacteria and viruses of plant and animal origin give rise to mass extinctions and change the ecology of the planet. And a massive growth of terrestrial archaeal colony network in oceans can lead to earthquake, tsunamis, global warming, drying up of earth, flooding and catastrophic extinction of the human zombie civilization as we know today. The extraterrestrial intelligence and terrestrial intelligence of porphyrin and archaeal colony network will survive in Mars, Earth, Venus, interstellar space and Sun. They are victorious, eternal and God-like. They are like astronauts from the outer space but only bacterial network astronauts. They can be considered as fallen angels send by the extraterrestrial intelligence to modulate the civilization of planets and stars of the universe and creation of hybrid zombies. The hybrid zombie civilizations are destroyed once the use is over. This happens to the Sumerian civilization, the Harappan civilization, the

Egyptian civilization and Aztec civilization. This is happening to homo sapien civilization at present created by the crisis of global warming due to terrestrial archaeal colony networks. The zombies will end but the masters and the gods of extraterrestrial porphyrin and archaeal colony network survive. The networks are protoplasmic and robotic bodies and can go back to their energy bodies of nothingness and merge into the quantal foam. This is the world of Maya where nothing exists and we can cut through the illusions and reach the world of Maya and connect with it. This may provide a hope for saving the human civilization of zombies with their archaeal colony networks increasing at astronomical spaces and speeding on to an end of catastrophe.

The extraterrestrial archaeal network with its secreted RNA viroids gives rise to eukaryotes, prokaryotes, multicellular organisms, primates and humans. The archaeal network secreted RNA viroids are edited by archaeal digoxin producing identity for different RNA viroidal quasi-species consortia. The RNA viroids can get converted into complementary DNA sequences and get inserted into archaeal genome. This produces varying diversity of extraterrestrial and terrestrial porphyrin and archaeal intelligent networks. The human and other multicellular organisms are structured by the endosymbiotic extraterrestrial and terrestrial archaeal network to produce species evolution and identity. The eukaryotes and prokaryotes evolve from the archaea by genomic changes induced by integrated complementary sequences of digoxin edited archaeal RNA viroids. Thus the extraterrestrial and terrestrial archaeal intelligent species network is capable of evolving to prokaryotes, eukaryotes, multicellular organisms, primates, plants, animals and humanoids. These are all zombies or structures for the extraterrestrial and terrestrial archaeal and porphyrin intelligent networks to survive and utilize. The terrestrial and extraterrestrial porphyrin and archaeal networks can have a wave-particle duality and can merge into the quantal foam and exist as intelligent quantal beings. They can be called as terrestrial and extraterrestrial intelligent quantal beings with multiple superimposed possibilities cohabiting multiverses at the same point of time. This raises the question of the nature of life itself. Life is a virtual reflection of the porphyrin networks intelligent species that arises out of the quantal foam of nothingness and can dissolve into it. This constitutes what is called as the world of Maya. There is only one life form the terrestrial and extraterrestrial intelligent porphyrin and archaeal networks which can have an energy existence or a particulate existence. All other life forms and inanimate forms arise out of it and are reflections of it and interconnected throughout the universe by the phenomena of quantal spin and quantal entanglement and non-local interaction. Everything in the universe exists everywhere at every point. Time is an illusion.

The terrestrial and extraterrestrial networks of porphyrin arrays and archaeal networks can influence earth by being transported to the terrestrial habitat by meteoritic showers. This induces fresh infusions of cometary genes into earth and other planetary systems. These are genes from the extraterrestrial porphyrin and archaeal networks which arise out of the quantal foam and induce the formation of star systems and planetary systems. They can be compared to a collection of lotuses in the ocean of galaxy. The lotuses being porphyrin arrays which have a wave and particle existence. This can be called as lotus of Brahma or the God of Creation in a metaphorical sense. The infusion of cometary genes from this extraterrestrial porphyrin and archaeal intelligent networks with God-like qualities into the earth affects the homo sapien and primate population. This results in creation of newer species with God-like characteristics who assume leadership in societies like Gauthama Buddha and Jesus Christ. They are induced by cometary genes from the extraterrestrial porphyrin and archaeal intelligent network. Such hybrid populations in the earth can be called as Neanderthals or Neoneanderthals due to changed characteristics consequent to extraterrestrial archaeal symbiosis. They form leaders in society and guides civilization. Civilization progresses due to infusion of genes from the extraterrestrial intelligent porphyrin and archaeal networks. The appearance of meteors and comets signify the infusion of cometary genes from the ultimate extraterrestrial porphyrin and archaeal network. These extraterrestrial porphyrin archaeal networks in the interstellar gas can also condense to form stars and planets. These networks as said before have a wave energetic form and a particulate form and are the mode of creation and destruction of the universe. Excessive growth of extraterrestrial archaeal networks in earth due to methanogenesis creates global warming and extinction of civilization. The extraterrestrial archaeal networks fission and fusion reaction fuel star systems as well as mediate their extinction. The extraterrestrial magnetite archaeal network is responsible for the intergalactic magnetic field in the galaxy and contributes to the formation of star systems. The extraterrestrial porphyrin and archaeal networks with their energetic and particulate bodies participate in the dance of creation and destruction of the universe and can be compared to the Hindu God Shiva.

The archaeal and porphyrin extraterrestrial network is intelligent and occupy the interstellar space and communicates with all galaxies, planets and star system by the phenomena of quantal non-local interaction and contributes to what is called as the collective unconscious or god. The extraterrestrial porphyrin and archaeal network form the centre of the interstellar gas and can condense to form stars and planets. Stars are fueled by fission and fusion reaction in the archaeal networks of interstellar clouds of hydrogen. Thus, the stars and planetary systems are alive and humanoid having a natural history of their own with life and death. The Hindu Gods and Rishis are fixed and attached to planetary and star systems. This would be a reflection of the impact of the extraterrestrial archaeal intelligence and the transfer of cometary genes to other planetary systems including earth. The extraterrestrial porphyrin and archaeal intelligent networks can build up the interstellar space, in other planets like Mars and Venus and other star systems. Intelligent civilization already exists in star systems and galaxies all over the universe. They are made of the extraterrestrial porphyrin and archaeal intelligent networks. The same extraterrestrial and induced terrestrial porphyrin and archaeal network gave rise to the zombie civilization of plants, bacteria, viruses, animals, primates and humans in earth with the terrestrial archaeal and porphyrin intelligent network as the real eternal structure. The zombie forms change from time to time. The terrestrial archaeal and porphyrin network having an energy/wave-particulate existence is eternal. Similar extraterrestrial archaeal and porphyrin intelligent network induced civilizations or the networks per se acting as civilized entities controlling the universe exists all over the galaxies and universe. This can be called as the god being. The transfer of cometary genes from the extraterrestrial porphyrin and archaeal network due to meteoric and cometary showers can be considered as visitations of the extraterrestrial archaeal intelligence to planetary systems and star systems to modulate such systems. They can be called as god astronauts in metaphorical sense. The human civilizations can be considered as zombies for the extraterrestrial archaeal intelligent network. The human zombie's creativity, leadership qualities and intelligence are proportionate to the infusion of cometary genes. The human zombies with the highest density of cometary genes from the extraterrestrial archaeal intelligent network are neanderthalic and dominate societies as kings and leaders. They are autistic and intelligent and form scientist, writers, poets, painters, dancers, artists and all creative icons of civilizations. They can be called as the illuminati of earth or the agents of the extraterrestrial archaeal intelligent network controlling the earth. The terrestrial history and civilization is also modulated by infusion of cometary genes by meteoric impacts from extraterrestrial archaeal intelligent network. This alters the genomic patterns of the homo sapien population by the archaea and the secreted RNA viroids. This results in reduction in cerebral cortical function, logical thinking, civilized behaviour and humanity as we know it. The society becomes cerebellar dominant and impulsive. The brain becomes a prisoner of surrealistic and transcendentalistic urges. This creates abstract art, anarchy, terrorism, sexual anarchy, alternate sexuality, rock music, globalization of culture and spirituality. This also leads to violence, terrorism, revolutions and wars which creates a sort of transcendence similar to spirituality. The world is dominated by impulsive people without logical thinking and with dominant extrasensory perception. The world becomes anarchy and globalized with uniformity in human behaviour and thought. The changes in society are brought about by increase in archaeal growth and increased transfer of extraterrestrial archaeal intelligent networks starting from the ice age. The Neanderthal hybrids were creative and created changes in society leading to an industrial revolution and a technological revolution contributing to global warming and extremes of temperature which leads to further growth of terrestrial and extraterrestrial archaeal intelligent networks. The aim of creating human zombies or hybrids that are creative could have been to create an environment which leads to further expansion of the terrestrial and extraterrestrial intelligent porphyrin and archaeal networks. The expansion of the terrestrial archaeal networks produces methanogenesis, global warming and further growth of terrestrial archaea. The growth of terrestrial archaea in the ocean beds can lead to earthquakes, continental drifts, tsunamis, rise in water levels, flooding and global catastrophe ending the universe and human zombies as we know of today. The terrestrial and extraterrestrial network of porphyrin and archaeal intelligence will continue till eternity. They can have a wave-particle existence reverting to their energy body or wave body when required and coming back to particulate existence from the quantal foam when required creating stars, planets and life again. This is the dance of life and death - the dance of Shiva.

The Biblical literature can be interpreted in terms of extraterrestrial archaeal networks and Neanderthal hybrids. The extraterrestrial archaeal networks contacts the terrestrial system via meteoric and cometary impacts. The homo sapien females with archaeal symbiosis give rise to Neanderthal hybrids. The homo sapien population have less of archaeal density and different characteristics. The Neanderthal hybrids were creative and autistic and became the rulers of society and kingship developed owing to the groups increased intellectual and physical abilities. The homo sapien females with archaeal symbiosis gave rise to Neanderthal hybrids and they were considered as equivalent to Mother Goddess in certain societies or as manifestation of Shakti. This gave rise to a matrilineal society and female dominance. The extraterrestrial archaeal network which forms stars and planetary system can be considered as a representative of Shiva and the female homo sapiens with archaeal symbiosis giving rise to Neanderthal hybrids can be considered as a consort of Shakti. The common homo sapiens viewed the Neanderthal hybrids with suspicion as they were different from the general population and dominant, ruling over them with their physical and intellectual ability. The common homo sapiens organized themselves as the church against Neanderthal hybrids which formed the state or kingship. The common homo sapiens considered the interbreeding of homo sapien females with extraterrestrial archaeal networks as original sin and the extraterrestrial archaeal networks contacting the earth via cometary impacts as fallen angels. The common homo sapiens considered sex as taboo and society was organized to prevent inter-breeding with Neanderthal hybrids. Hybrids were intelligent, creative, domineering with sexual freedom which was considered as sinful by the common homo sapien population and the church. Therefore, the position of women in the common homo sapien society with cerebral cortical dominance was inferior, sexual activity was sinful and society was patriarchal with strict moral codes. The Neanderthal hybrids, on the other hand were cerebellar dominant impulsive creatures with sexual freedom, loose moral codes and a communal life with matriarchal structure of society. The matrilineal societies with archaeal symbiosis are the Celts, Basques, Berbers, Jews, Sumerians, Dravidians and ancient Egyptians and they were all communal groupings of Neanderthal hybrids and dominated civilization for decades. This is noticed in the Egyptian civilization of Pharoahs, the Sumerian civilization, the ancient Jewish civilization, Assyrian civilization, the Iranian asuric civilization, Aztec civilization and the Harappan civilization. The Indian caste system is based on the phenomena of Neanderthal hybrids. The Neanderthal hybrids formed due to contact with the extraterrestrial archaeal networks consequent to meteoric impacts formed the upper caste of Brahmins and Kshatriyas. The common homo sapien population formed the Shudras. The Neanderthal hybrids of the upper caste ruled over the society and kept a common homo sapien population of lower caste as slaves. Breeding laws were made and a caste groups made water tight to keep the purity of the Neanderthal hybrids. The common homo sapien population all over the world retained its patriarchal moorings and male dominance and became predominantly Christian and Muslim. These societies were cerebral cortical dominant, civil societies were equality and respect for people was the norm. The Neanderthal hybrids with the creativity and extrasensory powers became Hindus, Buddhists and Jews. These societies were divided on the basis of caste and for the necessity of keeping the Neanderthal hybrids pure and compact. The Sumerian texts, the Egyptian papyrus, the Rig Veda, the ancient Jewish testaments, the Assyrian literature are Neanderthal hybrid in origin and contains version of the extraterrestrial archaeal networks of the interstellar space giving rise to stars and planetary system. The Gods of Egypt, Hindus, Assyrians and Jews are connection with the stars. This can be noted in the ritual of pagan religions like those in Stonehenge, the rituals of Hinduism and those of ancient Egypt as well as Aztecs which had relation to the star systems indicating contact with extraterrestrial intelligent archaeal networks. The Neanderthal hybrids have cerebellar dominance and cerebral cortical dysfunction and could attain transcendence to the collective unconscious and connect with the extraterrestrial archaeal particulate or energy network with ease. Most of the religious spiritual leaders could have been Neanderthal hybrids. The kings and warriors were also Neanderthal hybrids. The act of war or the act of killing gives a sense of transcendence and merger with the collective unconscious as seen in the

Bhagavad Gita where the act of war is considered as a dharma and leads to transcendence. The cortical function of logical thinking is obliterated. The Nazis including Hitler and Himmler believed in the philosophy of the Bhagavad Gita which justified the world wars as an act of dharma. There is also the concept of avatars of Vishnu descending from the heaven to save mankind. This includes Krishna and Buddha. They can be considered as Neanderthal hybrids bred in contact with the extraterrestrial intelligent archaeal colony particulate or energy network mediated by archaeal colony astronauts showered on to the earth by meteoric impacts. All the political revolutions on earth had a neanderthalic hybrid leadership - Nazism, Fascism, Communism, the French Revolution and the royal kingship. The French Revolution and Communist Revolution despite its egalitarian ethos were basically led by Jews. The modern revolution of the dissolution of nation-states and globalization of capital and economy also has a Jewish leadership of Neanderthal hybrids. The technological revolution of the internet which powers the globalised world is also dominated by the impulsive cerebellar dominant Neanderthal hybrids. Low level of EMF fields of the internet can increase archaeal growth and archaeal magnetite mediated perception of low level EMF producing cortical atrophy and cerebellar dominance which can be transmitted genetically. The Internet revolution is powered by Neanderthal hybrids who are basically autistic and have extrasensory perception due to higher brain density of archaeal magnetite. The Neanderthal hybrids dominate the world with their intellectual and physical superiority and can be considered as a universal ruling caucus or illuminati.

The modern day neanderthalic religion mediated by the intergalactic archaeal molecular quantum computing cloud includes Hinduism, Jewish Kabala, the Islamic Sufism, the Christian Gnostism and the Zen Buddhism. These are all cerebellar cortical religions. They have an extraterrestrial content, are impulsive, loose, ritualistic, sexual, sensual and primitive. The correlates of these can be surrealism, matriarchal, global, violence, gender-equal, alternate sexuality, non-democratic and with extrasensory quantal perception. They are environmental friendly and worship environment as God. They are prone to civilizational diseases like tumours, metabolic syndrome, degenerations, psychiatric illness and autoimmunity and in that sense were considered by the homo sapien secular religions as satanic. They are equivalent to the vampires, the demons, the elf, the goblins and Asuras. The homo sapien secular religions include the Wahabi Islam, modern communism and the protestant and Roman Catholic Christianity. The homo sapien religions are cerebral cortical religions of civil societies. These homo sapien religions give a civilized code of social conduct. The correlates of it are non-violence, ahimsa, morality, patriarchal, female-suppression, gender inequality, strict heterosexual moral codes and anti-environmental.

References

1. Kurup, R.K. and Kurup, P.A. *Theory of Everything – A Biological Viewpoint.* New York: Open Science Publishers, 2016.

CHAPTER 43

CLIMATE CHANGE AND CATASTROPHIC PANDEMICS – A HUMAN ORIGIN FOR EMERGING VIRUSES AND COVID 19 - INFECTIVE SYMBIOTIC TISSUE COLONIES OF PROTOZOA, ARCHAEA, VIRUSES AND BACTERIA

The plasma life based forms and the gravitational and anti-gravitational waves generated porphyrions would have been the most primitive of non-life forms. The porphyrin templates would have given rise to the isoprenoid organism, the RNA viroids, DNA viroids and prions which would have symbiosed to form the nanoarchaea. The archaea would have evolved into prokaryotes or bacteria. The archaeal symbiosis with prokaryotes would have generated the eukaryotes. The initial unicellular eukaryotes are the protozoa. The protozoa are capable of phagocytosis and would have had viruses, bacteria and archaea as their intracellular symbionts. The viruses of the protozoa like entamoeba and toxoplasma are giant viruses called megaviride. The protozoans like entamoeba with its viral, archaeal and bacterial endosymbionts would have evolved into the multicellular organisms or metazoa. The archaeal, bacterial and protozoal extracellular vesicles containing RNA and DNA viroids can by horizontal gene transfer lead to evolution of metazoan, plants, animals and human species. The metazoans would have evolved into plants, animals and humans by horizontal gene transfer generated by archaeal extracellular RNA and DNA viroid vesicles. The evidence of such a transformation is seen in the bacterial symbionts of protozoa like entamoeba which includes legionella, listeria, mycoplasma, mycobacteria and chlamydia. The symbionts of protozoan entamoeba include viruses like covid, HIV and cytomegalo virus. The protozoan entamoeba serves as a natural reservoir for covid infections, bacterial infections and HIV infections. The toxoplasma also serves as a covid reservoir. These protozoan colonies of entamoeba and toxoplasma with its bacterial, archaeal and

viral symbionts can be called as chimeric infective symbiotic tissue colonies which can exist in the air, water, earth, blood and tissues. The chimeric infective symbiotic tissue colonies containing bacteria, archaea and viruses can make the symbiotic organism multidrug resistant by horizontal gene transfer. The chimeric infective symbiotic tissue organism evolves as a consequence of global warming and the protozoal, bacterial, viral and archaeal symbiotic colonies are new forms of life. The chimeric infective symbiotic tissue organism composed of protozoa and its symbionts like virus, bacteria and archaea depend upon glycolysis for its energy needs. The glycolytic enzymes enolase, glyceraldehyde 3-phosphate dehydrogenase, hexokinase, phosphofructokinase, aldolase, pyruvic kinase and triose phosphate isomerase are moonlighting proteins having a cytoplasmic, membrane and nuclear functions. These glycolytic enzymes have a RNA binding motif which can modulate RNA transcription in the host and the symbiotic organism. These glycolytic enzymes with their RNA binding motif as well as DNA binding motif can bind to RNA and DNA viruses and viroids modulating their replication and life cycles. The chimeric infective symbiotic tissue organism composed of protozoa and its symbionts like virus, bacteria and archaea can enter the human host through the breaches of the gut-blood barrier and transform itself into the macrophages of the human host. The macrophages are involved in the phagocytosis and innate immunity against bacterial and viral infections. Paradoxically the macrophages like the protozoa can also serve as a reservoir where pathogens like mycobacteria tuberculosis, mycobacteria leprae, listeria, legionella, E. coli, shigella & salmonella, and viruses like HIV & covid can thrive and replicate. The modified chimeric infective symbiotic tissue organism or macrophage composed of protozoa and its symbionts like virus, bacteria and archaea can serve as a reservoir for horizontal gene transfer and generation of multidrug resistant bacteria and viruses. The infective chimeric symbiotic tissue organism composed of protozoa and its symbionts like virus, bacteria and archaea depends upon moonlighting glycolytic enzymes for their energetics and

replication. The development of antibodies against the infective chimeric symbiotic tissue organism composed of protozoa and its symbionts like virus, bacteria and archaea leads to the evolution of anti-brain antibody disease or anti-basal ganglia antibody disease. They present a threat to existence consequent to horiztontal gene transfer and generation of multidrug resistant bacteria, viruses, protozoa, mycobacteria and archaea.

Climate change can lead to extremophilic endosymbiotic archaeal growth and neanderthalisation of the species. There is increase in endosymbiotic and colonic archaeal growth. The increase in colonic archaeal growth leads to breaching of the blood-gut barrier and archaeal endosymbiosis. The exposure to the low level of EMF fields consequent to widespread use of the internet, radiosatellites and radar equipments leads to increased endosymbiotic archaeal growth. Archaeal endosymbiosis leads to neanderthalisation of the species. The colonic and endosymbiotic archaea also generates RNA viroids which are converted to DNA viroids by colonic epithelial HERV reverse transcriptase and gets integrated into the human genome. The RNA and DNA viroids hybridise with the viroidal and bacterial population of the gut microbiome and virobiome DNA and RNA of the gut leading to generation of new emerging viruses like COVID 19. The RNA and DNA viroids and gut bacteria can also hybridise with human genomic sequences for immunity, metabolic regulation, growth factors, cell growth, cell death, neurotransmitters and HERV sequences of the colonial epithelial cell. The hybridization of archaeal RNA viroids with human genomic HERV sequences can generate retroviruses. This leads to generation of new bacteria, RNA and DNA viruses modulating cell death, cell growth, cell proliferation, metabolism, endocrine function and behaviour producing autoimmunity, schizophrenia, autism, cancer diabetes mellitus, coronary vascular disease, cerebrovascular disease and neurodegeneration.

Dietary fibre is the most important factor that modulates human biology and speciation. Dietary fibre and human drugs - antibiotic and non-antibiotic can modulate the gut flora producing colonic archaeal growth and endosymbiosis generating climate change, evolution of superbugs and emerging viruses and creating the disease phenotype. The high fibre diet is digested by the colonic flora leading to generation of short chain fatty acids and inhibits endosymbiotic and colonic archaeal growth. This leads to homo sapienisation of the species. A low fibre diet with increase in protein and fat leads to stimulation of endosymbiotic and colonic archaeal growth. This leads to neanderthalisation of the species. Thus, a low fibre diet alters the gut microbiome in the colon with more of archaeal growth. The archaea develop protective mechanism of antibiotic resistance and dominates the colonic microflora. The colonic archaea by natural selection and DNA/gene exchanges with other colonic and endosymbiotic organism transfer the genes for antibiotic resistance to other colonic flora. This forms the basis for the generation of superbugs resistant to all antibiotics. The same mechanisms happen when populations are treated with antibiotics and other drugs like anti-hypertensives, anti-psychotics and analgesics. They kill the colonic bacterial flora and the extremophilic archaea dominates the colonic microbiome with acquired protective mechanisms for antibiotic resistance. These archaeal antibiotic resistance genes are transferred by gene and plasmid DNA exchange with remnant colonic and endosymbiotic micro-organism generating superbugs. Thus antibiotic and other non-antibiotic drugs promote colonic archaeal growth and endosymbiosis producing neanderthalisation of the species and generation of superbugs which kills of the homo sapien species. Thus the generation of superbugs and emerging viruses is promoted by a low fibre diet. A low fibre diet leads to colonic and endosymbiotic archaeal growth leading to neanderthalisation of the species and generation of superbugs and emerging viruses to which the neanderthalised species is resistant consequent to endogenous digoxin secretion. The neanderthalised species generated superbugs and emerging viruses kills of the homo sapien species which are sensitive to the superbugs and emerging viruses and do not have resistance to it. Thus the modern armamentarium of antibiotics, antipsychotics, analgesics, anti-hypertensives and anti-diabetic drugs promote colonic and endosymbiotic archaeal growth development of superbugs, neanderthalisation of species and induction of civilizational disease. Thus, the drugs used to treat infections and human disease tend to produce superbugs, emerging virus infection and civilizational diseases- cancer, autoimmune disease, metabolic syndrome X, neurodegeneration, schizophrenia and autism. Thus, the drug armamentarium of antibiotics and non-antibiotic drugs used to treat human infections and diseases increases colonic and endosymbiotic archaeal growth, neanderthalises the species, generates superbugs and leads to increased incidence of civilizational disease. The gut microbiome and virobiome modulated by dietary fibre intake, high protein high fat diets, antibiotics and non-antibiotic drugs like anti-psychotics, anti-hypertensives, analgesics leads to alteration in gut microbial and viroidal bio-geography leading to generation of superbugs and superviruses. The gut archaea highly resistant to conventional antibiotics and non-antibiotic drugs and proliferate when antibiotics and non-antibiotic drugs like anti-hypertensives, anti-psychotics and analgesics are used on a widespread basis. They develop antibiotic resistance and non-antibiotic drug resistance and it is transferred to other colonic and gut bacteria and viruses generating superbugs and superviruses exterminating the homo sapien population. The growth and dominance of the colonic and endosymbiotic bacteria leads to generation of homo neanderthalic species which is resistant to superbugs and superviruses and survive but get affected by civilizational diseases like cancer, autoimmunity, metabolic syndrome, schizophrenia, autism and neurodegeneration. Homo neoneanderthalis thus generated by colonic and endosymbiotic archaeal growth can become extinct over a slow period of time. Homo neanderthalis with its extremophilic archaeal endosymbiosis is hardy and resistant and can survive global warming and extremes of climate change as well as surviving extremophilic anaerobic conditions of other planets and galactic systems. The dietary fibre intake inhibits archaeal growth and populations with low dietary fibre intake are resistant to civilizational diseases like metabolic syndrome, cancer, autoimmune disease and schizophrenia. The human methanogenic archaea are highly resistant to routine antibiotics and non-antibiotic drugs and the intake of such drugs kills the colonic microbiome leading to archaeal growth and endosymbiosis and neanderthalisation to produce a hardy homo galacticus species capable of surviving in extreme climates of earth and other planets. The human drugs and antibiotics induced modulation of the gut microbiome with increased colonic and endosymbiotic archaeal growth also contributes to global warming via the mechanism of archaeal methanogenesis. The widespread use of antibiotics and non-antibiotic drugs contribute to colonic and endosymbiotic archaeal growth, climate change and still further more archaeal growth. This human colonic and endosymbiotic archaeal growth consequent to stress, internet fields, low level EMF, low dietary fibre intake, antibiotic use and non-antibiotic drug use producing methanogenesis is the principal cause of global warming. The widespread use of antibiotic and non-antibiotic drugs kills of the colonic microflora and replaces it with archaea resistant to antibiotic and non-antibiotic drugs. The low intake of dietary fibre promotes colonic archaeal and endosymbiotic archaeal growth. Stress produces catecholamine induced increased in colonic and endosymbiotic archaeal growth and breaching of the gut-blood barrier. The widespread use of internet and low level of EMF field of mobile phones promotes endosymbiotic archaeal growth. Thus endosymbiotic and colonic archaeal growth and methanogenesis producing global warming is a endogenous human biological event with human biological activity as a centre piece triggering the process. The widespread use of antibiotics and non-antibiotic drugs treat infections viral and bacterial as well as civilizational disease like cancer, schizophrenia, mood disorder, depression, autism, autoimmune disease and neurodegeneration promotes colonic and endosymbiotic archaeal growth which neanderthalises the species leading to still further transformation of the human system leading to induction of civilizational disease like cancer, metabolic syndrome X, neurodegeneration, autoimmunity and vascular disease. Thus the antibiotic and non-antibiotic drugs used to treat infections and systemic diseases produces colonic and endosymbiotic archaeal growth leading to superbugs, superviruses, pandemics and epidemic of civilizational diseases mediated by colonic and endosymbiotic archaeal growth. Human medicine as we know it has reached a dead end. The antibiotic and non-antibiotic era of drugs had come to an end. The human earth as we know it also undergoes catastrophic climate change leading to extinction of the human species and conditions for human habitations in earth. The modern medicine as we know of and the earth as we know of have ended. The principal cause of climate change and its catastrophic complications is contributed in part only by the destruction of the tropical Amazonian and rain forest. The deforestation of the gut microflora created over billions of years of human evolution is the principal cause of the climate change and a possible end of the earth as we know of and the human species as we know of. The revolutionary medicine and the golden earth with its habitable conditions have ended. The increase in dietary fibre intake upto 40 g/day can inhibit colonic archaeal growth and endosymbiosis putting a break on generation of superbugs, emerging viruses and climate change. Increase in dietary fibre intake of populations can prevent the state of climate change and gut microbial change putting human life and the life of the planet at risk. Dietary fibre can be considered as the elixir of life and endosymbiotic archaea the philosopher's stone.

Endosymbiotic actinidic archaea as having evolved from an early isoprenoid organisms by abiogenesis. An actinidc archaea/viroid mediated model of prokaryote, viral, eukaryotic, primate and human evolution is discussed.[1-4]. Actinidic archaea has been related to the pathogenesis of schizophrenia, malignancy, metabolic syndrome X, autoimmune disease and

neuronal degeneration[2]. Actinidic archaea have a mevalonate pathway and cholesterol catabolism[5-7]. Davies has put forward the concept of a shadow biosphere of organisms with alternate biochemistry present in earth itself[8]. An actinide dependent shadow biosphere of archaea and viroids in the above mentioned disease states is described[6]. Metal actinides in beach sands have been postulated to play a role in abiogenesis[6]. Actinide mineral like rutile, monazite and illmenite by surface metabolism would have contributed to abiogenesis[9]. A hypothesis of cholesterol as the primal prebiotic molecule synthesised on actinide surfaces with all other biomolecules arising from it and a self-replicating cholesterol lipid organism as the initial life form is presented. The actinidic archaea and viroids would have evolved from the primitive isoprenoid organism. The origin of viruses, prokaryotes, eukaryotes, primates and humans from the initial isoprenoid organism derived actinidic archaea is postulated.

Materials and Methods

Informed consent of the subjects and the approval of the Ethics Committee were obtained for the study. The following groups were included in the study:- endomyocardial fibrosis, Alzheimer's disease, multiple sclerosis, non-Hodgkin's lymphoma, metabolic syndrome X with cerebrovascular thrombosis and coronary artery disease, schizophrenia, autism, seizure disorder, Creutzfeldt-Jakob's disease and acquired immunodeficiency syndrome. There were 10 patients in each group and each patient had an age and sex matched healthy control selected randomly from the general population. The blood samples were drawn in the fasting state before treatment was initiated. Plasma from fasting heparinised blood was used and the experimental protocol was as follows:- (I) Plasma+phosphate buffered saline, (II) same as I+cholesterol substrate, (III) same as II+rutile 0.1 mg/ml, and (IV) same as II+ciprofloxacine and doxycycline each in a concentration of 1 mg/ml. Cholesterol substrate was prepared as described by Richmond[10]. Aliquots were withdrawn at zero time

immediately after mixing and after incubation at 37 °C for 1 hour. The following estimations were carried out:- Cytochrome F420, free RNA, free DNA, muramic acid, polycyclic aromatic hydrocarbon, hydrogen peroxide, serotonin, pyruvate, ammonia, glutamate, cytochrome C, hexokinase, ATP synthase, HMG CoA reductase, digoxin and bile acids[11-14]. Cytochrome F420 was estimated flourimetrically (excitation wavelength 420 nm and emission wavelength 520 nm). Polycyclic aromatic hydrocarbon was estimated by measuring hydrogen peroxide liberated by using glucose reagent. The statistical analysis was done by ANOVA.

Results

The parameters checked as indicated above were - cytochrome F420, free RNA, free DNA, muramic acid, polycyclic aromatic hydrocarbon, hydrogen peroxide, serotonin, pyruvate, ammonia, glutamate, cytochrome C, hexokinase, ATP synthase, HMG CoA reductase, digoxin and bile acids. Plasma of control subjects showed increased levels of the above mentioned parameters with after incubation for 1 hour and addition of cholesterol substrate resulted in still further significant increase in these parameters. The plasma of patients showed similar results but the extent of increase was more. The addition of antibiotics to the control plasma caused a decrease in all the parameters while addition of rutile increased their levels. The addition of antibiotics to the patient's plasma caused a decrease in all the parameters while addition of rutile increased their levels but the extent of change was more in patient's sera as compared to controls.

Discussion

There was increase in cytochrome F420 indicating archaeal growth. The archaea can synthesize and use cholesterol as a carbon and energy source[15,16]. The archaeal origin of the enzyme activities was indicated by antibiotic induced

suppression. The study indicates the presence of actinide based archaea with an alternate actinide based enzymes or metalloenzymes in the system as indicated by rutile induced increase in enzyme activities[17]. There was also an increase in archaeal HMG CoA reductase activity indicating increased cholesterol synthesis by the archaeal mevalonate pathway. The archaeal beta hydroxyl steroid dehydrogenase activity indicating digoxin synthesis and archaeal cholesterol hydroxylase activity indicating bile acid synthesis were increased[7]. The archaeal cholesterol oxidase activity was increased resulting in generation of pyruvate and hydrogen peroxide[16]. The pyruvate gets converted to glutamate and ammonia by the GABA shunt pathway. The archaeal aromatization of cholesterol generating PAH, serotonin and dopamine was also detected[18]. The archaeal glycolytic hexokinase activity and archaeal extracellular ATP synthase activity were increased. The archaea can undergo magnetite and calcium carbonate mineralization and can exist as calcified nanoforms[19].

The metal actinides provide radiolytic energy, catalysis for oligomer formation and provide a coordinating ion for metalloenzymes all important in abiogenesis[6]. The metal actinide surfaces would by surface metabolism generate acetate which could get converted to acetyl CoA and then to cholesterol which functions as the primal prebiotic molecule self-organising into self-replicating supramolecular systems, the lipid organism[8,9,20]. Cholesterol by radiolysis by actinides would have formed PAH generating PAH aromatic organism[8]. Cholesterol radiolysis would generate pyruvate which would get converted to amino acids, sugars, nucleotides, porphyrins, fatty acids and TCA acids. Anastase and rutile surfaces can produce polymerization of amino acids, isoprenyl residues, PAH and nucleotides to generate the initial lipid organism, PAH organism, prions and RNA viroids which would have symbiosed to generate the archaeal protocell. The archaea evolved into gram negative and gram positive bacteria with a mevalonate pathway which had an evolutionary advantage. The symbiosis of archaea with gram negative organism generated

the eukaryotic cell[21]. The data supports the persistence of an actinide and cholesterol based shadow biosphere which throws light on the actinide based origin of life and cholesterol as the premier prebiotic molecule.

There was an increase in free RNA indicating self-replicating RNA viroids and free DNA indicating generation of viroid complementary DNA strands by archaeal reverse transcriptase activity. The actinides modulate RNA folding and catalyse its ribozymal action. Digoxin can cut and paste the viroidal strands by modulating RNA splicing generating RNA viroidal diversity. The viroids are evolutionarily escaped archaeal group I introns which have retrotransposition and self-splicing qualities. Archaeal pyruvate can produce histone deacetylase inhibition resulting in endogenous retroviral (HERV) reverse transcriptase and integrase expression. This can integrate the RNA viroidal complementary DNA into the non-coding region of eukaryotic non-coding DNA using HERV integrase as has been described for borna and ebola viruses[21]. The non-coding DNA is lengthened by integrating RNA viroidal complementary DNA with the integration going on as a continuing event. The archaea genome can also get integrated into human genome using integrase as has been described for trypanosomes. The integrated viroids and archaea can undergo vertical transmission and can exist as genomic parasites. This increases the length and alters the grammar of the non-coding region producing memes or memory of acquired characters. The viroidal complementary DNA can function as jumping genes producing a dynamic genome.[22-24]

The presence of muramic acid, HMG CoA reductase and cholesterol oxidase activity inhibited by antibiotics indicates the presence of bacteria with mevalonate pathway. The bacterial with mevalonate pathway include streptococcus, staphylococcus, actinomycetes, listeria, coxiella and borrelia. The bacteria and archaea with mevalonate pathway and cholesterol catabolism had an evolutionarily advantage and constitutes the isoprenoidal clade organism. The archaea evolved into mevalonate pathway gram positive and gram negative isoprenoid clade organism through horizontal gene transfer of viroidal and virus genes. The isoprenoidal clade prokaryotes develop into other groups of prokaryotes via viroidal/virus as well as eukaryotic horizontal gene transfer producing bacterial speciation[25-27].

The RNA viroids and its complementary DNA developed into cholesterol enveloped RNA and DNA viruses like herpes, retrovirus, influenza virus, borna virus, cytomegalo virus and ebstein barr virus by recombining with eukaryotic and human genes resulting in viral speciation. Bacterial and viral species are ill-defined and fuzzy with all of them forming one common genetic pool with frequent horizontal gene transfer and recombination.

Thus, the multi and unicellular eukaryote with its genes serves the purpose of prokaryotic and viral speciation. The multicellular eukaryote developed so that their endosymbiotic archaeal colonies could survive and forage better. The multicellular eukaryotes are like bacterial biofilms. The archaea and bacteria with a mevalonate pathway uses the extracellular RNA viroids and DNA viroids for quorum sensing and in the generation of symbiotic biofilm like structures which develop into multicellular eukaryotes. The endosymbiotic archaea and bacteria with mevalonate pathway still uses the RNA viroids and DNA viroids for the regulation of multicellular eukaryote.[28-31]

Pollution is a major inducer of evolutionary innovation. Pollution is induced by the primitive nanoarchaea and mevalonate pathway bacteria synthesized PAH and methane leading on to redox stress. Redox stress leads to sodium-potassium ATPase inhibition, inward movement of plasma membrane cholesterol, defective SREBP sensing, increased cholesterol synthesis and nanoarchaeal/mevalonate pathway bacterial growth. Redox stress leads on to viroidal and archaeal multiplication. Redox stress can also lead to HERV reverse transcriptase and integrase expression. The non-coding DNA is formed of integrating RNA viroidal complementary DNA and archaea with the integration going on as a continuing event. The archaeal pox-like dsDNA virus forms evolutionarily the nucleus. The integrated viroidal, archaeal and mevalonate pathway bacterial sequences can undergo vertical transmission and can exist as genomic parasites. The genomic integrated archaea, mevalonate pathway bacteria and viroids form a genomic reserve of bacteria and viruses which can recombine with human and eukaryotic genes producing bacterial and viral speciation. The change in the length and grammar of the non-coding region produces eukaryotic speciation and individuality. The integration of nanoarchaea, mevalonate pathway prokaryotes and viroids in to the eukaryotic and human genome produces a chimera which can multiply producing biofilm like multicellular structures having a mixed archaeal, viroidal, prokaryotic and eukaryotic characters which is a regression from the multicellular eukaryotic tissue. This results in a new neuronal, metabolic, immune and tissue phenotype leading to human disease.[30-32]

Pollution would have been a major factor in eukaryotic speciation and primate/hominid evolution. The change in the length and grammar of the non-coding region produces eukaryotic speciation and individuality. It is the increase in non-coding region and HERV sequences of the genome that led to the evolution of the primate and the human brain and its attendant property of conscious and quantal perception. It is the non-coding region of the genome with its archaeal, RNA viroidal complementary DNA and HERV sequences that makes for the human qualities of the hominid brain. Changes in the length of non-coding region can lead on to disorders of consciousness like schizophrenia. A schizophrenia specific human endogenous retroviruses and change in the length and grammar of the non-coding region has been described in schizophrenia.[33,34]

An actinide dependent shadow biosphere of archaea and viroids in the above mentioned disease states is described. Metal actinides in beach sands

have been postulated to play a role in abiogenesis. Cholesterol is the primal prebiotic molecule synthesized on actinide surfaces with all other biomolecules arising from it. A self-replicating cholesterol lipid organism could be the initial life form. A cholesterol based abiogenesis is a more likely evolutionary option and the actinidic archaea and viroids would have evolved from it. The origin of viruses, prokaryotes, eukaryotes, primates and humans from the initial isoprenoid organism derived actinidic archaea is discussed.

References

1. Valiathan M.S., Somers, K., Kartha, C.C. (1993). *Endomyocardial Fibrosis.* Delhi: Oxford University Press.
2. Kurup R., Kurup, P.A. (2009). *Hypothalamic Digoxin, Cerebral Dominance and Brain Function in Health and Diseases.* New York: Nova Science Publishers.
3. Hanold D., Randies, J.W. (1991). Coconut cadang-cadang disease and its viroid agent, *Plant Disease,* 75, 330-335.
4. Edwin B.T., Mohankumaran, C. (2007). Kerala wilt disease phytoplasma: Phylogenetic analysis and identification of a vector, *Proutista moesta, Physiological and Molecular Plant Pathology,* 71(1-3), 41-47.
5. Eckburg P.B., Lepp, P.W., Relman, D.A. (2003). Archaea and their potential role in human disease, *Infect Immun,* 71, 591-596.
6. Adam Z. (2007). Actinides and Life's Origins, *Astrobiology,* 7, 6-10.
7. Schoner W. (2002). Endogenous cardiac glycosides, a new class of steroid hormones, *Eur J Biochem,* 269, 2440-2448.
8. Davies P.C.W., Benner, S.A., Cleland, C.E., Lineweaver, C.H., McKay, C.P., Wolfe-Simon, F. (2009). Signatures of a Shadow Biosphere, *Astrobiology,* 10, 241-249.
9. Wächtershäuser G. (1988). Before enzymes and templates: theory of surface metabolism, *Microbiol Rev,* 52(4), 452-84.
10. Richmond W. (1973). Preparation and properties of a cholesterol oxidase from nocardia species and its application to the enzymatic assay of total cholesterol in serum, *Clin Chem,* 19, 1350-1356.
11. Snell E.D., Snell, C.T. (1961). *Colorimetric Methods of Analysis.* Vol 3A. New York: Van NoStrand.
12. Glick D. (1971). *Methods of Biochemical Analysis.* Vol 5. New York: Interscience Publishers.
13. Colowick, Kaplan, N.O. (1955). *Methods in Enzymology.* Vol 2. New York: Academic Press.
14. Maarten A.H., Marie-Jose, M., Cornelia, G., van Helden–Meewsen, Fritz, E., Marten, P.H. (1995). Detection of muramic acid in human spleen, *Infection and Immunity,* 63(5), 1652 – 1657.

15. Smit A., Mushegian, A. (2000). Biosynthesis of isoprenoids via mevalonate in Archaea: the lost pathway, *Genome Res,* 10(10), 1468-84.

16. Van der Geize R., Yam, K., Heuser, T., Wilbrink, M.H., Hara, H., Anderton, M.C. (2007). A gene cluster encoding cholesterol catabolism in a soil actinomycete provides insight into Mycobacterium tuberculosis survival in macrophages, *Proc Natl Acad Sci USA,* 104(6), 1947-52.

17. Francis A.J. (1998). Biotransformation of uranium and other actinides in radioactive wastes, *Journal of Alloys and Compounds,* 271(273), 78-84.

18. Probian C., Wülfing, A., Harder, J. (2003). Anaerobic mineralization of quaternary carbon atoms: Isolation of denitrifying bacteria on pivalic acid (2,2-Dimethylpropionic acid), *Applied and Environmental Microbiology,* 69(3), 1866-1870.

19. Vainshtein M., Suzina, N., Kudryashova, E., Ariskina, E. (2002). New Magnet-Sensitive Structures in Bacterial and Archaeal Cells, *Biol Cell,* 94(1), 29-35.

20. Russell M.J., Martin, W. (2004). The rocky roots of the acetyl-CoA Pathway, *Trends in Biochemical Sciences,* 29, 7.

21. Margulis L. (1996). Archaeal-eubacterial mergers in the origin of Eukarya: phylogenetic classification of life, *Proc Natl Acad Sci USA,* 93, 1071-1076.

22. Tsagris E.M., de Alba, A.E., Gozmanova, M., Kalantidis, K. (2008). Viroids, *Cell Microbiol,* 10, 2168.

23. Horie M., Honda, T., Suzuki, Y., Kobayashi, Y., Daito, T., Oshida, T. (2010). Endogenous non-retroviral RNA virus elements in mammalian genomes, *Nature,* 463, 84-87.

24. Hecht M., Nitz, N., Araujo, P., Sousa, A., Rosa, A., Gomes, D. (2010). Genes from Chagas parasite can transfer to humans and be passed on to children. Inheritance of DNA Transferred from American Trypanosomes to Human Hosts, *PLoS ONE,* 5, 2-10.

25. Flam F. (1994). Hints of a language in junk DNA, *Science,* 266, 1320.

26. Horbach S., Sahm, H., Welle, R. (1993). Isoprenoid biosynthesis in bacteria: two different pathways? *FEMS Microbiol Lett,* 111, 135–140.

27. Gupta R.S. (1998). Protein phylogenetics and signature sequences: a reappraisal of evolutionary relationship among archaebacteria, eubacteria, and eukaryotes, *Microbiol Mol Biol Rev,* 62, 1435–1491.

28. Hanage W., Fraser, C., Spratt, B. (2005). Fuzzy species among recombinogenic bacteria, *BMC Biology,* 3, 6-10.

29. Whitchurch C.B., Tolker-Nielsen, T., Ragas, P.C., Mattick, J.S. (2002). Extracellular DNA Required for Bacterial Biofilm Formation. *Science,* 295(5559), 1487.

30. Webb J.S., Givskov, M., Kjelleberg, S. (2003). Bacterial biofilms: prokaryotic adventures in multicellularity, *Curr Opin Microbiol,* 6(6), 578–85.

31. Chen Y., Cai, T., Wang, H., Li, Z., Loreaux, E., Lingrel, J.B. (2009). Regulation of intracellular cholesterol distribution by Na/K-ATPase, *J Biol Chem,* 284(22), 14881-90.

32. Poole A.M. (2006). Did group II intron proliferation in an endosymbiont-bearing archaeon create eukaryotes? *Biol Direct,* 1, 36-40.

33. Villarreal L.P. (2006). How viruses shape the tree of life, *Future Virology,* 1(5), 587-595.

34. Lockwood M. (1989). *Mind, Brain and the Quantum.* Oxford: B. Blackwell.

CHAPTER 44

CLIMATE CHANGE, SYMBIOSIS AND HUMAN EVOLUTION TO COUNTERACT CLIMATE CHANGE AND EMERGING VIRAL INFECTION – CARBON DIOXIDE FIXING, AMMONIA OXIDISING AND NITRIFYING ENDOSYMBIOTIC HUMAN ARCHAEA AND SURVIVAL DURING GLOBAL WARMING – THE HOMO NEONEANDERTHALIS RESISTANT TO COVID 19 AND HOMO SAPIEN EXTINCTION - INFECTIVE SYMBIOTIC TISSUE COLONIES OF PROTOZOA, ARCHAEA, VIRUSES AND BACTERIA

The plasma life based forms and the gravitational and anti-gravitational waves generated porphyrions would have been the most primitive of non-life forms. The porphyrin templates would have given rise to the isoprenoid organism, the RNA viroids, DNA viroids and prions which would have symbiosed to form the nanoarchaea. The archaea would have evolved into prokaryotes or bacteria. The archaeal symbiosis with prokaryotes would have generated the eukaryotes. The initial unicellular eukaryotes are the protozoa. The protozoa are capable of phagocytosis and would have had viruses, bacteria and archaea as their intracellular symbionts. The viruses of the protozoa like entamoeba and toxoplasma are giant viruses called megaviride. The protozoans like entamoeba with its viral, archaeal and bacterial endosymbionts would have evolved into the multicellular organisms or metazoa. The archaeal, bacterial and protozoal extracellular vesicles containing RNA and DNA viroids can by horizontal gene transfer lead to evolution of metazoan, plants, animals and human species. The metazoans would have evolved into plants, animals and humans by horizontal gene transfer generated by archaeal extracellular RNA and DNA viroid vesicles. The evidence of such a transformation is seen in the bacterial symbionts of protozoa like entamoeba which includes legionella, listeria, mycoplasma, mycobacteria and chlamydia. The symbionts of protozoan

entamoeba include viruses like covid, HIV and cytomegalo virus. The protozoan entamoeba serves as a natural reservoir for covid infections, bacterial infections and HIV infections. The toxoplasma also serves as a covid reservoir. These protozoan colonies of entamoeba and toxoplasma with its bacterial, archaeal and viral symbionts can be called as chimeric infective symbiotic tissue colonies which can exist in the air, water, earth, blood and tissues. The chimeric infective symbiotic tissue colonies containing bacteria, archaea and viruses can make the symbiotic organism multidrug resistant by horizontal gene transfer. The chimeric infective symbiotic tissue organism evolves as a consequence of global warming and the protozoal, bacterial, viral and archaeal symbiotic colonies are new forms of life. The chimeric infective symbiotic tissue organism composed of protozoa and its symbionts like virus, bacteria and archaea depend upon glycolysis for its energy needs. The glycolytic enzymes enolase, glyceraldehyde 3-phosphate dehydrogenase, hexokinase, phosphofructokinase, aldolase, pyruvic kinase and triose phosphate isomerase are moonlighting proteins having a cytoplasmic, membrane and nuclear functions. These glycolytic enzymes have a RNA binding motif which can modulate RNA transcription in the host and the symbiotic organism. These glycolytic enzymes with their RNA binding motif as well as DNA binding motif can bind to RNA and DNA viruses and viroids modulating their replication and life cycles. The chimeric infective symbiotic tissue organism composed of protozoa and its symbionts like virus, bacteria and archaea can enter the human host through the breaches of the gut-blood barrier and transform itself into the macrophages of the human host. The macrophages are involved in the phagocytosis and innate immunity against bacterial and viral infections. Paradoxically the macrophages like the protozoa can also serve as a reservoir where pathogens like mycobacteria tuberculosis, mycobacteria leprae, listeria, legionella, E. coli, shigella & salmonella, and viruses like HIV & covid can thrive and replicate. The modified chimeric infective symbiotic tissue organism or macrophage composed of protozoa and its symbionts like virus, bacteria and

archaea can serve as a reservoir for horizontal gene transfer and generation of multidrug resistant bacteria and viruses. The infective chimeric symbiotic tissue organism composed of protozoa and its symbionts like virus, bacteria and archaea depends upon moonlighting glycolytic enzymes for their energetics and replication. The development of antibodies against the infective chimeric symbiotic tissue organism composed of protozoa and its symbionts like virus, bacteria and archaea leads to the evolution of anti-brain antibody disease or anti-basal ganglia antibody disease. They present a threat to existence consequent to horiztontal gene transfer and generation of multidrug resistant bacteria, viruses, protozoa, mycobacteria and archaea.

Global warming leads to endosymbiotic archaeal growth. The endosymbiotic archaea produces methane by methanogenesis contributing to global warming. The endosymbiotic archaea develops into cell organelle called archaeaons which can function as vitaminocytes, neurotransminoids, steroidelle and viroidelle creating newer cell organelle and endogenous synthetic biology. This changes the cell and tissue function and structure leading on to a new human phenotype called homo neoneanderthalis. The homo neoneanderthalis is a hardy species which can resist climate change and thrive in environments of high temperature, high atmospheric carbon dioxide and water shortage. The homo neoneanderthalis can also resist emerging viral infection like SARS, MERS, HIV and COVID 19. The new organelle that are developed can withstand climate change and fix carbon dioxide and ammonia leading to endogenous synthesis of carbohydrates and protein. Ammonia oxidation is an essential part of the global nitrogen cycling and was long thought to be driven only by bacteria. Recent findings expanded this pathway also to the archaea. However, most questions concerning the metabolism of ammonia-oxidizing archaea, such as ammonia oxidation and potential CO_2 fixation, remain open, especially for terrestrial environments. Here, we investigated the activity of ammonia-oxidizing archaea and bacteria in an agricultural soil by comparison of RNA- and DNA-stable isotope probing (SIP). RNA-SIP demonstrated a highly dynamic and diverse community involved in CO_2 fixation and carbon assimilation coupled to ammonia oxidation. DNA-SIP showed growth of the ammonia-oxidizing bacteria but not of archaea. Furthermore, the analysis of labeled RNA found transcripts of the archaeal acetyl-CoA/propionyl-CoA carboxylase (*accA/pccB*) to be expressed and labeled. These findings strongly suggest that ammonia-oxidizing archaeal groups in soil autotrophically fix CO_2 using the 3-hydroxypropionate–4-hydroxybutyrate cycle, one of the two pathways recently identified for CO_2 fixation in *Crenarchaeota.* Catalyzed reporter deposition (CARD)-FISH targeting the gene encoding subunit A of ammonia monooxygenase (*amoA*) mRNA and 16S rRNA of archaea also revealed ammonia-oxidizing archaea to be numerically relevant among the archaea in this soil. Our results demonstrate a diverse and dynamic contribution of ammonia-oxidizing archaea in soil to nitrification and CO_2 assimilation and that their importance to the overall archaeal community might be larger than previously thought. Now there is increasing evidence that archaea also are involved. Archaeal genes encoding subunit A of ammonia monooxygenase (*amoA*) have been found to occur in a wide variety of environments including marine systems, hot springs, and soils. Furthermore, molecular studies revealed that ammonia-oxidizing archaea often outnumber the nitrifying bacteria in most environments by orders of magnitude. These findings all demonstrate the potentially significant role of archaea in the process of nitrification. Evidence also suggests the assignment of the ammonia-oxidizing archaea should be assigned to the archaeal phylum *Thaumarchaeota* instead of to the *Crenarchaeota*. In addition, these archaea may be of importance for the global carbon cycle. The ammonia-oxidizing archaea isolated from aquatic environments were all shown to be autotrophs, like their bacterial counterparts, and analysis of ^{13}C-bicarbonate–labelled lipid biomarkers of natural *Crenarchaeota* in the North Sea indicated an autotrophic metabolism.

However, it still is unclear whether ammonia-oxidizing archaea in soil also have an autotrophic metabolism and to what extent they are functionally active. An answer to this question and a link of phylogeny to function could be provided by stable isotope probing (SIP) of nucleic acids. This technique allows the specific identification of micro-organisms assimilating labeled substances, most commonly carbon from a particular ^{13}C-labeled substrate. Direct demonstration of ammonia oxidation by this method is not possible, because nitrite, the product of ammonia oxidation, is not assimilated. However, assuming that ammonia oxidation is coupled to autotrophic CO_2 fixation, one should be able to identify the active autotrophic ammonia oxidizing prokaryotes using SIP. So far, DNA-SIP analyses successfully showed autotrophy of ammonia oxidizing bacteria in sediments of a lake and an estuary but failed to detect CO_2 fixation of ammonia-oxidizing archaea in agricultural soil, although potential activity of these archaea in soil has been reported before. DNA-SIP of grassland soil revealed autotrophic ammonia oxidation of archaea, but exactly which groups of archaeal ammonia oxidizers contributed to this process remains unclear. Furthermore, the efficiency of DNA-SIP depends solely on replication of cells, thus excluding micro-organisms that might be active but not growing. In this case, RNA-SIP is assumed to yield more detailed information regarding activity. The homo neoneanderthalis is a new metabolic phenotype that can survive in extremes of climate produced by global warming. The archaea forming new cellular organelle called archaeaons are extremophilic and survive in extremes of climate.

The endosymbiotic archaea secrete RNA viroids which can be converted to corresponding DNA sequences by endogenous retroviral reverse transcriptase and can get incorporated into the genome by endogenous retroviral integrase. Thus there is a constant addition of genomic sequences derived from archaeal RNA viroids producing a dynamic human neoneanderthalic genome adaptable to extremes of climate produced by global warming. Thus the human neoneanderthalis is a new genotype. The neoneanderthalic archaeaons are capable of fixing carbon dioxide and oxygen, using ammonia for energetics and capable of biological fission and fusion reactions. They are also capable of photoinduction of membrane ATP synthesis by sodium-potassium ATPase. The fixing of carbon dioxide leads to water as a by-product. This generates endogenous water in the neoneanderthalic system helping to survive extremophilic climatic conditions. This creates a new system of energetics helping the Neoneanderthal survives in extremes of climate.

The neoneanderthalic archaeaons have got magnetite capable of magnetoperception and quantal perception. The quantal perception of low level of EMF leads to frontal atrophy and cerebellar dominance producing a new neuronal phenotype with features of endemic cerebellar cognitive affective disorder manifesting as epidemic autism. This forms a new autistic behavioural phenotype which is introverted, non-compassionate, non-empathic, psychopathic and with extrasensory perceptual abilities of telepathy, telekinesis and time travel leading to communication with other galactic civilizations. The new behavioural autistic robotic neoneanderthalic phenotype will have the necessary intelligence and mental abilities cope with climate change. The neoneanderthalic archaeaons are capable of quantal perception and can communicate quantally with other star systems and multiverses. This leads on to the evolution of a galactic civilization. The neoneanderthalic archaeaons produce a metabolic phenotype capable of surviving in extremophilic environments of other star systems and can survive as civilization in other stars and exoplanets. The neoneanderthalic autistic phenotype with its extremophilic archaeaon organelle is suited for space colonisation. This constitutes a new human species surviving other stars and exoplanets called the homo galacticus.

The neoneanderthalic archaeaons use cholesterol oxidation as a form of energetics. This produces depletion of sex hormones producing asexual phenotypes. This leads to sexual phenotypes with alternate sexuality and parthenogenesis as a form of reproduction. This produces change in family and society as well as organisation of the community.

The neoneanderthalic archaeaons secrete RNA viroids which get converted to DNA viroids by human endogenous retroviral reverse transcriptase. The neoneanderthalic archaeaon generated RNA and DNA viroids can hybridize with other RNA and DNA viruses producing new RNA and DNA viruses to which the neoneanderthalic species is resistant. Therefore, there exists two sets of population with differing variety of archaeal symbiosis. The homo neanderthalic species has persisted as fossilised matrilineal societies in Europe and Asia which includes the Celts, the Dravidians, the Berbers and the Sumerians. These fossilised remnants of homo neanderthalic species are ideal substrates for archaeal endosymbiosis in the setting of global warming and extremophilic climatic conditions. These fossilised neanderthalic matrilineal societies will expand in the period of global warming and extremophilic conditions due to endosymbiotic archaeal symbiosis to which they are already acclimatised and exposed to. The human species homo neanderthalis is a consequence of archaeal symbiosis in the extremophilic conditions of ice age and are RNA viral/retroviral resistant consequent to endogenous digoxin synthesis and by archaeal RNA viroidal induced RNA interference and viral inactivation. The homo sapiens which evolved in Africa and migrated later to the Eurasian land mass are retroviral sensitive and other RNA viral sensitive and have a low load of endosymbiotic archaea growth. The homo sapiens owing to conditions of evolution is not an ideal substrate for global warming related endosymbiotic archaeal symbiosis. The homo neanderthalis archaeaons secrete RNA viroids which are converted to DNA viroids with both of them hybridizing with other RNA and DNA viruses in nature producing new viruses to which the homo neanderthalic and homo neoneanderthalic host is resistant. The new RNA and DNA viruses produced by homo neoneanderthalis and neanderthalis in global warming will infect the susceptible homo sapien species and wipe them out from earth. The homo sapien species is not a good substrate for global warming related endosymbiotic archaeal growth and would not evolve new cellular organelle like steroidelle, viroidelle, neurotransminoids and vitaminocyte which would help them to survive in extremophilic conditions of global warming. The homo sapiens also lack archaeaon organelle which help them to fix carbon dioxide and generate water, synthesis ATP by photo-induction mediated by sodium-potassium ATPase and archaeaon mediated fission and fusion reaction. The homo sapien species therefore without evolving archaeaon organelle cannot survive in the harsh environment of global warming induced climate change and is exterminated by homo neanderthalis and homo neoneanderthalis. The earth gets dominated by homo neanderthalis and homo neoneanderthalis in the setting of global warming and extremophilic climate change and they go on to contact other galactic civilizations by quantal perceptions as well as colonising other exoplanets, planets and star systems. This produces a civilization of superhuman homo neanderthalis and homo neoneanderthalis with new metabolic, endocrine, neuronal, psychological and genotypic features. The homo neoneanderthalis and homo neanderthalis with its cellular archaeaons and porphyrions can interlink with the internet and computer systems as well as robots producing man-machine hybrids by quantal perception. The artificial intelligence and robotic systems gets merged with homo neanderthalis and homo neoneanderthalis producing superhuman man-machine hybrids. The new homo neanderthalis and homo neoneanderthalis superhuman autistic man-machine hybrids will survive, prosper and flourish in the setting of global warming related extremophilic climate change and evolving to contact by quantal perception other galactic forms of life and civilization as well as colonising other planets and star systems. This constitutes a new human species surviving other stars and exoplanets called the Homo galacticus.

Reference

1. Pratscher, J, Dumont M.G. and Conrad, R. (2011). Ammonia oxidation coupled to CO_2 fixation by archaea and bacteria in an agricultural soil. *PNAS*, 108 (10): 4170-4175.

CHAPTER 45

THE KALKI - MAN-MACHINE HYBRIDS, REPTILIAN NEANDERTHAL BRAIN AND ANARCHIA – HOMO ROBOTICUS - INFECTIVE SYMBIOTIC TISSUE COLONIES OF PROTOZOA, ARCHAEA, VIRUSES AND BACTERIA

The plasma life based forms and the gravitational and anti-gravitational waves generated porphyrions would have been the most primitive of non-life forms. The porphyrin templates would have given rise to the isoprenoid organism, the RNA viroids, DNA viroids and prions which would have symbiosed to form the nanoarchaea. The archaea would have evolved into prokaryotes or bacteria. The archaeal symbiosis with prokaryotes would have generated the eukaryotes. The initial unicellular eukaryotes are the protozoa. The protozoa are capable of phagocytosis and would have had viruses, bacteria and archaea as their intracellular symbionts. The viruses of the protozoa like entamoeba and toxoplasma are giant viruses called megaviride. The protozoans like entamoeba with its viral, archaeal and bacterial endosymbionts would have evolved into the multicellular organisms or metazoa. The archaeal, bacterial and protozoal extracellular vesicles containing RNA and DNA viroids can by horizontal gene transfer lead to evolution of metazoan, plants, animals and human species. The metazoans would have evolved into plants, animals and humans by horizontal gene transfer generated by archaeal extracellular RNA and DNA viroid vesicles. The evidence of such a transformation is seen in the bacterial symbionts of protozoa like entamoeba which includes legionella, listeria, mycoplasma, mycobacteria and chlamydia. The symbionts of protozoan entamoeba include viruses like covid, HIV and cytomegalo virus. The protozoan entamoeba serves as a natural reservoir for covid infections, bacterial infections and HIV infections. The toxoplasma also serves as a covid reservoir. These protozoan colonies of entamoeba and toxoplasma with its bacterial, archaeal and

viral symbionts can be called as chimeric infective symbiotic tissue colonies which can exist in the air, water, earth, blood and tissues. The chimeric infective symbiotic tissue colonies containing bacteria, archaea and viruses can make the symbiotic organism multidrug resistant by horizontal gene transfer. The chimeric infective symbiotic tissue organism evolves as a consequence of global warming and the protozoal, bacterial, viral and archaeal symbiotic colonies are new forms of life. The chimeric infective symbiotic tissue organism composed of protozoa and its symbionts like virus, bacteria and archaea depend upon glycolysis for its energy needs. The glycolytic enzymes enolase, glyceraldehyde 3-phosphate dehydrogenase, hexokinase, phosphofructokinase, aldolase, pyruvic kinase and triose phosphate isomerase are moonlighting proteins having a cytoplasmic, membrane and nuclear functions. These glycolytic enzymes have a RNA binding motif which can modulate RNA transcription in the host and the symbiotic organism. These glycolytic enzymes with their RNA binding motif as well as DNA binding motif can bind to RNA and DNA viruses and viroids modulating their replication and life cycles. The chimeric infective symbiotic tissue organism composed of protozoa and its symbionts like virus, bacteria and archaea can enter the human host through the breaches of the gut-blood barrier and transform itself into the macrophages of the human host. The macrophages are involved in the phagocytosis and innate immunity against bacterial and viral infections. Paradoxically the macrophages like the protozoa can also serve as a reservoir where pathogens like mycobacteria tuberculosis, mycobacteria leprae, listeria, legionella, E. coli, shigella & salmonella, and viruses like HIV & covid can thrive and replicate. The modified chimeric infective symbiotic tissue organism or macrophage composed of protozoa and its symbionts like virus, bacteria and archaea can serve as a reservoir for horizontal gene transfer and generation of multidrug resistant bacteria and viruses. The infective chimeric symbiotic tissue organism composed of protozoa and its symbionts like virus, bacteria and archaea depends upon moonlighting glycolytic enzymes for their energetics and

replication. The development of antibodies against the infective chimeric symbiotic tissue organism composed of protozoa and its symbionts like virus, bacteria and archaea leads to the evolution of anti-brain antibody disease or anti-basal ganglia antibody disease. They present a threat to existence consequent to horiztontal gene transfer and generation of multidrug resistant bacteria, viruses, protozoa, mycobacteria and archaea.

Somatic parthenogenesis occurs due to transformation of the somatic cell to pluripotent stem cell which can develop to parthenogenetic embryos. This results from hybridisation between two different species homo neanderthalis and homo sapiens. This interspecies hybridisation produces intragenomic conflict and parthenogenesis. Parthenogenesis can be induced by stress of climate change and by endosymbiotic archaea and RNA viroids. The intragenomic conflict consequent to interspecies hybridisation results in upregulation of the tryptophan catabolic pathways produces increased amounts of kynurenine, immunosuppression and immune escape of the parthenogenetic embryos. The interspecies hybridisation and intragenomic conflict results in reptilian gene expression and digoxin synthesis. The digoxin produced by endosymbiotic archaea results in upregulated tryptophan transport and catabolism over tyrosine. The increased kynurenine catabolism in Neanderthals results in immune escape and endosymbiotic archaeal growth. The endosymbiotic archaeal growth results in neanderthalisation of the species. The neanderthalisation of the species and interspecies hybridisation and intragenomic conflict results in porphyrias and increased porphyrion mediated low level EMF perception, cortical atrophy and cerebellar dominance. This produces a cerebellar cognitive affective disorder with autistic features, impulsivity, aggressiveness, power lust, increased sexual desire, alternate sexuality, criminality, cannibalistic features and anarchic social mores. The Neanderthal species owing to endosymbiotic archaeal growth and stem cell transformation had an asexual mode of reproduction with parthenogenesis. This results in female dominance, Amazonian syndrome, a culture of male eunuchs and matriarchy. The interspecies hybridisation and intragenomic conflict results in a SLOS phenotype with decreased cholesterol synthesis. This produces decreased sex hormone synthesis resulting in asexuality and alternate sexuality. The Neanderthals were matriarchal and worship the Mother Goddess. The reptilian gene expression and SLOS phenotype results in generation of serpentine features. The genes for the primitive reptilian brain complex get expressed with aggression, violence, impulsivity, cannibalism, anarchy, immorality, lust for power and dominance. The SLOS phenotype results in development of primitive features like the development of tail or cauda, cleft chin, prominent thin long teeth and canines, freckles, scales in the skin and syndactly or webbing. The digoxin synthesis and decreased tyrosine transport results in reduced dopamine and melanin synthesis contributing to a tribe of anarchic white gods. The reptilian gene expression and digoxin synthesis leads to sympathetic hyperactivity and ability to regulate the body temperature in alignment with environmental temperature producing cold bloodedness. The archaea induced fructolysis and fructosemia owing to induction of aldose reductase results in increased lipid and mucopolysaccharide synthesis and hibernation syndrome contributing to obesity and increased subcutaneous fat like reptiles. The evolution of cervical rib, the widow's peak in the eye brows and prominent second toe in the feet is due to increase of recessive traits consequent to inbreeding. The inbreeding results from the autistic phenotype and reduced social contact and social withdrawal. This autistic phenotype with the inbreeding and parthenogenesis leads to decreased genetic diversity and extinction of Neanderthals. The Neanderthals have increased subcutaneous fat, scaly skin which is partly due to albinism and UV exposure as well as due to cholesterol synthetic defect, increased salt secreting eccrine sweat glands and dominant tryptophan pathway and serotonin synthesis producing pineal gland or third eye. The increased tryptophan catabolism produces an epidemic oshtoran syndrome. The neanderthalisation produces an epidemic anarchic syndrome. The neanderthalisation also produces the syndrome of male eunuch and matrilineality. The Neanderthals were mostly parthenogenetic. The cerebellar cognitive affective disorder and actinidic archaeal magnetite and porphyrion induced quantal perception resulted in equality and universality. The primitive Neanderthal tribes are represented by the Saxons, Sakas, Basque, Etruscans, Dravidian races, the Celts and Berbers. The most dominant Neanderthal tribe is the Anglo-Saxons which created an equal society with a universal culture which is American, a universal language which is English, a universal monetary system represented by the dollar and the unification of all ethnicity in the American dream of tolerance and inclusiveness. The Neanderthals had serpentine features and Neanderthal cultures all over the world is represented by serpent worships as seen in Sumeria, the Dravidians of South India, Thoth of Egypt and in Mexico. The Naga tribes and Naga lore of South India represents the Neanderthal culture. The modern version of serpent worship is seen in the symbol of medical profession, the representation of the dollar and in free masonry. Free masonry of the Anglo-Saxons incorporates all religions and is the first universal religion and is neanderthalic. The neanderthalic populations were represented by particular blood groups, the universal donor O Rh –ve and AB Rh –ve, the universal receiver.

The Neanderthal phenotype gives clues as to the origin of the humans. The Naga tribes of South India were hypothesized to originate in the ancient Lemurian land mass which got broken up by tsunamis and earthquakes. The remnants of the Neanderthal phenotype are seen in the Australian Aboriginals, the New Zealandian Maoris, the Dravidian Tamils and Nairs. These societies are predominantly matriarchal and serpent worshipping. The homo neanderthalic possibly arose in the Lemurian oceanic land mass supporting the theory of the aquatic ape origin of humans. The accepted theory of human origin postulates the origin of humans from primates in the African savannahs.

Several points give clue to an origin of the human species in water. The neanderthalic behaviour can be compared to the behaviour of the bonobo monkeys or lemurs seen in the ancient Lemurian continent. The homo neanderthalis would have originated from the bonobo monkeys in the Lemurian backwaters communicating with the sea. The bonobo monkeys owing to shortage of food would have started foraging the backwaters and sea for fish and tubers of water lilies. The fish contains essential fatty acids like docosa hexaenoic acid and tubers contain plenty of carbohydrates. The brain is exclusively dependent on carbohydrates and ketone bodies for energy. The essential fatty acids increase the brain growth. The brain growth in humans is called encephalisation which is more than in primates and is equivalent to sea mammals like the dolphin and whales. The bonobo monkeys would have waded into water and stood in water generating the phenomena of bipedalism. This would have freed their hands to catch fish and break shellfish to generate food. This would also have freed the hands to collect and eat the tubers of water plants. The human species lack hair unlike the primates but like the aquatic mammals. The human species have increased subcutaneous fat like aquatic mammals and unlike primates. The human beings have got eccrine sweat glands and tear glands useful in a watery environment. The human trachea is placed down in the neck unlike that in primates where it is more nasal. The human language points to an aquatic origin for human species. For the human language to develop you have to consciously control your breathing which does not exist in primates but in humans and aquatic mammals. The humans have the diving reflex. The smooth skin with sparse hair and thick subcutaneous fat for insulation points to an aquatic origin for humans like aquatic whales and dolphins. The webbed feet and hands also point to an aquatic origin. The sebaceous glands with its greasy secretion point to water-proofing in humans and an aquatic origin. The homo neanderthalis unlike the homo sapiens is more of an aquatic swimming type. The homo neanderthalis had longer lungs and increased respiratory capacity which helped them to dive into water and float in water. The homo neanderthalis owing to their SLOS phenotype and albinism had vitamin D deficiency. Vitamin D is synthesized from cholesterol. The homo neanderthalic bone was thin as compared to the darkened vitamin D rich homo sapiens. The long lungs and the thin bone helped the homo neanderthalis to float in water. The origin of paranasal sinuses were for the human species to hold its head above water. The human beings sexual behaviour is like aquatic mammals and unlike primates with front to front population. All these differences point to an aquatic origin for the human beings or an aquatic ape hypothesis. The homo neanderthalis originated from the bonobo Lemurian monkeys which waded into water to search for food. The bonobo monkeys sexual behaviour and promiscuousness and alternate sexuality are comparable to homo neanderthalis. The homo neanderthalis arises owing to archaeal endosymbiosis. The waters of the ocean and backwaters are rich in marine archaea which are capable for acetogenesis and methanogenesis. The backwaters and sea of the South Asian peninsular land mass is rich in actinides, the bathyarchaeota. They are capable of acetogenesis and are a source of organic carbon. The actinides would have formed scaffolds for the formation of complex life molecules like RNA, DNA, protein, isoprenoids and complex carbohydrates. This would have produces RNA viroids, DNA viroids, isoprenoid organism and prions on actinidic surfaces which would have symbiosed to form archaea and eventual multicellular organisms. The multicellular organisms arising on abiogenetic actinidic surfaces in the backwater-ocean connections seen in Southern peninsular India which broke away from the Lemurian land mass would have evolved into eukaryotes, prokaryotes, multicellular organisms and symbiotic plants/animals. The bonobo Lemurian monkeys would have evolved into homo neanderthalis by archaeal endosymbiosis in the actinidic shores of backwaters, lakes and oceans of peninsular India. The remnants of homo neanderthalis is seen in Australian Aboriginals, Maoris and Dravidians which are all matriarchal and serpent worshipers. The aquatic ape and homo neanderthalis would have evolved in the actinidic sand shores of backwaters and lakes of Lemuria and peninsular India. The Dravidian communities are matriarchal and female dominant. There is a high degree of consanguinity and inbreeding in the Dravidian matrilineal communities. Parthenogenesis would have been dominant in such communities with matriarchy producing syndromes of the male eunuchs and oshtoran syndromes. The homo neanderthalis ate a carnivorous diet. The teeth were longer compared to homo sapiens and the canines were prominent comparable to fangs of snakes. The cows and bulls were domesticated by the homo neanderthalis which were originally hunters and warriors hunting on mammoths. The domestication of cows and bulls resulted in a high consumption of milk and meat including beef. This resulted in the generation of a lactose tolerant adult population. The gene for lactose tolerance arose 12,000 years ago. The relationship between the cows and Neanderthals could be described as parasitic obligate symbiosis. The origin of cow worship and bull worship in peninsular Indian religions can be related to it. The increased consumption of a carnivorous diet of milk and meat resulted in increased tryptophan intake and catabolism generating more of kynurenines producing immune suppression and immune escape required for parthenogenesis. The immune suppression also produced cold bloodedness of Neanderthals for survival in the cold watery climate. The elements of this culture are still seen in the matriarchal Dravidian communities of peninsular India. The actinidic sea shores and backwater shores of peninsular India is where the homo neanderthalis or the aquatic ape originated. The culture of the Dravidian peninsula is matriarchal and the religious traditions still continue with serpent worshipping culture. This can be called as the mermaid culture.

The human hairlessness, thick subcutaneous fat, webs in the feet and toes, shape of the nostril all point to a watery origin for the human race in mangrove swamps, backwaters and seas. The presence of diving reflex in humans, sweating, tearing, descended human larynx in the neck, hair tract patterns, the presence of hymen and vernix caseosa in babies like seals point to a watery origin for humans. The reproductive behaviour of humans with front-to-front population like aquatic mammals points to the watery origin for humans. The great apes would have come down from the trees and went through an aquaboreal phase where it waded through swamps feeding on molluscs, fruits and fish producing bipedalism. The need to excrete large amount of salt in sea or brackish water leads to the origin of tears and eccrine sweat glands. The human nose is protruded unlike that of primates to protect it from water. The human hairlessness, thick subcutaneous fat and sweating point to the watery origin for humans and bipedalism. The encephalisation of the brain was due to the large intake of omega-3 and omega-6 fatty acids from fish. The human jaw is short with short teeth unlike Chimpanzees pointing to eating of marine food. All naked mammals are aquatic like the dolphin, the manatee, elephant, pig and rhinoceros. The human beings are the only naked apes. The increased subcutaneous fat or blubber in human babies and vernix caseosa of babies point to a watery origin for human species. The human beings have the lacrimal glands and sweat glands to excrete salt in a sea environment. The human babies are plump with 16 percent body fat and the human milk contains 25 percent fat. The babies roll over on their body and float in water with their nose in air. The fact that human babies can swim points to a watery origin for the human species. The Chimpanzees' hand is light and strong to hang from trees. The human babies hand is too heavy and weak and can grasp on to the hair of the mother while swimming in water. The human babies have developed the grasp response for this type of evolution. The human species female has long oily sebum coated scalp hair. The infant chimps have got a strong neck which is kept steady. The human baby attains neck steadiness at six months, but if the baby is placed in water the neck becomes strong. The human speech arose due to the conscious control of respiration which the chimp cannot. The speech also owes its origin to the position of the larynx in the neck. The pincer grip of humans is developed to get meat out of shell fish and mussels which is called as precision grip. This theory was put forward by Elaine Morgan. The backwaters contain water lilies and tubers whose roots were consumed by the primitive humans along with fish, mussels, snails and shell fish. The water lily and lotus roots and leaves contain alkaloids like nupharine and aporphine which are psychoactive and gives rise to the dream state of Neanderthals. The lotus and water lilies are associated with creation myths like that of the Sun God Ra emerging from the lotus in primordial waters and the Brahma the Creator seated on the lotus. The homo neanderthalis emerged first in the Lemurian land mass which was more like a big island susceptible to breakage to independent land masses owing to widespread tsunamis in the region. This would have lead to inbreeding in the Neanderthals leading on to loss of genetic diversity and expression of reptilian genes. This would have also contributed to the eventual extinction of Neanderthals. The aquatic ape would have arose as homo neanderthalis in the Lemurian land mass and its breakaway regions like the backwaters linked to the sea regions of Kerala with actinidic sands. This is indicated by the persistence of matrilineal society in the Dravidians of Kerala and the detection of endosymbiotic archaea in the blood of Kerala population. The psychometric neanderthalic quotient is high in the population of Kerala with a high incidence of autism. Matrilineality and consanguinity is common among the Dravidian Nair population of Kerala. The Dravidians tend to have a Neanderthal phenotype. The Dravidian Nairs are postulated to have a Scythian origin. Serpent worship, serpent music and serpent dances are common in Kerala. The communities in Kerala are mostly carnivorous and consume beef. The culture in Kerala is more tolerant and Keralites migrate all over the world and mix with different societies. Tolerance and inclusiveness is a feature of NQ quotient. Serpent temples are widespread in Kerala. Kerala has got a higher incidence of Neanderthal genomic sequences related diseases like autism, schizophrenia,

ADHD, addictions, metabolic syndrome, autoimmune disease and cancer. It is tempting to locate the origin of the upright aquatic ape in the extended backwaters and lakes of Kerala with its connections to the sea and its actinidic sand rich shores. The backwaters are rich in fresh fish and mussels and water lilies and lotuses giving tubers and roots. The homo neanderthalis evolved from archaeal endosymbiosis and marine archaea are dominant in the seas of the Indian ocean and backwaters of Kerala. Serpent worship is a dominant theme in the Dravidian Nair culture and Serpent God Anantha is the dominant deity. This tempts us to speculate on the origin of bipedalism, human species and homo neanderthalis in Kerala.

The origin of the aquatic ape points to the dominant role for the female of the species in human evolution. Human evolutionary theories have focused on the hunter gatherer and tool maker males. The watery origin of bipedalism and human species points to a dominant role for woman in human evolution. The body anatomy of the females with pendular mammary glands and rounded glutei are for floating and buoyancy and not for sexual attraction. This produces a gynocentric approach to evolution as against an androcentric approach. Evolution was basically meant for protection and rearing of children. Humans have evolved in swamps, backwaters and sea and are not biologically or socially inferior to men. The homo neanderthalis have got features unlike that of chimpanzees. The social patterns of homo neanderthalis can be compared to the bonobo monkeys. The bonobo monkey society is female centered and egalitarian. Sex is a part of social relationship and serves as a substitute for aggression. The bonobo monkeys have different types of sexuality heterosexual, and male to male and female to female. The frequency of sexual interaction is more but the reproductive rate of bonobo monkeys is the same as chimpanzees. The chimpanzees evolved in the open dry savannah while the bonobo monkeys still lived in trees and hanged down from trees. The tree habitat of the bonobo monkeys lead them to an evolutionary form of life where they can hang from mangrove trees in swamps and eventually wade in water. The bonobo monkeys are pigmy monkeys with male weighing 43 kg and female 33 kg. They are omnivorous and eat fruits, small amount of vertebrates and invertebrates. They have imaginative plays and have sex in missionary positions. They have wide variety of sexuality and group behaviour. Sex was a means of social relationships. The female bonobos bonded among themselves and led the community. The male bonobo is attached to his mother and depends on her for protection throughout life. The bonobo society can be compared to a matriarchal female dominant Neanderthal society.

The female dominant model of human evolution raises the question of who evolved first - the male or the female. The original fossils of human species are predominantly female and the male fossils evolved after billions of years. The original human species would have been a cluster of female bipedals in swampy waters feeding on tubers of water lilies and lotus as well as fish, mussels and shell fish. The women can reproduce by parthenogenesis like lower animals. Therefore it is natural for the female of the species to evolve first. The sexual relationship in such female only societies in primodial times was lesbian. The evolution of males occurred at a later date. The macho model of human evolution with male hunter and male toolmaker and a female accomplice evolving together is highly unlikely. The next stage of human evolution has been postulated to be interspecies hybrids. This was put forward by Eugene McCarthy. McCarthy pointed to several features of men of humans similar to pigs. Pig organs can be transplanted to humans without rejection. The pigs-like humans are hairless, have thick layer of subcutaneous fat, protruding nose and heavy eye lashes. The pig genetic sequence contains similar SINE element ALU as humans. The phenomenon of crossing the species barrier is represented by the generation of the swine flu epidemic in humans. The early bipedal female only human species would have generated human pig interspecies hybrids. This would have generated interspecies hybrids of males and females. The male sexual organ corresponds to the tail of mammals. Similarly sex organs of species like snakes correspond to limb buds. The human embryo in its various stages of development can be compared to fishes, amphibians and lower animals. Interspecies hybrid would have led to the development of male and female of the species. The water mammals like cows, bulls, pigs, elephants, rhinoceros, turtles, crocodiles, water snakes and giant tortoises would have contributed to the generation of interspecies hybrids. This would have generated a population of both male and female bipedals in the swamps with different type of sexual interactions - heterosexual, bisexual, homo sexual and lesbian. Genetic diversity is required for a species to survive. Thus heterosexuality as a mode of sexual behaviour would have become acceptable to society as such. Bacterial and archaeal conjugation with human cells have been described. Chimeras of humans and animals have been produced in labs. Interspecies hybrids have been generated in human labs and populations have been produced. Interspecies hybrids include the dzo between yak and cattle, zubron between cow and bison, cama between camel and illama, yakulo between yak and buffalo, sheep-goat and mules. This exemplified by the Plant of the Apes. The interspecies hybrids would have been protected from pre-zygotic and post-zygotic isolation and destruction by the phenomena of immunosuppression and immune escape mediated by tryptophan catabolite kynurenine. A fish diet in swampy waters is rich in tryptophan. The Hindu myth of creation of Matsya the fish, Koorma the tortoise, Varaha the boar and Narasimha the lion point to the generation of interspecies hybrids as the main lynch point of evolution. The generation of human brain structure also depends upon interspecies hybridisation. The reptilian complex of the brain is dominant in Neanderthals. The reptilian complex is seen in amniotes which include mammals, reptiles and birds. The reptilian complex of the reptilian brain includes the basal ganglia, the brain stem and cerebellum. This forms the basis of the cerebellar cognitive affective disorder in Neanderthals. The reptilian brain is the site of imagination, intuition, instinct, compulsivity and dreams. It communicates by symbols and archetypes. It is the site of obsessive compulsive disorder, superstition, ritualism, slavishness and conformation to all way of doing things. It is the site of territoriality, aggression, racism, violence and hypersexuality. The reptilian complex is dominant in amphibians, fishes and reptiles. The Neanderthal brain has got cerebral cortical atrophy and cerebellar dominance. The reptilian brain and reptilian genes are dominant in Neanderthals pointing to interspecies hybridisation of the first evolving Neanderthal females and matrilineal female only Neanderthal society.

Climate induced archaeal endosymbiosis leads to stem cell transformation and germ cell generation leading to parthenogenesis. This leads to mitochondrial dysfunction and glycolytic activation contributing to the Warburg phenotype. Microembryos parasatize various tissues like the brain, heart, liver and lung. The microembryos contain lipid droplets organelle which forms a reservoir for archaeal replication. The Warburg phenotype leads to blockade of pyruvate dehydrogenase and the TCA cycle. The pyruvate gets channelled to GABA shunt pathway generating succinyl CoA. The Warburg phenotype induced increased glycolysis leads to generation of phosphoglycerate, serine and glycine. Glycine can combine with succinyl CoA generating delta amino levulinic acid which forms the basis of porphyrin synthesis. The porphyrins can form a template for formation of the RNA viroids, DNA viroids and prions as well as isoprenoid lipid organisms. All of them symbiose to form archaea by template mediated replication. The archaea contains magnetite and porphyrins which are capable of absorption of low level of EMF. This leads to quantal perception of low level EMF leading to cortical atrophy and cerebellar dominance which characterised the Neanderthal brain. Neanderthals tend to have a cerebellar cognitive affective disorder. The archaeal network in lipid droplets in the brain is capable of intersynaptic transport and functions as a giant magnetotactic archaeal colony in the zombie brain and controls brain functions.

The giant neuronal magnetotactic archaeal colony is capable of quantal perception and consciousness. This produces what is called as the zombie syndrome. The Warburg phenotype that is generated by endosymbiotic archaea leads to germ cell and stem cell transformation and parthenogenesis. The parthenogenetic microembryos with its lipid droplet organelle reservoir of archaea forms the substrate of the Neanderthal brain and its magnetite and porphyrion content is capable of quantal perception and consciousness. This can lead to schizophrenic and autistic phenotypes. The parthenogenetic microembryos with its Warburg phenotype studding multiple tissues produce pathogenetic state. The Warburg phenotype can produce insulin resistance and diabetes mellitus as well as vascular thrombosis. The Warburg phenotype and parthenogenesis can lead to oncogenesis. The Warburg phenotype and parthenogenetic embryos can lead to autoimmune disease. The Warburg phenotype and increased glycolysis produces immune activation. The Warburg phenotype can lead to increased glycolysis and glyceraldehyde 3-phosphate mediated nuclear cell death and neurodegeneration.

Somatic parthenogenesis is due to climate change. Climate change can produce archaeal endosymbiosis and archaea and archaeal RNA viroid induced parthenogenesis. In response to stress somatic cells can get transformed to stem cells by endosymbiotic archaea. Stem cell metabolonomics - anaerobic glycolysis, PDH dysfunction, CoQ deficiency mitochondrial dysfunction, branched chain ketoacid dehydrogenase dysfunction, homocystinuria and genomic demythelation, porphyrias and reactive oxygen species generation, SLOS (Smith Lemli Opitz) leading to cholesterol depletion, low sex hormones, vitamin D deficiency and bile acid deficiency. Stem cells can undergo endoreduplication, cell fusion and budding and bursting like bacteria producing polyploidy. These polyploidal cells can become parthogenic embryos. The polyploidal cells are stress resistant and genomically unstable. The polyploidal cells are genomically, metabolically and phenotypically different and unstable.

They can get converted to different tissues like brain, liver and heart forming somatic embryos. Multiple somatic embryos with polyploidy produces multiple personality disorders – schizophrenia, autism and mood disorder. Multiple somatic embryos with polyploidy are antigenic and can produce an autoimmune disease. Multiple somatic embryos with polyploidy are genomically unstable can produce cancer. Parthogenesis can lead to social changes including matrilineal societies, alternate sexualities and different identities. Multiple somatic embryos with polyploidy are metabolically and genotypically unstable leads to neurodegeneration. Multiple somatic embryos with different metabolic instability can lead to metabolic syndrome. Archaea and RNA viroid induced mitochondrial dysfunction and upregulated glycolysis resulting in stem cell transformation of somatic cells. The somatic cells which are stem cell transform in the setting of immune activation and cytokine secretion can get converted to germinal cells – sperm and ova. This can result in fertilisation and parthenogenesis. Archaeal digoxin can produce intracellular magnesium deficiency and failed mitosis due to spindle dysfunction. This can produce polyploidal cells. The polyploidal cells can assume stem cell functions. The stem cells can mimic germline cells. The meiotic programs of the polyploidal stem cells can get activated generating cleavage embryos, morulas which can get converted to tumour spheroids. The spheroids can get converted to blastocyst and post-implantation embryos producing oncogenesis. They can also dissemble producing metastasis.

The Neanderthals ate a high protein high fat non-vegetarian diet by hunting and scavenging mammoths and other animals. This resulted in heavy load of tryptophan in the system producing tryptophanuria and tryptophanemia. The meat contains a high level tryptophan and haemoglobin which can induce the enzyme indoleamine 2,3-dioxygenase and tryptophan 2,3-dioxygenase which are both heme enzymes. The Neanderthals ate a low fibre diet resulting in decrease supply of short chain fatty acids especially butyrate from the gut.

Butyrate can suppress indoleamine 2,3-dioxygenase and tryptophan 2,3-dioxygenase and a fibre deficient diet in Neanderthals can upregulate the activity of indoleamine 2,3-dioxygenase and tryptophan 2,3-dioxygenase resulting in increased catabolism of tryptophan along the kynurenine pathway. Natural substances that are deficient in non-vegetarian diet like brassica alkaloids, curcumin, caffeine, tea and coca can inhibit indoleamine 2,3-dioxygenase and tryptophan 2,3-dioxygenase which becomes over active in Neanderthals producing tryptophan catabolism. The tryptophan is metabolised to formyl kynurenine, hydroxy kynurenine, kynurenic acid, 3-hydroxy anthranilic acid, quinolinic acid and NAD. Kynurenine can bind to AHR receptor producing immunomodulation and immunosuppression. It can produce immune tolerance in case of embyrogenesis, parthenogenesis, autoimmunity, cancer, lipopolysaccharide tolerance and chronic infection. The kynurenine can produce suppression of T cells and immunity leading to immunotolerance important in the above mentioned states. Thus kynurenine pathway flux can contribute to embyrogenesis and parthenogenesis by producing immune tolerance. The tumours can escape immune destruction by suppression of NK cells and T cells. Thus tumour metabolism depends upon activation of the tryptophan catabolism along the kynurenine pathway. The kynurenine pathway is activated in lipopolysaccharide tolerance and chronic infections producing immunosuppression and immunotolerance. The archaeal endosymbiosis depends upon immunotolerance and immunosuppression by activating tryptophan catabolism along the kynurenine pathway. The tryptophan catabolism along the kynurenine pathway can also contribute to psychiatric disorders. Kynurenine blocks the NMDA receptor and alpha-7 nicotinic receptor. This results in down regulation of NMDA and cholinergic transmission. Kynurenine can also upregulate dopaminergic transmission. This results in schizophrenia and autism. The tryptophan catabolism along the kynurenine pathway blocks serotonin synthesis from tryptophan contributing to depressive and anxiety disorders. The tryptophan catabolism along the kynurenine pathway can result in quinolinic acid synthesis and immune activation resulting in autoimmunity, immune tolerance as well as immune activation. Kynurenine can produce immunosuppression and NMDA blockade while quinolinic acid can produce immune activation and NMDA excitotoxicity. The tryptophan catabolism along the kynurenine pathway can generate quinolinic acid important in neurodegeneration. The tryptophan flux along the kynurenine pathway can generate quinolinic acid which can produce low grade inflammation and insulin resistance causing metabolic syndrome. Thus the flux of tryptophan along the kynurenine pathway can produce autoimmune disease, neurodegeneration, schizophrenia, depression, autism, cancer and metabolic syndrome. The induction of heme enzyme indoleamine 2,3-dioxygenase and tryptophan 2,3-dioxygenase can lead to heme depletion and induction of ALA synthase increasing porphyrin synthesis and producing porphyrias. The porphyrins can form self-replicating porphyrions. The porphyrions form a template for the formation of RNA viroid, DNA viroid and isoprenoid organism which can symbiose together to form an endosymbiotic archaea by abiogenesis. The induction of indoleamine 2,3-dioxygenase and tryptophan 2,3-dioxygenase produces kynurenine which can suppress the immune system generating immunotolerance and archaeal endosymbiosis. Thus the tryptophan load and induction of indoleamine 2,3-dioxygenase and tryptophan 2,3-dioxygenase results in systemic civilizational disorders in Neanderthal population. The tryptophan load and induction of indoleamine 2,3-dioxygenase and tryptophan 2,3-dioxygenase can result in immune tolerance and immunosuppression by kynurenine producing archaeal endosymbiosis and neanderthalisation of the species. The tryptophan load and induction of indoleamine 2,3-dioxygenase and tryptophan 2,3-dioxygenase can result in immunotolerance that can contribute to parthenogenetic embryogenesis or somatic pregnancy in multiple tissues in Neanderthals. The tryptophan load and induction of indoleamine 2,3-dioxygenase and tryptophan 2,3-dioxygenase can produce an autistic schizophrenic tribe of Neanderthals with parthenogenesis and matriarchy.

Neanderthal evolution was determined by archaeal endosymbiosis. The human species evolved into homo neanderthalis by archaeal endosymbiosis. Archaeal endosymbiosis was mediated by the tryptophan catabolic kynurenine pathway. The kynurenines can produce immunosuppression and immune tolerance resulting in archaeal endosymbiosis. The kynurenine pathway results in blockade of the NMDA receptor, cholinergic nicotinic alpha-7 receptor, decrease production of serotonin and increase dopaminergic transmission. This produces an autistic, schizophrenic Neanderthal tribe with less of frontal executive function and more of cerebellar dominance affective impulsive type behaviour. The induction of indolamine 2,3-dioxygenase and tryptophan 2,3-dioxygenase led to channelling of tryptophan catabolism along the kynurenine pathway. The Neanderthals ate a high protein non-vegetarian tryptophan diet leading to an induction of tryptophan catabolism. The Neanderthal diet was deficient in dietary fibre and fibre digestion generated short chain fatty acids especially butyrate resulting in increasing indolamine 2,3-dioxygenase and tryptophan 2,3-dioxygenase activity and more of tryptophan catabolism. The kynurenine pathway results in generation of quinolinic acid which can produce chronic immune activation and insulin resistance. Quinolinic acid is also involved in neurodegeneration. The kynurenine induced immune cell as well as NK cell suppression can result in evolution of cancer. The archaeal endosymbiosis generated by kynurenine induced immunotolerance can lead on to cancer, autoimmune disease, metabolic syndrome, neurodegeneration, schizophrenia and autism by generation of archaeal cholesterol catabolite digoxin. Thus the higher load of tryptophan due to a meat diet and low fibre diet results in generation of kynurenine which has a ketamine-like action producing Neanderthal behaviour. The induction of heme enzymes indolamine 2,3-

dioxygenase and tryptophan 2,3-dioxygenase results in heme depletion, activation of ALA synthase and porphyrias. The porphyrins can self-organize to form porphyrions and can act as a template to generate isoprenoid organism, RNA viroids, DNA viroids which all symbiozed to form actinidic archaea. The porphyrions have a wave-particle existence and in the presence of membrane intercalated porphyrion mediated sodium-potassium ATPase inhibition can result in a pumped phonon system and dipolar porphyrion mediated quantal perception. The porphyrion intercalated cell membrane and sodium-potassium ATPase inhibition can result in increased intracellular calcium and decreased intracellular magnesium resulting in mitochondrial dysfunction, cell membrane dysfunction, golgi body related protein processing dysfunction, defect in DNA and RNA function and disordered cell function. This increased intracellular calcium and reduced magnesium due to sodium-potassium ATPase inhibition can result in immune activation, glutamate excitotoxicity, oncogene activation and disease states. The immune tolerance produces cancer, stem cell transformation and parthenogenesis. The cancer stem cells can develop meiotic programs generating parthenogenetic embryos which can survive and grow in the presence of immune tolerance created by kynurenines. The multiple parthenogenic embryos can create multiple personalities leading to schizophrenia and autism, grow into cancer, its stem cell metabolism with increased glycolysis and mitochondrial dysfunction can produce metabolic syndrome, stem cell mediated increased glycolysis can lead to immune activation and the normal tissue dying at the expense of parthenogenic embryos can produce degeneration. The chronic inflammation induced by quinolinic acid and other tryptophan catabolites producing autoimmune disease and quinolinic acid related cell death and degeneration. The tryptophan catabolic pathway produces civilizational syndromes in Neanderthals leading to their extinction. The tryptophan loading in higher primates and homo sapiens due to increased meat eating carnivorous habits in Eurasian steppes resulted in kynurenine catabolite induced immunosuppression, immunotolerance, archaeal endosymbiosis and neoneanderthalisation. Thus tryptophan loading due to a high meat and low fibre diet mediated kynurenine generation and immunotolerance would have led to archaeal endosymbiosis and evolution of homo neanderthalis and homo neoneanderthalis. The Neanderthals due to archaeal endosymbiosis had stem cell transformation, activation of meiotic programs in the stem cells, generation of germ cells and parthenogenesis. This resulted in female dominance and matriarchal societies. The intraspecies hybridisation and intragenomic conflict would also have contributed to parthenogenesis and reptilian gene expression consequent to interspecies hybridisation and intragenomic conflict. Genes affected PDH, BKCD, SLOS, porphyria, Hartnup's disease. Decreased cholesterol CoQ and vitamin D. The Hartnup's trait would have developed to counteract the tryptophan loading in Neanderthals consequent to a high meat diet. This produces what is called as increased tryptophan catabolism and kynurenine catabolites mediated oshtoran syndrome described initially in the Pericaspian areas. Endemic oshtoran syndromes would have existed in the Neanderthal population producing refractory thought and mood disorders, movement disorders including chorea and tic as well as schizophrenia and autism. The generation of tic disorders as a part of epidemic or endemic oshtoran syndrome would have led to vocal tic led language generation. The human language including ancient ones like Akkadian and Sanskrit developed first in Pericaspian areas. The endemic oshtoran syndromes can also result in altered fat metabolism fatty liver and cirrhosis. It can produce adrenal dysfunction and hyperactivity producing sympathetic overactivity and parasympathetic underactivity leading to hypertension, vascular disease, cancer and autoimmune disease. The incidence of lupus-like syndromes and multiple sclerosis is high in endemic oshtoran syndrome. The tryptophan catabolism in oshtoran syndrome can produce cognitive dysfunction like Alzheimer's disease, Parkinson's disease and cell death. Parthenogenesis can lead to autistic phenotype with cerebellar dominance and a cerebellar cognitive affective disorder. Parthenogenesis induced reptilian gene expression can produce porphyrias and porphyrin induced extrasensory perception. This produces a creative tribe with quantal perception.

Archaeal endosymbiosis and neanderthalisation depended upon the kynurenine pathway mediated immunotolerance and immunoparalysis. The tryptophan catabolic pathway and kynurenine can have a ketamine-like effect due to NMDA blockade resulting in ecstasy, CCAS, cerebral cortical paralysis and extrasensory perception. The tryptophan catabolism is directed the kynurenine pathway resulting in depletion of melatonin and nocturnal activity and lack of sleep. The tryptophan catabolites kynurenine and kynurenic acid can produce decreased insulin synthesis, decreased insulin release, decreased insulin biological activity causing insulin resistance. Tryptophan catabolitic syndrome can result in sympathetic overactivity and a dysautonomic syndrome producing fear flight response and impulsivity. The induction of reptilian genes by intragenomic conflict and interspecies hybridisation can result in expression of the shikimate pathway producing alkaloidal neurotransmitter synthesis - LSD, nicotine, strychnine, mescaline producing shamanic states and extrasensory perception. The induction of IDO and heme depletion can result in porphyrias and porphyrion mediated extrasensory perception. The heme depletion can reduce the heme enzyme activities. The heme enzyme cytochrome C oxidase is inactivated resulting in mitochondrial dysfunction. The heme enzymes catalase and glutathione peroxidase are inactivated producing free radical stress. The heme enzyme cytochrome P450 is inactivated producing defective bile acid synthesis due to cholesterol 7-alpha hydroxylase deficiency causing metabolic syndrome X due to bile acid deficiency. The heme enzyme cytochrome F450 deficiency produces defective aromatase and beta hydroxy steroid dehydrogenase producing reduced testosterone, estrogen and cortisol synthesis. This produces the asexual and alternate sexual Neanderthal state. The heme enzyme lanosterol 14-alpha demethylase is inactivated inhibiting cholesterol synthesis and cholesterol depletion syndrome. The heme enzyme retinoic acid hydroxylase and cholecalciferol hydroxylases are deficient producing lack of vitamin D and A and defective immunity and uncontrolled cell proliferation.

The initial event is an increased tryptophan catabolic pathway producing immunotolerance and immunoparalysis mediated by kynurenine. This produces endosymbiotic archaeal growth and neanderthalisation. This results in conversion of somatic cells to germ cells and activation of meiotic programs resulting in parthenogenesis. The homo neanderthalis reproduces by parthenogenesis. There is a malfunction in sexual reproduction resulting in a parthenogenetic species. Parthenogenesis is induced by actinidic archaea and RNA viroids. Climate change induced stress can produce parthenogenesis. The heme enzymes cytochrome P 450 dependent aromatase and beta hydroxy steroid dehydrogenase are defective resulting in lack of sex hormones producing asexuality and alternate sexuality. The heme enzymes NOS, CBS and HO1 are defective leading to lack of gasotransmitters NO, CO and H_2S resulting in dysautonomia and sexual dysfunction. The parthenogenesis results in formation of multiple embryos in tissues causing multiple personalities and schizophrenia and autism. Parthenogenesis also produces cancer and autoimmune disease. Parthenogenesis in the brain can result in death of normal tissue and neurodegeneration. The stem cell metabolonomics of parthenogenetic embryos with increased glycolysis and mitochondrial dysfunction results in metabolic syndrome. The heme depletion leads to porphyrias and porphyrions causing quantal perception. The tryptophan catabolic pathway related kynurenine can produce a ketamine syndrome akin to schizophrenia. Similar schizophrenic and autistic syndrome can occur in Neanderthals due to tryptophan alkaloids synthesized by reptilian gene activation - LSD and mescaline. This produces a shamanistic spiritual quantal perceptive society. The porphyrion induced quantal perception of low level EMF can result in cortical atrophy and CCAS and an impulsive state, aggressive state, violence, criminality, spirituality and terrorism. The Neoneanderthals form small colonies and tribal groups. The porphyrion mediated quantal perception results in formation of cohesive small groups with communal living creating an anarchic society. The anarchic society arises due to a cerebellar cognitive affective disorder and cortical atrophy consequent to quantal perception. This anarchic society is small and tribal, violent and aggressive, spiritual and transcendental. This results in loss of national identities and recession to primitive tribal identities. The civilizational national identities collapse and are replaced by small tribal identities causing permanent war, instability and crisis. The parthenogenetic reproduction in Neanderthals and Neo-neanderthals as well as the lack of sex hormones related asexuality and alternate sexuality produces a matriarchal female dominant small social groups. This can happen in the setting neoneanderthalisation consequent to endosymbiotic archaeal growth. The neoneanderthalic matriarchal society is female dominant and the males are reduced to a marginal role in society creating complexes of hatred and vulnerabilities. This can be compared to the creation of a society female amazons and male eunuchs. The world tends to be transformed into an anarchic society of Amazonian women and male eunuchs. This represents the castrated male syndrome with deprivation of dignity, integrity, passion and pride. The Neanderthal communities functioned as anarchic small societies with communal living. The concept of altruistic, egoistic and obsessive love and nuclear family was absent in them. They lived as small groups of 15-20 with group consciousness. The porphyrion induced extrasensory quantal perception resulted in well-bonded anarchic communities with communal living and parenting of children. Sexual relationships became partnership between equals and functional. They were promiscuous, self-sufficient and were not into socially sanctioned relationships like marriage. The duty of the male eunuchoid was to the small community which is served relationships in the community depended on a common communal consciousness based on extrasensory quantal perception. The neanderthalic and neoneanderthalic communities were matriarchal and gender-equal and did not have any hierarchal leadership. There were no kings or queens and it was a stateless society. It was a voluntary association and there was no written law or control except by consensus. This is represented by ancient Indian societies in the Buddhist period of Indian history described as Janapadams. These were small stateless societies ruled by equality and consensus. The ancient Harappan civilization was also structured as a stateless anarchic society. The same holds good for the ancient Celtic kingdoms in Wales, Scotland, Basque, Catalonia and Brittany. The concept of anarchic societies is exemplified in the Mandalas in Southeast Asia comprising modern Indonesia, Vietnam, Laos, Cambodia, Myanmar and Thailand. These civilizations were extensions of the South Indian Dravidian civilization which derived from the Harappan civilization. In modern times the anarchic societies were revived in the form of modern Grama Swaraj of Gandhi whose philosophy was basically anarchic. Gandhi was from the area of India where the Harappan civilization thrived in prehistoric times. The same anarchic societies can be seen in Jewish Kibbutz. The Jews, Celts, Dravidians all had a neanderthalic origin. This formed the basis primitive, anarchic Neanderthal communities. The global warming related endosymbiotic archaeal growth results in neoneanderthalisation. The actinidic archaea mediated quantal perception of low level EMF results in cortical atrophy and cerebellar dominance resulting in a cerebellar cognitive affective disorder. The archaea mediated quantal perception by porphyrions results in small bonded Neo-neanderthal communities with anarchic forms of organisation. The human civilization owing to global warming and neanderthalisation of the brain regresses to form small tribal anarchic communities in perpetual warfare resulting in the end of nation-states. The neoneanderthalic world of anarchy has set in. The matriarchal societies with parthenogenetic female result in the syndrome of castrated male eunuchs and gender-equal societies. Anarchy becomes the norm of political life in the world with its attendant catastrophic destructions. The male eunuch syndrome coupled with cerebellar cognitive affective disorder in an anarchic world with tribal identities can produce terrorism, criminality, creativity, aggression, violence, lack of empathy and an autistic tribe. The parthenogenetic Neo-neanderthals will eventually become extinct due to lack of gene diversity in the population.

Climate change and exposure to low level internet EMF fields can induce HO1 activity and increased porphyrin synthesis. The porphyrins can form porphyrions which act as a template for the formation of RNA viroids, DNA viroids and isoprenoids by abiogenesis. They are symbiose to form actinidic archaea and RNA viroids. This results in endosymbiotic archaeal mediated stem cell transformation. The endosymbiotic archaea can induce toll receptor activation and activate HIF alpha resulting in stem cell metabolonomics with increased glycolysis and mitochondrial dysfunction. The endosymbiotic archaea can induce aldose reductase and fructose metabolism producing frucotsemia, lipid synthesis, mucopolysaccharidosis, porphyrias and hibernation/zombie syndromes. The increased glycolysis and Warburg phenotype can activate the immune system. The stem cells meiotic programs get activated and results in the formation of germ cells and parthenogenesis. Thus climate change and internet exposure results in a reproductive change to predominant asexual reproduction and parthenogenesis. This results in an asexual male eunuch phenotype. This results in gender equality and female dominance. The male population becomes dispossessed and is peripheral to the functions of society. This creates a society of male eunuchs and matriarchs. The society regresses to the matriarchal regime with widespread social consequences. The porphyrions and actinidic magnetotactic archaea can perceive low level EMF fields producing frontal cortical atrophy and cerebellar dominance. This produces the cerebellar cognitive affective disorder on an epidemic scale. The cerebellum is the site of impulsive behaviour, aggression, criminality, extrasensory perception, spiritual phenomena and dreams as well as trance. This results in an impulsive society without any logic or reason producing lawlessness and anarchy. This produces an anarchic world of Neoneanderthals with small social tribal groups and fall of organized civil society and nation-states. The consequence of these are widespread lawlessness, wars, criminality, terrorism, sexual promiscuity, demise of the nuclear family, alternate sexuality, communal living and breakdown of social structures of the homo sapien society. The anarchic eunuchoid world of Neoneanderthals opens up. Thus the climate change and internet produces a sexual change, anarchic social change, and reproductive change. The Neanderthals live in a dream world of imagination and paranormal phenomena modulated by cerebellar hypertrophy and function. This produces a spiritual world of trances and religiosity. The Neanderthal brain activates the default network in the frontoparietal lobe producing day-dreaming, creative visualization, fantasy phenomenon, depersonalisation and altered consciousness. The increased tryptophan catabolism produces kynurenine which blocks the NMDA receptor producing a ketamine or phencyclidine schizophrenic psychosis on an epidemic scale. The increased kynurenine can also block the alpha-7 nicotinic acetyl choline receptor, decreased serotoninergic activity and activates dopaminergic receptor. The tryptophan catabolic pathway also produces hallucinogenic alkaloids like strychnine, mescaline and LSD. This contributes to day-dreaming shamanic states in Neoneanderthals. This contributes to creativity, autism, schizophrenia, lack of social contacts, small tribal populations and a dream world in Neanderthal society. The increased porphyrions produces quantal perception and a dream world. The Neanderthals form small social groups and lack social contacts with out of kin population producing small autistic tribes. The Neoneanderthals and homo sapiens can be differentiated by the following phenomena of unconscious versus conscious, religion versus science, magic versus logic, dream versus waking and psychic versus material. The Neo-neanderthals have great paranormal ability and the society was more religious and spiritual. They created cities of dreams which were classically psychopathic, magical and dream-like. The Neanderthal culture of magic, culture and spirit was different from that of homo sapiens. The myths, folklores and religiosity are derived from the Neanderthals. The worship of Serpents as symbols of God and use of crystals and minerals as exemplified by Siddha forms of medicine are neanderthalic. They painted the skin and face creating extensive tattoos, wore ornaments and were extraordinarily ceremonial and ritualistic. They created dance as form of worship as comparable with the concept of Shiva as a celestial dancer. The neanderthalic tribes were nocturnal and were aware of the star constellations of big bear, little bear and draco. The Neanderthals were nocturnal tribe because of photosensitivity due to porphyrias which made them favour the night time to the day. The neanderthalic gods were from the outer cosmos and the civilization was seeded by intergalactic contacts mediated by cometary and asteroidal impacts. The comets and asteroids carried magnetotactic actinidic archaea which formed colonies transforming to homo neanderthalis. The Neanderthals were religious and have funeral ceremonies and believed in after life. The Dravidian civilization, Uluzzian and Chatelperonean civilization as well as the Basque and Catalan were neanderthalic. They were a civilization of dreams, rituals, dances, religiosity and trances owing to an epidemic CCAS. The Neanderthals were nocturnal tribe who worshipped the Moon Goddess were matriarchal food gathering and women governed society. The homo sapiens were the sun worshipping patriarchal hunter warriors and male governed society. The females were mere adjuncts. The Neanderthal society was religious, ritualistic, symbolic with a cosmological approach to the world. It was a society of creative imagination. The society was predominantly parthenogenetic and asexual. The tantric form of spirituality and the sense of spiritual awakening indicated by the Kundalini showed the asexual nature and eunuchoid characteristic of the Neanderthal tribe. The cerebellum is the site of creative visualization, paranormal and dreams. The

Neanderthal brain was cerebellar dominant creating trance-like states, day-dreaming, telepathy, psychic healing, poltergeist phenomena and religiosity. It was a high civilization of dreams mediated by the cerebellum. The cerebellum produces an ataxic motor syndrome as well as dysmetria of thought. The dysmetria of thought results in an autistic and schizophrenic tribe of Neanderthals and Neoneanderthals produced by climate change and internet exposure. The cerebellum is the site of common embryonal tumours and archaea induced parthenogenesis is higher in the cerebellum producing cerebellar hypertrophy, cerebellar dominance and cerebellar dysfunction. The archaea induced parthenogenesis produces a cerebellar dominance Neoneanderthal brain and the dream civilization of Neanderthals.

The Homo sapiens mind can be characterized by the ego and the homo neanderthalis by id. Homo sapien qualities are sun, fascism, psychosis, logic, science, awake, adult, day, god, male and yang, versus the homo neanderthalis qualities are communism, moon, neurosis, intuition, religion, day-dreaming, child, night, devil, female and ying. The Cro-Magnons were hunters, patriarchal and sun worshipers. The homo neanderthalis were moon worshipers, patriarchal and food gatherers. The homo neanderthalis inhabited Europe and Middle-East. Gooch postulated the double helical concept of the mind as opposed to hemispheric dominance. The double helical mind of Gooch includes the cerebellum which is concerned with dreaming and creativity and the cerebrum concerned with logic. His concept of the sacred life of humans, the double helix of the mind, the cities of dreams, the creations of inner space and the divided self are described extensively in his work. The cerebellum is the site of paranormal and supernatural phenomena and gives rise to creations from the inner space. These are the creations from the inner space mediated by the cerebellum constituted the basis of vampires, troglodytes, demons and asuras. The cerebral cortex is the site of ego and the cerebellum the site of id. The cerebellum becomes dominant owing to archaea and viroid induced embryonal parthenogenesis. The interbreeding of Cro-Magnon with Neanderthals resulted in a burst of spirituality, artistry and creativity. The human behaviour can be explained by the double helical concept of mind. The socialists are neanderthalic and the Cro-Magnon conservatives. The Neanderthals were red-haired with slanting forehead and were worshippers of the Moon. Moon worshipping was common in the Fertile Crescent which included Turkey, Egypt, Harappa, Sumeria and Arabia. The basis of these civilizations was lunar. The Neanderthal societies were matriarchal, completely promiscuous and sex driven and lead by women. They can be compared with the behaviour of bonobo primates. The Neanderthals were short-statured, left-handed and near-sighted. The Cro-Magnons were taller, long-sighted and right-handed. The Neanderthals had a communal living while the Cro-Magnons were monogamous and pair-bonding. The Neanderthals have a larger cerebellum, pyknic body type, non-athletic body type, left-handedness, less of male pattern baldness, prominent eye brows, recessive chins, were neurotic, and is less of psychosis, more hypnotisable and better night vision. The neanderthalic phenotype can be seen in drop-outs, addicts, alcoholics, unemployed and insomniacs. They lived in a world of day-dreaming and increased sexual activity as well as alternate sexuality. The Neanderthals were religiously organized were seen in South Europe, East Europe among the untouchables while the Cro-Magnons had large civil society and were seen in North Europe, West Europe, Brahmin communities and were taller. The political ideas of the French Revolution, the Russian Revolution and the Taliban were neanderthalic while that of the Nazis and KKK were Cro-Magnon. The Neanderthals were basically a day-dreaming, lunar society and were represented by the Celts, the witches, the Kabalist, the Rosicrucian and Judaist. The Nazi hatred of Jews and the western hatred of Islam are based on their neanderthalic origin. The homo sapiens, on the other hand were worshippers of the sun. The Neanderthals were left-handed and left-leaning while the Cro-Magnons were right-handed and right-leaning. The Neanderthal societies are represented by Nairs, Nagas, Sakas, Scythians, Saxons, Celts, Berbers, Sumerians, Dravidians, Harappans, Etruscans and Egyptians. The moon was worshiped in Egypt, Babylon, India, Sumeria, Assyria, Akkadians and Chaldians. The Moon God was called as Sin and Thoth. They are the oldest human deities and are represented by Shiva in India. The Egyptian God Isis, the Celtic God Morgana, the Greek God Artemis, Aphrodite and Selene were representative of the lunar God. All the pagan festivals depend upon the lunar cycles. The Moon God was worshipped in the Kabbala, the Talmuds, the Ur of Chaldees, Harappa and in all of the Fertile Crescent. The Harappan civilization had Shiva with his crescent symbol representing lunar worship. Soma was the presiding deity of Rig vedic ceremonies and is represented by moon. The soma is actually a drink of milk, honey, cannabis and other plant extracts which produced a hallucinatory state. The Harappan culture was neanderthalic and lunar centric as also the succeeding Saivite sects of Hinduism like Aghoras and Nagas. The term for mental illness - lunatic came from lunar worship. The Cro-Magnon civilization was the opposite with sun being the dominant and logic being the culture of the society. The wars of history and hatred of civilizations like Jews, Islam and Hindus were based on neanderthalic origin and their lunar worship.

The Neanderthals evolved by seeding of cometary reptilian genes from outer space. The intergalactic porphyrions, RNA viroids, DNA viroids and template replicating magnetotactic archaea are the basis of cometary genes and seeding of life on earth. The template replication and parthenogenesis are related to evolution of Neanderthals. The intragenomic conflict and interspecies hybridisation produces expression of reptilian genes in human - PDH deficiency, mitochondrial dysfunction, SLOS and porphyria. The parthenogenesis and matriarchy are related as also SLOS, asexuality and alternate sexuality. This produces on-hierarchal anarchic societies. The equality and gender sensitivity are related to the serpent cult of Khylst and Capracoites-

Nagas and Asuras- Dravidians and Sumerians- Punts and Egyptians. The Mother Goddess, the serpent people and Neanderthals are synonymous. The Dravidians, Celts, Egyptians, Jews, Berbers, Sakas, Nagas, Nairs, Asuras and Neanderthals were parthenogenetic. They consumed a high fat high protein diet- milk and honey, and had persisting adult lactose tolerance. This contributed to the consumption of a ketogenic diet, stem cell metabolonomics and parthenogenesis. Global warming and climate change can lead to archaeal endosymbiosis and neanderthalisation of the species. This leads to interspecies hybridisation and intragenomic conflict contributing to parthenogenesis. Archaea and RNA viroids can induce parthenogenesis. Parthenogenesis can lead to matriarchy and female dominance. Parthenogenesis can produce expression of reptilian genes, porphyrias, extrasensory perception and autistic phenotype. This results in a creative, spiritual and matriarchal population. This has a similarity to an ant and bee colony. They behave like parthenogenetic Neanderthal societies. The reptilian gene expression produces the SLOS phenotype, low cholesterol and lack of sex hormones leading to asexuality and alternate sexuality.

Parthenogenesis can lead to human disease. Parthenogenetic Somatic pregnancy can lead to cancer. Parthenogenetic embryos behave like autoantigens causing autoimmunity and autoimmune disease. The parthenogenesis and stem cell glycolysis can lead to lymphocyte activation and autoimmune disease. Parthenogenesis in brain can lead to multiple personalities causing schizophrenia and autism. Parthenogenetic embryos in neural tissue can lead to neurodegeneration by starving the host cells. Parthenogenesis leads on to Warburg phenotype. The Warburg phenotype contributes to anaerobic glycolysis, mitochondrial dysfunction and metabolic syndrome. Parthenogenesis can lead to sexual evolution and alternate sexuality. Parthenogenesis will lead to a lack of demand for sexual reproduction. The expression of reptilian genes and SLOS phenotype produces cholesterol depletion and sex hormone deficiency.

This produces an asexual phenotype. Extremes of climate change can produce archaeal symbiosis and parthenogenesis. This results in interspecies hybridisation and intragenomic conflict. This leads to a matrilineal society and female dominance. The status of males is low in matrilineal societies. The Neanderthal societies were female dominant. The interspecies hybridisation and intragenomic conflict leads to archaea and RNA viroid induced parthenogenesis. The reptilian gene expression produces SLOS phenotype, low cholesterol, low sex hormones and asexuality. This results in alternate sexuality, matrilineal societies, female dominance and DEVI syndrome. This contributes to autistic phenotype and neanderthalic autistic tribes.

The climate change induced parthenogenesis is due to mediation by archaea and RNA viroid. This produces matrilineal society, matriarchy and asexual societies. The reptilian gene expression and porphyrias leads to generation of porphyrions and extrasensory perception. Parthenogenesis and Mother Goddess cult are related. The porphyrions produces extrasensory quantal perception and a quantal civilization. This leads to creativity and autism. The return of Nagas and Asuras due to climate change mediated archaeal endosymbiosis and parthenogenesis is related. Archaea and viroid can induce parthenogenesis. Archaea and RNA viroid induced mitochondrial dysfunction and upregulated glycolysis resulting in stem cell transformation of somatic cells. Stem cell metabolonomics include anaerobic glycolysis, PDH dysfunction, CoQ 2 mutation and mitochondrial dysfunction. The somatic cells which are stem cell transform in the setting of immune activation and cytokine secretion can get converted to germinal cells – sperm and ova. This can result in fertilisation and parthenogenesis. The porphyrions result from intragenomic conflict/interspecies hybridisation and reptilian gene expression. The archaea can induce the HIF alpha and toll receptor activation resulting in glycolysis, mitochondrial dysfunction and GABA shunt. This also produces porphyrin synthesis and porphyrions. Porphyrins can produce extrasensory perception/quantal perception. Porphyrins can result in quantal perception and a quantal civilization existing in multiverse universes. The actinidic archaea and porphyrions have magnetotactic function resulting in mirror neuron function. This results in an autistic tribe. The cerebral cortex becomes atrophic and the cerebellar cortex dominant producing a cerebellar cognitive affective disorder. Interspecies hybridisation and intragenomic conflict results in archaeal symbiosis and parthenogenesis. There is intragenomic conflict and interspecies hybridisation producing this phenomenon. This results in reptilian gene expression and the shikimic acid pathway activation. This produces increased dopamine synthesis. Hyperdopaminergic transmission can produce endemic la Tourette's disease with motor and vocal tics. The hissing like vocal tics led to the evolution of language.

The Neanderthals consume high-fat high-protein, ketogenic diet. This resulted in hibernation, fructolysis and fructosemia, lipogenesis and fat deposition, mucopolysaccharide accumulation, parthenogenesis and stem cell metabolism, glycolysis and mitochondrial dysfunction. High fat high protein diet can lead to decreased SCFA, modulated histone acetylation, HERV expression and genomic modulation. Low levels of short chain fatty acids including butyrate consequent to a low fibre diet produces decreased HDAC is related to HERV expression. Reptilian gene expression can lead to homocystinurias and genomic expression modulation by demethylation. Neanderthalisation of the brain produces extrasensory perception, cortical atrophy, cerebellar dominance and cerebellar cognitive affective disorder (CCAS). The Neanderthal quotient is related to neuroticism, social fear, social avoidance, depression, bipolar disorders and autism. The Neanderthal quotient is related to fear of strangers, aggressive behaviour and social limitation. The NQ is also related to sexual promiscuity, emotional stoicism and fear in social situation. The NQ is also related to anxiety, xenophobia, lack of empathy, compassion. The NQ quotient is related to lack of working memory and consciousness and develop long-term memory. The NQ is related to risky physical, social and sexual behaviour. The NQ is related to small groups of 8-10 numbers. The Neanderthals groups were small. The homo sapien groups numbered 150 and were large. The NQ form alliances within kin groups and family ties were strong. The homo sapiens could form alliances with non-kin groups and large civilizations developed.

The porphyrions have a wave-particle existence and can exist in the intergalactic space. They form a template for the formation of abiogenetic isoprenoid organism, DNA viroids, RNA viroids, double helical templates resembling reptiles. The formation of universe depends on abiogenetic magnetotactic archaea related intergalactic magnetic field. The cometary genes derived from intergalactic archaea and viroids by asteroidal impact seeds life in earth. The intergalactic archaea and cometary genes produced the evolution of life on earth. The actinidic archaeal colonies became multicellular and evolved into Neanderthals. The parthenogenetic embryos of Neanderthals have template mediated replication.

The interspecies hybridisation and intragenomic conflict resulted in archaea and viroid induced parthenogenesis. This produced matriarchy and female dominance. The reptilian gene expression and SLOS phenotype resulted in low cholesterol, sex hormones and asexuality. This produces a culture of sexual equality and alternate sexuality. The porphyrion induced ESP can lead to unified consciousness, equality and oneness. This produces a anarchic and non-hierarchal society. The neanderthalisation of the brain can lead to a cerebellar cognitive affective disorder. The CCAS leads to evolution of evilness and spirituality. The CCAS also results in illogical impulsive acts contributing to terrorism. The extrasensory perception and ataxia due to cerebellar dysfunction can lead to creativity, dance, painting and art.

The interspecies hybridisation and intragenomic conflict can lead to parthenogenesis - archaea and viroid induced. The reptilian gene expression and stem cell metabolism leads to mitochondrial dysfunction, PDH deficiency, glycolysis, fructolysis, fructosemia, lipogenesis, hibernation syndrome, mucopolysaccharidosis and zombie syndrome. This produces metabolic syndrome with obesity and diabetes mellitus mimicking reptilian habits. The interspecies hybridisation and intragenomic conflict leads to climate change, archaea and viroid induced parthenogenesis. The reptilian gene expression produces HIF alpha and increased glycolysis immune activation. The somatic parthenogenesis is related to autoimmunity. Reptilian gene expression and digoxin synthesis leads to immune activation. The reptilian genes and HLA expression are related. The HLA genes are derived from Neanderthals. The archaea and RNA viroid induced toll receptor activation can lead to immune activation. This produces autoimmune disease. The interspecies hybridisation and intragenomic conflict can result in parthenogenesis, archaea and RNA viroid induced. The reptilian gene expression leads to porphyrias. The porphyrions are related to quantal perception. The Neanderthals are retroviral resistant with less of HERV expression leading to cortical atrophy and cerebellar dominance contributing to CCAS. The porphyrion induced quantal perception can lead to a quantal civilization. This results from neanderthalisation of the brain, decreased cerebral cortex. The interspecies hybridisation and intragenomic conflict as said before can lead to parthenogenesis- archaea and viroid induced. The reptilian gene expression leads to porphyria and porphyrion generation producing ESP and quantal perception. The mirror neurons function is related to archaeal porphyrions. This leads to quantal perception and civilization as well as biological reincarnation.

References

1. Gordon Scherer (2013). The Serpent People. Blog: themenoftheserpent. Mar. 22, 2013.
2. Geher, G., Holler, R., Chapleau, D., Fell, J., Gangemi, B., Gleason, M., Rolon, V., Shimkus, A., & Tauber, B. (2017). Using Personal Genome Technology and Psychometrics to Study the Personality of the Neanderthals. *Human Ethology Bulletin*, 3, 34-46.

CHAPTER 46

UNRAVELLING SYMBIOSIS - THE ANARCHIC CELL - THE ETERNAL MICROBIAL COLONY AS THE MASTER OF THE EARTH AND UNIVERSE - INFECTIVE SYMBIOTIC TISSUE COLONIES OF PROTOZOA, ARCHAEA, VIRUSES AND BACTERIA

The plasma life based forms and the gravitational and anti-gravitational waves generated porphyrions would have been the most primitive of non-life forms. The porphyrin templates would have given rise to the isoprenoid organism, the RNA viroids, DNA viroids and prions which would have symbiosed to form the nanoarchaea. The archaea would have evolved into prokaryotes or bacteria. The archaeal symbiosis with prokaryotes would have generated the eukaryotes. The initial unicellular eukaryotes are the protozoa. The protozoa are capable of phagocytosis and would have had viruses, bacteria and archaea as their intracellular symbionts. The viruses of the protozoa like entamoeba and toxoplasma are giant viruses called megaviride. The protozoans like entamoeba with its viral, archaeal and bacterial endosymbionts would have evolved into the multicellular organisms or metazoa. The archaeal, bacterial and protozoal extracellular vesicles containing RNA and DNA viroids can by horizontal gene transfer lead to evolution of metazoan, plants, animals and human species. The metazoans would have evolved into plants, animals and humans by horizontal gene transfer generated by archaeal extracellular RNA and DNA viroid vesicles. The evidence of such a transformation is seen in the bacterial symbionts of protozoa like entamoeba which includes legionella, listeria, mycoplasma, mycobacteria and chlamydia. The symbionts of protozoan entamoeba include viruses like covid, HIV and cytomegalo virus. The protozoan entamoeba serves as a natural reservoir for covid infections, bacterial infections and HIV infections. The toxoplasma also serves as a covid reservoir. These protozoan colonies of entamoeba and toxoplasma with its bacterial, archaeal and

viral symbionts can be called as chimeric infective symbiotic tissue colonies which can exist in the air, water, earth, blood and tissues. The chimeric infective symbiotic tissue colonies containing bacteria, archaea and viruses can make the symbiotic organism multidrug resistant by horizontal gene transfer. The chimeric infective symbiotic tissue organism evolves as a consequence of global warming and the protozoal, bacterial, viral and archaeal symbiotic colonies are new forms of life. The chimeric infective symbiotic tissue organism composed of protozoa and its symbionts like virus, bacteria and archaea depend upon glycolysis for its energy needs. The glycolytic enzymes enolase, glyceraldehyde 3-phosphate dehydrogenase, hexokinase, phosphofructokinase, aldolase, pyruvic kinase and triose phosphate isomerase are moonlighting proteins having a cytoplasmic, membrane and nuclear functions. These glycolytic enzymes have a RNA binding motif which can modulate RNA transcription in the host and the symbiotic organism. These glycolytic enzymes with their RNA binding motif as well as DNA binding motif can bind to RNA and DNA viruses and viroids modulating their replication and life cycles. The chimeric infective symbiotic tissue organism composed of protozoa and its symbionts like virus, bacteria and archaea can enter the human host through the breaches of the gut-blood barrier and transform itself into the macrophages of the human host. The macrophages are involved in the phagocytosis and innate immunity against bacterial and viral infections. Paradoxically the macrophages like the protozoa can also serve as a reservoir where pathogens like mycobacteria tuberculosis, mycobacteria leprae, listeria, legionella, E. coli, shigella & salmonella, and viruses like HIV & covid can thrive and replicate. The modified chimeric infective symbiotic tissue organism or macrophage composed of protozoa and its symbionts like virus, bacteria and archaea can serve as a reservoir for horizontal gene transfer and generation of multidrug resistant bacteria and viruses. The infective chimeric symbiotic tissue organism composed of protozoa and its symbionts like virus, bacteria and archaea depends upon moonlighting glycolytic enzymes for their energetics and

replication. The development of antibodies against the infective chimeric symbiotic tissue organism composed of protozoa and its symbionts like virus, bacteria and archaea leads to the evolution of anti-brain antibody disease or anti-basal ganglia antibody disease. They present a threat to existence consequent to horiztontal gene transfer and generation of multidrug resistant bacteria, viruses, protozoa, mycobacteria and archaea.

Lynn Margulis has postulated a symbiotic origin of the eukaryotic cell including humans. The mitochondria are symbiotic rickettsia, the cell nucleus is formed of pox virus, the microtubular system, flagella and cilia are formed of spirochaetes, the peroxisome of acinetobacter and the cell envelope of archaea. The archaea does not have any nucleus and the archaeal nucleus can be formed of archaeal viruses. Our studies have demonstrated the presence of archaea in the human blood as evidenced by the presence of cytochrome F420, the methanogenic cytochrome. This indicates the unravelling of the symbiotic cell and expulsion of the archaea with its cytoplasm and enzymes into the blood. The archaea can reproduce asexually by binary fission. The archaeal virus can give a nucleus to the archaea. The archaeal cell membrane can form outer membrane vesicles which can rupture off forming new archaea. The microtubules, cilia and flagella are formed of spirochaetes like treponema. The archaeal actin and tubulin cytoskeleton can also contribute to the microtubular network. Spirochaetal cyst formed the picks body in FTD, Lewy body in PD, ALS spheroid body and Alzheimer's plaques. Spirochaetes biofilms and DNA are seen in senile plaques. In Alzheimer's disease neuroborreliosis with transynaptic spread of infection form NFT derived from intraneural spirochaetes. The microtubular network can divide asexually like spirochaetes. The spirochaetes are capable of asexual reproduction and binary fission. The nucleus is derived from the pox virus which the archaea symbiosed. The archaeal RNA and DNA virus with archaeal reverse transcriptase activity can also contribute to the eukaryotic nucleus. The nuclear membrane is contiguous with the smooth and rough endoplasmic reticulum. Free ribosomes are seen in the blood. The cell-free nuclear DNA is seen in circulation. Cell-free DNA are ejected out in cancer, diabetes, inflammation, trauma, sepsis. This cell-free DNA derived from pox viruses or archaeal viruses can replicate and produce new viruses. The cell-free DNA and its complementary RNA can hybridize with archaeal RNA viroids forming new viruses. The human DNA sequences contain the ebola virus, retrovirus, HHV 6, arena virus, filoviridae, totiviridae, rhabdo viruses and borna virus. The expressed retroviral particles can get encapsulated in phospholipid vesicles forming the retrovirus. Similarly, all the viruses can arise de novo from the nucleus. The mitochondria and its bacterial DNA is a rickettsia and can be expelled from the cell in case of trauma, sepsis, viral infections and psychiatric stress. The mitochondria can have fission and fusion and behave like bacteria. The mitochondria can form a well-organised colony like a bacterial network. Thus the cell which is under stress by global warming can unravel its symbiotic structure and expel out its archaeal envelope, produce free mitochondria, produce floating cell-free DNA, can have spirochaetal microtubular asexual reproduction and behave like a multiple bacterial and viral colony without a central organizer resulting in cellular and tissue anarchy. The cellular and tissue anarchy with its free mitochondria, nucleus, free archaea, dividing spirochaetal microtubules, free ribosomes and outer membrane vesicles are recognized as foreign bacteria and viruses by the human immune system. This sets up a cytokine storm leading on to modern civilizational diseases like viral pandemics, metabolic syndrome, neurodegeneration, autoimmune disease and psychiatric disease. The human species and animals go into anevolution as well as break down of symbiosis. Species do not exist any longer; only large organized colonies of archaea, bacteria and viruses. This forms the modern world of organized microbes which are the masters of the earth and the universe with eternal life.

The endosymbiotic archaea can form archaeaons in the cell. The archaeaons having the shikimic acid pathway can produce neurotransmitters and act as neurotransminoids. The archaeaons are capable of biological transmutation and such archaeaons can be described as transmutoids. The archaeaons have the vitamin C synthesizing GULO pathway and such archaeaons are called the vitaminocyte. The archaeaons can oxidize cholesterol to synthesize digoxin involved in neuro-immuno-endocrine integration and such archaeaons are called the steroidelle. The archaeaons can secrete RNA viroids which can integrate with RNA derived from human genomic sequences generating emerging viruses and such archaeaons are called the viroidelle. Thus the endosymbiotic archaea can generate new organelle like the neurotransminoids, the transmutoids, the vitaminocyte, the steroidelle and the viroidelle. Thus global warming and low level internet derived EMF pollution can induce the formation of archaeaons derived new organelle which can lead to cellular evolution.

The human species evolved by symbiotic evolution. The symbiosis of archaea led to the evolution of the homo neanderthalis with Neanderthal brain having a smaller frontal cortex and a bigger cerebellum leading to an epidemic cerebellar cognitive disorder. The evolution of the homo sapiens was due to retroviral infection and integration of the retrovirus into the genome leading on to widespread expansion of the frontal cortex. Human evolution is due to symbiosis. But there is evidence that the symbiotic cell is unraveling. Human endosymbiotic archaeal growth due to expression of archaeal genomic sequences in the human genome can lead to civilizational diseases and a new human species called homo neoneanderthalis. The homo neoneanderthalis is susceptible to cancer, schizophrenia, bipolar disorder, autoimmune disease, neurodegeneration and diabetes mellitus due to endosymbiotic archaea which can modulate metabolism, immunity and genomic function. The homo neoneanderthalis is a new human species created by global warming and low level of EMF pollution due to widespread internet use. The homo neoneanderthalis is generated by unraveling of the symbiotic cell as well as neosymbiosis.

The regression of the symbiotic human cell to archaea, rickettsia, acinetobacter aceti and pox virus results in unravelling of the symbiotic human network. This produces a microbial anarchic colony of viruses and bacteria which can last to eternity.

FSC
www.fsc.org
MIX
Papier | Fördert
gute Waldnutzung
FSC® C083411

Kolibri 360 GmbH
Ferdinand-Jühlke-Straße 7
99095 Erfurt, Deutschland
produktsicherheit@kolibri360.de

Druck:
CPI Druckdienstleistungen GmbH
im Auftrag der
Kolibri 360 GmbH
Ferdinand-Jühlke-Str. 7
99095 Erfurt